FUNGAL DIFFERENTIATION

MYCOLOGY SERIES

Edited by

Paul A. Lemke

Department of Botany,
Plant Pathology, and Microbiology
Auburn University
Auburn, Alabama

Volume 1 Viruses and Plasmids in Fungi, *edited by Paul A. Lemke*

Volume 2 The Fungal Community: Its Organization and Role in the Ecosystem, *edited by Donald T. Wicklow and George C. Carroll*

Volume 3 Fungi Pathogenic for Humans and Animals (in three parts), *edited by Dexter H. Howard*

Volume 4 Fungal Differentiation: A Contemporary Synthesis, *edited by John E. Smith*

Volume 5 Secondary Metabolism and Differentiation in Fungi (in preparation), *edited by Joan W. Bennett and Alex Ciegler*

Other Volumes in Preparation

FUNGAL DIFFERENTIATION

A Contemporary Synthesis

Edited by John E. Smith
Strathclyde University
Glasgow, Scotland

MARCEL DEKKER, INC. New York and Basel

Library of Congress Cataloging in Publication Data

Main entry under title:

Fungal differentiation.

(Mycology series ; v. 4)
Includes indexes.
1. Fungi--Morphogenesis. I. Smith, John E.
II. Series. [DNLM: 1. Fungi--Growth and development.
2. Morphogenesis. QK 601 F981]
QK601.F84 1983 589.2'043 83-5136
ISBN 0-8247-1734-1

MARCEL DEKKER, INC.
270 Madison Avenue, New York, New York 10016

Current printing (last digit):
10 9 8 7 6 5 4 3 2 1

PRINTED IN THE UNITED STATES OF AMERICA

Series Introduction

Mycology is the study of fungi, that vast assemblage of microorganisms which includes such things as molds, yeasts, and mushrooms. All of us in one way or another are influenced by fungi. Think of it for a moment — the good life without penicillin or a fine wine. Consider further the importance of fungi as pathogens to plants and to humans. Yes, fungi are ubiquitous and important.

Mycologists study fungi either in Nature or in the laboratory, and at different experimental levels ranging from descriptive to molecular and from basic to applied. Since there are so many fungi and so many ways to study them, mycologists often find it difficult to communicate their results even to other mycologists, much less to other scientists or to Society in general.

This series establishes a niche for publication of works dealing with all aspects of mycology. It is not intended to set the fungi apart, but rather to emphasize the study of fungi and of fungal processes as they relate to mankind and to science in general. Such a series of books is long overdue. It is broadly conceived as to scope, and should include textbooks and manuals as well as original and scholarly research works and monographs.

The scope of the Series will be defined by, and hopefully will help define, progress in mycology.

Paul A. Lemke

Preface

Fungi exhibit a bewildering array of forms, ranging from simple unicellular types such as *Blastocladiella* and yeast to the complex and macroscopic Basidiomycetes. Between these extremes are the filamentous mold fungi that exert such major influences on mankind by way of biodeterioration, toxicology, and industrial mycology.

Morphogenesis can be considered as the development of an organism to its specific form. How this form is achieved at a biochemical and cellular level in a wide range of fungal types is the subject of this volume.

To cover the field of fungal morphogenesis would require the writing of several voluminous treatises. However, by the skill and application of the contributors, a volume has been produced which is a concise and advanced analysis of this absorbing subject.

The contributors were asked to consider the state of the art of their specific subject and also to indulge in creative speculation for the future. In this way it is hoped that this volume will act not only as a source of reference for what has already been accomplished, but also as a guideline for future directions of research.

John E. Smith

Contributors

Knud Allermann, M.D. Senior Lecturer, Institute of Plant Physiology, University of Copenhagen, Copenhagen, Denmark

David R. Berry, Ph.D., M. Sc., M.A. Senior Lecturer, Department of Bioscience and Biotechnology, University of Strathclyde, Glasgow, Scotland

Jean-Paul Bret, M.D. Laboratoire de Physiologie Végétale, Université Lyon 1, Villeurbanne, France

Edward C. Cantino, Ph.D., A.B. Professor, Department of Botany and Plant Pathology, Michigan State University, East Lansing, Michigan

Mariette Carels, Research Assistant, Linor-Food Products Development Center, Nestlé Products Technical Assistance Company, Ltd., Orbe, Switzerland

Roderic Cooke, Ph.D., D.Sc. Reader in Mycology, Department of Botany, University of Sheffield, Sheffield, England

Kurt R. Dahlberg, Ph.D., M.S., B.S.* Postdoctoral Research Associate, Department of Plant Pathology, University of Nebraska - Lincoln, Lincoln, Nebraska

Roger Durand, M.D. Laboratoire de Physiologie Végétale, Université Lyon 1, Villeurbanne, France

Present affiliation:

*Project Leader, Microbiology Department, Ralston Purina Company Central Research, St. Louis, Missouri

Ruth Emyanitoff, Ph.D.* Research Associate, Department of Developmental Biology, Boston Biomedical Research Institute, Boston, Massachusetts

Karl Esser, Dr. Phil., Dr. h.c. Professor, Lehrstuhl für Allgemeine Botanik, Ruhr-Universität, Bochum, Federal Republic of Germany

Michel Fèvre, D.Sc. Laboratoire de Physiologie Végétale, Université Lyon 1, Villeurbanne, France

Graham W. Gooday, Ph.D. Reader, Department of Microbiology, Marischal College, University of Aberdeen, Aberdeen, Scotland

David Lloyd, D.Sc, Ph.D., B.Sc. Professor and Head of Department, Department of Microbiology, University College, Cardiff, Wales

James S. Lovett, Ph.D., A.B. Professor of Biology, Department of Biological Sciences, Purdue University, West Lafayette, Indiana

Gèrard Manachère, D.Sc. Professor, Laboratoire de Physiologie Végétale, Université Lyon 1, Villeurbanne, France

Friedhelm Meinhardt, Dr. rer. nat., Dipl. Biol. Hochschulassistent, Lehrstuhl für Allgemeine Botanik, Ruhr-Universität, Bochum, Federal Republic of Germany

Gary L. Mills, Ph.D. Visiting Assistant Professor, Department of Botany and Plant Pathology, Michigan State University, East Lansing, Michigan

Peter C. Newell, M.A., D. Phil. University Lecturer, Department of Biochemistry, University of Oxford, Oxford, England

Jørgen Olsen, M.D., Cand. scient. Senior Lecturer, Institute of Plant Physiology, University of Copenhagen, Copenhagen, Denmark

John F. Peberdy, Ph.D., B.Sc. Reader in Plant Microbiology, Department of Botany, University of Nottingham, Nottingham, England

Gerard Pierron† Research Associate, Zoologisches Institut, Universität Würzburg, Würzburg, Federal Republic of Germany

Present affiliations:

*Senior Staff Scientist, Technology Assessment, Genex Corporation, Rockville, Maryland
†Biology Department, Texas A & M University, College Station, Texas

J. I. Prosser, Ph.D., B.Sc. Department of Microbiology, Marischal College, University of Aberdeen, Aberdeen, Scotland

Jean-Claude Robert, D.Sc. Laboratoire de Physiologie Végétale, Université Lyon 1, Villeurbanne, France

Peter J. Rogers, Ph.D., M.Sc., B.Sc. Senior Lecturer, School of Science, Griffith University, Brisbane, Queensland, Australia

Geraldine M. Russo, M.S., B.S. Research Assistant, Department of Plant Pathology, University of Nebraska-Lincoln, Lincoln, Nebraska

Helmut W. Sauer, Dr. rer. nat.* Professor, Zoologisches Institut, Universität Würzburg, Würzburg, Federal Republic of Germany

David Shepherd, Ph.D., M.Sc., B.Sc. Section Manager, Linor-Food Products Development Center, Nestlé Products Technical Assistance Company, Ltd., Orbe, Switzerland

John E. Smith, Ph.D., D.Sc., M.Sc., B.Sc., F.R.S.E., F.I. Biol. Professor and Chairman, Department of Bioscience and Biotechnology, Strathclyde University, Glasgow, Scotland

Peter R. Stewart, Ph.D., B. Agr. Sc. Reader in Biochemistry and Microbiology, Department of Biochemistry, Faculty of Science, Australian National University, Canberra, Australia

Gilbert Turian, D.Sc. Professor, Laboratory of General Microbiology, Department of Plant Biology, University of Geneva, Geneva, Switzerland

Herman van den Ende, M.D. Professor, Department of Plant Physiology, University of Amsterdam, Amsterdam, The Netherlands

James L. Van Etten, Ph.D., M.S., B.A. Professor, Department of Plant Pathology, University of Nebraska - Lincoln, Lincoln, Nebraska

Barbara E. Wright, Ph.D.† Director, Department of Developmental Biology, Boston Biomedical Research Institute, Boston, Massachusetts

Present affilations:

*Biology Department, Texas A & M University, College Station, Texas
†Research Professor, Department of Microbiology, University of Montana, Missoula, Montana

Contents

FUNGAL DIFFERENTIATION

1

Concepts of Fungal Differentiation

GILBERT TURIAN *University of Geneva, Geneva, Switzerland*

I. GENERAL DEFINITIONS

Morphogenesis can be defined as the development of an organism to its specific form. This implies differentiation of its cells into types differing in their shape, size, structure, and chemical composition. In unicellular forms, internal differentiation also results in morphogenetic processes leading to a change in form. Thus endospores in bacteria or asci in yeasts are obviously different in form from the normal vegetative cells (Smith and Berry, 1974). Such cases of individual cells exhibiting differences in structure and function at different stages of their life cycle are examples of time-dependent or *temporal* differentiation, in contrast to the more usual *spatial* differentiation typical of conidiogenous fungi, not to mention embryos (Bonner, 1974).

The connection between cell division and cell differentiation is not only a considerable theoretical problem, but also quite a technical nuisance, for it is often difficult to distinguish between events which are really a consequence of differentiation and those that are a consequence of the cell cycle when both sets of events are occurring simultaneously. These difficulties can be overcome with the cellular slime mold, because in its life cycle growth and differentiation are mutually exclusive phenomena (Ashworth, 1973). Another attractive feature of this eukaryotic microbe is that it can separate cellular differentiation (spores plus stalk cells) from morphogenesis (culmination into a sorocarp), as has recently been shown in an aggregateless mutant (Ishida, 1980).

It is generally believed that the process of differentiation is associated with the activation of different genes in different cells at different stages of development. Controls of gene functioning can intervene at *both* transcriptional *and* translational levels; in the latter case the process would proceed only until a specific "inducer" (or stimulator; see Gross, 1968) arrived which would induce at the translational level. Indeed, where hormones appear to control differentiation by influencing transcription, translation-level controls have been shown to play an important role in the orderly expression of hormone-induced changes, as stated by Gross (1968), who provided a few examples drawn from animal cells, as well as from the evolution of the patterns of synthesis and decay during differentiation of cellular slime mold (timing of relevant transcriptional and translational events).

Experimental facts generally suggest that cells become committed to a particular differentiated state before the existence of that state is detectable. The high degree of fixity in visible, final differentiated states would thus contrast with the lability, gradually reduced with time, of the initial determination process. However, there is no evidence, experimental or logical, in favor of the idea that commitment to a cause of change involves molecular mechanisms distinct from the final realization of that change. Hence, according to Gross (1968), the biochemical analysis of determination must be a part of the analysis of differentiation, which may differ from determination only in that some change in affected cells becomes more easily detectable and appears to be reversible only with difficulty.

The question arises as to whether even a complete biochemical description of the pathway to terminal chemodifferentiation could contribute to an understanding of the shape changes and movements of cell groups implicated in morphogenesis. Analysis of the supra- and inframolecular forces acting on cells with that of biosynthetic controls could contribute to such understanding, but "the visible events of morphogenesis might be a spontaneous consequence of biochemical differentiation, creative of differential cell surface properties, adhesiveness, motility, response to the environmental composition" (Gross, 1968).

A most restrictive and operational definition of differentiation depends upon foreknowledge of the adult condition of cell types. The study of their cytodifferentiation requires identification of a particular biochemical pathway, a catalytic or structural macromolecule, or a morphological feature uniquely characteristic of the adult cell type (Gross, 1968). A relevant case in fungi is that of the male cells in the sexual species of *Allomyces*, which elicit the bright yellow color of γ-carotene indicative of a selective deviation of terpenoid biosyntheses toward unsaturated carotenoids at their level (Turian, 1969).

Assuming that whole differentiation embracing the succession of determining, differentiating, and terminally differentiating systems is the result of differential gene action permitted Gross (1968) to establish the following categories of the process of gene action at differentiation: replication, transcription, intranuclear processing, and translation, beyond which is "epigenetic control."

More recently, a differentiated cell has been considered as poised in a state of dynamic equilibrium between the nucleus and cytoplasm, in which signals from the cytoplasm are necessary for the maintenance of the nuclear gene expression which characterizes that state (Maclean, 1977). The limited number of possible types of differentiated cells in a complex organism implies that a very limited number of "tissue-master" genes may exist, each one responsible for the release of a particular genetic program. This conception of differentiation (Maclean, 1977) provides an easy understanding of the distinction between determination and differentiation. The first is the assumption of the stable state between the nucleus and cytoplasm, which involves choice of one master gene, while the second is the actual expression of the program under the direction of that gene. Clearly, additional factors may often be necessary in order to release the activity of an already selected master gene. The factors controlling and determining the activity or inactivity of tissue-master genes are presumably often cytoplasmic, may well be partitioned during embryonic cell division, and are really the decisive factors in cell differentiation. Their identity remains a mystery (Maclean, 1977).

Essential as selective gene activation and sequential genome readout are for differentiation, they necessarily interplay with other equally important extranuclear components and processes. Wright (1973) has explored the consequences of metabolite availability on the process of differentiation and has proposed that in primitive systems such as slime molds, changes in the rate of availability, or flux, of endogenous metabolites could control the activity of essential metabolic pathways essential to differentiation. This would involve, for example, the rate of substrate-limited reactions critical to differentiation; the concentration of enzymes, by changing their rate of synthesis (induction) or degradation (stabilization); enzyme activity, by allosteric modulation and end-product inhibition; and the duration and stability of differentiated structures. From such critical variables, Wright (1973) has been led to construct kinetic models which simulate long-term change in metabolism (e.g., carbohydrates in *Dictyostelium*) such as occur during differentiation.

Differentiation can be studied at several levels of structural and functional organization. Even in prokaryotic organization it is possible to recognize some anticipation of the process of eukaryotic differentiation. Thus the switching on and off of bacterial operons constitutes an altered commitment on the part of the cell and its metabolic machinery in response to the changing environment. By contrast, both division and sporulation demand an even closer commitment to a particular pathway.

Two cells are differentiated from each other if they have the same genome but the pattern of other molecules which they possess and synthesize is different. It follows from this statement that very simple cells may have a greatly reduced potential for differentiation. Indeed, a population of cells might exist in which the genome was so simple that no cell could possibly survive without constantly expressing the full genetic complement. Such cells probably do not exist, but certainly within populations of bacteria the possibilities of differentiation are severely limited. Some bacteria may be induced by a substrate to transcribe a particular operon, while their sister cells, not exposed to the substrate, fail to express these genes. But such differences in metabolism are temporary and do not involve gross changes of cellular morphology. We must therefore try to distinguish between changes of an adaptive type, which are simply part of the normal cell cycle, and changes which lead to more permanent differences between cells in the same population.

As selective gene activation might not be the only way to control differentiation, a scheme of progressive repression of previously active genes has appeared as most consistent, with observations at least from experimental embryology, as well as from more recent biochemical experimentation (Caplan and Ordahl, 1978). How far the model proposed, in which each initial pluripotent embryo cell progressively becomes irreversibly restricted to the expression of a single phenotype, can be transposed to sporulating fungi remains conjectural. As yet, the "activation-repression" rather than the "irreversible repression" mechanism seems to fit most generally to the several progressive sporulation types of differentiation. According to the first mechanism, genes that have not previously been transcribed are activated (or derepressed), and their transcripts and subsequent translational products appear for the first time in the newly differentiated spore; part, at least, of the previously transcribed "vegetative genes" are then repressed. The newly available plasmid technology or "genetic engineering" to obtain phenotype specific probes should help to further assess the value of that conceptual framework concerned with the role of genes in differentiation (Paul, 1971). Such an approach involving the cloning of selected pieces of DNA (specific gene isolation) has been aimed at a better understanding of the temporal control of genes turned on during the yeast cell cycle (Halvorson, 1977) and is currently being further tested in sporulating filamentous fungi (Dutta et al., 1981).

Reversibility of differentiation is evident in microorganisms. Thus spores do change back into normal, dividing, vegetative cells by the process of germination. Should spore formation, then, be considered a process of adaptation rather than of differentiation? Either case could be argued according to Pasternak (1970), who takes the following point to consider sporulation (or sporogenesis) as a differentiation process:

> One of the stimuli for the formation of bacterial spores is the exhaustion of nutrients. If nutrients are restored to the culture immediately after depletion, spore formation does not take place; the culture is in the reversible, adaptive state. But if nutrients are added some time after spore formation has already begun, sporulation continues despite the presence of nutrients. In other words, cells are now *committed* to sporulation. Likewise spores become committed to germination on the return to more favourable conditions. It is the presence of a commitment stage, also referred to as *determination* during embryological development, which allows one to distinguish between differentiation and adaptation.

Differential gene expression, although no doubt important in cell differentiation, does not, of itself, explain how the initial commitment is made. This initial commitment need only involve a choice between perhaps a hundred or so separate programs of gene expression. The striking similarity of different cells of the same differentiated type probably reflects the identical nature of the program selected within them (Maclean, 1977).

From the above considerations, we can consider differentiation as the stable development of altered structure and function. Its stimulus is sometimes environmental, as in the formation of spores. Agents that cause it are known as initiators or inducers and their action as *initiation* or *induction* (Pasternak, 1970). Biochemical differentiation is a manifestation of an altered synthesis of specific proteins; these function either as structural units, such as actomyosin of muscle or other motile cells, or as enzymes catalyzing the synthesis and transport of a variety of molecules.

Differentiation is generally irreversible in animals (liver cells cannot change back to an ovum, etc.). It is, however, not always based on the loss of various potentialities by the cell; these particular potentialities can merely be masked by others. This is well exemplified in higher plants, in which certain cells in the phloem of the carrot are capable of *dedifferentiation*, with eventual production of a complete new plant; moreover, many differentiated plant tissues lose much of their diagnostic characters when grown in tissue culture (Gautheret, 1966). In germinating spores of fungi, dedifferentiation first occurs during the isometric growth (Turian, 1969), followed by redifferentiation into a germ tube that becomes a new vegetative hypha.

Regarding this question of the *stability* of differentiation, we could follow the suggestion of Abercrombie (1967), that with his kind of approach in terms of the epigenotype, "one is relieved of the necessity of postulating any sort of stability of differentiation: the stability is in the epigenotype and the differentiation is the phenotype; and this can fluctuate and change." The complicated term *epigenotype* could, however, be avoided if we use Gurdon's (in Abercrombie, 1967) following simple assumption:

> ...in any differentiated cell the products of its genes continually influence these same genes to be active; consequently, any cell will necessarily stay the same until something changes it. This would account in early development for how the different cytoplasmic substances cause nuclei in different parts to behave differently, and we have many examples of induction where we have only to suppose that the influences coming from outside interfere in some way with this continual process.

II. FUNGAL DIFFERENTIATION

Fungi as simple eukaryotic organisms have been considered as favorable for studies of differentiation (Turian, 1969; Smith and Berry, 1974), as they provide relatively simple morphological systems which are genetically controlled and amenable to biochemical investigation.

A. Dimorphism

Dimorphism corresponds to the easily reversible alternative between a mycelial (M) and a yeast (Y) phase, as shown by many fungal types exposed to certain physical (temperature, etc.), chemical (O_2-CO_2 tensions, sugar concentration, etc.), or environmental factors (Stewart and Rogers, 1978). It does not, however, fully meet the basic criteria of differentiation. At the limit, it could be viewed as an adaptation in Pasternak's sense or environmental modulation (Smith and Berry, 1974). It could rather be considered as a relatively simple morphogenetic event, as it involves a contrasted change in the shape of growth, from the normally anisometrical, polarized, cylindrical type of the standard fungal hyphae, to an isometrical, spherical or subspherical yeast type with its unpredictable peripheral budding of new spherical or ellipsoidal cells.

The changed vegetative shape has initially been explained by Nickerson's group (Baldwin and Rusch, 1965) as being due to chemical changes in the macromolecular components of the cell wall (disulfide cross-linking of a manno-protein complex and/or changes in the polysaccharide components). However, primary disturbance in the polarity of growth has more recently been envisaged: The wall-making vesicles could no longer be polarly directed toward the apex of a normally elongating hypha (Bartnicki-Garcia, 1981). A newly differentiated structure found during the isodiametrical growth phase of spore germination is the vegetative wall, which is deposited anew under the existing spore wall around the germ sphere. In this reconversion Y → M, the initial appearance of a hyphal tip from the yeast form can be considered as homologous to the appearance of the germ tube from a spore. It then becomes crucial to investigate further the transitory processes initiating polarized growth of the germ tube emerging from swollen (initial spherical growth) spores.

B. Polarity and Gradients

There is no differentiation without *polarity* (Bloch, 1943; Bünning, 1952; Sinnott, 1960). Such a statement, that differentiation is impossible without protoplasmic polarity, is based not only on theoretical considerations, but also on experimental facts. Polarity is indeed the first step in the differentiation of an egg cell or a spore. The eggs of *Fucus* and other marine algae that are discharged into the surrounding medium, as well as the spores of mosses and ferns, are unpolarized at first. Their polarization is the result of the unilateral influence of external factors, such as light, gravity, or substratum (Bonner, 1974).

A specific protoplasmic structure providing the necessary physical asymmetry has been assumed to be the ultimate cause of polarity (Bünning, 1952). Such structural asymmetry may then lead to chemical gradients and electrical potential differences. This last has also been ascribed to the selective entrance of positive ions (Ca^{2+}) across the plasma membrane of the prospective pole (Jaffe and Nuccitelli, 1977). Stability of polarity during dormancy periods can be ensured by the asymmetry of protoplasm, despite momentary cessation of all the metabolic energy-yielding processes necessary for the maintenance of the gradients. As a consequence, it can be shown that visible polarity can only be temporarily reversed by its overcompensation by changing external (centrifugal, electrical, hormonal, etc.) factors. The concentration gradients produced by polarity offer the requisite ground for intercellular differentiation following unequal division, provided that the new cell wall is at right angles to the axis of polarity. A well-known example is the division in pollen grains, with their mitotic spindle polarity oriented along a gradient of protoplasmic density (Bünning, 1952) probably corresponding to the gradient distribution of ribonucleoprotein (Browder, 1980).

The origin of *spatial* patterns therefore appears to be related to the existence of physiological gradients (Child, 1941) providing a basis for difference in gene action. The recognition of such an important role of gradients finally led to the concept of positional information developed by Wolpert (1971), which implies the existence of a coordinate system within which all positions are specified. This results in a spatial differentiation that precedes, and is independent of, molecular differentiation at the subcellular level. The latter takes place in a second stage, during which positional information is interpreted by the individual cells (Nicolis and Prigogine, 1977). The positional differentiation produced might result from the diffusion of a morphogen between a source and a sink, as illustrated by the "French flag problem" (Goodwin, 1976).

Processes other than diffusion (Turing, 1952) are likely to influence pattern formation, such as active transport, cellular contacts, time oscillations, and cell sorting. However, in the development of a polar pattern, there are specific conditions giving rise by diffusion to a source and a sink along a field. Onset of such polarity therefore occurs in a previously unpolarized field and results in a stable gradient. The symmetry-breaking instability being induced by diffusion implies that this primary pattern is necessarily a spatial dissipative structure (Nicolis and Prigogine, 1977).

In addition to polarity, such *primary* pattern endows the field with well-defined boundary values of morphogens and with positional information for the individual cells. How this information is interpreted by individual cells and induces a *secondary* pattern of differentiation has been analyzed by diffusion gradients of the morphogens. Turing's (1952) two-morphogen linear model based on diffusivity waves of the morphogens is relevant here. Its application has been suggested (Lacalli and Harrison, 1979) for the stability of the simple binary pattern in slime molds. Such a quantitative model that was also applied to the alga *Acetabularia* (Nicolis and Prigogine, 1977) could well be considered for fungi. In such organisms, the polarized distribution of hyphal organelles (tip vesicles, back-positioned mitochondria, etc.) intervenes very early, at the emergence stage of spore germination, and grows in the elongating hyphal apex. It can be regarded as a model of simple, vegetative differentiation involving spatial pattern formation.

C. Germination Models

According to Ross (1979), once the germ tube makes its appearance, it is obvious that a change in orientation of the logistics of the cell has to occur. The swelling or spherically growing spore is developing in a nonpolarized manner, while the germ tube and all subsequent hyphal growth develops in a polarized manner. This change in polarity has important consequences, as quoted by Ross (1979):

> Without such a change, if accumulation continued in a uniform, nonpolarized manner, the thallus would develop as a spherical cell, as in the case of holocarpic chytrids and in single-celled yeasts. Only if polarity is imposed on the accumulation of wall materials can there be long, cylindrical, apically elongating

hyphae that constitute the fundamental element of the fungal thallus be formed.

The question arises, then, of what initiates polarity first. In other words (Ross, 1979), what induces the transition from the vesicles arriving at all parts of the plasma membrane in the nonpolarized, uniformly swollen cells to a situation in which all the vesicles arrive at the same place in order to initiate the germ tube?

Autoradiographical evidence obtained by Bartnicki-Garcia (1973) by means of radioactive precursors of wall polymers has shown that polarization of wall synthesis occurs during the spore germination of *Mucor rouxii*. It had previously been suggested from a study of budding mutants of *Mucor* that the functioning of mitochondria is required because such mutants deficient in such organelles grow isometrically, as yeastlike cells (Haidle and Storck, 1966). The abnormal forms obtained only returned to the hyphal state if their mitochondrial activity was restored, therefore strongly involving electron transport and oxidative phosphorylation in the establishment of polarity.

Being intimately associated with the establishment of primary differentiation, polarity has then to be studied at its origin. We were led to take this novel approach by our observations of a low pH in the hyphal tips (Turian, 1979a) having its origin back at the outgrowing stage of the germinating spores (Turian, 1980). We thought that the source of this decreasing pH gradient in the emergent and elongating apexes could be related to the well-known chemiosmotic theory (Mitchell, 1977) of proton extrusion from mitochondria into the cytosol. Our hypothesis (Turian, 1979b) was that, following some type of aleatory event in the peripheral distribution of mitochondria in the initially isodiametrical, swollen cell, protons having escaped from strategically back-positioned mitochondria (Fig. 1) could initiate the formation of a proton sink self-amplified at the proximate site of the plasma membrane by reciprocal events such as ionic exchange with K^+ and phosphate anions. The localized acidification resulting from the formation of the proton sink would lead to local gelification of the cytosol; the resulting cytogel, presumably constituted of an actomyosin highly cross-linked in its microfibrillar network, would prevent entrance of mitochondria. Such a restrictive, or exclusive, amitochondrial zone becomes the tip of the outgrowing germ tube and then that of the elongating hypha known to contain only vesicles and microfilaments associated with the microvesicles of the "Spitzenkörper," at least in the Septomycetes.

According to our working hypothesis, no oxidative phosphorylation would be necessary for the initiation of the proton sink ensuring outgrowth of the germ tube; rather, it would depend upon a burst of hydrolytic ATPase activity vectorially liberating the H^+, as suggested by the stimulated emergence of germ tubes from conidia of *Neurospora* in the presence of 2,4-dinitrophenol (Turian, 1980). Oxidative phos-

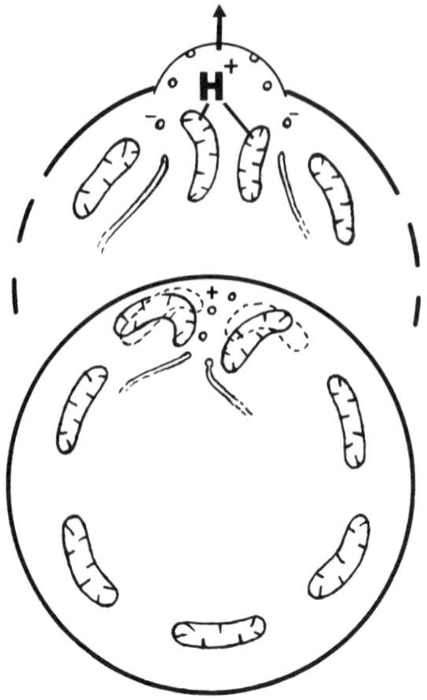

FIG. 1 Model of primary, apical differentiation based on studies (Turian, 1979a,b, 1980, 1981) of the onset of germ tube or bud outgrowth in fungal conidia (type *Neurospora*) or vegetative yeasts (type *Saccharomyces*): Polar sorting out of cytosol acidified by protons (+) vectorially dissipated from a few mitochondria contingently back-positioned. Proton sink (H^+) self-expanded below the depolarized membranar site by further vectorial dissipation of mitochondrial protons and transmembranous ionic exchanges. Parallel gelation of the acidified polar cytosol pushes back proximal mitochondria into an axially oriented position, while cytomembranous vesicles (-) are electrophoretically attracted toward the emerging tip bud (arrow).

phorylation would start in the second hour, with protons reintruded into all mitochondria except the back-front longitudinally oriented ones (uncoupled); such back-positioning of the mitochondria leaving a free cytoplasmic front would then be favorable to the further anisotropic dissipation of protons toward the hyphal tip thus maintained in its exclusive cytogel state (Turian, 1979a,b).

Given such conditions of polarity induction, the question arises again of what causes the change in migration pattern of the vesicles. These last do not appear to have any self-contained motile system, and microtubules often associated with intracellular movements of orientation are virtually absent in spores and hyphal ultimate tips

(G. Turian and T. C. Ton-That, unpublished observations, 1981). The fact that an electrical protential could be measured between the apex and the subapical region of *Neurospora crassa* hyphae (Slayman and Slayman, 1962) suggests that the movement of the vesicles may be caused by electrophoresis (Jaffe, 1968; Harold, 1977). For Ross (1979), all that is necessary, then, to switch from nonpolarized to polarized vesicle movement is for a change in electrophoretic fields to occur at some stage during spore swelling. This, of course, merely pushes the unknown one step back. What causes the change in electric current?

Our proton sink model could be a first answer to that question, as it becomes evident that the emergence site being positive (Fig. 1), the presumably negatively charged vesicles would therefore polarly migrate to fill the needs of the extending apical plasma membrane and its acid-softened thin wall. The maintenance of the electrochemical gradient and therefore the self-maintenance of apical differentiation would be insured by the continuous, unilateral provision of protons that had vectorially escaped from the front of the back-positioned mitochondria through a mechanism which would thus fit in with the thermodynamical concept of "spatial dissipation" proposed by Nicolis and Prigogine (1977).

Emergence of germ tubes can be prevented by the addition of cyclic AMP, a depolarizer of cytoplasmic membranes, to germinating spores of *Mucor racemosus* during the spherical growth stage; interestingly, this leads to an indefinite "swelling" of the germ spheres into giant cells (Paznokas and Sypherd, 1975). This suggests that for the onset of polarity, the depolarization at the plasmalemma level must be *locally* provoked and that in this case it can normally only occur from the inside of the spore. This is what is proposed by our model, which involves a flux of protons vectorially directed to the presumptive site of germ tube outgrowth; such a proton flux would originate from mitochondria which could be considered as the "creators" of first cytoplasmic heterogeneity and therefore of primary differentiation at germination.

In the polarly growing thallus of *Blastocladiella emersonii*, circumstantial evidence suggests that protons also carry much of the electric current, which appears to play an important role in the spatial localization of its development. In growing cells, positive current enters the rhizoid and leaves from the thallus, whereas upon sporulation, the current pattern is reversed (Stump et al. 1980). As the origin of these currents, both extracellular and intracellular, is not known, it will be interesting to reinvestigate the initial positioning of the mitochondria, which we described as frontal to the emergence point of the rhizoid in the germlings of the closely related aquatic fungus *Allomyces* (Turian, 1969).

In the first phase of germination, preceding that of mitochondrial polar positioning, zygotes or zoospores of *Allomyces* show a dedifferentiation of their compartimentalized organization, marked by disintegration of their ribosomal nuclear cap and symmetrical distribution of mitochondria (Turian, 1969). The encystment of the symmetrical cell which follows can be considered as a new differentiation involving a complicated subcellular reorganization highlighted by the cyst coat and wall syntheses from chitin synthetase-containing microvesicles originating from the γ particles and fusing with the plasma membrane (Bartnicki-Garcia, 1981). In the last phase, the bipolarization process intervenes, directing the vesicles first to the point of rhizoid outgrowth and then to the opposite, hyphal point of emergence. Such bipolarized differentiation is paralleled by opposite mitochondrial axial displacements (Turian, 1969).

D. Other Models

In the slime mold *Dictyostelium*, we see another most impressive exploitation of differential organization. This has made it possible to isolate mutants of this organism in which the mutant gene is only expressed in the differential phase, thus providing further evidence for implication of differential gene activity in the differentiation process. Three classes of intercellular regulatory substances or "morphogens" have been described in that slime mold (Newell, 1978): The first is NH_3, which can switch developing aggregates of cells from the sorocarp- (stalk and spore) forming program to the alternative "slug" program; the second class is cyclic AMP, the inductive signal of development, in general without regard to the cell types formed; and the third class inducing the differentiation of specific cell types includes a diffusible glycopeptide as a stalk-specific factor and a nondiffusible factor for sporogenesis. These last endogenous compounds appear to redirect the function of cells in an integrated way and in an interaction with genes involved in the achievement of differentiation.

Our understanding of the molecular changes in chromatin involved in the receiving and processing of differentiation *signals* is still vague. Environmental changes (O_2 tension, pH, etc.) could be viewed as primary events, dictating the altered pattern of gene transcription (Caplan and Ordahl, 1978). Internal events such as those involved in random cytotopological reorganization, as exemplified by mitochondrial redistribution leading to the formation of a low-pH proton sink in the thus gelled cytosol (Fig. 1, top), might be a primary signal to change gene transcription. Sensor molecules (polyADP-ribose), more recently guanylate polyphosphates in bacteria and yeasts, have been proposed to link the signals (NAD concentration in Caplan and Ordahl's model for the mesenchymal cell alternate differentiation into muscle or cartilage cells) to the chromatin. Such molecules covalently bind to a discrete subfraction of chromatin, thus specifically repressing certain gene sequences either related to vegetative growth or to alternate differentiation choices.

Cytodifferentiation in yeasts includes budding, their symbolic morphological characteristic, and an irreversible process in their life cycle (Hayashibe, 1977). In *Saccharomyces* budding implies a periodic differentiation in intracellular organization with the restrictive apical region containing only membrane-limited vesicles but excluding mitochondria as the most characteristic feature (Matile et al., 1969). Such a local back-positioning of the mitochondria is an aleatory, periodic event in multipolar yeasts (limited in its possibilities by the number of birth and bud chitinous scars) and a relatively permanent event in bipolarly budding yeasts. It might be the determining event in the choice of the budding site (primarily to the positioning of the spindle plaque) by allowing the vectorial escape of protons from these few oriented mitochondria toward the mother-cell plasma membrane, as evidenced by the low pH cytochemically measured in the bud initials (Turian, 1981). The proton sink thus formed below the bulging membrane could provide the electrochemical power necessary for the apical transport of the vesicles into the expanding bud. Such a tentative proposal rejoins that made above for apical differentiation of hyphae (Fig. 1) which elongate through a process of continuous budding, in contrast with the rhythmicity of yeast budding. Both processes appear, however, to be functionally homologous in their requirement for the onset of polarity through initial mitochondrial positioning.

The polarized rhythmic outgrowth of buds is interrelated with the periodic events of the yeast cell cycle such as ordered genome transcription and enzyme synthesis (Halvorson, 1977). Into this conceptual framework of a relatively simple type of cell differentiation can be inserted the chitin synthesis intimately associated with the very beginning stage of the bud formation, implying periodic activation of a chitin synthetase zymogen through a proteinase (Cabib, 1975).

Both types of reproduction, asexual and sexual, involve the most complex type of hyphal differentiation achieved in fungi (Turian, 1969; Smith and Berry, 1974; Ross, 1979). In its asexual aspects, it is generally apical and, as such, corresponds to a new differentiation secondarily imposed upon the primary, vegetative one. The apices of conidiogenous hyphae first undergo cytodedifferentiation, both the polar organization and its generated electrochemical gradient vanishing when mitochondria and the vesicular elements spread isometrically around the enlarging tip to fulfill the needs of spherical conidial morphogenesis (Najim and Turian, 1979). Biochemical differentiation then intervenes, expressing itself by the progressive accumulation of secondary metabolites (pigments, etc.).

Fungal hyphae must, however, obtain *competence* before normally differentiating conidia or other types of spores. Such developmental competence is genetically regulated in *Aspergillus nidulans* (Champe et al., 1981). Premature differentiation of conidia on very short hyphae-conidiophores can be induced following initial heat treatment of conidia both in *Aspergillus niger* (Anderson and Smith, 1971) and *N. crassa* (Cortat and Turian, 1974). In such cases, premature competence is somewhat paralleled by commitment to differentiation or determination, acquired already during the phase of isodiametrical growth induced by the heat treatment. In macroconidia of *N. crassa*, such commitment is visibly expressed by dense spots of RNA protein material crowning of nucleoli (Ton-That et al., 1981). The shift down at the permissive temperature (to 25°C after 16 h at 46°C for *N. crassa*) then allows, with the disappearance of the dense spots, the premature expression of the conidiogenous genes soon after the transition from spherical to cylindrical growth. Because of its heat-induced commitment, the emerging germ tube is therefore determined as a presumptive conidiophore. Its capacity to synthesize vegetative protein becomes restricted during conidiogenous differentiation (Van Tuinen and G. Turian, work in progress). Such restriction must mean that certain mechanisms are operating to select portions of the genome to be utilized. This confirms that the variability in protein composition in differentiated cells must result from differential gene expression (Browder, 1980).

Reproductive differentiation of the hyphae of *Allomyces* is always *apical* and, still elusively, leads to different organs, depending upon the sequential doubling of the chromosomal set: N-2N-4N leads to the double apical differentiation of couples of gametangia, with DNA-controlled inverse positioning of the sexual organs; an acridinelike endogenous compound has been hypothesized to be polarly directed by positioning DNA to the mitochondria thus induced to a partial respiratory deficiency in the presumptive male cytoplasmic zone (Turian, 1975). Positioning DNA might also be involved in the significant loss of ribosomal cistrons measured in masculinized cultures (Ojha and Turian, 1978). The 2N-4N-8N series leads to single apical formation of either zoosporangia or resistant sporangia.

Lateral differentiation is the alternative to apical differentiation; it is usually associated with sexualization and produces, for example, phialides basipetally budding microconidia in *Neurospora* (microcyclical induction) (Rossier et al., 1977), coiled ascogonia in Euascomycetes, zygophores in Zygomycetes or antheridial hyphae in Oomycetes. These last are hormonally produced in *Achlya* by antheridiol, which induces the localized synthesis of specific male products, thus providing an outstanding fungal model for deeper study of the interaction hormone differentiation (Horgen, 1980).

Ultimately, chemostructural differentiation of the several types of functional hyphae shaping the elaboration of reproductive stromas in the higher Ascomycetes and Basidiomycetes raises the most provoking and mostly unanswered questions pertaining to the visual achievement of macrofungal morphogenesis in Nature.

REFERENCES

Abercrombie, M. (1967). General review of the nature of differentiation. In *Cell Differentiation*, CIBA Foundation Symposium. V. S. De Renck and J. Knight (Eds.). Churchill, London, pp. 3-17.

Anderson, J. G., and Smith, J. E. (1971). The production of conidiophores and conidia by newly germinated conidia of *Aspergillus niger* (microcycle conidiation). *J. Gen. Microbiol. 69*:187-197.

Ashworth, J. M. (1973). *Cell Differentiation*, Outline Studies in Biology. Chapman and Hall, London.

Baldwin, H. H., and Rusch, H. P. (1965). The chemistry of differentiation in lower organisms. *Annu. Rev. Biochem. 34*:565-594.

Bartnicki-Garcia, S. (1973). Fundamental aspects of hyphal morphogenesis. In *Microbial Differentiation*, Symposium 23, Society of General Microbiology. J. M. Ashworth and J. E. Smith (Eds.). Cambridge University, Cambridge, pp. 245-267.

Bartnicki-Garcia, S. (1981). Cell wall construction during spore germination in Phycomycetes. In *The Fungus Spore, Morphogenetic Controls*, Third Spore Symposium, Gwatt, Switzerland. Academic, London, G. Turian and H. R. Hohl (Eds.) pp. 533-556.

Bloch, R. (1943). Polarity in plants. *Bot. Rev. 9*:261-310.

Bonner, J. T. (1974). *On Development: The Biology of Form*. Harvard University, Cambridge, Mass.

Browder, L. W. (1980). *Developmental Biology*. Saunders College, Philadelphia, Pa.

Bünning, E. (1952). Morphogenesis in plants. In *Survey of Biological Progress*. J. S. Avery, Jr. (Ed.). Academic, New York, pp. 105-140.

Cabib, E. (1975). Molecular aspects of yeast morphogenesis. *Annu. Rev. Microbiol. 29*:191-214.

Caplan, A. I., and Ordahl, C. P. (1978). Irreversible gene repression model for control of development. *Science 201*:120-130.

Champe, S. P., Kurtz, M. B., Yager, L. N., Butnick, N. J., and Axelrod, D. E. (1981). Spore formation in *Aspergillus nidulans*: Competence and other developmental processes. In *The Fungus Spore, Morphogenetic Controls*, Third Spore Symposium, Gwatt, Switzerland. G. Turian and H. R. Hohl (Eds.). Academic, London, pp. 255-276.

Child, C. M. (1941). *Patterns and Problems of Development*. University of Chicago, Chicago, Ill.

Cortat, M., and Turian, G. (1974). Conidiation of *Neurospora crassa* in submerged culture without mycelial phase. *Arch. Microbiol. 95*: 305-309.

Dutta, S. K., Frederick, L., and Austin, W. L. (1981). Gene level control. In *The Fungus Spore, Morphogenetic Controls*, Third Spore Symposium, Gwatt, Switzerland. G. Turian and H. R. Hohl (Eds.). Academic, London, pp. 229-254.

Gautheret, R. J. (1966). Factors affecting differentiation of plant tissues grown in vitro. In *Cell Differentiation and Morphogenesis*. North-Holland, Amsterdam, pp. 55-95.

Goodwin, B. C. (1976). *Analytical Physiology of Cells and Developing Organisms*. Academic, London.

Gross, P. R. (1968). Biochemistry of differentiation. *Annu. Rev. Biochem. 37*:631-660.

Haidle, C. W., and Storck, R. (1966). Control of dimorphism in *Mucor rouxii*. *J. Bacteriol. 93*:1236-1244.

Halvorson, H. O. (1977). Microorganisms as model systems for differentiation. *ASM News 43*:470-476.

Harold, F. M. (1977). Ion currents and physiological functions in microorganisms. *Annu. Rev. Microbiol. 31*:181-203.

Hayashibe, M. (1977). Cytodifferentiation in the cell cycle of yeasts. In *Growth and Differentiation in Microorganisms*, NRI Symposium on Modern Biology. T. Ishikawa, Y. Maruyama, and H. Matsumiya (Eds.). University Park, Tokyo, pp. 165-191.

Horgen, P. A. (1980). The role of the steroid sex pheromone antheridiol controlling the development of male sex organs in the water mold, *Achlya*. In *Sexual Interactions in Eukaryotic Microbes*. D. H. O'Day and P. A. Horgen (Eds.). Academic, New York, pp. 155-174.

Ishida, S. (1980). A mutant of *Dictyostelium discoideum* capable of differentiating without morphogenesis. *Dev. Growth Differ. 22*: 143-152.

Jaffe, L. F. (1968). Localization in the developing *Fucus* egg and the general role of localizing currents. *Adv. Morphog. 7*:295-328.

Jaffe, L. F., and Nuccitelli, R. (1977). Electrical controls of development. *Annu. Rev. Biophys. Bioeng. 6*:445-476.

Lacalli, T. C., and Harrison, L. G. (1979). Turing's conditions and the analysis of morphogenetic models. *J. Theor. Biol. 76*:419-436.

Maclean, N. (1977). *The Differentiation of Cells*. Edward Arnold, London.

Matile, P., Moore, H., and Robinow, C. F. (1969). Yeast cytology. In *The Yeasts, Biology of yeasts*, Vol. 1. A. Rose and J. S. Harrison (Eds.). Academic, London and New York, pp. 219-302.

Mitchell, P. (1977). Vectorial chemiosmotic processes. *Annu. Rev. Biochem. 46*:996-1005.

Najim, L., and Turian, G. (1979). Conidiogenous loss of structurofunctional polarity in the hyphal tips of *Sclerotinia fructigena*. *Eur. J. Cell. Biol. 20*:24-27.

Newell, P. C. (1978). Progress on slime mould morphogens. *Nature 271*:302-303.

Nicolis, C., and Prigogine, I. (1977). *Self-Organization in Nonequilibrium Systems*. Wiley-Interscience, New York.

Ojha, M., and Turian, G. (1978). Loss of rRNA genes during a phenotypic, masculinizing mutation in *Allomyces arbuscula*. *Mol. Gen. Genet.* *166*:225-227.

Pasternak, C. A. (1970). *Biochemistry of Differentiation*. Wiley-Interscience, New York.

Paul, J. (1971). Transcriptional regulation in mammalian chromosomes. In *Control Mechanisms of Growth and Differentiation*. D. D. Davies and M. Balls (Eds.). *Symp. Soc. Exp. Biol.* *25*:117-126.

Paznokas, J. L., and Sypherd, P. S. (1975). Respiratory capacity, cyclic adenosine 3', 5'-monophosphate and morphogenesis of *Mucor racemosus*. *J. Bacterial* *124*:134-139.

Ross, I. K. (1979). *Biology of the Fungi*. McGraw-Hill, New York.

Rossier, C., Ton-That, T. C., and Turian, G. (1977). Microcyclic microconidiation in *Neurospora crassa*. *Exp. Mycol.* *1*:52-62.

Sinnott, E. W. (1960). *Plant Morphogenesis*. McGraw-Hill, New York.

Slayman, C. L., and Slayman, C. W. (1962). Measurement of membrane potentials in *Neurospora*. *Science* *136*:876-877.

Smith, J. E., and Berry, D. R. (1974). *An Introduction to Biochemistry of Fungal Development*. Academic, London.

Stewart, P. R., and Rogers, P. J. (1978). Fungal dimorphism: A particular expression of cell wall morphogenesis. In *The Filamentous Fungi, Developmental Mycology;* Vol. 3. J. E. Smith and D. R. Berry (Eds.). Edward Arnold, London, pp. 164-196.

Stump, R. F., Robinson, K. R., Harold, R. L., and Harold, F. M. (1980). Endogenous electrical currents in the water mold *Blastocladiella emersonii* during growth and sporulation. *Proc. Nat. Acad. Sci. USA* *77*:6673-6677.

Ton-That, T. C., Turian, G., Fakan, J., and Gautier, A. (1981). Ultrastructural cytochemistry of perinucleolar dense spots in heat-treated macroconidia of *Neurospora crass*. *Eur. J. Cell Biol.* *24*: 317-319.

Turian, G. (1969). *Différenciation fongique*. Masson, Paris.

Turian, G. (1975). Differentiation in *Allomyces* and *Neurospora*. *Trans. Br. Mycol. Soc.* *64*:367-380.

Turian, G. (1979_a). Cytochemical gradients and mitochondrial exclusion in the apices of vegetative hyphae. *Experientia* *35*:1164-1166.

Turian, G. (1979_b). Polarity of elongation growth generated and sustained by anisotropic distribution of the protons ejected from mitochondria into the cytosol of hyphal apices (Neurospora model)? *Arch. Sci. Geneve* *32*:251-254.

Turian, G. (1980). Germination of conidia (*Neurospora, Trichoderma*): Evidence of an electrogenic sink of protons mitochondrially-extruded into emerging tubes. *Ber. Schweiz. Bot. Ges.* *90* (3/4):203-212.

Turian, G. (1981). Low pH in fungal bud initials. *Experientia* *37*: 1378-1379.

Turing, A. M. (1952). The chemical basis of morphogenesis. *Philos. Trans. R. Soc. Ser. B.* 237:37-72.

Wolpert, L. (1971). Positional information and pattern formation. In *Current Topics in Developmental Biology*, Vol. 6, A. A. Moscona and A. Monroy (Eds.). Academic, New York, pp. 183-224.

Wright, B. E. (1973). *Critical Variables in Differentiation*, Concepts of Modern Biology Series. Prentice-Hall, Engelwood Cliffs, N.J.

2

Metabolic Organization During Differentiation in the Slime Mold

BARBARA E. WRIGHT* and RUTH EMYANITOFF *Boston Biomedical Research Institute, Boston, Massachusetts*

I. INTRODUCTION

Among the most difficult phenomena to analyze in living organisms are intracellular and intercellular metabolic compartmentation. However, such analyses are crucial, since subcellular organization and the interdependence of specialized cells are the very essence of differentiating systems. One approach for understanding metabolic organization is through the construction of kinetic models, which represent one of the

**Present affiliation*: University of Montana, Missoula, Montana few noninvasive techniques available for analyzing compartmentation. As our biochemical knowledge grows, models will become increasingly essential to its integration and comprehension. By organizing the data, models serve to clarify which facts are compatible and pinpoint those which are not. Models help us ask questions and think in realistic, dynamic terms which relate directly to the living organism. They serve as a realistic framework within which to judge the relevance of in vitro data to metabolism in vivo, make specific predictions to be tested experimentally, and suggest analytical approaches that would be difficult to imagine otherwise. In effect, the construction and exploration of kinetic models represents a powerful *analytical* technique.

Over the past few decades, progress in understanding the biochemical basis of differentiation has been achieved primarily through the isolation and analysis of individual cellular components. Progress in the future will depend upon the extent to which this in vitro information can be used to understand the functioning of an intact organism. Experimental advances in this direction have been made with the study of enzymes in vivo (Wright and Bard, 1963; Kim and Miller, 1969; Ferrari and Varner, 1970) and in situ (Srere and Mosbach, 1974), with the isolation or reconstitution of enzyme complexes (Halper and Srere, 1977; Pramanik et al., 1979) and with the analysis of immobilized enzyme systems in which metabolite-enzyme relationships approach conditions existing in vivo (Srere and Mosbach, 1974; Ottaway and Mowbray, 1975). Other investigations have been concerned with the intercellular (Rutherford, 1976; Rutherford and Harris, 1976; Wilson and Rutherford, 1978) or intracellular distribution of enzymes and metabolites (Sols and Marco, 1970; Oaks and Bidwell, 1970; Davis et al., 1978), with the influence of this distribution on enzyme activity (Lowry and Passonneau, 1964; Masters, 1977) and on the ratio of soluble to bound enzymes (Koch-Schmidt et al., 1977; Wilson, 1978). A further step toward understanding metabolism in the intact cell has been the study of pathways in isolated organelles such as mitochondria, peroxisomes, or glyoxysomes (Kobr et al., 1969; Stucki and Walter, 1972; Tischler et al., 1977). The long-term goal in analyzing all of these systems is to understand the relationships and control mechanisms of metabolism under the dynamic conditions of living organisms. Each experimental approach has its particular advantages and disadvantages.

Although the ultimate goal is to understand metabolism in intact cells, the complexity therein is enormous, and we are very limited in approaches to the analysis of metabolism in vivo. By contrast, elegant and definitive investigations can be carried out with isolated, well-characterized enzyme systems; however, the experimental conditions used must differ from those existing in vivo, and the relevance of such work is therefore uncertain (Sols and Marco, 1970; Srere and Mosbach, 1974; Masters, 1977). The more complex the experimental system, the more difficult it is to study, and yet the closer we are to the living cell. Hopefully, in the future, the in vitro and in vivo approaches will converge, and the extensive information obtained by using purified components isolated from differentiating cells will be incorporated into realistic kinetic models which simulate metabolism within these cells. Until such models are developed and demonstrated to have predictive value, we do not fully understand metabolism in vivo (Wright, 1979; Wright and Davison, 1980).

It should be pointed out that modeling is not esoteric; some sort of model or hypothesis is used as the basis for most experimentation. A mathematical (computer) model is really a hypothesis formulated in quantitative detail. If the outcome of an experiment is consistent with the model, the validity of the model becomes more probable. Thus model building is an essential part of any scientific pursuit. This article will deal with two steady state models of metabolic organization in *Dictyostelium*, based on extensive experimental data: (1) intra- and extramitochondrial compartments of the citric acid cycle intermediates and (2) intercellular compartments of specific saccharides and their precursors in the spore and stalk cells. Before these models can be discussed it is necessary to describe the life cycle of *Dictyostelium discoideum* and the computer program used for data analysis.

A. The Life Cycle

The life cycle of the ameba *Dityostelium discoideum* is characterized by a free-living vegetative phase in the presence of an adequate food source and, under starvation conditions, a phase in which the amebas aggregate, age, and differentiate into one of two cell types. Differentiation is completed in about 24 h, with the formation of a fruiting body composed of a cellulose stalk supporting a mass of spore cells. The stalk cells die and the spore cells enter a state of dormancy. Since differentiation in the cellular slime mold occurs in a starving system; endogenous reserves of protein, carbohydrate, RNA, and lipids must supply all the precursors for the synthesis of new materials (e.g., cellulose, trehalose) and for the generation of energy over the 24 h of differentiation. During this period, total carbohydrate remains essentially constant (Rosness and Wright, 1974) and total lipid changes are relatively small (Long and Coe, 1974). However, protein decreases by approximately one-half (Wright and Anderson, 1960; White and Sussman, 1961). A quantity of ammonia compatible with the amount of protein degraded is released from the differentiating organism (Sussman et al., 1977; Wilson and Rutherford, 1978). In addition, there is evidence that the conversion of amino acids to carbohydrates is not significant (Cleland and Coe, 1969). Taken together, these results strongly suggest that carbon supplied by protein degradation is the major energy source for differentiation and aging in this organism.

B. The TFLUX Program

Since compartmentation in the intact cell exists at the intercellular, organellar, multienzymatic, and microenvironmental levels, it is highly desirable to obtain biochemical information in vivo without affecting cellular integrity. To assess the rate or flux of a reaction in vivo, cells can be exposed to tracer (nonperturbing) levels of a radioactive metabolite over brief periods of time under steady state conditions. The cell samples are then removed, the metabolites isolated, and their specific radioactivities determined. A computer program called TFLUX has been developed to simulate the changes in the specific radioactivity of each metabolite pool as a function of time (Sherwood et al., 1979).

The systems which TFLUX can simulate consist of chemically distinct species (pools) with time-varying specific radioactivities. Chemical reactions or transport mechanisms provide an interchange of tracers among pools. The metabolic system must be in steady state; that is, all pool sizes must be constant (the pool size is the sum of labeled and unlabeled material), and the total (labeled plus unlabeled) fluxes between pools must also be constant. TFLUX solves a set of simultaneous first-order differential equations to simulate the specific radioactivities for all pools as a function of time. To use TFLUX, the user specifies a number of pools, giving the name, size (millimolar cell volume), and initial specific radioactivity for each pool. The interconnections between pools are specified by giving the (constant) flux between any pair of pools. FLUX is expressed as the millimolar concentration per minute, which is equivalent to the number of micromoles per minute per milliliter cell volume. It is also possible to specify "external pools, whose properties include infinite size and constant specific radioactivity. For more information regarding this program see Sherwood et al. (1979).

II. INTRACELLULAR COMPARTMENTATION IN *DICTYOSTELIUM*: THE CITRIC ACID CYCLE

A. Experimental Basis for Model Construction

Flux through the citric acid cycle has been estimated from net protein degradation, NH_3 production, and O_2 uptake (Kelly et al., 1979a). These estimates were in fair agreement; the O_2 uptake data was used for the model, as it was considered the most accurate indication of cycle activity. Flux through the cycle, about 0.4 mM/min, had to be matched by an equivalent total flux of cycle precursors (amino acids) into the cycle. Thus 0.4 mM/min was broken down into several flux rates into the cycle at various points by factoring in the amino acid composition of typical cell proteins (Kelly et al., 1979a,b). Figure 1 indicates the points of entry into the cycle, as well as the compart-

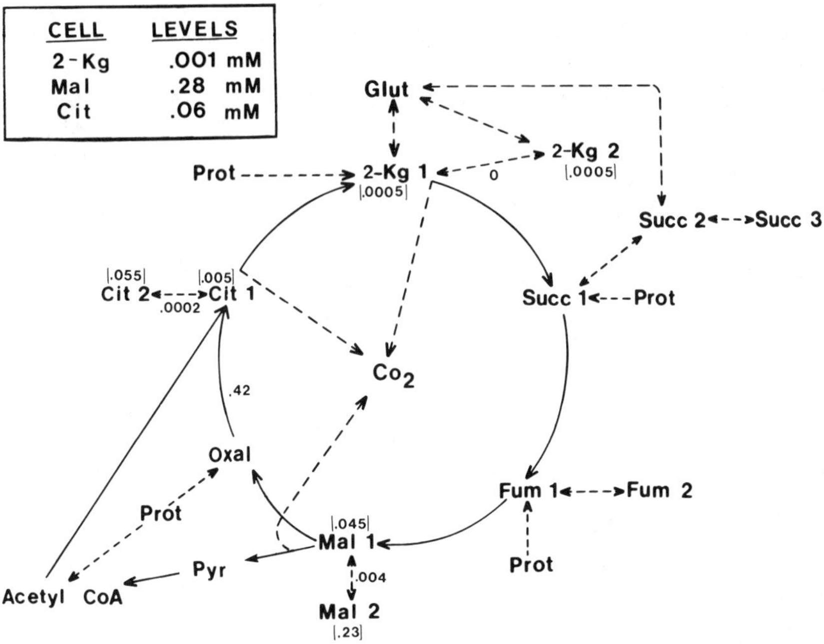

FIG. 1 Model used in the initial simulations of the [^{14}C]glutamate data. Total cellular levels of 2-ketoglutarate (2-Kg), malate (Mal), and citrate (Cit) are indicated in the table insert. The intramitochondrial levels of these pools are given in brackets associated with 2-Kg 1, Mal 1, and Cit 1, respectively; extramitochondrial levels are given in brackets associated with 2-Kg 2, Mal 2, and Cit 2, respectively. Selected flux values are shown near arrows indicating reactions between metabolite pools. See text for further discussion. (From Kelly et al., 1979a).

mentation of cycle intermediates into intra- and extramitochondrial pools, which will be discussed below.

A problem arises in a system like this, a system which uses only amino acids as the source of cycle intermediates. More amino acids are converted to four- and five-carbon intermediates than to acetyl CoA. For the citrate synthetase reaction to operate and maintain steady state levels of cycle intermediates, the flux of acetyl CoA into the cycle must equal that of oxaloacetate. Therefore a pathway is required which converts "excess" four- or five-carbon intermediates to acetyl CoA. Isotopic and enzymatic analyses have demonstrated that malic enzyme (E.C.1.1.1.40; malate + NAP = pyruvate + CO_2 + NADH + H^+), indicated in the lower left corner of Fig. 1, serves this function (Kelleher et al., 1979). Furthermore, this enzyme is allosteric, being positively affected by aspartate and, to a lesser extent, glutamate. Thus amino acid production from protein degradation stimulates a reaction essential to the efficient utilization of these amino acids for energy.

The total cellular levels of cycle intermediates and related amino acids were determined; three of these are shown in Fig. 1 (table insert). These levels, of course, represent the combined intra- and extramitochondrial pools, which mix upon cell rupture. The specific radioactivities of these intermediates over time were determined following brief exposure of intact cells to tracer levels of [^{14}C]glutamate, [^{14}C]aspartate, or [^{14}C]alanine. Figure 2 shows the results of exposing culminating cells to [^{14}C]glutamate. As this experiment was done under conditions of metabolic steady state, pool sizes were constant and fluxes into and out of each pool were equal. Inspection of the data led to some initial conclusions: [^{14}C]glutamate equilibrates immediately with the glutamate pool and the intramitochondrial cycle intermediates. This must be the case, since all cycle intermediates are labeled rapidly, and the whole cycle only functions within the mitochondria. Mitochondrial malate, fumarate, and citrate then exchange with their counterpart cytoplasmic pools, accounting for the slow rise

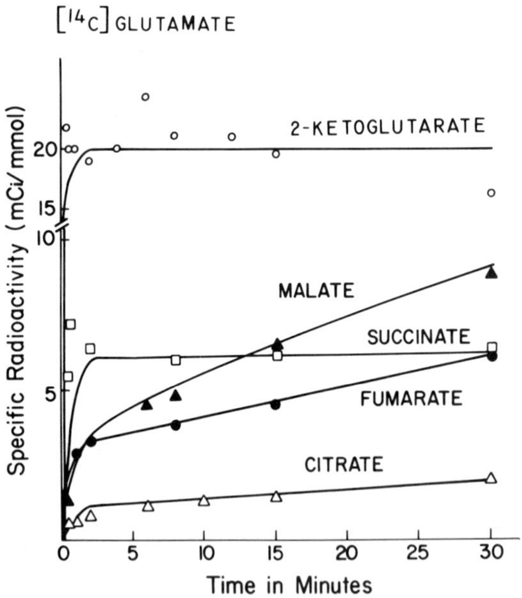

FIG. 2 Simulation of the [^{14}C]-glutamate data. The solid lines represent computer output superimposed on the experimental data (symbols) for the metabolites as indicated. (From Kelly et al., 1979a).

in their (average) specific radioactivities. Since 2-ketoglutarate and succinate remain constant in specific radioactivity, either there is only one pool of each or their cytoplasmic pools are labeled immediately or no exchange occurs between cytoplasm and mitochondria. The relative sizes of the intra- and extramitochondrial pools and the fluxes between them cannot be measured, but they can be predicted using TFLUX. These parameters are considered as unknowns, and their values are varied until the computer specific radioactivity curves match the data. The TFLUX program starts with the input specific radioactivity of [^{14}C]-glutamate at t=0 and calculates the specific radioactivities over time of all the metabolites shown in Fig. 1, as well as the average values for the two pools of a given metabolite. Selected pool concentrations and fluxes from the model giving output matching the experimental data are indicated in Fig. 1. Note that no exchange occurs between the two 2-ketoglutarate pools, and that the rate of exchange between the malate pools is higher than that between the citrate pools. This circumstance was necessitated by the more rapid rise in the (average) malate specific radioactivity compared to that of citrate (see Fig. 2). The final model, capable of simulating all the labeling data, is even more complicated (Kelly et al., 1979b). Output from this model matching the [^{14}C]-glutamate data (symbols) is actually represented by the curves (solid line) in Fig. 2. Figure 3 shows the match to data obtained using [^{14}C]-aspartate as the tracer.

B. Conclusions and Model Predictions

These analyses have led to the following conclusions.

1. Flux through the citric acid cycle is approximately 0.4 mM/min. It can be calculated from the model that an ATP flux of about 5 mM/min would be produced if all of the $FADH_2$ and NADH produced were oxidized by the cytochrome chain. Rutherford and Wright (1971) have calculated a value of 12 mM/min as a minimum flux through the ATP pool. These calculations were derived from the labeling kinetics of the adenosine phosphate pools by [^{3}H]-adenine. One explanation for this discrepancy is that the calculation by the latter authors included ATP flux due both to tricarboxylic acid cycle activity and to an active adenylate kinase present in *Dictyostelium*, whereas the value derived from the model represents only the contribution to ATP turnover from the tricarboxylic acid cycle. Calculation of ATP yield from the model is consistent with the earlier estimate of ATP turnover if the activity of adenylate kinase is considered. An ATP flux of 5 mM/min is approximately four times that required by the known metabolic demands of the system, such as protein and RNA turnover and the synthesis of new carbohydrate end products of differentiation (Wright and Gustafson, 1972).

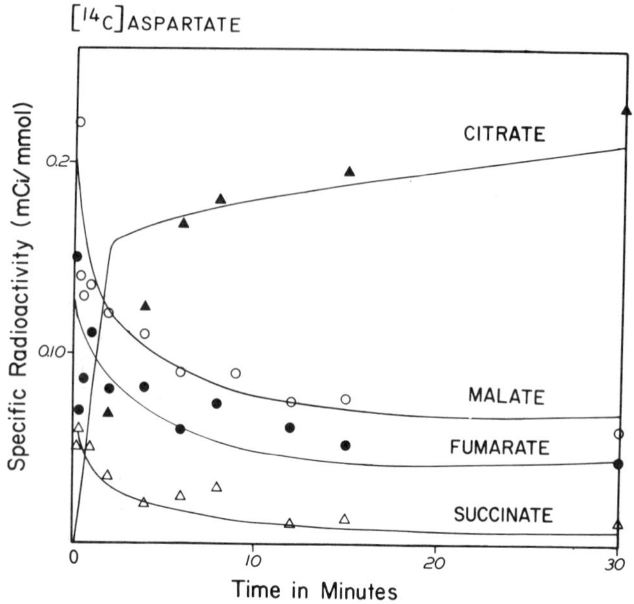

FIG. 3 Simulation of the [^{14}C]-aspartate data. The solid lines represent computer output superimposed on the experimental data (symbols) for the metabolites as indicated. (From Kelly et al., 1979a).

2. Both fumarate and malate are compartmentalized between cycle and noncycle pools at a ratio of approximately 1:5. Citrate is compartmentalized at a 1:10 ratio between cycle and noncycle pools. Succinate appears to exist in three compartments, two of which become labeled by [^{14}C]-glutamate and only one by [^{14}C]-aspartate.
3. Two pools of aspartate with two associated pools of oxaloacetate are necessary for simulation.
4. Exchange between the cycle and noncycle pools of both citrate and fumarate occurs at very low rates of about 0.0003 mM/min, whereas exchange between the malate pools is about 0.004 mM/min. The exchange reaction glutamate $\rightleftharpoons$ 2-ketoglutarate occurs at approximately 15 times the rate of the cycle flux.
5. A reaction catalyzed by malic enzyme is included in the model, as this reaction is necessary for complete oxidation of amino acid substrates.

One prediction resulting from this analysis concerns malic enzyme, which presumably is rate limiting to cycle flux because of the need in this system for acetyl CoA. If this is correct, the addition of metabolites which can be rapidly transformed to pyruvate or acetyl CoA might

TABLE 1 Stimulation of O_2 Uptake by Exogenous Metabolites

Metabolite	Percentage of stimulation
Glucose	100
Glutamate	30
Alanine (→pyruvate)	30
Cysteine (→pyruvate)	16
Histidine (→2-ketoglutarate)	2
Phenylalanine (→fumarate)	0
Pyruvate	2
Succinate	0
Malate	0
Citrate	0

Source: Modified from Table 3 of Liddel and Wright (1961).

stimulate oxygen uptake. About 20 years ago we had tested the effect of certain metabolites on oxygen uptake (Liddel and Wright, 1961), but did not understand the results at that time (Table 1). The following interpretation is now possible: Glutamate stimulates O_2 consumption as an allosteric effector of malic enzyme, whereas alanine and cysteine stimulate because of their conversion to pyruvate (and hence acetyl CoA). The poor stimulation by added pyruvate may be due to its lability or poor permeability as an acid. If these interpretations are correct, acetate and pyruvate might be more effective at a lower pH. Also, aspartate should stimulate more than glutamate, as the former is the more effective in enhancing the activity of malic enzyme.

Another prediction from the model is based on the observation that toward the end of differentiation, the cycle intermediates and associated amino acids accumulate (Kelly et al., 1979a). This circumstance may have been selected for in evolution to provide dormant spores with a supply of energy sources which can be used immediately in times of stress or under circumstances suitable to germination. One might further speculate that intra- and extramitochondrial compartments evolved so that during dormancy, cycle intermediates could be stored predominantly in cytoplasmic pools, thus preventing their rapid oxidation. Unknown mechanisms would then allow their reentry into the cycle in a time of need. Thus the prediction arises that the cytoplasmic pools should become relatively larger during dormancy.

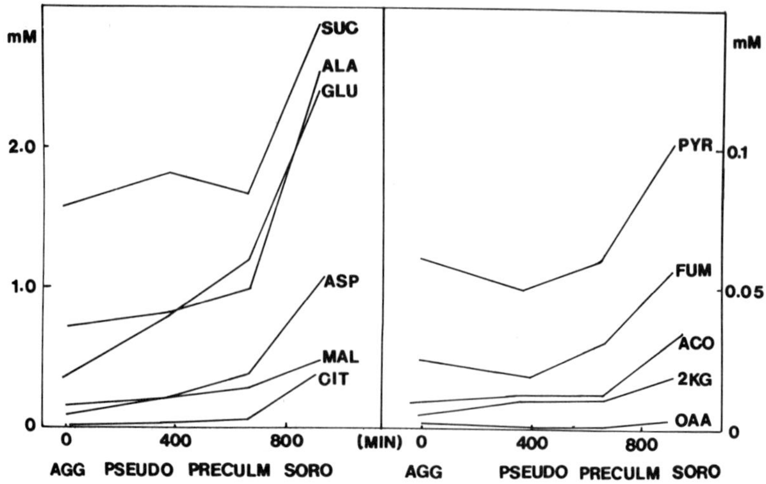

FIG. 4 The accumulation during differentiation (AGG, aggregation; PSEUDO, pseudoplasmodium; PRECULM, preculmination; SORO, sorocarp) of eight citric acid cycle intermediates and three amino acids (SUC, succinate; ALA, alanine; GLU, glutamate; ASP, aspartate; MAL, malate; CIT, citrate; PYR, pyruvate; FUM, fumarate; ACO, acetyl CoA; 2 Kg, 2-ketoglutarate; OAA; oxaloactetate). Concentration is expressed in terms of cell volume. (Data from Kelly et al., 1979b.)

C. The Steady State Model as a Basis for Constructing a Transition Model

The steady state model of the citric acid cycle is serving as the basis for constructing a transition model of the cycle during differentiation. As mentioned above, the cycle intermediates and related amino acids accumulate during differentiation. The exact patterns of accumulation are shown in Fig. 4. We have chosen, initially, to simulate the accumulation of these metabolites over the last 300 min of differentiation, as flux through the cycle (i.e., O_2 consumption) is constant during this period (Kelly et al., 1979a). This transition model must be consistent with the constancy of O_2 consumption over this period and with the intra- and extramitochondrial compartmentation and flux relationships characterizing our steady state model. During the construction of the transition model, it has become clear that two general conditions must be met in order to achieve the observed accumulation of metabolites while keeping the cycle flux constant: (1) flux from protein to amino acids must increase about six fold over the steady state conditions and (2) reactions in which amino acids are converted to cycle intermediates, as well as reactions within the cycle, must be inhibited to approximately the same extent (Wright, 1981).

Simulation of this transition period requires a more complicated program called METASIM. The METASIM program utilizes the mathematical expressions for the kinetic mechanism of the enzymes involved and incorporates the effects of known inhibitors and effectors. A more detailed explanation of METASIM can be found in the review by Wright and Kelly (1981). In order to simulate the observed accumulation of intermediates, a physiologically feasible mechanism for inhibiting reactions feeding into and within the cycle is required in the model. We speculated that both NADH, a product of several cycle reactions, and NH_3, which accumulates during differentiation (Sussman et al., 1977; Wilson and Rutherford, 1978) might serve as inhibitors of dehydrogenases within the cycle, as well as dehydrogenases and deaminases providing sources of cycle intermediates. As shown in Table 2, several key enzymes were inhibited by NADH or NH_3 at levels at or below intracellular concentrations of those compounds (Emyanitoff et al., 1981). These inhibition constants are being incorporated into a transition model. Other data being used, in addition to that described for the steady state model, are kinetic constants and mechanisms for several key enzymes in the model. These include malic enzyme (Kelleher et al., 1979), glutamate dehydrogenase (Landridge et al., 1977), citrate syntetase (Porter and Wright, 1977), aspartate and alanine transaminases (Komuniecki and DeToma, 1979), malate dehydrogenase (Emyanitoff and Kelly, 1982), and isocitrate dehydrogenase (Emyanitoff, 1982). Thus results from the steady state model, along with data from additional experimental analyses, will serve as a basis for developing a model of metabolism in an organism undergoing a transition from one metabolic steady state to another, such as occurs during differentiation and aging.

TABLE 2 Effect of NADH and NH_3 on Selected Citric Acid Cycle-Associated Enyzmes, Comparison of K_i with Intracellular Concentration

Enzyme	Inhibitor	K_i (mM)	Intracellular concentration of inhibitor (mM)
Glutamate dehydrogenase	NADH	0.03	0.04
	NH_3	3.00	11.00
Malate dehydrogenase	NADH	0.02	0.04
Isocitrate dehydrogenase	NADH	0.01	0.04

III. INTERCELLULAR COMPARTMENTATION: CARBOHYDRATE METABOLISM

A. Preliminary Evidence

In connection with the analysis of transition models of carbohydrate metabolism in *Dictyostelium*, predictions were made that glucose and the hexose phosphates must exist in at least two metabolic compartments during the culmination process (for a review, see Wright and Kelly, 1981). It was also predicted that the glucose pool in the pathway leading to glycogen synthesis was very low in concentration (compared to total cellular levels), and that it was this pool which was perturbed by exogenous glucose, resulting in higher glycogen levels. In an independent investigation, Wilson and Rutherford (1978) found that the glucose measured in extracts of culminating organisms was derived from the stalk cells; glucose in spores was undetectable by their assay. Hence glucose perturbation resulting in enhanced glycogen levels must occur in the spore cells. In order to further analyze the intercellular compartmentation of glucose and the hexose phosphates and to investigate the metabolic relationship between the two cell types, steady state models were constructed.

B. Experimental Basis for Model Construction

Figure 5 indicates the metabolites and reactions included in the model. As in the case of the citric acid cycle intermediates, the total cellular levels of all species is known (Table 3). However, unlike the citric acid cycle, a number of the compartmented metabolites and many flux relationships are known. The rates of synthesis of UDP-glucose (UDPG), glycogen (GLY), and the saccharide end products of differentiation (cellulose, trehalose, cell wall glycogen, and glucosaminoglycan) are also known, as well as the contribution of net glycogen and RNA degradation to saccharide biosynthesis (Table 4). The model is thus highly constrained by the extensive data which is available; that is, only a few compartmentation ratios and fluxes could be varied while constructing the model.

In order to label the relevant metabolites, *Dictyostelium* was exposed for 40 min to tracer levels of [U-^{14}]-glucose (EXGLU) at various times during the culmination process. A perchloric acid extract was prepared, and all of the metabolites indicated in Fig. 5 were isolated to determine their specific radioactivities. A combination of differential precipitation, absorption, and column and thin layer chromatographic techniques were used, and the purity of each compound was verified with specific chemical and enzymatic analyses (Wright et al., 1982). A total of 10 such experiments were carried out. This discussion will describe one in which the synchrony of culminating organisms was excellent, and in which the

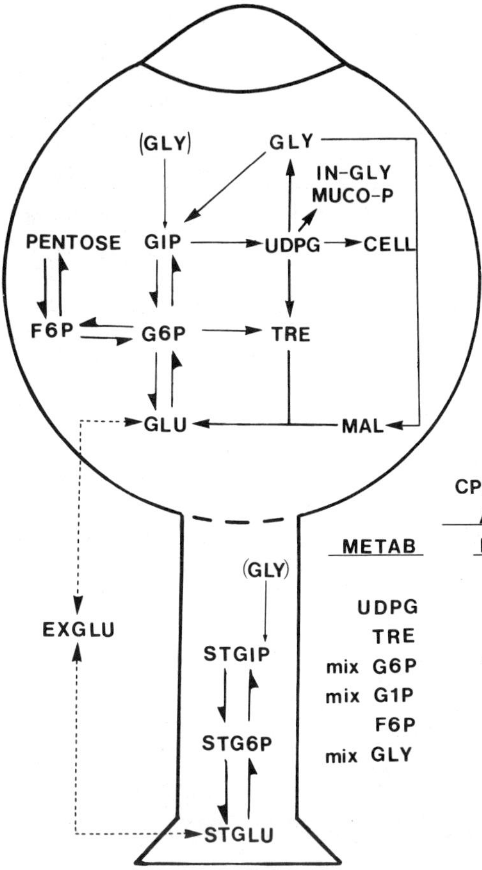

METAB	CPM x 10^{-4}/µmol AT 40 min DATA	MODEL
UDPG	6.7	7.2
TRE	5.0	5.7
mix G6P	164.0	144.0
mix G1P	1.2	2.1
F6P	3.3	3.1
mix GLY	0.63	0.65

FIG. 5 Reactions included in the steady state model of spore and stalk cells in *Dictyostelium* [EXGLU, exogenous radioactive glucose; (GLY), spore and stalk glycogen which is degraded; GLY, spore glycogen participating in the glycogen cycle; G1P, spore glucose-1-phosphate; G6P, spore glucose-6-phosphate; UDPG, UDP-glucose; IN-GLY, insoluble cell wall glycogen; MUCO-P, mucopolysaccharide or glucosaminoglycan; CELL, cellulose; TRE, trehalose; MAL, maltose; GLU, spore glucose; F6P, fructose-6-phosphate; STG1P, stalk glucose-1-phosphate; STG6P, stalk glucose-6-phosphate; STGLU, stalk glucose]. The table insert compares model with data specific radioactivities at 40 min. "Mix" refers to metabolite pools existing in both spore and stalk cells; they mix upon cell rupture and are isolated as a single pool. See Tables 3 and 4 for model parameters.

TABLE 3 A Comparison of Metabolite Levels Determined Experimentally[a] with Values Used in the Model

Metabolite	Concentration of packed cell volume of total organism (mM)	
	Experimental	Model
Exogenous glucose (EXGLU)	≅0.1	0.17
UDP-glucose (UDPG)	0.1-0.4	0.10
Fructose-6-phosphate (F6P)	0.04	0.04
Maltose (MAL)	0.14	0.14
Total cellulose[b] (CELL)	5.0	5.0
Total wall glycogen (IN-GLY)	Represented as external pool in model[d]	
Glucosaminoglycan (MUCO-P)	Represented as external pool in model[d]	
Pentose	Represented as external pool in model[d]	
Total trehalose[b] (TRE)	1.0-3.0	1.0
Total glycogen [GLY+(GLY)]	17.0	16.5
Degradation glycogen	–	15.0
Turnover glycogen	–	1.5
Total glucose-6-phosphate (G6P+STG6P)	0.02-0.1	0.05
Spore G6P (G6P)	–	0.0001
Stalk G6P[c] (STG6P)	–	0.05
Total glucose-1-phosphate (G1P+STGLU)	0.003-0.02	0.05
Spore G1P (G1P)	–	0.005
Stalk G1P[c] (STG1P)	–	0.05
Total glucose (GLU+STGLU)	0.04-0.1	0.03
Spore glucose (GLU)	–	0.005
Stalk glucose[c] (STGLU)	–	0.03

[a]For review of the experimental data and original references see Wright and Kelly (1981).
[b]Although these exist in both cell types, they are shown only in spore cells where they are synthesized.
[c]If stalk pools were expressed in terms of stalk volume, they would be 10 times greater, as stalk cells represent about 10% of the total cellular volume of the organism.
[d]See text.

TABLE 4 A Comparison of Reaction Rates Determined Experimentally[a] with Values Used in the Model

Reaction	Flux (mM/min)[b] Experimental	Model
UDPG → GLY	≅0.16	0.14
GLY → G1P	≅0.13	0.14
G1P → UDPG	≅0.16	0.21
(GLY) → G1P	0.05 (average)	0.06
UDPG → CELL	0.03	0.03
UDPG → TRE	0.01-0.03	0.01
UDPG → IN-GLY+MUCO-P	≅0.02	0.02
GLY → MAL	–	0.0002
G1P → G6P	–	0.08
G6P → G1P	–	0.09
G6P → TRE	0.01-0.03	0.01
TRE → GLU	0.0	0.0
MAL → GLU	–	0.0002
G6P → F6P	–	0.02
F6P → G6P	–	0.05
GLU ↔ G6P	–	0.01
EXGLU ↔ GLU	–	0.008
EXGLU ↔ STGLU	–	0.5
STGLU ↔ STG6P	–	0.04
STG6P ↔ STG1P	–	0.0009
(GLY) → STG1P	0.05 (average)	0.08
Pentose → F6P	0.02	0.05
F6P → Pentose	–	0.02

[a]For a review of the experimental data and original references see Wright and Kelly (1981).
[b]Flux to or from external pools not shown; see text.

data were the most compelling with respect to constraining the model parameters; that is, the model approached a unique solution. The specific radioactivities of the metabolites isolated in this experiment are indicated in the table of Fig. 5, which compares these data with the model values at 40 min. The kinetics of labeling over this 40-min period in the model are shown in Fig. 6; qualitatively, this may be compared to Fig. 7, which is a composite of two experiments. Glycogen and the saccharide end products are labeled in a fairly linear manner compared to the small, rapidly turning over pools of glucose-6-phosphate (G6P), glucose-1-phosphate (G1P), and fructose-6-phosphate

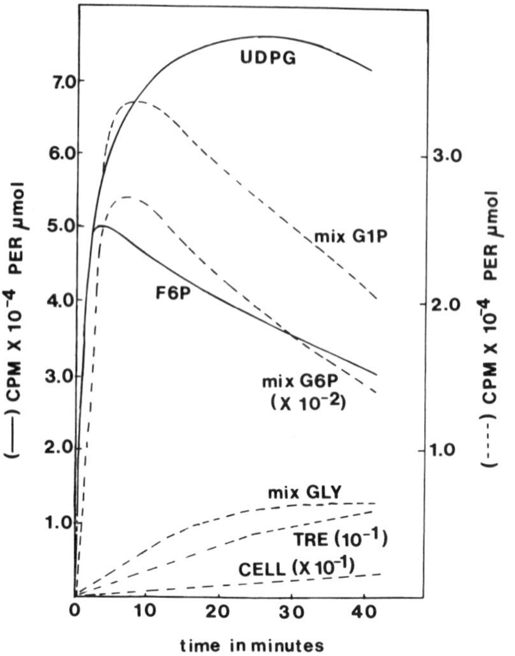

FIG. 6 Simulation of the labeling kinetics of the pools isolated. The final specific radioactivities are given in Fig. 5. "Mix" refers to metabolites existing in each cell type in vivo. Model parameters are given in Tables 3 and 4.

(F6P); the UDPG pool shows intermediate labeling kinetics. The word *mix* in Figs. 6 and 7 refers to a metabolite existing in both spore and stalk cells; the two pools will mix upon cell rupture, and the specific radioactivity of that metabolite will equal the total number of counts in both pools divided by the total number of micromoles in both pools.

Table 3 compares the experimental and model values for total metabolite concentrations and indicates the distribution in the model of those metabolites which are divided into spore and stalk cell compartments. The higher glucose concentrations in stalk as compared to spore cells has been demonstrated experimentally (Wilson and Rutherford, 1978) and predicted by modeling (Wright et al., 1977). Compartmentation of glycogen has also been demonstrated experimentally (Rutherford, 1976), and the compartmentation of G6P and G1P has been predicted (Wright et al., 1977; Wright and Kelly, 1981). The specific concentrations of these metabolites in the spore and stalk cell compartments are dictated by their specific radioactivities, known fluxes, and other model constraints. As indicated in Table 4, many reaction rates

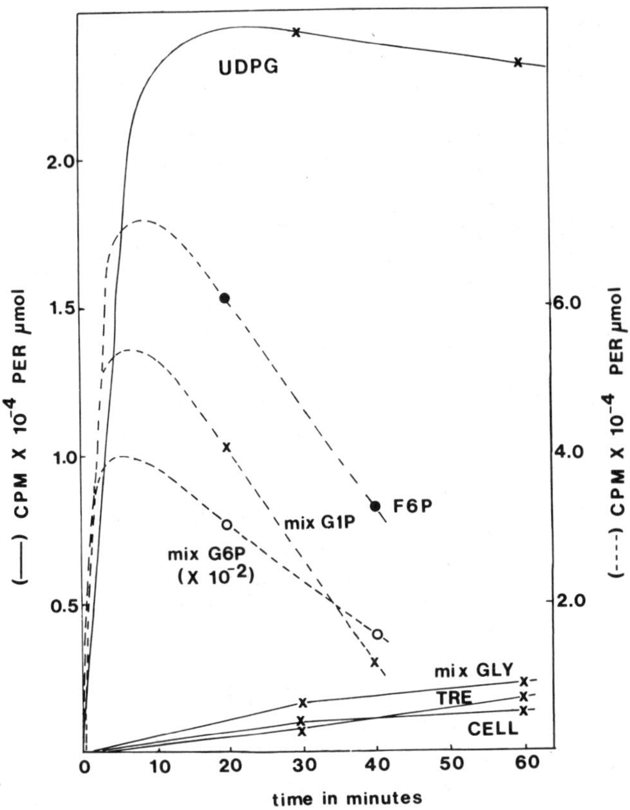

FIG. 7 A composite of two experiments indicating the shape of the labeling patterns for the metabolites isolated.

have been determined experimentally (for a review, see Wright and Kelly, 1981); others can be accurately deduced from the specific radioactivity data and/or the assumption of steady state, as will be discussed below.

In the experiment being analyzed (and in most of the others performed), the hexose phosphates had specific radioactivities one to two orders of magnitude higher than that of the UDPG isolated. We assume that a species of UDPG with high specific radioactivity comparable to the hexose phosphates does not exist. Apparently there is only one UDPG pool, in spore cells, based on the following considerations. The UDPG isolated at all stages of differentiation in these (and previous) investigations must be the precursor, in vivo, of soluble glycogen and all end-product saccharides synthesized during differentiation. This is so because the rates of synthesis of UDPG, soluble glycogen, cellulose, and trehalose calculated from, and based upon, the specific radioactivity of this UDPG are consistent with the rates of accumulation (cellulose and trehalose) and turnover (glycogen) of these saccharides (see Table 4). If the rates of synthesis of these saccharides were calculated from the specific radioactivity of a separate pool of UDPG derived from the high-specific-radioactivity hexose phosphates isolated, they would be one to two orders of magnitude too low to account for their rates of turnover or accumulation. It follows that there must be a single pool of UDPG and that the hexose phosphates and glucose must exist in two metabolically independent compartments which mix during their isolation: one species of low specific radioactivity, acting as a precursor to UDPG, and one species of high specific radioactivity, not associated with the UDPG isolated. The simplest assumption is that the latter hexose phosphates are located in stalk cells, whereas the low-specific-radioactivity hexose phosphates are in the spore cells, with the UDPG which serves as a precursor to soluble glycogen and the end-product saccharides. This interpretation is compatible with the fact that permeability to exogenous metabolites peaks during sorocarp construction (Wright, 1963), when dying stalk cells are breaking down and becoming vacuolated and presumably more permeable. As discussed earlier, perturbation studies with phosphorylated metabolites also indicate that stalk cells are more permeable than spore cells. The lack of trehalose synthesis in stalk cells under the nonperturbed conditions of the present study is apparently due to substrate limitation (B. E. Wright, 1981 unpublished data).

C. Model Constraints

The model is highly constrained by the extensive data available. For example, flux through the glycogen cycle is fixed by the experimental data; this flux and the specific radioactivity of the isolated glycogen determine the fraction of glycogen present which is actively turning over in the cycle. The rate of exchange between spore G6P and F6P, which arises from unlabeled pentose phosphate, is determined by the rate of F6P formation from pentose and the specific radioactivity of spore G6P. The rate of exchange between stalk G6P and G1P and the rate of net (unlabeled) glycogen degradation in the stalk is dictated by the relative specific radioactivities of mix G6P and mix G1P. The rates of exchange between the exogenous labeled glucose (EXGLU) and the spore (GLU) and stalk (STGLU) glucose pools are determined by the kinetics of labeling and the final specific radioactivities of all the metabolites. Many simulation analyses were carried out in which the specific fluxes or the relative sizes of pool compartments were varied singly or together. It was found that very few compensatory effects are possible among the parameters involved, as so many are fixed by the data available.

D. Conclusions and Model Predictions

Analyses of the steady state models have led to the following conclusions: (1) Metabolically, two cell types can be distinguished, the presumptive spore plus stalk cells (thus far called spore cells) and the mature stalk cells (thus far called stalk cells). (2) During normal, unperturbed differentiation, there is a single UDPG pool that is in the presumptive stalk and spore cells and which is the precursor of glycogen and the saccharide end products of differentiation. Saccharides are not synthesized in stalk cells. (3) Glycogen and the end products are synthesized in these cell types and thereafter distributed into spore or stalk cells, where they accumulate (cellulose, cell wall glycogen), are degraded (glycogen), or both (trehalose). (4) Separate glucose and hexose phosphate pools are in each metabolic cell type. (5) The fraction of glycogen present which turns over increases from 3 to 9% during the culmination process. (6) Net glycogen degradation occurs in both cell types; unlabeled glycogen must be degraded in stalk cells. (7) In young sorocarps, stalk cells are approximately 10 times more permeable than presumptive cell types. (8) Metabolically, the presumptive cells and stalk cells behave like two independent cell populations. This last conclusion is interesting in view of the observation that under artificial experimental conditions, isolated amebas can transform into stalk or spore cells in the absence of cell-to-cell contact (Bonner, 1970; Kay and Trevan, 1978). A number of the predictions implicit in the above conclusions could theoretically be tested by independent examination of the two cell types. For example, stalk G6P (STG6P) should have a higher specific radioactivity than spore G6P, while stalk glycogen should be unlabeled and spore glycogen labeled. Although techniques are available for separation of the two cell types (Tsang and Bradbury, 1981; Killick, 1981), it is questionable whether stalk metabolites could be isolated in sufficient quantities for analysis.

E. The Steady State Model as a Basis for Constructing Transition Models of the Two Cell Types

The conclusions from the present analyses provide a solid basis on which to develop independent transition models of each metabolic cell type over the course of differentiation. Actually, the most recent transition model, including both cell types (Wright et al., 1979), necessarily incorporated certain features of what must now become the transition model of only the presumptive cells (see Sec. III.A). It is interesting to note that these features were actually predictions of the two-cell type transition model and served in part as the basis for constructing the steady state models of each cell type. For example, the measured (total) cellular level of glucose during culmination was, of course, one parameter of the transition model of the two cell types. However, in order to simulate certain data, it was necessary to lower the gluose level in the model about 20-fold (Wright et al., 1977). This low glucose level is now known to exist in presumptive cells, and the high glucose level in stalk cells (Wilson and Rutherford, 1978).

Based on the results of our steady state analyses, the synthesis of saccharides will only be depicted in the transition model of presumptive cells. Trehalose degradation will only occur in stalk cells, while glycogen degradation will occur in both metabolic cell types (Rutherford, 1976). Glycogen in the presumptive cell model will be represented as two pools, as we now know the fraction of the glycogen present which is actively engaged in turnover. Our new understanding of the flux relationships, metabolite compartmentation, and particularly the metabolic independence of the two metabolic cell types should allow a realistic analysis of their individual fates during the course of the differentiation process.

Finally, three cell types should really be modeled, rather than two. In general, the literature supports the interpretation that the apical tip is metabolically a very active region of the culminating organism. Presumably the end-product saccharides are synthesized in this small area and thereafter distributed by cell migration into stalk cells, where they are accumulated, degraded, or both. Hence the model of the spore cells in the present analysis should be considered as including both prespore and apical tip cells. The latter treatment was possible because of the similarity of metabolism in these cell types, as compared to mature stalk cells.

ACKNOWLEDGMENTS

This work was supported by Public Health Service grants AG00260 and AG00433 from the National Institutes of Health.

REFERENCES

Bonner, J. T. (1970). Induction of stalk cell differentiation by cyclic AMP in the cellular slime mold *Dictyostelium discoideum*. *Proc. Nat. Acad. Sci. USA 65*:110-113.

Cleland, S. V., and Coe, E. L. (1969). Conversion of aspartic acid to glucose during culmination in *Dictyostelium discoideum*. *Biochim. Biophys. Acta. 192*:446-454.

Davis, R. H., Bowman, B. J., and Weiss, R. L. (1978). Intracellular compartmentation and transport or metabolites. *J. Supramol. Struc. 9*:473-488.

Emyanitoff, R. G., DeToma, F. J., and Wright, B. E. (1981). Product inhibition of TCA cycle enzymes in mitochondrial extracts from *Dictyostelium discoideum*. *Abstr. Annu. Met. Am. Soc. Microbiol. 1981*:110.

Emyanitoff, R. G. (1982). Purification and characterization of NAD-dependent isocitric dehydrogenase from *Dictyostelium discoideum*. *Exp. Mycol.* *6*:274-282.

Emyanitoff, R. G., and Kelly, P. J., (1982). Kinetic characterization of mitochondrial malate dehydrogenase from *Dictyostelium discoideum*. *J. Gen. Microbiol.* *128*:1767-1771.

Ferrari, T. E., and Varner, J. E. (1970). Control of nitrate reductase activity in barley aleurone layers. *Proc. Nat. Acad. Sci. USA* *65*:729-736.

Halper, L. A., and Srere, P. A. (1977). Interaction between citrate synthase and mitochondrial malate dehydrogenase in the presence of polyethylene glycol. *Arch. Biochem. Biophys.* *184*:529-534.

Kay, R. R., and Trevan, D. J. (1981). *Dictyostelium* amoebae can differentiate into spores without cell-to-cell contact. *J. Embryol. Exp. Morphol.* *62*:369-375.

Kelleher, J. K., Kelly, P. J., and Wright, B. E. (1979). Amino acid catabolism and malic enzyme in differentiating *Dictyostelium discoideum*. *J. Bacteriol.* *138*:467-474.

Kelly, P. J., Kelleher, J. K., and Wright, B. E. (1979a). The tricarboxylic acid cycle in *Dictyostelium discoideum*: Metabolite concentrations, oxygen uptake and [^{14}C]-amino acid labeling patterns. *Biochem. J.* *184*:581-588.

Kelly, P. J., Kelleher, J. K., and Wright, B. E. (1979b). The tricarboxylic acid cycle in *Dictyostelium discoideum*: A model of the cycle at preculmination and aggregation. *Biochem. J.* *184*:589-597.

Killick, K. A. (1981). Isolation of spores from *Dictyostelium discoideum* by centrifugation in percoll density gradients. *Anal. Biochem.* *114*:46-52.

Kim, J. H., and Miller, L. L. (1969). The functional significance of changes in activity of the enzymes, typtophan pyrrolase and tyrosine transaminse, after induction in intact rats and in the isolated, perfused rat liver. *J. Biol. Chem.* *244*:1410-1416.

Kobr, M. J., Vanderhaeghe, F., and Combépine, G. (1969). Particulate enzymes of the glyoxylate cycle in *Neurospora crassa*. *Biochem. Biophys. Res. Commun.* *37*:640-45.

Koch-Schmidt, A. C., Mattiasson, B., and Mosbach, K. (1977). Aspects on microenvironmental compartmentation. *Eur. J. Biochem.* *81*:71-78.

Komuniecki, P. R., and DeToma, F. J. (1979). The mitochondrial alanine aminotransferase and aspartate aminotransferase enzymes of *D. discoideum*. Their role during growth and morphogenesis. *Abstr. Annu. Meet. Am. Soc. Microbiol.* *1979*:103.

Landridge, W. H. R., Komuniecki, P., and DeToma, F. J. (1977). Isolation and regulatory properties of two glutamate dehydrogenases from the cellular slime mold *Dictyostelium discoideum*. *Arch. Biochem. Biophys.* *178*:581-587.

Liddel, J. U., and Wright, B. E. (1961). The effect of glucose on respiration of the differentiating slime mold. *Dev. Biol.* 3:265-276.

Long, B. H., and Coe, E. L. (1974). Changes in neutral lipid constituents during differentiation of the cellular slime mold *Dictyostelium discoideum*. *J. Biol. Chem.* *249*:521-529.

Lowry, O. H., and Passonneau, J. V. (1964). The relationships between substrates and enzymes of glycolysis in brain. *J. Biol. Chem.* *239*:31-42.

Masters, C. J. (1977). Metabolic control and the microenvironment. *Curr. Top. Cell. Regul.* *12*:75-105.

Oaks, A., and Bidwell, R. G. S. (1970). Compartmentation of intermediary metabolites. *Annu. Rev. Plant Physiol.* *21*:43-66.

Ottaway, J. H., and Mowbray, J. (1975). The role of compartmentation in the control of glycolysis. *Curr. Top. Cell. Regul.* *12*:107-208.

Porter, J. S., and Wright, B. E. (1977). Partial Purification and characterization of citrate synthase from *Dictyostelium discoideum*. *Arch. Biochem. Biophys.* *181*:155-163.

Pramanik, A., Pawar, S., Antonian, E., and Schulz, H. (1979). Five different enzymatic activities are associated with the multienzyme complex of fatty acid oxidation from *Escherichia coli*. *J. Bacteriol.* *137*:469-473.

Rosness, P. A., and Wright, B. E. (1974). *In vivo* changes of cellulose, trehalose and glycogen during differentiation of *Dictyostelium discoideum*. *Arch. Biochem. Biophys.* *164*:60-72.

Rutherford, C. L. (1976). Glycogen degradation during migration of presumptive cell types in *Dictyostelium discoideum*. *Biochim. Biophys. Acta.* *451*:212-222.

Rutherford, C. L., and Harris, J. F. (1976). Localization of glycogen phosphorylase in specific cell types during differentiation of *Dictyostelium discoideum*. *Arch. Biochem. Biophys.* *175*:453-462.

Rutherford, C. L., and Wright, B. E. (1971). Regulation of nucleotide metabolism during differentiation in *Dictyostelium discoideum*. *J. Bacteriol.* *108*:269-275.

Sherwood, P., Kelly, P.J., Kelleher, J. K., and Wright, B. E. (1979). TFLUX: A general purpose program for the interpretation of radioactive tracer experiments. *Comp. Programs Biomed.* *10*:66-74.

Sols, A., and Marco, R. (1970). Concentrations of metabolites and binding sites. Implications in metabolic regulation. *Curr. Top. Cell. Regul.* *2*:227-273.

Srere, P. A., and Mosbach, K. (1974). Metabolic compartmentation: Symbiotic, organellar, multienzymatic, and microenvironmental. *Annu. Rev. Microbiol.* *28*:61-83.

Stucki, J. W., and Walter, P. (1972). Pyruvate metabolism in mitochondria from rat liver. Measured and computer-simulated fluxes. *Eur. J. Biochem.* *30*:60-72.

Sussman, M., Schindler, J., and Kim, H. (1977). Toward a biochemical definition of the morphogenetic fields in *Dictyostelium discoideum*. In *Development and Differentiation in the Cellular Slime Molds*. J. M. Ashworth (Ed.), Elsevier/North-Holland, New York, pp. 31-50.

Tischler, M. E., Hecht, P., and Williamson, J. R. (1977). Determination of mitochondrial/cytosolic metabolite gradients in isolated rat liver cells by cell disruption. *Arch. Biochem. Biophys.* *181*:278-292.

Tsang, A., and Bradbury, E. M. (1981). Separation and properties of prespore cells of *Dictyostelium discoideum*. *Exp. Cell Res.* *132*:433-441.

Wilson, J. B., and Rutherford, C. L. (1978). ATP, trehalose, glucose and ammonium ion localization in the two cell types of *Dictyostelium discoideum*. *J. Cell. Physiol.* *94*:37-45.

Wilson, J. E. (1978). Ambiquitous enzymes: Variation in intracellular distribution as a regulatory mechanism. *Trends Biochem. Sci.* *3*:124-125.

White, G. J., and Sussman, M. (1961). Metabolism of major cell components during slime mold morphogenesis. *Biochim. Biophys. Acta.* *53*:285-293.

Wright, B. E. (1963). Endogenous activity and sporulation in slime molds. *Ann. N.Y. Acad. Sci.* *102*:740-754.

Wright, B. E. (1979). Causality in biological systems. *Trends Biochem. Sci.* *4*:N110-N111.

Wright, B. E., (1981). A transition model of the citric acid cycle in *Dictyostelium*. *Abstr. Annu. Meet. Am. Soc. Microbiol.* *1981*:111.

Wright, B. E., and Anderson, M. L. (1960). Protein and amino acid turnover during differentiation in the slime mold. I. Utilization of endogenous amino acids and proteins. *Biochim. Biophys. Acta.* *43*:62-66.

Wright, B. E., and Bard, S. (1963). Glutamate oxidation in the differentiating slime mold. I. Studies *in vivo*. *Biochim. Biophys. Acta.* *71*:45-49.

Wright, B. E., and Davison, P. F. (1980). Guest editorial: Mechanisms of development and aging. *Mech. Aging Dev.* *12*:213-219.

Wright, B. E., and Gustafson, G. L. (1972). Expansion of the kinetic model of differentiation in *Dictyostelium discoideum*. *J. Biol. Chem.* *247*:7875-7884.

Wright, B. E., and Kelly, P. J. (1981). Kinetic models of metabolism in intact cells, tissues, and organisms. *Curr. Top. Cell. Regul.* *19*:103-188.

Wright, B. E., Tai, A., and Killick, K. A. (1977). Fourth expansion and glucose perturbation of the *Dictyostelium* kinetic model. *Eur. J. Biochem.* *74*:217-225.

Wright, B. E., Tai, A., Killick, K. A., and Thomas, D. A. (1979). The effect of exogenous glucose, uracil, and inorganic phosphate on differentiation in *Dictyostelium discoideum*. *Arch. Biochem. Biophys.* *192*:489-499.

Wright, B. E., Thomas, D. A., and Ingalls, D. W. (1982). Metabolic compartments in *Dictyostelium discoideum*. *J. Biol. Chem.* *257*:7587-7594.

3

Attraction and Adhesion in the Slime Mold *Dictyostelium*

PETER C. NEWELL *University of Oxford, Oxford, England*

I. INTRODUCTION

The striking characteristic of the cellular slime molds is the way in which they can change in the course of a 24-h period from a population of independent amebas to a multicellular organism. Such a transition requires intercellular communication over distances of many cell diameters and aggregative movement that is finely coordinated.

In their natural habitat of decaying forest litter the slime molds chemotactically seek and then ingest bacteria around them until this food source becomes exhausted. Their problem, then, is to survive and, if the food source has become depleted because of overpopulation, to move away to unspoiled feeding grounds. Formation of the multicellular cell mass allows formation of a motile body (the "slug" or "grex") capable of moving several centimeters over the substratum. The slug responds positively to both heat and unidirectional light signals that direct it to the surface of the substratum and away from the original site of formation. Once out in the open the organism can produce heat- and drought-resistant spores on top of a delicate stalk and can then remain dormant until better times.

This chapter reviews progress made into the different (but probably related) chemotactic systems used for food seeking and aggregation and summarizes current thinking about the cellular adhesion mechanism used when the amebas have aggregated to form the motile slug.

II. CHEMOTAXIS TO FOLIC ACID

The notion that amebas of *Dictyostelium* employed folic acid as a chemotactic stimulus in food seeking came from the work of Pan et al. (1972, 1975), who found that amebas in the growth phase were strongly attracted toward small drops of folic acid placed on the agar adjacent to them. By placing the amebas on a small cellophane square and determining the rate at which they left it the authors were able to produce an index of attraction to different substances. They found that only compounds possessing the pteridine ring were active for the species *Dictyostelium discoideum*, *Polysphondylium pallidum*, and *Polysphondylium violaceum*. Although vegetative cells rapidly showed a positive response to folic acid, starving cells in the aggregation phase were almost inactive, clearly demonstrating that folic acid is not an aggregation phase attractant. *Dictyostelium* requires folic acid as a vitamin (Franke and Kessin, 1977), so that liberation of this compound by bacteria would serve as a very suitable indicator of a nearby food source.

Recent work by Wurster and Butz (1980) has demonstrated the existence of cell surface receptors that can reversibly bind folic acid. These receptors are specific for this compound and its closely related derivatives such as aminopterin and amethopterin. The concentration of folic acid for half-maximal binding ($K_{0.5}$) is 150 nM and in the growth phase there are 6×10^4 receptors per cell. It is of interest that Scatchard plots of the binding data are curvilinear (concave upward), suggestive of negative cooperativity, as previously found for the cAMP receptors formed during development (see below). Cyclic AMP and cGMP do not compete for the folic acid receptors, however, and the two receptor types seem to be distinct. An unexpected finding of Wurster and Butz was that although the sensitivity to folic acid decreased greatly during development (amebas only responding to concentrations in excess of 5 μM after 4 h of starvation), the number and affinity of folic acid receptors dropped only slightly, and even after 10 h of starvation amebas possessed over half the vegetative number and these showed a $K_{0.5}$ of 24 nM. This finding implies that the folic acid receptors are somehow uncoupled from the signal-processing system rather than being removed.

Insight into the folic acid-coupled signal-processing system has come from the work of Wurster et al. (1977) and Mato et al. (1977b), who have found that in *D. discoideum*, folic acid binding to amebas can trigger an increase in activity of the enzyme guanylate cyclase and can thereby produce a brief pulse of cGMP. This again resembles the effect of cAMP binding to its cell surface receptors during the developmental phase and suggests that folic acid and cAMP may share a common membrane-transducing system linking them to the internal chemosensory mechanism.

In sensory systems, for every signal source there must be a signal-destroying system to prevent the receptors from becoming saturated and inactive. In the case of folic acid this may be the enzyme folate deaminase, which is excreted by amebas and causes the hydrolytic deamination of folic acid to 2-hydroxy-2-deamino folic acid (Pan and Wurster, 1978). For *D. discoideum* and *P. violaceum* this derivative is chemotactically inactive, although curiously Wurster and Butz (1980) found that for *D. discoideum* the 2-hydroxy-2-deamino derivative still binds to the folic acid receptors with similar affinity to folic acid. Presumably such binding must differ from that of folic acid and be incapable of triggering the signal-transducing mechanism, a notion that is confirmed by the finding of Kakebeeke et al. (1980a) that 2-hydroxy-2-deamino folic acid does not induce the increase in intracellular cGMP that is seen with folic acid binding. The difference in the binding mechanism between folic acid and the 2-hydroxy-2-amino derivative is clearly a subtle one, and this subtlety is emphasized by the finding that for some species (such as *Dictyostelium minutum*) 2-hydroxy-2-deamino folic acid is chemotactically active and such species must inactivate the folic acid by breaking the link between the pteridine ring and p-aminobenzoyl glutamate (Kakebeeke et al., 1980b).

III. CHEMOTAXIS TO cAMP

Possibly as an evolutionary extension of chemotaxis to folic acid, starving amebas have developed the ability to move toward sources of cAMP. This behavior, which results in aggregation to form multicellular slugs, may be analyzed visually, genetically, and biochemically.

A. Visual Observations

For the first 4-6 h after the initiation of starvation no directed movement of the amebas is observable, but then during the next 4-5 h the cells begin to move toward collecting centers. Their progress is not continuous but occurs in rhythmic steps lasting about 100 s each, separated by quiescent phases of 3-4 min during which the amebas remain comparatively still and await the next signal to move forward. In dense populations this periodic movement results in the appearance of dark and light bands under dark-field illumination (Fig. 1). Such bands, which have been studied by a number of observers over many years (Arndt, 1937; Bonner, 1944; Shaffer, 1957; Gerisch, 1968;

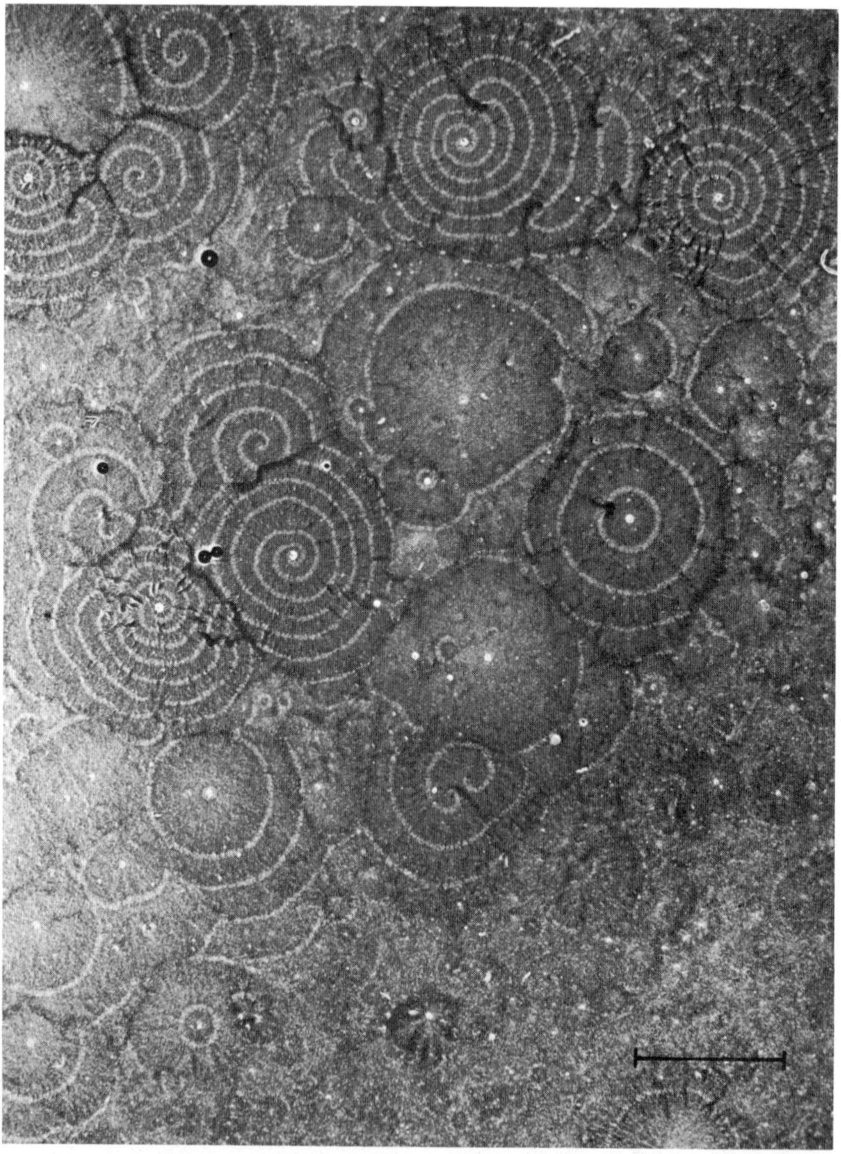

FIG. 1 Concentric and spiral waves of aggregating fields of *Dictyostelium discoideum* on an agar surface, seen using dark-field optics. The signalling centers which appear as bright dots are emitting pulses of cAMP at variable frequency. The bands of amebas moving toward the centers appear bright, and the stationary ones dark. In a number of cases the concentric banding pattern has given way to a spiral

Alcantara and Monk, 1974; Gross et al., 1976), result from the slight differences in optical properties between moving and stationary amebas that make moving cells look bright and stationary cells dark under carefully aligned dark-field optics. At population densities of approximately 10^5 amebas per square centimeter the bands (and hence the relayed signals) move outward at approximately 300 μm/min. The width of the bright bands is determined by the duration of the movement period. In wild-type populations the width is roughly constant, the movement period being about 100 s, during which the amebas move about 20 μm (approximately two cell diameters) at a velocity of about 12 μm/min. In certain "streamer" mutants, however, this movement phase is greatly prolonged and in such strains the bright band continues until the next chemotactic signal wave abruptly stops the cells and starts them once again (Fig. 2) (Ross and Newell, 1981).

The width of the dark stationary band is determined in normal wild-type populations by the frequency of pulsations emanating from the particular collecting center. The higher the frequency, the narrower the dark bands. This frequency changes during the course of aggregation, starting at roughly one pulse every 10 min, then decreasing to one pulse every 5 min, and finally to one every 2-3 min, before becoming indistinct and (possibly) continuous during the later stages of aggregate formation (Fig. 3 a-d).

The light and dark bands seen under dark-field optics may either be concentric or spiral (Fig. 1). Detailed observations by Durston (1974) and Gross et al. (1976) suggest that spiral aggregation territories may arise from breakage of a concentric ring, followed by propagation of the signal from this break point as a continuous two-dimensional spiral. Once initiated, such a spiral wave no longer depends on the center for its periodic initiation, but propagates itself continuously. The apparent frequency of the wave experienced by a cell positioned in the field is only limited by the refractory period for signal relay. Because of this effect, the apparent frequency of spiral waves is able to increase faster than that of concentric ones, and spirals commonly swamp concentrics when centers are competing for territory.

The inward motion of the amebas toward the collecting centers, which is initially radial, eventually becomes unstable and streams of cells rather than a uniform field are produced. The instability of the radial motion arises because populations of amebas show slight inhomogeneity of distribution and consequently the signal amplitude will not be uniform in the expanding wave. Any inhomogeneity will be rein-

(FIG. 1 Continued) pattern. Such spirals are produced from a broken concentric band. A double spiral wave in the process of being formed from a concentric wave may be seen in the lower middle portion of the photograph. The bar marker represents 1 cm.

FIG. 2 Signalling patterns of aggregating amebas of a wild-type strain (left) and a streamer mutant (right) seen by dark-field optics. Note that in the streamer mutant the bright (movement) band is wide and the dark (stationary) band narrow, which is the converse of the pattern seen with the wild-type strain. The bar marker represents 1 cm.

forced by chemotactic movement causing amebas to collect into streams, leaving areas devoid of cells (Fig. 3 e,f).

B. Genes for Aggregation

The existence of genes concerned with aggregation may be inferred by mutagenesis of amebas and isolation of clones that are normal during the growth phase but aberrant during aggregation. A number of different phenotypes have been isolated, the most common being the simple "aggregateless" type that completely fails to aggregate in clones grown on bacterially inoculated agar plates. Because multicellular development is dependent on aggregation, such mutants completely fail to form fruiting bodies and remain as a flat or rippled lawn of starving amebas until death overtakes them (Fig. 4). A rather different mutant phenotype is that of the "aggregation pattern" mutant that can aggre-

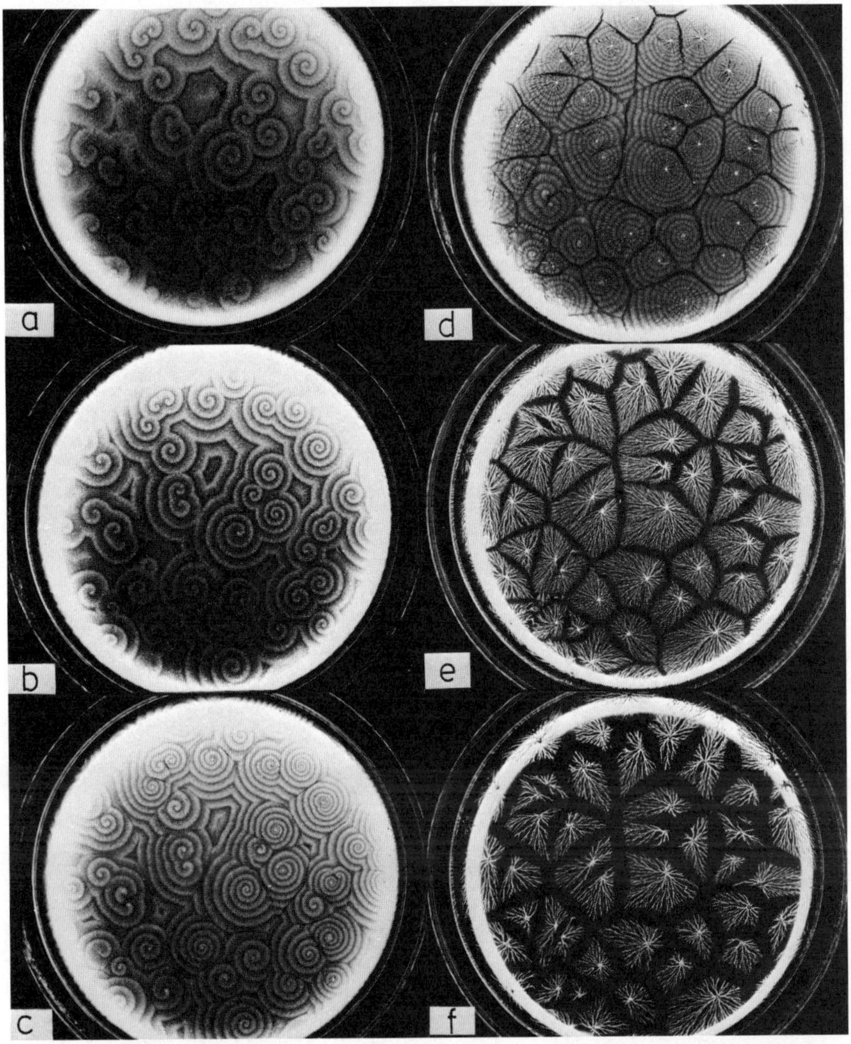

FIG. 3 Signalling patterns of *D. discoideum* showing the progressive increase in frequency of signalling, leading to the formation of territories and finally to aggregating streams. The photographs, taken at approximately 10-min intervals, show a field of 5×10^7 amebas of the streamer mutant NP377 in a 50-mm-diameter Petri dish. The pattern seen with the wild type is similar, but with narrower bright bands.

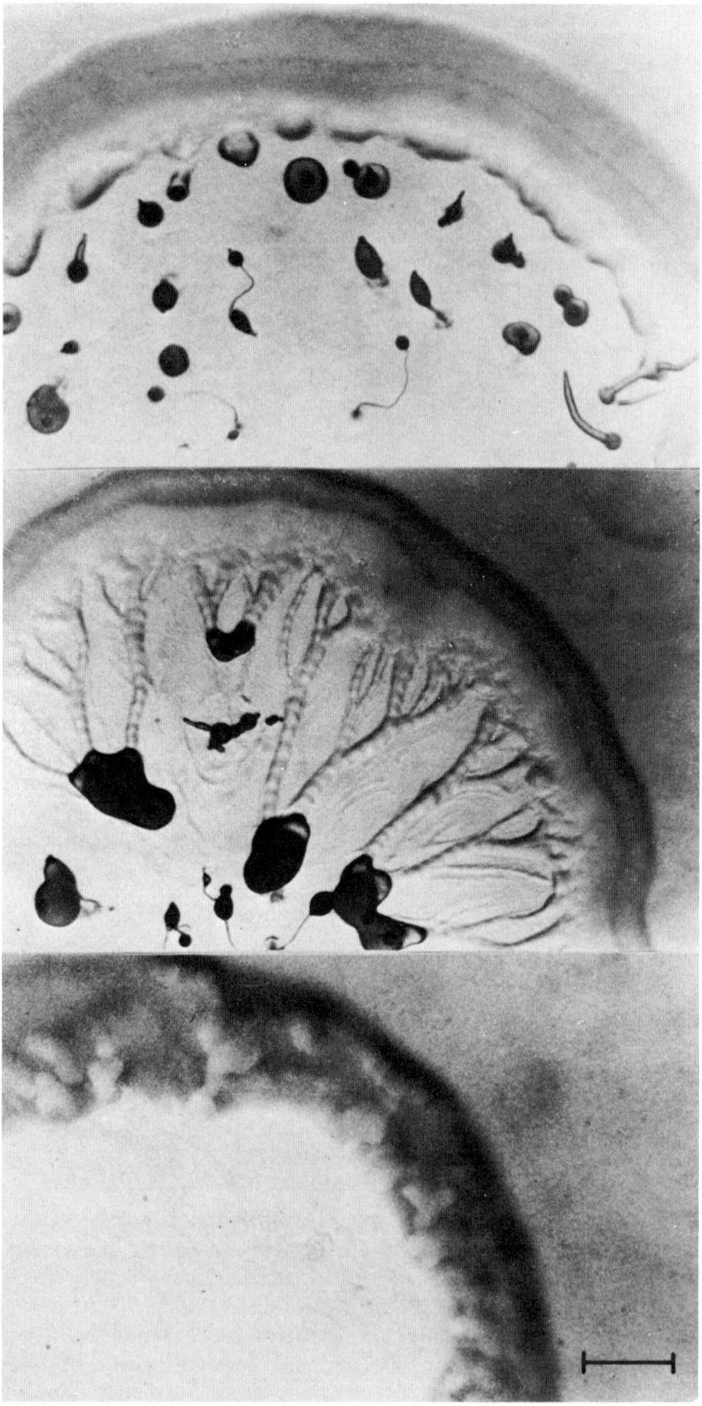

gate but does so abnormally. The streamer mutants mentioned above fall into this category, forming huge streams under conditions where the wild type forms scarcely observable streams (Fig. 4). Under conditions of lower population density, where waves and streams are seen in the wild type, other types of mutants are observable that form no waves or streams at all and aggregate from tiny territories to form miniature fruiting bodies (Fig. 5).

The number of genes specifically required for the aggregation process and their position on the seven linkage groups has been investigated in *Dictyostelium* by Williams and Newell (1976) and Coukell (1975, 1977) and in the genus *Polysphondylium* by Warren et al. (1975, 1976). For these genetic studies, use was made of the parasexual cycle. In this cycle haploid amebas occasionally fuse to form diploids which can be selected and isolated. Subsequent spontaneous haploidization of these diploids leads to segregation of chromosomes and allows identification of linkage groups (see Newell, 1978, for a review of slime mold genetics). Haploid mutants were fused under conditions of starvation, and the diploids which formed (at a frequency of about 2×10^{-5} in 24 h) were selected using temperature sensitivity for growth. By analysis of the aggregational abilities of the diploids isolated in this way, it was calculated that approximately 50-100 genes were required specifically for the aggregation process.

One of the objectives of genetic analysis is to correlate genetic mutations with lesions in known components of the chemotactic process. Because of the pleiotropic nature of many mutations, such a correlation has proved difficult. However, Brachet and colleagues have found that mutants bearing mutations in the gene *pdsA* (located on linkage group IV) fail to produce the phosphodiesterase enzyme needed to destroy the cAMP signal (see below). The mutant phenotype is aggregateless, but it will aggregate if exogenous phosphodiesterase enzyme is added to the amebas starving on agar (Brachet et al., 1979; Barra et al., 1980). It remains to be determined whether the mutation is in a structural gene for the enzyme or at a regulatory locus concerned with phosphodiesterase production. Another correlative example is the *cbpA* gene. A mutant found by Ray et al. (1979) that bears a mutation in this gene was shown to be unable to aggregate and to lack

---

FIG. 4 Patterns of aggregation and development seen in clones of *D. discoideum* growing on lawns of bacteria. Top: Wild type showing an expanding ring of amebal growth behind which the starving cells proceed through development. Middle: A streamer mutant which forms huge streams leading into the aggregates. Bottom: An aggregation mutant that grows normally but which does not aggregate or develop. The scale bar represents 2 mm. [From Newell (1981).]

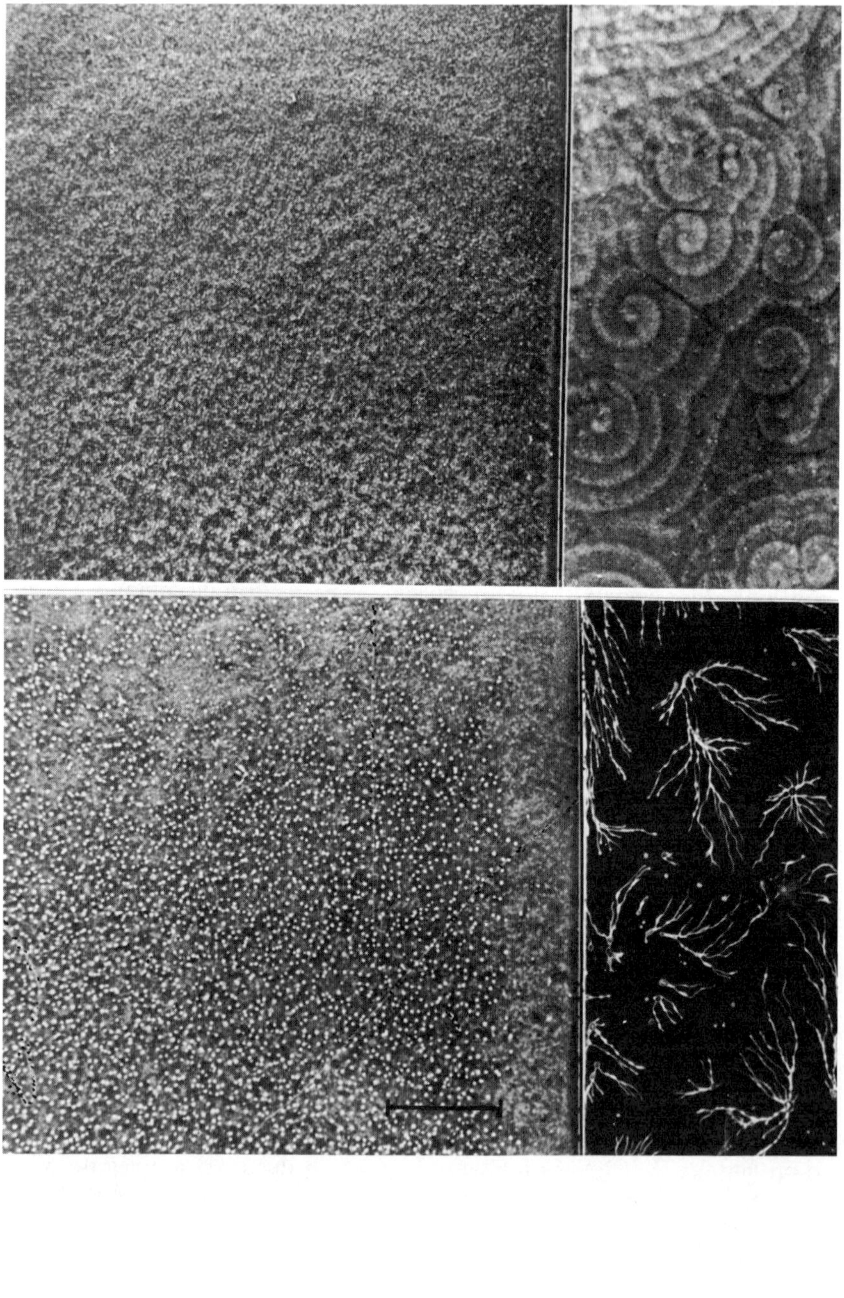

normal discoidin I lectin (thought to be involved in cell adhesion; see below). Because of the method of isolation used for this mutant, a simple relationship between the aggregation phenotype and loss of discoidin was not certain. More recent studies by Shinnick and Lerner (1980) confirm such a correlation, as they found that a large number of revertants at this locus regained functional discoidin and the ability to aggregate simultaneously. Whether or not *cbpA* is the structural gene for discoidin has yet to be established.

Streamer mutants have been analyzed genetically by Ross and Newell (1979). A total of 24 haploid mutants were fused together in pairs and the pattern of complementation groups determined. The mutants were found to occur in six complementation groups located on three linkage groups. From their phenotypic behavior, the mutants appeared to possess a greater than normal ability to suppress further initiation of centers once a local center had formed in a growing clone. Because of the continuous outward growth of the clone on a lawn of bacteria and the continuous production of newly starving amebas as the growth ring expands, aggregation centers once formed tend to dominate the newly starving amebas and entrain them into the center. In this way very large streams can be formed in the mutant clones. In wild-type clones this dominance is somehow regulated to allow new centers to be formed. The biochemical basis of the dominance effect is as yet incompletely understood. The mutants appear to resemble the wild type in every respect tested, except that they seem to persist in the movement phase for longer than the wild type. This leads to the formation of an abnormally wide (bright) movement band seen with dark-field optics of lawns of starving amebas (Fig. 2) (Ross and Newell, 1981). (The reason for this prolonged movement seems to be connected with aberrant cGMP metabolism, as discussed in the next section below.) A possible connection between this prolonged movement response and the domination of one center by another in clones could be that cells might not be able to initiate center formation while chemotactically responding to a signal from another center. This would have the effect of decreasing the tendency of the streamer mutants to initiate new centers in clones, compared to the wild type,

FIG. 5 Aggregating field of amebas of a mutant that fails to relay compared to a relay-competent strain. Using dark-field optics, two blocks of agar were observed: the left-hand one bearing the relayless mutant, and the right-hand one a relay-competent mutant. In the upper photograph the relayless mutant is forming tiny territories, and in the lower photograph, taken a few hours later, a large number of small fruiting bodies may just be seen. In contrast, the relay-competent strain forms the large normal territories using its signal relay system. The bar marker represents 1 cm.

which has a period without chemotaxis (the wide dark band) following each chemotactic movement response.

One question about aggregation in *Dictyostelium* that has often been asked is, What is the nature of the cell that initiates aggregation? The early work of Sussman and Noel (1952) and Ennis and Sussman (1958) suggested that the initiator or "I" cell was genetically distinct from the general population and present at a frequency of roughly 1 in 2000. From time-lapse film, it appeared to be larger and faster moving. Later work by Konijn and Raper (1961) and Gerisch (1961) produced evidence against the I cell hypothesis, as they found that small populations of amebas (down to 100 cells) could generally aggregate and such populations would not be expected to contain an I cell. However, as populations of less than 100 amebas normally do not aggregate, it was still possible that I cells were present at 1 in 100 of the population. At this level statistical analysis by Raman (1976) and Raman et al. (1976) could not distinguish between models in which I cells were present as a genetically distinct subpopulation and those in which initiator cells arose by differentiation from the general population during starvation. More recent evidence strongly favors the latter model. Glazer and Newell (1981) used mixed populations of wild-type amebas at very low numbers mixed with an excess of a mutant (NP160). This mutant could not itself make or relay aggregation signals, but it could respond chemotactically with high sensitivity to those signals produced by spatially separated individual wild-type cells and produce fruiting bodies. In this way the ability of isolated wild-type cells which were outnumbered by up to 10,000 to 1 by the mutant could be assessed for their ability to initiate aggregation. It was found that at least one in five of the wild-type amebal population could initiate aggregation in this way, making a genetically distinct subpopulation extremely unlikely. It seems more likely that starvation causes amebas to develop a signal-amplifying system that becomes increasingly excitable until some cell just exceeds a threshold, setting up spontaneous and self-sustaining signal pulses. The theoretical aspects of such a model have been presented by Keller and Segal (1970) and Durston (1973).

C. Biochemistry of Aggregation

The molecule that is rhythmically emitted by the aggregation centers of the four species *D. discoideum*, *D. rosarium*, *D. purpureum*, and *D. mucoroides* is cAMP (Konijn, 1972). Other species use different attractants that have been only partially characterized. Starving amebas of *Polysphondylium violaceum* and *P. pallidum* are attracted to a small peptide. This molecule is either circular or blocked at both ends, but its detailed structure is unknown (Wurster et al., 1976). The species *D. lacteum* is attracted to a negatively charged aromatic heterocyclic compound of unknown structure (Mato et al., 1977a), and *D. minutum* is specifically attracted by a different aromatic moiety bound to a glycine molecule (Kakebeeke et al., 1978).

Only the cAMP chemotactic system used by *D. discoideum* has been studied in detail, but considerable progress has been made with this system.

1. *Signal Production*

The enzyme that produces the pulses of cAMP is membrane bound, but not easily isolated. Evidence has been presented that it is located on the inner face of the plasma membrane (Farnham, 1975), but more recent work suggests that it may also be contained in vesicles (Hintermann and Parish, 1979). The latter observation is interesting because apparent fusion of vesicles with the cell surface in phase with cAMP release has been reported by Maeda and Gerisch (1977). It seems possible, although not yet proven, that the cAMP formed may be released from vesicles in a similar manner to the release of neurotransmitters from the preganglionic cells at nerve synapses.

The mechanism of adenylate cyclase activation by starvation is unknown. At some point a few hours after the food supply has been depleted a small number of the cells in a starving population activate their enzyme in an oscillatory manner and thereby produce pulses of cAMP (Roos et al., 1977; Klein, 1977). Very few of a normal population of starving amebas spontaneously produce such pulses of cAMP, but nearly all the cells become excitable and can act as relays for the pulses produced by the center-forming cells.

2. *Cell Surface cAMP Receptors*

The cAMP signal diffuses only 50-60 μm from the pulsatile source before it is destroyed. During its brief existence, however, a small fraction can bind to highly specific cAMP receptors situated on the outside of the plasma membrane of nearby starving amebas. These cell surface cAMP receptors have been extensively studied (Malchow and Gerisch, 1974; Green and Newell, 1975; Mato and Konijn, 1975; Henderson, 1975, Klein and Juliani, 1977; Mullens and Newell, 1978; Klein, 1979). Their properties closely resemble hormone receptors in higher cells and show, for example, curvilinear Scatchard plots indicating multiple classes of receptor, or variable receptor affinity dependent on the cAMP concentration. Evidence for such a variable-affinity receptor is the finding that increasing the concentration of cAMP increases the rate of dissociation of previously bound cAMP under conditions that favor rapid dissociation. The requirement by the organism for such a variable-affinity receptor may lie in the need to respond to a wide range of cAMP concentrations such as may be found in the slime molds' natural habitat. The number of these receptors has been found to increase rapidly from a basal level during the starvation period, with a maximum number present at 8-10 h. After this the receptors disappear equally rapidly and it is not clear that they play a role in later development.

These receptors have recently been isolated on polyacrylamide gels by Wallace and Frazier (1979) using the technique of photoaffinity labeling. Intact amebas were treated with 8-N_3-cyclic[^{32}P]AMP, which binds to cAMP receptors. Subsequent exposure to short-wavelength light caused covalent bonding of the molecules to the receptor protein, which was then isolated and purified, its presence during purification being monitored by its ^{32}P label. It was found that the only band on electrophoretic gels that was specifically labeled in this way was a 40,000-dalton protein with a high specificity for cAMP and a time of formation that identified it as the cAMP receptor.

3. *Signal Transduction System*

An important and interesting area of current research is concerned with the connecting system that links cAMP receptor binding with the chemotactic and relay responses. The chemotactic response of amebas is fast, occurring within 15 s of the stimulus reaching the cell, as shown by the speed with which amebas will change direction if they receive stimuli from different directions (Alcantara and Monk, 1974; Gerisch et al., 1975). Cellular events that are observed to be temporarily related over this short time scale (i.e., occur within a few seconds of cAMP binding to receptors) are the following:

1. A very rapid increase in intracellular cGMP (Mato et al., 1977c; Wurster et al., 1977; Mato and Malchow, 1978)
2. A very rapid influx of Ca^{2+} ions that occurs within a few seconds of cAMP binding, followed by a slower efflux over several minutes (Wick et al., 1978)
3. A rapid release of protons, about 10^3 being released per molecule of cAMP added, followed by an increase of the internal pH to the initial value over about 4 min (Malchow et al., 1978).

Which of these events might represent the link between cAMP receptor stimulation and adenylate cyclase activation is not yet known, but all three occur over the appropriate time scale.

It is noteworthy with regard to cGMP metabolism that recent experiments with the streamer mutants by Ross and Newell (1981) have shown that members of the complementation group *stmF* show an abnormal cGMP response to pulses of cAMP. Instead of the normal brief pulse of cGMP that is formed and degraded within about 25 s, these mutants form a large peak of cGMP that is only slowly degraded (Fig. 6). The reason for this large peak may be due to increased synthesis of cGMP or a deficiency of the specific cGMP phosphodiesterase, as found by Dicou and Brachet (1980). (The non-

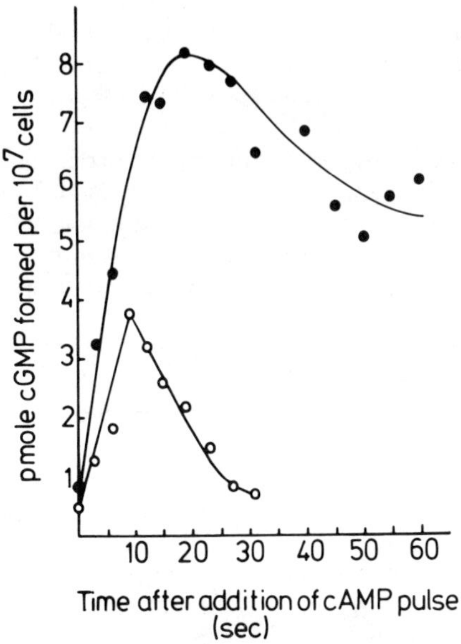

FIG. 6 Formation of intracellular cGMP in starving amebas of parental strain XP55 (○) and a streamer mutant (●) in response to a 100-nM pulse of cAMP at time zero. Note the rapid formation and degradation of the cGMP peak in the parental strain and the slow degradation of the large peak of cGMP formed in the streamer mutant. (From Ross and Newell, 1981.)

specific cAMP/cGMP phosphodiesterase is normal in these mutants.) The mutants are clearly abnormal in their chemotactic and signal relay response and should be useful in further defining the important role of cGMP in these processes.

4. *The Movement Response to cAMP*

The change in optical properties that occurs when cells move may also be correlated with cAMP binding. Such changes are best studied quantitatively, with the amebas starving in oxygenated suspension in a spectrophotometer cuvette (Gerisch and Hess, 1974). After addition of cAMP to amebas that have been starving for a few hours, a rapid decrease of extinction is observed within seconds of cAMP binding. The extinction then rapidly rises toward the original value, followed by a slower wave of decreased extinction. These changes in optical properties may be correlated with those seen with lawns of starving amebas on agar plates and probably represent an initial rapid shock response to the cAMP followed by a slower movement response. This correlation is supported by the observation that streamer mutants, which show a wide movement band on agar, demonstrate an extended slow wave of decreased extinction in oxygenated suspensions (Ross and Newell, 1981).

The way that movement results from the cAMP binding is little understood. However, with Ca^{2+}, cGMP, and proton changes occurring rapidly after cAMP binding it may be envisaged that the actomyosin system is activated by the changing concentrations of these entities. It is of interest in this connection that a protein factor that confers Ca^{2+} sensitivity to the activation of myosin ATPase by actin has been isolated from *Dictyostelium* amebas (Mockrin and Spudich, 1976). What is not clear is the way the direction of movement is established by the amebas after receipt of a relayed signal from the aggregation center. In some manner they find their "chemical compass bearing." Although it is conceivable that this might occur through differential cAMP receptor occupancy from the front to the back of the amebas, any cAMP gradient that exists would be very shallow, and it seems more likely that the amebas are able to recognize which side of a cell was first exposed to a wave of signal reaching it.

5. *Signal Relay and the Problem of Unidirectional Movement*

Besides moving toward a signal source, starving amebas also develop the ability to relay the signal, amplifying it in the process so that the signal wave can expand from a center without a decrease in amplitude. The speed of the relay response is currently subject to conflicting ideas. According to the theoretical analysis of Cohen and Robertson (1971), the relay response must be very fast. In order for the relay system to sustain the observed velocity of wave propagation on agar surfaces, the release of cAMP must start to occur within 5-15 s of the binding of the incoming signal. If the "relay delay time" were slower than this, either the signal range of each pulse would be very much larger than that reported (about 60 μm, according to Alcantara and Monk, 1974) or the wave velocity would be much slower than that observed. Although some measurements support this prediction (Grutsch and Robertson, 1978), other detailed studies show a slower buildup of cAMP that extends over a period of minutes (Roos et al., 1975; Chung and Coe, 1978; Devreotes and Steck, 1979). The significance of the gradual accumulation of cAMP is presently uncertain; however, from the work of Devreotes and Steck (1979) and Futrelle et al. (1980) it appears to be possibly related to the maintenance of movement of stimulated amebas toward the signal center. It has always been a problem with the model of Cohen and Robertson (1971) that amebas could respond to a pulse of cAMP liberated further out from the center, causing them to move outward to the relayed pulse instead of moving consistently toward the aggregation center. In Cohen and

Robertson's model the problem was obviated by making the pulse duration extremely brief (a few seconds) and assuming a refractory period (longer than a few seconds) after stimulation during which the amebas would not respond chemotactically to further stimulation. In this way the amebas would move inward for 100 s once briefly stimulated and ignore the relayed signal briefly regenerated just behind them. In fact, Alcantara and Monk (1974) and Gerisch et al. (1975) found that there is virtually no refractory period for amebal chemotaxis, amebas responding to a new signal within 12-15 s, and more recently Futrelle et al. (1980) reported that chemotactic movement toward an electrode producing artificial pulses lasts only as long as the duration of the signal. It seems probable, therefore, that a signal of extended duration (over approximately 100 s) rather than a brief signal (of less than 15 s) is used by the amebas to maintain the 100 s of movement observed as the bright band on agar plates. With such a broad band of cAMP-secreting cells the problem of outward movement to the relayed signal would not arise, as the amebas would be experiencing either a steadily increasing cAMP concentration while the front of the wave came over them, or a decreasing concentration as the back of the broad wave moved further outward. Devreotes and Steck (1979) have convincingly demonstrated that amebas only respond to an increment in the cAMP concentration for signal relay. A declining wave would not, therefore, stimulate further cAMP synthesis or, probably, chemotactic movement, and amebas would not be induced to follow the wave as it moved outward.

6. *Adaptation to cAMP Signals*

The ability of *D. discoideum* to adapt to continuously maintained concentrations of cAMP has been investigated in an interesting series of experiments by Dinauer et al. (1980). Their technique employed a specially designed perfusion apparatus in which starving amebas (prelabeled with [^{3}H]adenosine during the growth phase) were placed on support filters and perfused with buffer containing various concentrations of cAMP. The release of tritiated cAMP in response to such external cAMP was monitored by collection of the perfusate and estimation of ^{3}H by scintillation counting. In this manner the authors were able to observe the effects on the perfused amebas of fixed concentration levels of extracellular cAMP. As mentioned above, Devreotes and Steck (1979) found that amebas secrete cAMP only in response to a relative increase in the extracellular concentration. They found, moreover, that the response is graded rather than "all or none," the size of the relay response being proportional to the increment in extracellular cAMP concentration. The duration of the cAMP secretion depended on the continued presence of extracellular cAMP, and removal of the stimulus at any time terminated the response. However, even in the continued presence of the cAMP stimulus the response ceased after approximately 2 min. Such cessation of response is indicative of the phenomenon of adaptation which has been extensively studied in other systems, including bacterial chemotaxis. If the cAMP concentration was increased in tiny incremental steps over a period of 30 min, then the authors were able to demonstrate a continuous secretion of cAMP over this period. In any other than this abnormal situation of continuous incremental stimuli, the response is clearly of limited duration owing to the adaptation phenomenon. Although the molecular mechanism of adaptation is not yet understood, analogy with the bacterial chemotactic adaptation system (Goy and Springer, 1978) suggests that methylation of membrane proteins could be involved and such methylation has recently been reported by Mato and Marin-Cao (1979) in starving amebas in response to cAMP signals.

7. *Signal Destruction*

Unless cAMP were rapidly destroyed after its formation and binding to the cell receptors, the chemosensory system would become saturated and would cease to function. In normal cells the cAMP is rapidly removed by the phosphodiesterase enzymes. These enzymes have been extensively studied (see reviews by Gerisch and Malchow, 1976, and Newell, 1977, for references). There are two major types of phosphodiesterase; one of these is bound to the outside of the plasma membrane and the other is excreted as soluble enzyme. A protein inhibitor of the soluble enzyme is also formed and helps to regulate its activity, the inhibitor being repressible by excess cAMP (Tsang and Coukell, 1977, 1979; Klein and Darmon, 1977). Some of the apparent complexity concerning the various forms and their different substrate affinities has been clarified by the recent finding of Kessin et al. (1979), that the inhibitor binds to the soluble form by a dithiothreitol-sensitive bonding system. These authors were able to show that in the presence of the inhibitor the soluble enzyme has a Km in the millimolar range, but in the presence of dithiothreitol or in mutants lacking inhibitor the enzyme has 20- to 100-fold higher activity, with a Km of approximately 10 μM.

The different roles of the soluble and plasma membrane-bound enzymes and their structural relationship is not clear, but each may be needed under particular environmental conditions to ensure rapid breakdown of the cAMP signal.

IV. CELL COHESION

By the time aggregating amebas make contact with each other, they have on their surfaces certain macromolecules that enable them to cohere. In shaken, starved suspensions the onset of cellular cohesiveness may be gauged by counting the number of single amebas left in the suspension. Studies by Gerisch and coworkers (reviewed in Gerisch, 1980) have shown that as cells starve, they produce adhesion molecules called contact sites A on their cell surface that make cell interactions resistant to EDTA. These molecules seem to be active preferentially at the two opposite ends of the cell and chains of amebas are formed adhering end to end. In the absence of EDTA, side-to-side adhesion is stabilized by other molecules (not specifically formed during development) called contact sites B. Other groups of workers have implicated slime mold lectins (carbohydrate-binding proteins) as candidates for cell cohesion (see Rosen and Barondes, 1978, for references). In *D. discoideum* these molecules are the discoidins, and in *P. pallidum* they are the pallidins. There seems to be no evidence to suggest that contact sites A and lectins are related in their structure or mode of action, and their study, outlined below, has proceeded along rather different avenues.

A. Contact Sites A

Beug and Gerisch (1969) and Beug et al. (1970) injected whole starving, cohesive amebas into rabbits and raised antisera against their cell surfaces. The antibodies active against growth phase components were then absorbed out with excess growth phase amebas. The resultant antiserum caused agglutination of cohesive but not of vegetative amebas. Monovalent antibody (Fab) prepared by papain treatment of the antiserum completely inhibited amebal agglutination in suspension culture. It seems unlikely that this inhibition was due to a general covering of the amebal surface by the antiserum, as the contact sites represented only a small fraction (about 2%) of the cell surface (about 3×10^5 contact sites A per cell). Use of fluorescein- or ferritin-labeled anticontact sites A immunoglobulin showed no specific arrangement over the amebal surface (Gerisch et al., 1974), suggesting that the observed end-to-end contacts must have arisen by redistribution or activation of uniformly situated sites.

The contact sites A have been purified by Huesgen and Gerisch (1975), Eitle and Gerisch (1977), and Muller and Gerisch (1978) and have the characteristics of a single species of glycoprotein of about 70,000-80,000 daltons. This single species completely absorbed the Fab molecules that bound contact sites A, implying that the glycoprotein contains the complete set of antigenic determinants to which the Fab molecules bind.

Mutants deficient in contact sites A have been studied and found to be of three types (Gerisch, 1980): (1) A number of mutants showed pleiotropic loss of many aggregation functions, including formation of contact sites A (shown by their failure to bind anticontact sites A Fab); (2) one mutant bound anticontact sites A Fab incompletely, leaving about 30% of the Fab in an unbound state and capable of binding to and inhibiting aggregation in wild-type cells; and (3) two mutants failed to aggregate or form EDTA-stable contacts, but could completely absorb anticontact sites A Fab. Mutants of the third type are of particular interest, in that they appeared to be defective in pseudopodal activity and failed to elongate properly during amebal movement, apparently extending pseudopodia in different directions simultaneously. Such uncontrolled pseudopodial activity may be connected with the failure to produce active contact sites A, since such activation may occur in elongated cells forming end-to-end contacts during chemotaxis.

B. Cell Surface Lectins

The evidence for the involvement of discoidins and pallidins in cell surface adhesion may be summarized as follows:

1. They are found on the cell surface. Studies by Rosen et al. (1974) found that erythrocytes formed rosettes around cohesive slime mold amebas by binding to their cell surfaces. Using ferritin- or fluorescein-labeled antibody directed against rabbit antidiscoidin antibodies, Chang et al. (1975) found that the surface of cohesive but not vegetative cells became ferritin or fluorescein labeled. Later, Siu et al. (1976) reported that cohesive amebas treated with ^{125}I by lactoperoxidase labeled a protein that could be precipitated from detergent-solubilized preparations by specific rabbit antidiscoidin antiserum in conjunction with goat antirabbit immunoglobulin. The two predominantly labeled proteins so obtained had molecular weights of approximately 56,000 and 26,000, the latter corresponding approximately to discoidin.

Estimates by Siu et al. (1976) of the relative amount of discoidin on the cell surface compared to the amount found intracellularly indicated that in amebas starved for 10 h, discoidin can constitute 1% or more of the soluble cytoplasmic protein (about 10^6 molecules per cell). However, only a small minority were estimated as present on the amebal surface. The significance of this large intracellular pool for cell cohesion requires explanation.

2. They are developmentally regulated. Siu et al. (1976), using the ^{125}I surface-labeling technique, found that bacterially grown vegetative cells of *D. discoideum* showed undetectable amounts of cell surface or internally soluble discoidin, but over the course of development on filter supports the amount of cell surface discoidin rose dramatically and formed a plateau at 10 h. The time course of this increase was very similar to the increase in cohesiveness of the developing cells.

For reasons not yet fully understood, axenically grown cells formed discoidin during the exponential growth phase. Another protein that is formed by axenic amebas in this way and which is developmentally regulated in bacterially grown amebas is the enzyme N-acetyl glucosaminidase. It has been suggested that the untimely formation of such proteins by axenically grown cells may be due to such cells being in a semistarvation state (the mean doubling time being 9-10 h instead of 3-4 h, as with bacteria). However, recently Longmore and Watts (1980) have produced evidence indicating that in the case of formation of N-acetyl glucosaminidase, the enzyme is derepressed by the absence of gram-negative bacteria in the axenic cultures and lipopolysaccharide fractions from such bacteria could repress the synthesis of axenically growing cells. Whether the discoidin protein is under similar control would seem a possibility worth testing.

3. Species-specific cell cohesion correlates with the sugar-binding specificity of lectins from different species. In a study with six species of cellular slime molds, Rosen et al. (1975) found that various monosaccharides had different inhibitory abilities in an erythrocyte agglutination assay with lectins from the six different species.

Surprisingly, studies using slime mold aggregation assays give much less interpretable results. For example, purified discoidin does not of itself inhibit reassociation of dissociated aggregates of *D. discoideum* in suspension. Dissociated aggregates of *Polysphondylium pallidum* amebas are, however, inhibited from reassociating by purified pallidins (Frazier, 1976) but antipallidin Fab has only a weak effect in inhibiting agglutination by *P. pallidum* unless the adhesiveness is attenuated by 0.4 M glucose or 2.5 mM 2,4-dinitrophenol (Rosen et al., 1976, Bozzaro and Gerisch, 1978).

Although the evidence above suggests a cell surface role during development for the lectins, it is not clear that this role is simply that of cellular cohesion. Confirmatory evidence that discoidin I (the main discoidin in *D. discoideum*) is essential for aggregation comes from the work of Ray et al. (1979) and Shinnick and Lerner (1980), with the mutant of *D. discoideum* bearing a mutation at the locus *cbpA1*. As mentioned above, this mutant fails to aggregate or become cohesive and fails to form functional discoidin I. A study of reversion at the *cbpA* locus indicated that the degree of aggregation and cohesiveness that reappeared in such revertants correlated well with the reappearance of agglutination of sheep erythrocytes.

In studies by Yamada et al. (1980) and Aramaki et al. (1980) it was found that in a certain strain of *D. discoideum* the production of discoidin was influenced by the growth medium. If grown on media containing a proteose peptone of a particular composition, the strain failed to produce discoidin (as judged by agglutination assay with rabbit erythrocytes) yet aggregated and developed normally. Under these conditions, however, the cells produced a lectin with properties different from those of the discoidins. (It was, for example, extractable by Tris-HCL at pH 6.0 rather than pH 7.3, and agglutinated sheep but not rabbit red blood cells.) It appears that in some way the change in growth medium was influencing the lectin produced by the starving amebas. From molecular-weight studies using SDS gel electrophoresis, it seemed possible that the lectin produced in proteose peptone-grown cells was closely related to one of the subunits of the discoidin I protein.

While the function of the discoidins still remains a mystery, significant progress has recently been made in understanding the molecular biology of the discoidin genes. Gene-cloning experiments by Roewekamp et al. (1980) and Tsang et al. (1981) have isolated the DNA coding for discoidin I in a bacterial plasmid. A family of structurally related and closely linked genes coding for this (probably tetrameric) protein have been found. Very recently Williams et al. (1982), using a plasmid bearing a discoidin gene as a probe, have made the exciting finding that cAMP inhibits the transcription of the mRNA coding for discoidin I during the first few hours of development. As the natural rise in cAMP concentration in amebas occurs in development during the time when transcription of the discoidin I gene ceases, the authors suggested that cAMP acts at the level of transcription for regulation of discoidin synthesis.

ACKNOWLEDGMENTS

I would like to thank Frank Caddick for his untiring help with the photographic developing and printing, and Fiona Ross for her help and advice.

REFERENCES

Alcantara, F., and Monk, M. (1974). Signal propagation during aggregation in the slime mold *Dictyostelium discoideum*. *J. Gen. Microbiol.* *85*:321-334.

Aramaki, Y., Yamada, H., and Miyazaki, T. (1980). Isolation and characterization of a carbohydrate-binding protein from *Dictyostelium discoideum* NC4 deficient in normal discoidin synthesis. *J. Biochem. Tokyo* *87*:1145-1151.

Arndt, A. (1937). Untersuchungen über *Dictyostelium mucoroides* Brefeld. *Wilhelm Roux Arch. Entwicklungswech. Org.* *136*:681-747.

Barra, J., Barrand, P., Blondelet, M. H., and Brachet, P. (1980). *pdsA*: A gene involved in the production of active phosphodiesterase during starvation of *Dictyostelium discoideum* amebas. *Mol. Gen. Genet.* *177*:607-613.

Beug, H., and Gerisch, G. (1969). Univalente Fragments von Antikorpern zur Analyse von Zelimembran-Funktionen. *Naturwissenshaften* *56*:374.

Beug, H., Gerisch, G., Kempff, S., Riedel, V., and Cremer, G. (1970). Specific inhibition of cell contact formation in *Dictyostelium* by univalent antibodies. *Exp. Cell Res. 63*:147-158.

Bonner, J. T. (1944). A descriptive study of the development of the slime mold *Dictyopstelium discoideum. Am. J. Bot. 31*:175-182.

Bozzaro, S., and Gerisch, G. (1978). Contact sites in aggregating cells of *Polysphondylium pallidum. J. Mol. Biol. 120*:265-279.

Brachet, P., Dicou, E. L., and Klein, C. (1979). Inhibition of cell differentiation in a phosphodiesterase defective mutant of *Dictyostelium discoideum. Cell Differ. 8*:255-265.

Chang, C. M., Reitherman, R. W., Rosen, S. D., and Barondes, S. H. (1975). Cell surface location of discoidin: A developmentally regulated carbohydrate-binding protein from *Dictyostelium discoideum. Exp. Cell Res. 95*:136-142.

Chung, W. J. K. and Coe, E. L. (1978). Correlations among responses of suspensions of *D. discoideum* to pulses of 3'5' cAMP. *Biochim. Biophys. Acta 544*:29-44.

Cohen, M. H., and Robertson, A. (1971). Wave propagation in the early stages of aggregation of cellular slime molds. *J. Theor. Biol. 31*:101-118.

Coukell, M. B. (1975). Parasexual genetic analysis of aggregation deficient mutants of *Dictyostelium discoideum. Mol. Gen. Genet. 142*:119-135.

Coukell, M. B. (1977). Evidence against mutational "hot-spots" at aggregation loci in *Dictyostelium discoideum. Mol. Gen. Genet. 151*:269-273.

Devreotes, P. N., and Steck, T. L. (1979). Cyclic 3',5'AMP relay in *Dictyostelium discoideum*. II. Requirements for the initiation and termination of the response. *J. Cell. Biol. 80*:300-309.

Dicou, E. L., and Brachet, P. (1980). A separate phosphodiesterase for the hydrolysis of cGMP in growing *Dictyostelium discoideum* ameba. *Eur. J. Biochem. 109*:507-514.

Dinauer, M. C., Steck, T. L., and Devreotes, P. N. (1980). Adaptation of the cAMP signalling response during cAMP stimulation. *J. Cell Biol. 86*:554-561.

Durston, A. J. (1973). *Dictyostelium discoideum* aggregation fields as excitable media. *J. Theor. Biol. 42*:483-504.

Durston, A. J. (1974). Pacemaker activity during aggregation in *Dictyostelium discoideum. Dev. Biol. 37*:225-235.

Eitle, E., and Gerisch, G. (1977). Implication of developmentally related concanavalin A binding proteins of *Dictyostelium* in cell adhesion and cAMP regulation. *Cell Differ. 6*:339-346.

Ennis, H. L., and Sussman, M. (1958). The initiator cell for slime mold aggregation. *Proc. Nat. Acad. Sci. USA. 44*:401-411.

Farnham, C. J. M. (1975) Cytochemical localization of adenylate cyclase and 3',5' nucleotide phosphodiesterase in *Dictyostelium discoideum. Exp. Cell. Res. 91*:36-46.

Franke, J., and Kessin, R. H. (1977). Defined minimal medium for axenic strains of *Dictyostelium discoideum*. *Proc. Nat. Acad. Sci. USA*. *74*:2157-2161.

Frazier, W. A. (1976). The role of cell surface components in the morphogenesis of the cellular slime molds. *Trends Biochem. Sci.* *1*:130-133.

Futrelle, R. P., McKee, W. G., and Traut, J. (1980). Response of *Dictyostelium discoideum* to localized cAMP stimuli: Computer analysis of cell action. *J. Cell Sci.* *87*:57a.

Gerisch, G. (1961). Zellfunktionen und Zellfunktionswechse in der Entwicklund von *Dictyostelium discoideum*. *Dev. Biol.* *3*:685-724.

Gerisch, G. (1968). Cell aggregation and differentiation in *Dictyostelium*. In *Current Topics in Developmental Biology*, Vol. 3. A. A. Moscona and A. Monroy (Eds.). Academic, New York, pp. 157-197.

Gerisch, G. (1980). Univalent antibody fragments as tools for the analysis of cell interactions in *Dictyostelium*. In *Current Topics in Developmental Biology*, Vol. 14. A. A. Moscona and A. Monroy (Eds.). Academic, New York, pp. 243-270.

Gerisch, G., and Hess, B. (1974). Cyclic-AMP controlled oscillations in suspended *Dictyostelium* cells: Their relation to morphogenetic cell interactions. *Proc. Nat. Acad. Sci. USA* *71*:2118-2122.

Gerisch, G., and Malchow, D. (1976) Cyclic AMP receptors and the control of cell aggregation in *Dictyostelium*. In *Advances in Cyclic Nucleotide Research*, Vol. 7. P. Greengard and G. A. Robison (Eds.). Raven Press, New York, pp. 49-68.

Gerisch, G., Beug, H., Malchow, D., Schwarz, H., and Stein, A. V. (1974). In *Biology and Chemistry of Eucaryotic Cell Surfaces*, Vol. 7. Miami Winter Symposium. E. Y. C. Lee and E. E. Smith (Eds.), Academic, New York, pp. 49-66.

Gerisch, G., Malchow, D., Huesgen, A., Nanjundiah, V., Roos, W., and Wick, U. (1975) Cyclic AMP reception and cell recognition in *Dictyostelium discoideum*. In *ICN-UCLA Symposium on Developmental Biology*. D. MacMahon and F. Fox (Eds.). Benjamin, New York, pp. 76-88.

Glazer, P. M., and Newell, P. C. (1981) Initiation of aggregation by *Dictyostelium discoideum* in mutant populations lacking pulsatile signalling. *J. Gen. Microbiol.* *125*:221-232.

Goy, M. F., and Springer, M. S. (1978) In search of the linkage between receptor and response: The role of a protein methylation reaction in bacterial chemotaxis. In *Taxis and Behaviour, Receptors and Recognition*, Vol. B5. G. L. Hazelbauer (Ed.). Chapman and Hall, London, pp. 1-34.

Green, A. A., and Newell, P. C. (1975) Evidence for the existance of two types of binding sites in aggregating cells of *Dictyostelium discoideum*. *Cell* *6*:129-136.

Gross, J. D., Peacey, M. J., and Trevan, D. J. (1976). Signal emisssion and signal propagation during early aggregation in *Dictyostelium discoideum*. *J. Cell Sci.* *22*:645-656.

Grutsch, J. F., and Robertson, A. (1978). cAMP signal from *Dictyostelium discoideum* amebas. *Dev. Biol.* *66*:285-293.

Henderson, E. J. (1975). The cyclic adenosine 3'5' monophosphate receptor of *Dictyostelium discoideum*. *J. Biol. Chem.* *250*:4730-4736.

Hintermann, R., and Parish, R. W. (1979). Intracellular location of adenylyl cyclase in the cellular slime molds *Dictyostelium discoideum* and *Polysphondylium pallidum*. *Exp. Cell Res.* *123*:429-434.

Huesgen, A., and Gerisch, G. (1975). Solubilized contact sites A from cell membranes of *Dictyostelium discoideum*. *FEBS Lett.* *56*:46-49.

Kakebeeke, P. I. J., Mato, J. M., and Konijn, T. M. (1978). Purification and preliminary characterization of an aggregation-sensitive chemoattractant of *Dictyostelium minutum*. *J. Bacteriol.* *133*:403-405.

Kakebeeke, P. I. J., de Wit, R. J. W., and Konijn, T. M. (1980a). Folic acid deaminase activity during development in *Dictyostelium discoideum*. *J. Bacteriol.* *143*:307-312.

Kakebeeke, P. I. J., de Wit, R. J. W., and Konijn, T. M. (1980b). A novel chemotaxis regulating enzyme that splits folic acid into 6-hydroxymethylpterin and p-aminobenzoylglutamic acid. *FEBS Lett.* *115*:216-220.

Keller, E. F., and Segal, L. A. (1970). Initiation of slime mold aggregation viewed as an instability. *J. Theor. Biol.* *26*:399-415.

Kessin, R. H., Orlow, S. J., Shapiro, R. I., and Franke, J. (1979). Binding of inhibitor alters kinetic and physical properties of extracellular cAMP phosphodiesterase from *Dictyostelium discoideum*. *Proc. Nat. Acad. Sci. USA* *76*:5450-5454.

Klein, C. (1977). Changes in adenylate cyclase during differentiation of *Dictyostelium discoideum*. *FEMS Microbiol. Lett.* *1*:17-19.

Klein, C. (1979). Slowly dissociating form of the cell-surface cAMP receptor of *Dictyostelium discoideum*. *J. Biol. Chem.* *254*:12573-12578.

Klein, C., and Darmon, M. (1977). Effects of cAMP pulses on adenylate cyclase and the phosphodiesterase inhibitor of *Dictyostelium discoideum*. *Nature* *268*:76-78.

Klein, C., and Juliani, M. H. (1977). cAMP induced changes in cAMP-binding sites on *D. discoideum* ameba. *Cell* *10*:309-335.

Konijn, T. M. (1972). Cyclic AMP and cell aggregation in the cellular slime molds. *Acta Protzool.* *11*:137-143.

Konijn, T. M., and Raper, K. B. (1961). Cell aggregation in *Dictyostelium discoideum*. *Dev. Biol.* *3*:725-756.

Longmore, K., and Watts, D. J. (1980). Control of N-acetylglucosaminidase specific activity in myxamoebae of *Dictyostelium discoideum*. *Dev. Biol. 78*:104-112.

Maeda, Y., and Gerisch, G. (1977). Vesicle formation in *Dictyostelium discoideum* cells during oscillations of cAMP synthesis and release. *Exp. Cell Res. 110*:119-126.

Malchow, D., and Gerisch, G. (1974). Short-term binding and hydrolysis of cyclic 3':5' adenosine monophosphate by aggregating *Dictyostelium* cells. *Proc. Nat. Acad. Sci. USA 71*:2423-2427.

Malchow, D., Nanjundiah, V., Wurster, B., Eckstein, F., and Gerisch, G. (1978). cAMP-induced pH changes in *Dictyostelium discoideum* and their control by Ca^{++}. *Biochim. Biophys. Acta 538*:473-480.

Mato, J. M., and Konijn, T. M. (1975). Chemotaxis and binding of cyclic AMP in cellular slime molds. *Biochim. Biophys. Acta 385*:173-179.

Mato, J. M., and Malchow, D. (1978). Guanylate cyclase activation in response to chemotactic stimulation in *Dictyostelium discoideum*. *FEBS Lett. 90*:119-122.

Mato, J. M., and Marin-Cao, D. (1979). Protein and phospholipid methylation during chemotaxis in *Dictyostelium discoideum* and its relationship to calcium movements. *Proc. Nat. Acad. Sci. USA 76*:6106-6109.

Mato, J. M., Van Haastert, P. J. M., Krens, F. A., and Konijn, T. M. (1977a). An acrasin-like attractant from yeast extract specific for *Dictyostelium lacteum*. *Dev. Biol. 57*:450-453.

Mato, J. M., Van Haastert, P. J. M., Krens, F. A., Rhijnsburger, E. H., Dobbe, F. C. P. M., and Konijn, T. M. (1977b). Cyclic AMP and folic acid mediated cyclic GMP accumulation in *Dictyostelium discoideum*. *FEBS Lett. 79*:331-336.

Mato, J. M., Krens, F. A., Van Haastert, P. J. M., and Konijn, T. M. (1977c). 3',5'cAMP dependent, 3',5'cGMP accumulation in *Dictyostelium discoideum*. *Proc. Nat. Acad. Sci. USA 74*:2348-2351.

Mockrin, S. C., and Spudich, J. A. (1976). Calcium control of actin-activated myosin adenosine triphosphatase from *Dictyostelium discoideum*. *Proc. Nat. Acad. Sci. USA 73*:2321-2325.

Mullens, I. A., and Newell, P. C. (1978). cAMP binding to cell surface receptors of *Dictyostelium*. *Differentiation 10*:171-176.

Muller, K., and Gerisch, G. (1978). A specific glycoprotein as the target site of adhesion blocking Fab in aggregating *Dictyostelium discoideum* cells. *Nature 274*:445-449.

Newell, P. C. (1977). Aggregation and cell surface receptors in cellular slime molds. In *Receptors and Recognition, Microbial. Interactions*, Vol. B3. J. L. Reissig (Ed.). Chapman and Hall, London, pp. 1-57.

Newell, P. C. (1978). Genetics of the cellular slime molds. *Annu. Rev. Genet. 12*:69-93.

Newell, P. C. (1981). In *The Biology of the Chemotactic Response*, J. M. Lackie and P. C. Wilkinson (Eds.). Cambridge University, Cambridge.

Pan, P., and Wurster, B. (1978). Inactivation of chemoattractant folic acid by cellular slime molds and identification of reaction product. *J. Bacteriol. 136*:955-959.

Pan, P., Hall, E. M., and Bonner, J. T. (1972). The bacterial attracting substance is folic acid. *Nature London New Biol. 237*: 181-182.

Pan, P., Hall, E. M., and Bonner, J. T. (1975). Determination of active portion of folic acid molecule in cellular slime mold chemotaxis. *J. Bacteriol. 122*:185-191.

Raman, R. K. (1976). The autonomous cell in *Dictyostelium discoideum. J. Theor. Biol. 59*:491-495.

Raman, R. K., Hashimoto, Y., Cohen, M. H., and Robertson, A. (1976). Differentiation for aggregation in the cellular slime molds: The emergence of autonomously signally cells in *Dictyostelium discoideum. J. Cell Sci. 21*:243-259.

Ray, J., Shinnick, T. M., and Lerner, R. A. (1979). Mutation altering the function of carbohydrate-binding protein blocks cell-cell cohesion in developing *Dictyostelium discoideum. Nature 279*:215-221.

Roewekamp, W., Poole, S., and Firtel, R. A., (1980). Analysis of members of the multi-gene family coding the developmentally regulated carbohydrate binding protein discoidin I in *Dictyostelium discoideum. Cell 20*:495-505.

Roos, W., Nanjundiah, V., Malchow, D., and Gerisch, G. (1975). Amplification of cyclic-AMP signals in aggregating cells of *Dictyostelium discoideum. FEBS Lett. 53*:139-142.

Roos, W., Scheidegger, C., and Gerisch, G. (1977). Adenylate cyclase activity oscillations as signals for cell aggregation in *Dictyostelium discoideum. Nature 266*:259-261.

Rosen, S. D., and Barondes, S. H. (1978). Cell adhesion in the cellular slime molds. In *Receptors and Recognition*, Vol. 4B. D. R. Garrod (Ed.). Chapman and Hall, London, pp. 233-264.

Rosen, S. D., Simpson, D. L., Rose, J. E., and Barondes, S. H. (1974). Carbohydrate-binding protein from *Polysphondylium pallidum* implicated in intercellular adhesion. *Nature 252*:128-151.

Rosen, S. D., Reitherman, R. W., and Barondes, S. H. (1975). Distinct lectin activities from six species of cellular slime molds. *Exp. Cell Res. 95*:159-166.

Rosen, S. D., Haywood, P. C., and Barondes, S. H. (1976). Inhibition of intercellular adhesion in a cellular slime mold by univalent antibody against a cell surface lectin. *Nature 263*:425-427.

Ross, F. M., and Newell, P. C. (1979). Genetics of aggregation pattern mutations in the cellular slime mold *Dictyostelium discoideum*. *J. Gen. Microbiol.* *115*:289-300.

Ross, F. M., and Newell, P. C. (1981). Streamers: Chemotactic mutants of *Dictyostelium* with altered cGMP metabolism. *J. Gen. Microbiol.* *127*:339-350.

Shaffer, B. M. (1957). Aspects of aggregation in cellular slime molds. *Am. Nat.* *91*:19-35.

Shinnick, T. M., and Lerner, R. A. (1980). The *cbpA* gene: Role of the 26,000 D carbohydrate binding protein in intercellular cohesion of developing *Dictyostelium discoideum*. *Proc. Nat. Acad. Sci. USA* *77*:4788-4792.

Siu, C. H., Lerner, R. A., Ma, G., Firtel, R. A., and Loomis, W. F. (1976). Developmentally regulated proteins of the plasma membrane of *Dictyostelium discoideum*. The carbohydrate-binding protein. *J. Mol. Biol.* *100*:157-178.

Sussman, M., and Noel, E. (1952). An analysis of the aggregation phase in the development of the slime molds *Dictyosteliaceae*. I: Populational distribution of the capacity to initiate aggregation. *Biol. Bull. Woods Hole Mass.* *103*:259-268.

Tsang, A. S., and Coukell, M. B. (1977). The regulation of cAMP phosphodiesterase and its specific inhibitor by cAMP in *Dictyostelium discoideum*. *Cell Differ.* *6*:75-84.

Tsang, A. S., and Coukell, M. B. (1979). Direct evidence for extracellular cAMP phosphodiesterase induction and phosphodiesterase inhibitor repression by exogenous cAMP in *Dictyostelium purpureum*. *Eur. J. Biochem.* *95*:419-425.

Tsang, A. S., Devine, J. M., and Williams, J. G. (1981). The multiple subunits of discoidin I are encoded by different genes. *Dev. Biol.* *84*:212-217.

Wallace, L. J., and Frazier, W. A. (1979). Photoaffinity labeling of cAMP binding and AMP binding proteins of differentiating *Dictyostelium discoideum* cells. *Proc. Nat. Acad. Sci. USA* *76*:4250-4254.

Warren, J. A., Warren, W. D., and Cox, E. C. (1975). Genetic complexity of aggregation in the cellular slime mold *Polysphondylium violaceum*. *Proc. Nat. Acad. Sci. USA* *72*:1041-1042.

Warren, J. A., Warren, W. D., and Cox, E. C. (1976). Genetic and morphological study of aggregation in the cellular slime mold *Polysphondylium violaceum*. *Genetics* *83*:25-47.

Wick, U., Malchow, D., and Gerisch, G. (1978). cAMP stimulated Ca^{++} influx into aggregating cells of *Dictyostelium discoideum*. *Cell Biol. Int. Rep.* *2*:71-79.

Williams, J. G., Tsang, A. S., and Mahbubani, H. (1982). A change in the rate of transcription of a eukaryotic gene in response to cyclic AMP. *Proc. Nat. Acad. Sci. USA*. In press.

Williams, K. L., and Newell, P. C. (1976). A genetic study of aggregation in the cellular slime mold *Dictyostelium discoideum* using complementation analysis. *Genetics 82*:287-307.

Wurster, B., and Butz, U. (1980). Reversible binding of the chemoattractant folic acid to cells. *Eur. J. Biochem. 109*:613-618.

Wurster, B., Pan, P., Tyan, G. G., and Bonner, J. T. (1976). Preliminary characterization of acrasin of the cellular slime mold *Polysphondylium violaceum. Proc. Nat. Acad. Sci. USA 73*:795-799.

Wurster, B., Schubiger, K., Wick, U., and Gerisch, G. (1977). Cyclic GMP in *Dictyostelium discoideum*: Oscillations and pulses in response to folic acid and cAMP signals. *FEBS Lett. 76*:141-144.

Yamada, H., Aramaki, Y., and Miyazaki, T. (1980). Effect of different growth media on the synthesis of carbohydrate binding protein from *Dictyostelium discoideum* NC4. *J. Biochem. Tokyo 87*:333-338.

4

Morphogenesis and Differentiation in *Physarum*

HELMUT W. SAUER* and GERARD PIERRON* *Zoologisches Institut, Universität Würzburg, Würzburg, Federal Republic of Germany*

I. INTRODUCTION

Every living organism needs to grow and reproduce as it interacts with other organisms that do the same. All living things, therefore, can be described as undergoing a life cycle. In simple organisms like bacteria, the cell division cycle is the life cycle. More complex organisms, the eukaryotes, show specializations. A unicellular organism contains a cell nucleus. Only after its genome has completely doubled does it divide, and only then will the cell split in two. Therefore the mitotic cycle equals the essential life cycle in protists, which potentially live forever. The cell cycle does not suffice to describe the life of a multicellular organism. During their life time, cells become different, they stop dividing and serve specific functions, and normally die. Complex

*Present affiliation: Texas A & M University, College Station, Texas

organisms survive because of a special cell type, the germ cells (eggs and sperm). Consequently, cell differentiation becomes a prominent element in the life cycle and an important concern of the developmental biologist as he asks the simple question of how to build an animal.

The division of labor is an important feature – either as a prerequisite or a consequence, when a protist became a multicellular system in the time course of evolution, just as an egg cell has the potential to turn into an organism as complex as us. We do not understand how this is achieved. No protist has turned into a multicellular organism under the watchful eye of a biologist. Obviously, the function essential for higher organisms is missing in a protist. A serious problem in experimental analysis of an egg cell is that it is too complex; it is the respective organism in disguise. Eggs are not undifferentiated cells, although often named as such. The human mature egg is so specialized, or terminally differentiated, that it lives only one day unless it is fertilized by an equally short-lived sperm cell.

For the higher organism, becoming multicellular is just one event among the processes of growth and differentiation which constitute embryonic development.

Many facts of life have been learned from simple systems. The structure and function of the genome have been found to be remarkably conserved in all organisms, making them all relatives to each other, by degrees. Also, the mitotic mechanism which assures accurate segregation of the chromosomes is quite similar in all eukaryotic organisms. The mechanism for becoming multicellular, although unknown in detail, may be a prerequisite for cell type-specific differentiation and could be a third conserved trade characteristic of all higher organisms.

Despite such conserved features, organisms are very different. What makes them different is their morphology. Changes in form occur at every level of biological organization, from molecules to societies and, in part, they result from environmental constraints (Sauer, 1980).

In short, mitotic cycles, multicellularity, morphogenesis, and differentiation are the key notes of biological development. *Physarum polycephalum* may be a system of choice in which to analyze these general yet central questions which have been probed for a quarter of a century. However, this true slime mold deserves to be studied in its own right to unravel the life cycle strategies of a Myxomycete, a class of "lower eukaryotes."

In the following we shall focus on some relevant developments of *Physarum*, the Myxomycete, and the model.

II. THE LIFE CYCLE OF *PHYSARUM*

Physarum polycephalum can be found only in the New World, where it lives in various geographical locations as two distinct macroscopic forms. One is a colony of black fruiting bodies, each about 1-2 mm tall and made up of a thin stalk with a head on top. These structures are sedentary and exposed to the light. The other form is a slimy, bright yellow mobile mass, the plasmodium, containing millions of nuclei in a common cytoplasm. It is not readily seen, as it hides in the dark under the rotting bark of fallen trees. This duality in shape and habitat is a favorable adaptation of the mold, as the sporangia harbor many spores, which can be better distributed from an exposed site. It is also useful for the experimenter to work with both forms. Indeed, when a plasmodium is kept on the window sill, it can change completely into many sporangia, whereas a sporangium cannot turn directly back into a plasmodium. This then describes a unidirectional irreversible change. Two conditions must be met for the process to occur: The plasmodium must not be fed and it must not be kept in the dark. When a plasmodium is starved in the dark, it becomes a sclerotium, a dormant form that can easily be revived with food. As long as a plasmodium is fed with organic substrate and microorganisms collected at its habitat, or – more conveniently – with oat flakes, it grows vigorously and stays around its food, migrating only after the food source is exhausted.

Spores are the microscopic form of *Physarum*, little round balls about 10 μm in diameter. They can be germinated in water. Little amebas hatch out and grow on a lawn of bacteria in a Petri dish. They divide like any other soil amebas. In addition, they can give rise to plasmodia, something no ordinary soil amebas can do. This cell transformation is a trait which is typical for Myxomycetes.

Much work has been done with these impure cultures and references can be obtained on the biology (Martin, 1940) and developmental physiology (von Stosch, 1965). More recently, the biosystematics of the true or acellular slime molds has been coherently treated (Olive, 1975), and the reproductive strategy of two related Myxomycetes, *Didymium* and *Physarum*, has been analyzed (Collins, 1979).

Biochemical work on *Physarum* was initiated after an axenic medium became available (Daniel and Rusch, 1961) that could support the growth of vegetative plasmodia. Later, a procedure to obtain sporulation under controlled conditions was derived (Daniel, 1966). The many descriptive results, concerning the composition and synthesis of biomolecules, and the outcome of some elegant experiments have been summarized by Rusch (1970). In the mean time genetic procedures have been adapted to an almost unique situation. Immense numbers (10^{10}) of amebas, cloned from a single spore, have been mutagenized. Mutants have been obtained that are deficient for functions required for amebal and/or the plasmodial state. Furthermore, any gene of interest can be analyzed and transferred into another strain by Mendelian crossing (Dee, 1973). Recombinant DNA techniques have also been applied, although so far this work has been limited for structural analysis of DNA.

The combination of these various experimental procedures can turn the humble slime mold *Physarum* into a model system, at least in the eyes of some "Physarologists."

III. *PHYSARUM*: THE MODEL SYSTEM

The complete life cycle of *Physarum* can be studied under defined conditions in the laboratory from spore to spore, revealing 15 distinct stages (Fig. 1).

Spores (1) release amebas (2), which can multiply (3) or become swarm cells (4), microcysts (5), or gametes (6). Under normal conditions, two gametes of an appropriate mating type can fuse and form a diploid zygote (7); in some strains, the "Selfers," plasmodia can be derived from a single ameba (8). The multinucleated plasmodium (diploid after sexual fusion of gametes or haploid in the selfer) can be grown as a submerged, agitated suspension of microplasmodia (9), or as macroplasmodia (10), after natural fusion of many microplasmodia on top of a filter paper. The microplasmodia can break up into clusters of macrocysts (11). In contrast to microcysts, which contain one nucleus, macrocysts contain 0-15 nuclei; they are also called spherules. A macroplasmodium forms a sclerotium (12), which consists of a large collection of macrocysts held together by dried up masses of slime.

Macrocysts are formed in response to unfavorable environmental conditions in the dark. Alternatively, starved macroplasmodia (but not submerged microplasmodia) can sporulate if they are illuminated. They differentiate into many fruiting bodies (13). Within the head, prespores are formed following a mitotic nuclear division (14). After meiosis (or a pseudomeiosis in the selfer) uninucleated spores are formed (15) and the life cycle has come full round. A concise review of the structural-functional aspects has recently been published (Goodman, 1980).

We can distinguish two forms of vegetative proliferation (ameba, plasmodium), three forms of motility (swarm cell, ameba, plasmodium), three dormant forms (macrocyst, microcyst, spore), three examples of reversible differentiation (macrocyst formation, microcyst formation, ameba-flagellate transition), and two developmental pathways leading to irreversible commitment, that is, sporulation (plasmodium-ameba transition) and plasmodium formation (ameba-plasmodium transformation). Their analysis may unravel two distinct developmental programs, if such programs exist. If not, we can describe in detail the history of developmental changes of a simple, but not too simple, eukaryotic organism as a network of gene action, and the interaction of gene products as an "epigenetic system."

Furthermore, *Physarum* can be viewed as a truly multipotent system. The plasmodium has the option to grow or to differentiate into either macrocysts or sporangia. The ameba has four choices: to grow

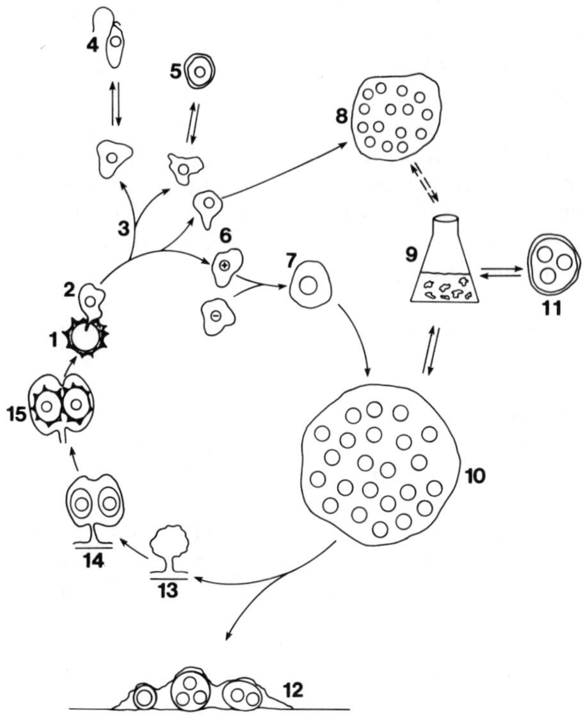

FIG. 1 The life cycle of *Physarum* (for details see text).

as an ameba or a swarm cell, to stop growth and encyst, or to become a gamete. Here we may be in a unique position to detect how an organism selects one pathway over more than just one alternate pathway.

Some other aspects of plasmodial life have received some attention. Two are concerned with endogenous rhythmic processes associated with subcellular transient morphogenetic changes: the shuttle streaming and the naturally synchronous nuclear divisions. In addition to studies of the actomyosin complex or the mitotic apparatus as such, the synchrony of both processes allows one to probe for the signals that cause mitosis or contractions. Also, the phenomenon of recognition of self has been described, as plasmodia can either recognize or reject one another if they meet on an agar dish. Finally, after an extended period of culture, plasmodia grow old and eventually die, thus displaying senescence.

The main reason for the success of *Physarum* in the laboratory was at first the availability of homogeneous material that could be analyzed in ample quantities in separate test tubes, because the growth and differentiation of the plasmodium are mutually exclusive. Contrary to embryos, which grow and differentiate, the plasmodium either grows, displaying freewheeling proliferative mitotic cycles, or changes into either macrocysts or sporangia. In the latter case, even though there is no trace of growth, there is at least one mitotic division which may be an essential step in the differentiation pathway of building spores.

Lately, as mutants of the transformation of a single ameba (the haploid selfer) into a plasmodium have become available, it has been suggested that a single gene, *mat A*, may be responsible for plasmodium formation. This offers a good opportunity to lay hands on a major regulatory gene.

As a reminder, the purpose of establishing *Physarum* as a model system was to find an anticancer agent. Rusch's (1970) aim was to extract a factor from a sporulating mold and put it on a growing plasmodium, thus forcing it to sporulate and abandon proliferative mitotic cycles. This goal has not yet been reached, nor has the cancer puzzle been solved. However, the approach is quite remarkable for two reasons: Its hypothetical basis is not so much that the cancerous cell has a defect in its proliferation control, but that it cannot differentiate properly. Perhaps every cancerous cell, if it were forced to differentiate, would cease to be a cancer cell. Consequently, we need to learn all we can about normal differentiation before we can understand cancer. The other presumption is that differentiation mechanisms may be basically the same in all organisms beyond protists, from *Physarum* to man.

Whether both presumptions, the ubiquitous differentiation mechanism and cancer as defective differentiation, hold true is not known, but a system that can either grow (a clone of an ameba or a plasmodium) or completely and irreversibly change (i.e., differentiate into a plasmodium or sporangia) must yield valuable information, as in most other experimental organisms mitosis and differentiation occur in the same system at the same time. This last point raises the chicken and egg situation, as we have introduced the plasmodial state both as the result of amebal transformation, that is, differentiation, and as the growing (i.e., undifferentiated) form to be differentiated into spores. Indeed, the term *differentiation* is ambivalent, as discussed in detail by Weiss (1973). It is applied to developmental changes, both as a dynamic process and a rather stable result.

To illustrate this problem with just one example from the life of *Physarum*, is the transitory assembly of the motile apparatus in the ectoplasm tube of the plasmodium a case of differentiation or not? We shall not pursue this question further, nor shall we repeat a coherent analysis of the life cycle of *Physarum*, since this has been done elsewhere (Dove and Rusch, 1980; Aldrich and Daniel, 1981). Instead, we shall select a few topics about which something has been learned and much more can be questioned. A book of current problems, dealt with at the Fifth European *Physarum* Meeting in Cracow, Poland, has been edited by Rakoczy (1981).

A. Intracellular Morphogenesis

Microtubuli (MTs) make up the transient mitotic spindle in the closed mitosis of the plasmodium, as well as in the open mitosis of amebas. They are also the major structural component of the ameba-flagellate transition. The α and β tubulins from amebas copolymerize into MTs with tubulins of pig brain and of *Paramecium*. However, the antitubular drug colchicine does not affect mitosis in *Physarum*. In plasmodia the tubulins accumulate in the nucleus, in part following de novo synthesis in the prophase of mitosis, but they do not guarantee MT formation. The microtubulus-associated proteins (MAPs) are involved in the regulation of MT growth, as seen from the inhibitory effect of griseofulvin on the in vitro assembly of MTs in the amebal extracts and on in vivo mitotic spindle formation of plasmodia (Roobol et al., 1980; Bernstam et al., 1981). These associated proteins may constitute the bridges connecting MTs and the fibrillar network within the complex flagellar apparatus (the cone) and an asymmetric rigid set of arrays of MTs (Wright et al., 1979) and perhaps explain the existence of many mutants specifically affecting flagellate formation. These MAPs are candidates for true morphogenic substances.

The main factor regulating MT arrangement is the appearance, activation, division, and translocation of the RNA-containing microtubulus-organizing centers (MTOC), which function in the initiation and orientation of MTs of the intranuclear spindle (Laane and Haugli, 1974). In plasmodia, a heat shock in the prophase of mitosis does not prevent formation of MTs, but does effect their proper orientation. In the amebal spindle each MTOC is associated with two typical centrioles which function as spindle poles (Wright et al., 1981). The two centrioles are also the initiation sites of the two flagella in swarm cells. The isolation of the MTOC, which appears close to the nucleolus at the prophase in the plasmodium, its transformation into the cytoplasmic kinetosome of amebas, mutant analysis, and knowledge of what activates MTOC would greatly increase our understanding of the many cellular processes involving MTs, not only in *Physarum*.

Actomyosin fibrils are another conserved cellular structure. *Physarum* actin has an amino acid sequence very similar to that of mammalian cytoplasmic actin (Vandekerckhove and Weber, 1978). Much work has been done on the oscillatory synchronous contractions of the ectoplasm tube, which serves as the motor to shuttle the endoplasm and probably keeps the whole plasmodium in synchrony, as separate plasmodia display different contraction and mitotic rhythms (Hatano et al., 1979; Pepe et al., 1979). At least some of the actin filaments (F actin) disaggregate and reaggregate with each contraction cycle. Therefore the transformation of globular actin (G actin) into F actin is an important prerequisite for the organization of the motile apparatus in *Physarum*. A protein of 41,000 daltons which copurifies with actin (43,000 daltons), has been shown to keep *Physarum* actin from polymerizing into filaments in vitro. It may be responsible for preventing F actin formation in the endoplasm, and it may do the same in *Amoeba proteus*, whose migration is completely upset after injection of the protein from *Physarum* (Gawlitta et al., 1980).

As discussed for the MTs, the orientation of the muscle fibrils is essential for their function. They are attached to certain sites of the plasmalemma invaginations which pervade the whole ectoplasm, perhaps through an actininlike protein which has been isolated in plasmodial extracts. A new protein with elastic properties, connectin, has recently been isolated (Ozaki and Maruyama, 1980). It may prevent the plasmalemma from herniating upon contraction.

Of course, calcium is involved in actomyosin activity, and its ubiquitous receptor molecule, calmodulin, is also present in *Physarum*. As a regulatory mechanism, a means for sequestering calcium would be required. In functional terms there is a sarcoplasmic reticulum in *Physarum*. It consists of vesicles that have recently been isolated as a microsome fraction (Zubrzycka-Gaarn et al., 1979). These vesicles contain a calcium-dependent ATPase of 105,000 daltons in their membrane and resemble the endoplasmic reticulum (ER) of normal cells. They are capable of accumulating calcium. Electron micrographs have not demonstrated an ER in the dynamic protoplasm of *Physarum*; however, during macrocyst formation a typical rough ER, studded with ribosomes, can be seen (Goodman and Rusch, 1970). Presumably the vesicles in the fluid growing plasmodium have fused and thus created the reticulum, as the cytoplasm becomes dessicated during encystment, when shuttle streaming also stops. Interestingly, the ER of normal tissue, like liver, has been isolated in the form of microsomes which arise artifactually during cell fractionation. Therefore the transformation of sarcoplasmic vesicles into the respective reticulum in *Physarum* may be the opposite and more natural process than that seen in standard cell fractionation. Furthermore, the sarcoplasmic vesicles may be derived from the nuclear membrane.

As an elaborate reticulum does not exist in the plasmodium, its nuclei can be isolated as clean spheres, without the need for breaking cytoplasmic connections like the ER. These nuclei may be uniquely suited for placement into test tubes and division in vitro.

The nuclear membrane disintegrates in amebal mitosis (hence open mitosis), but not in the plasmodium. However, the membrane seems to be extensively deformed during the anaphase and telophase of plasmodial mitosis, perhaps as MTs and MFs (microfibrils) are pushing and pulling on it. On the other hand, the MT apparatus of the flagellate is so tightly fixed to the nucleus that both can be isolated as one complex containing the nucleus, the cone of MT arrays, and the flagella, ready for fine structure analysis (Wright et al., 1979). The metaphase nucleus of plasmodia, though round, is not completely homogeneous, as there are holes in the membrane at the two spindle poles (Lord et al., 1977), potential gates for calcium ions to flow in and dismantled tubulin to escape through as the spindle shortens and thus drives daughter chromosomes apart.

The nucleolus, in the ameba and the plasmodium, shows the typical fibrillar core surrounded by a granular mass made up of ribonucleoprotein. As a unique situation for somatic cells, the core is made up of many minichromosomes, about 125 per haploid genome, each containing a piece of DNA which consists of two ribosomal genes (each 4.2 μm long) separated by a very large nontranscribed spacer (about 11 μm long) (Grainger and Ogle, 1978; Johnson et al., 1979; Gubler et al., 1980). Presently it is not known whether a single ribosomal gene exists in the chromosome or whether all of them are permanently "episomes." Under normal conditions the nucleolus disaggregates and reaggregates in concert with the chromosomes during mitosis.

The essential structural component of the nucleus is the chromatin, which is built up of a string of nucleosomes, as in all eukaryotic cells, composed of a long DNA molecule, probably one per chromosome, and octomers of four conserved histone proteins. However, a portion of the chromatin, roughly 15%, is not beaded but shows smooth contours, and there is good though circumstantial evidence that that fraction contains the portion of the genome which can be transcribed into RNA (Scheer et al., 1981). Regular condensation and decondensation marks the chromatin cycle, and the morphogenetic changes have been explained by two very different views. According to one, an overall dehydration of the chromatin, perhaps caused by an increase in free magnesium ions, leads to the compact chromosome structure (Jerzmanowski and Staron, 1980). The other view holds that phosphorylation of histone H_1 triggers condensation (Matthews, 1980). Recently, it has been confirmed, after delaying mitosis by irradiation, that phosphorylation of H_1 histone, but even more of the core histones, correlates well with chromosome condensation (Brewer and Oleinick, 1980). However, in a separate investigation it was found that decondensation of the chromosome after mitosis was not correlated with H_1 histone dephosphorylation (Fischer and Laemmli, 1980). Therefore, while the mechanism remains unsolved, the cyclic morphogenetic changes of the chromatin cycle are obvious. Furthermore, the nuclear contents, in particular the many chromosomes (40 in a haploid nucleus), are not free floating inside the nucleus. They are held in a proteinaceous and carbohydrate-containing network, a cage that may bind loops of chromatin such that they can be replicated in an orderly fashion in the S phase, as after isolation, nascent DNA is predominantly found in that cage (Mitchelson et al., 1979). If released from the cage, replicating chromatin displays typical "eye" structures which contain nucleosomes, which seem to be distributed in one piece (i.e., in a conservative fashion) to one or the other new DNA strand (Prior et al., 1980), as it duplicates bidirectionally. The genome is made up of 10 to 20-μm-long replicons that are replicated in clusters (Funderud et al., 1979). Activated replicons can be distinguished even as naked DNA, because they appear as an array of "microbubbles," although the exact meaning of that morphological feature is not known (Hardman and Gillespie, 1980). As a final comment on nuclei, there are two different kinds. Most get pushed around with the shuttle streaming, while many are much more stable within the ectoplasm sheet of a plasmodium.

The Golgi apparatus is another structure involved in intracellular morphogenesis. It is a source of more vesicles in the cytoplasm. Its activity, associated with typical lamellar stacks and the budding of dictyosomes, is very obvious during transition from growth to starvation, when massive slime production occurs in the plasmodium which is exported by exocytosis, and when the extracellular galactosamine-containing membranes are built. The Golgi apparatus is hardly seen during growth, where endocytosis dominates and brings in food particles which become digested after fusion of food vacuoles with lysosomes. What is left of the food is then expelled by exocytosis.

This leads us to the cell surface of a plasmodium, the plasmalemma. It shows extensive invaginations, as more than 80% of it is used in lining the reticulated ectoplasm.

The term *invagination* is deceptive, as it implies that the plasmalemma folds inwards. The formation of the "invaginations" has been analyzed in an ingenious simple experiment (Achenbach et al., 1979). After pricking a plasmodium with a needle, a drop of endoplasm gushes out and quickly gets covered with a membrane. Instant membrane formation is a trademark of *Physarum*, which makes it difficult to inject any foreign material. As an example, it can survive a shot of saturated RNase or α-amanitine solution. The drop of endoplasm "differentiates" within 20 min into ectoplasm and endoplasm. Actomyosin fibrils show up in the ectoplasm, contractions set in, pushing the endoplasm back into the vein where it came from. After about 30 min the drop is gone and has been fully integrated into the plasmodium again. Applying appropriate markers, the morphogenesis of plasmalemma invagination has been analyzed with the electron microscope and it was found that the major part of the invagination resulted from fusion of cytoplasmic vesicles with initial indentations of the outer membrane; that is, the invaginations are built from the inside.

In this particular experiment, all the membrane was derived from the endoplasm droplet and, maybe, this was not a fair test. However, a simple modification of the experiment has shown a remarkable result: Instead of leaving the endoplasm droplet sitting on the plasmodium, it can be put on agar and within a couple of hours it becomes a tiny yet healthy plasmodium. This little experiment convincingly demonstrates that the ectoplasm and the surface membrane of a proper plasmodium contain nothing that cannot be made from pure endoplasm.

Another way of looking at that tiny plasmodium is that a cell has been created. Indeed, the plasmodium can make cells in a similar manner, in contrast to the amebas, which are constricted after each nuclear division by cytokinesis, like any animal cell would be too. Cellularization in the plasmodium occurs by an alignment and fusion of vesicles, as has been clearly shown in macrocyst formation and as happens in the creation of prespores (Zaar and Kleinig, 1975). In the former case, cells arise in the absence of nuclear division. Their size is variable, as is the number of nuclei contained therein. Spore cells are carved out more accurately because most of them are uninucleate. Although cellularization does not resemble cytokinesis, a similarity to plant cell division may exist.

As a final comment to intracellular morphogenesis, after having recognized dynamic yet distinct compartments in the plasmodium, one wonders what would happen if the different vesicles accidentally fused, mixing up what should be separated. This happens regularly in association with starvation, as autophagic vacuoles cannibalize the protoplasm, which, in biochemical terms, results in a decrease of all macromolecules, including DNA, during macrocyst formation and sporulation.

B. Cellular Morphogenesis

Transient shape changes of mobile amebas, which use pseudopodia, are quite different from the change in shape seen in growing or migrating plasmodia.

Microplasmodia, which are kept in submerged, agitated suspension in growth medium, are round balls of 1-2 mm in diameter, therefore they must break up, as their mass increases with time to maintain a fairly homogeneous size. Some myxomycete plasmodia have a tendency to break in two naturally, by plasmotomy. This is a means of vegetative proliferation and of restricting the size of a plasmodium. In *Physarum* this trait is reversed, which is utilized when microplasmodia are fused on a filter paper to create the macroplasmodium on which most work is done. They are a hybrid structure, at least in terminology: a syncytial plasmodium.

The macroplasmodium grows as a flat disk, a coherent sheet of protoplasm which expands at the periphery. In the field *Physarum* looks different, more like a fan, hence it has been called a phaneroplasmodium (Alexopoulos, 1960). In contrast to well-fed laboratory cultures, the natural plasmodium is polarized. The reason for this difference is that the well-fed, "balanced" plasmodium is sedentary, while a natural plasmodium, always hungry, is on the move. Even the best-fed macroplasmodium is not homogeneous, as the oscillatory contractions push endoplasm into the periphery, where it gets in part transformed into ectoplasm. Thus there is an age gradient extending from the old center all the way to the edge. As long as enough material is synthesized to compensate for the newly deposited ectoplasm, the disk remains coherent. If not, the plasmodium breaks up into a network in the center as the endoplasm-ectoplasm transformation continues at the periphery. Distinct veins appear as a result of a patterning process. Such an unbalanced macroplasmodium finally breaks apart by a kind of delayed plasmotomy and each portion turns into a fanlike structure and begins to migrate. Migration results when more ectoplasm and plasmalemma is dismantled at the rear end and is moved up with each shuttle into the frontal edge. Then the front keeps more of the newly acquired mass as the remaining endoplasm shuttles backwards. This polarization process occurs also in tiny plasmodia, like microplasmodia. They turn into a wedgelike structure and their random migration also continues in the absence of food as the mobile mass gets smaller and smaller, but eventually manages to create cells as described in the previous section. This process starts at the "old" end, that is, the rear end (Lewis, 1942) and results in the formation of a colony of macrocysts. These have a thick double-layered wall, composed of a net of fibrils which consist of a galactosamine polymer, containing phosphate and sulfate (McCormick et al., 1970).

Macrocysts are resistant and stable structures and their formation is a simple case of cytodifferentiation. The state of dormancy is maintained as long as the environment is generally "unfavorable." However, these cysts may have come under this specific control of an extracellular factor, preventing germination of the established cysts. Only some of them are released from that negative control at a time, which may be of adaptive value to *Physarum*, as a sclerotium could thus probe the environment and manage to survive if favorable conditions are only transient, as after a rain shower (Holt and Hüttermann, 1980).

An interesting observation has been made in the white mutant, which was allowed to encyst as a suspension of microplasmodia. While these mutants make spherules in the dark, a distinct morphology of the starved microplasmodia has been observed in the light (Schreckenbach et al., 1981). The "light cells" look like little porcupines and they "round up" again in the dark. Moreover, if they are kept in the light, they do not make macrocysts at all. They omit the cellularization step and after a few days they all lyse completely and die.

Illumination of a starved plasmodium exposed to air evokes a different case of morphogenesis, sporulation. Again a patterning phenomenon is observed, as plasmodial strands break up into regular droplets of cytoplasm 1-2 mm in diameter. Each one is transformed into a pillar which is segregated into a stalk and a head with the help of contraction waves. Thereby most of the endoplasm is pushed into the head, where cellularization and spore differentiation take place. The remarkable morphogenetic process of sporulation has been described in detail (Guttes et al., 1961) and can be viewed in time-lapse films but has not been analyzed experimentally. Formally, cellularization can be seen as a reversal of plasmodial fusion. It may be an indication of incompatibility between some internal membrane systems that arises in an aged plasmodium.

It has been stressed by Bonner (1967) that many different lower organisms, myxobacteria, cellular slime molds, and small fungi, produce similar-looking fruiting bodies, although the morphogenetic processes may be very different. As an example, in *Dictyostelium* this process involves oriented mass movements of amebas, described as a "fountain running backwards" (Bonner, 1967). It seems that the typical form of fruiting bodies is of adaptive value for spore dispersal.

Typically, two meiotic divisions occur in the prespore of *Physarum* within 24 h and lead to a uninucleate haploid spore. However, this process shows great flexibility in morphology and timing (Laane and Haugli, 1976). Meiosis can occur as early as in the plasmodial state (making spores, some kind of meiocysts) and several nuclei can become enclosed by the spore membrane, which can lead up to eight nuclei in a single spore giving birth to several amebas at once.

As a last comment on cellular morphogenesis in *Physarum*, it seems that the many fruiting bodies need to be physically connected with each other by the remnants of the plasmodium, the hypothallus, devoid of shuttling endoplasm, not only to allow the synchrony of this morphogenesis but to generally sustain its progression. The plasmodium turns into a multiheaded individual and not a number of separate fruiting bodies.

C. Extracellular Signalling

Knowing about the elaborate intercellular signalling by cAMP pulses in the cellular slime mold *Dictyostelium* (Loomis, 1975), it has been asked whether a similar communication works between *Physarum* plasmodia. Several have been placed on both sides of a filter (suspended in a moist chamber) which allowed the passage of any molecules in both directions (Tso and Wong, 1978). As the individual plasmodia on both sides migrate at random, they seem to neither attract nor repel each other. They ignore one another until they meet; then they may or may not fuse (see below).

This does not mean that acellular slime molds do not communicate with each other. Extracellular slime is not an adequate description of the elaborate glycocalyx of slime molds. Only about 30% of its polysaccharides consist of the known galactose polymer (Hüttermann, 1973). Among the signal molecules suspected so far there is the inhibitor of germination already mentioned and an inhibitor of mitosis in *Physarum* amebas (as well as in bacteria) (Henney and Asgari, 1975). These factors could be classified among the still elusive chalones. In addition, there is a factor that promotes sporulation (Wilkins and Reynolds, 1979) and several hydrolytic enzymes are also present.

Furthermore, *Physarum* has a well-developed sensory mechanism, as the plasmodia are attracted or repelled by various chemical substances. Chemotaxis has been first analyzed by a simple behavioral test, as plasmodia will crawl over a period of 10-20 h toward an attractant if it is present above threshold concentration (Carlile, 1970). Elaborate biophysical methods, aided by the enormous size of plasmodia (as compared to, say, a lymphocyte) have provided much data (Ueda and Kobatake, 1979). As a generalized view, a conformational change of the surface membrane occurs at chemoperception, which is somehow transduced to the motile apparatus. Consequently, it is that membrane change, and not so much the change in motive force, that directs the movement of a plasmodium. The membrane becomes softened where an attractant hits threshold level first. Calcium also seems to play a role in chemotaxis, as its absence can revert positive to a negative chemotaxis (Ishida et al., 1979; Satoh et al., 1981).

A similar reversal of behavior occurs naturally as a plasmodium sporulates. As long as it is incompetent, it shows a phobic reaction which leads to avoidance of bright light. When it is ready to sporulate, a plasmodium no longer avoids light. As for all photobiological reactions in *Physarum*, blue light is most effective (Rakoczy, 1980), as it is also in the unphysiological situation already alluded to, where microplasmodia turn into "light cells." Therefore it is important to know whether the spikes radiating from the illuminated microplasmodia are tails (as the endoplasm is concentrated in the interior, hiding from the light) or whether they represent abortive sporangia, which cannot fully differentiate under water.

Recently, it has been shown that the endoplasm can transmit the the signal given by blue light (Wohlfarth-Bottermann and Block, 1981). The oscillatory contractions of protoplasmic strands display a phase shift after illumination. Even if only one end was illuminated, this shift could be observed at the other end of the strand, a couple of centimeters away. However, this occurred only if the endoplasm flowed in that direction during illumination, thus transporting the signal to the contractile ectoplasm. As a new observation, propagated waves have been detected that travel at 1 mm/min over a plasmodium engaged in chemotaxis (Hejnowicz and Wohlfarth-Bottermann, 1980).

At first sight, a colony of amebas grazing on a lawn of bacteria seems to suggest intercellular communication between individual amebas. But this is not so. Amebas stay together and form a ring-shaped colony that expands, leaving a "plaque" of bacteria-free space in the center. The colony form is dependent on the pH of the medium and whether bacteria are alive or dead, and there are mutants which alter the behavior of the amebas, leading to spreading growth of an amebal population. Careful analysis (Jacobson, 1980) has shown a close-range interaction where not motility as such (i.e., speed) is affected but, rather, the orientation of pseudopodium formation. As a result, an ameba stays close to bacteria and only moves on as it eats up the ones close by. As an abundant food supply – like a lawn of bacteria – is not likely in nature, this behavior is better adapted than unrestrained spreading growth, although the amebas must crowd in a ring on the Petri dish. Generally, such an interaction with signals from target cells may provide a useful means with which to study the phenomenon of contact guidance known in animal embryonic development.

Although the plaque is devoid of bacteria, it is inhabited by microcysts, because the encysting individuals stay behind the expanding ring of growing amebas. Whether cysts are protected in the free space of the plaque or are of any value for survival under unfavorable conditions in the field is not clear, Microcysts may be a dead end of amebal life, as plasmodia also arise in the plaque and engulf cysts and amebas alike as they grow up (Collins, 1979).

Nevertheless, cysts provide a very convenient selection procedure for screening for mutants which cannot germinate, and functions for interrupting dormancy may thus be discovered.

There is more to the colonial life of *Physarum* amebas. If spores of one fruiting body are germinated together, plasmodia will eventually arise. It takes a rather long time and no less than 500,000 individual amebas seem to be engaged in making only a few gametes. One signal in transforming an ameba into a gamete comes from an extracellular factor which is produced during the group effect until it reaches sufficient concentrations to induce gametes (Youngman et al., 1977). This factor can be provided by a large colony of amebas kept on one side of a filter, thus stimulating just a few amebas on the other side to become gametes. The factor has been enriched and can be added to a few amebas, and then most of them turn into gametes, fuse into pairs, and create plasmodia, as shown in a recent film (Holt and Hüttermann, 1980). This factor is not a sex hormone, as it is likewise made by the asexual selfer, and turns amebas into both male and female gametes, according to the genetic makeup at the mating-type gene locus.

Several temperature-sensitive (TS) mutants are available that interfere with gamete formation, and this process can be synchronized in a suspension culture of amebas kept from making plasmodia at the restrictive temperature. Afterwards, at a permissive temperature most individuals rapidly become gametes (Palotta et al., 1979). This is an important result for biochemical work yet to be done, as only a few plasmodia ever form on a Petri dish among millions of amebas. One reason for this is that young plasmodia secrete a still uncharacterized inhibitor of plasmodium formation and later on they feed on amebas. Thus it is obvious that of the many spores of a fruiting body only a few give rise to new plasmodia.

A remarkable experiment has been done by seeding out spores from a variety of different strains of *Didymium* (a close relative of *Physarum*), each collected in a separate geographical location and having its own alternate set of mating-type alleles. Consequently, just as many different plasmodia could be expected to arise in this mass culture of amebas (Clark, 1980); however, only a single strain managed to make plasmodia. A strong selection process must have occurred which can thus be analyzed in the Petri dish, and this brings us to the next section, on immediate communication by contact.

D. Cell Contact

Plasmodial fusion is a natural behavior that is used as a tool in making time heterokaryons, yet people began to wonder when a subline of the usual *Physarum* strain started killing its partner. The killer phenotype results from a single mutation, and the killing of the sensitive partner occurs in a programmed sequence of events, beginning with fusion and ending with lysis of nuclei and the whole plasmodium. At 6 h after fusion nuclei are committed to die and cannot be rescued by fusion with a big healthy plasmodium. Comparable studies with many strains of *Physarum* and *Didymium* have revealed several fusion genes, which can either be dominant or recessive. Plasmodial fusion can only be productive (i.e., the resulting heterokaryons are only stable) if identity at all loci is maintained. Consequently, *Physarum* plasmodia utilize a multigenic system to recognize themselves. Incompatibility occurs as a series of graded reactions if the various genotypes are put together pairwise. The plasmodia can completely avoid fusion, the nuclei of one partner can become slowly and selectively eliminated, or both plasmodia can kill each other. These results have been rationalized in two opposite arguments: Either the killer is almost compatible with its partner and those plasmodia which do not fuse at all are the most different individuals, or the killer is the most incompatible partner (see Carlile, 1973, and Carlile and Gooday, 1978, for review). To solve this problem we need to learn the functions which the fusion genes control and whether the mixed surface membranes signal changes in the cytoplasm of the mixed plasmodia directly, or indirectly after first altering the gene expression.

As we have seen above, there are many membrane boundaries in the plasmodium, and while several kinds of vesicles need to fuse to function, others must be kept apart. Cellularization, patterning of the plasmodial network, and segregation of the fruiting bodies may be explained by altered membrane fusion behavior, leading to local lysis of the plasmodial sheet and limited cell digestion.

One membrane compartment is the nucleus and normally somatic nuclei are not allowed to fuse. Sometimes, in aged strains, big nuclei have been observed (Lott and Clark, 1980). They may have been created by endopolyploidy or by nuclear fusion. In addition, some pycnotic nuclei are always seen in a plasmodium and many more occur during sporulation, leading to massive nuclear elimination (Guttes et al., 1961).

Highly specific cell recognition is required for sexual fusion. In contrast to the multigenic fusion control of plasmodia, gametes utilize a multiallelic single-gene complex to achieve this specificity: the *mat B* locus (Kirouac-Brunet et al., 1981).

A diploid plasmodium has two different alleles of the *mat B* gene and after meiosis half of its progeny receive one of the two alleles. Consequently, just a single allele of *mat B* is not enough to allow sexual fusion of two gametes. Maybe amebas turn into two kinds of gametes (+ and –) because they express the two different alleles of *mat B*. A surprise came when amebas from other *Physarum* strains were tested: Each strain had a different pair of alleles of the same gene *mat B*. Consequently, and in contrast to plasmodia – which must have identical sets of fusion genes in order to fuse successfully – gametes obtained from different strains do fuse with each other: They are cross-fertile. This may be useful for spreading genes through various populations of *Physarum*.

In summary, the *mat B* gene is a multiallelic gene, and since different strains have different sets of alleles, this genetic unit must be quite variable. Indeed, amebas from a single fruiting body need not necessarily harbor different alleles of *mat B* and they still become plasmodia. Only, this happens rarely, about 10^{-3} times the rate detected among amebas from heterothallic parents. It is assumed that a mutation has occurred, resulting in a different yet active allele of *mat B*, thus allowing sexual fusion in a single clone of amebas (Dee, 1978).

A brief look at yeast cells and their highly flexible determination of either *a* or α gametes tells us of a very frequent specific transposition of DNA, turning a male into a female, as often as every cell generation, by insertion of a piece of DNA stored in a "cassette" into the mating gene locus, which functions as a "cassette deck" (Strathern et al., 1980).

It is possible that similar alterations on the genome level occur in *Physarum* in the determination of gametes, although, other than in the bipolar yeast system, more than only two types of mates are created in *Physarum*. This is only the first half of the story of ameba-plasmodium transformation, that is, sexual fusion. But this speculation has turned our attention back inside the plasmodium, where most biochemical work has been done.

E. Intraplasmodial Communication

This is a very wide topic and can be treated only superficially. It is also the main area where molecular mechanisms to explain differentiation processes have been proposed with the theory of differential gene activation (Davidson, 1976).

The chromosomal genome of *Physarum*, 0.25 pg DNA per haploid genome with a sequence complexity of 1.8×10^8 base pairs of single-copy DNA, is quite complex and the unique DNA is interspersed with repetitive sequences as in other higher organisms (Whittaker et al., 1981).

The nucleolar genome is transcribed by RNA polymerase A, and restriction analysis of rDNA and comparison with mature rRNA products has indicated that in addition to processing of the 40-S rRNA precursor, the large subunit of the 26-S rRNA must be spliced in two places (Gubler et al., 1980). Control of rDNA expression seems to exist, as less rRNA accumulates during starvation as compared to exponential growth, thus limiting the concentration of available ribosomes. The reduced ribosome level is probably a consequence of both a decrease in RNA polymerase A activity and an increase in degradation by nucleases (see Turnock, 1980, for a review).

RNA polymerase B has been isolated and characterized. This enzyme activity, if analyzed as an endogeneous nuclear enzyme, is higher in the S phase than in the G_2 phase of the mitotic cycle and generally lower during starvation (see Seebeck and Braun, 1980, for review). The soluble enzyme, however, has been claimed to become phosphorylated during starvation and even to increase in activity (Freed and Zimmerman, 1980). Also, the largest subunit of 215,000 daltons is missing in enzyme B from starved cultures, probably due to proteolytic degradation (Smith and Braun, 1981). Indirect evidence for stimulatory factors for that enzyme has been obtained and acetylation of core histones may also be required for general activation of RNA polymerase B (Chahal et al., 1980).

The nascent RNA products, where visible as RNP fibril gradients, suggest that the transcripts are about 1-2 μm long, which corresponds to the size of polyA RNA after pulse labeling in vivo and the in vitro products obtained from isolated nuclei (Scheer et al., 1981). Only one paper has shown that mRNA is present in total RNA extracts from *Physarum* (Melera et al., 1979) and not one single polypeptide coding gene has been identified.

The sequence complexity of total polyA RNA has been estimated at about 10^7 nucleotides. This number is both too high, as the total polyA RNA and nonpolysomal mRNA has been analyzed, but also too low, as only stable RNA molecules can be studied by hybridization of cDNA to polyA RNA (Baeckmann, 1980).

The issue of information transfer to polysomes in *Physarum* is complicated by the important observation that poly(A)RNA in putative storage particles in the cytoplasm contains a long poly(A)stretch (pA_{65}), whereas mature mRNA contains only a short tail (pA_{15}), which may be lost in standard isolation procedures (Adams et al., 1981).

This maturation event could be utilized in translation control. It has been shown that mRNA is fed at an increasing rate into polysomes as plasmodia grow denser, followed by the slowing down of polysome formation and a decrease of their translation capacity (Brown and Hardman, 1980; Jeffery, 1979). Interestingly, cysts contain mRNA which is required for germination. This fraction may be functionally similar to the maternal mRNA in oocytes (Davidson, 1976).

With the exception of sporulation, where indirect evidence for new genetic activity has been gathered with the techniques available in the late sixties (Sauer, 1973), not one case of selective gene expression has been demonstrated in *Physarum* by RNA analysis, and the modern techniques must be applied before we begin to understand how genetic information is called upon in *Physarum* development.

One hypothesis (Sauer, 1978), if substantiated further, could be valuable in defining the freewheeling mitotic cycle of *Physarum* and perhaps proliferative mitotic cycles in general. As metaphase chromosomes do not contain RNA polymerase B, as the activity of this enzyme after mitosis depends on DNA replication, and as DNA replication follows a temporarily sequential order, it is possible that "replication-dependent RNA transcription," for which some evidence exists, is a prerequisite for the genetic contribution to the cell cycle program (Pierron and Sauer, 1980). Results with recombinant DNA probes will have to show whether this temporal correlation does provide for a controlled expression of genes as they are replicated.

We now turn to protein synthesis and search for specific changes during the mitotic cycle of *Physarum*. No results with high-resolution two-dimensional gels have been published, but a 30-fold increase in tubulin synthesis in the late G_2 phase has recently been claimed (Dove et al., 1980).

For some time, abrupt enzyme activity changes have been believed to indicate fluctuations in gene expression relevant to the cell cycle. In most cases these changes were caused by unbalanced growth (see Mitchison, 1977, for review). In *Physarum* most enzymes follow a linear synthesis pattern (Wegener and Sauer, 1981), which makes the few exceptions potential candidates for mitotic cycle control. Here are two examples, but both turn out to be cases of posttranslational control.

Thymidine kinase (TK) is the classic case and still a good marker for mitosis, as it increases in activity around mitosis (Wright and Tollon, 1979). However, the main activity in the S phase is caused by a posttranslational modification, probably a dephosphorylation (Gröbner, 1979) and many TK-minus mutants have normal mitotic cycles (Gröbner and Mohberg, 1980).

Histone kinase increases over 10-fold in activity in the late G_2 phase, just as the H_1 histone becomes phosphorylated. This is also due to enzyme activation (Mitchelson et al., 1978). We have already alluded to the putative signficance of histone phosphorylation and chromosome condensation. In this respect, the autonomous mitotic cycle of *Physarum* offers a unique control experiment that has rarely if ever beeen done. If any changes in the cycle are cell cycle specific, they should recur each cycle. Many alterations do not, like the general phosphorylation of DNA-binding proteins that have been followed for two mitotic cycles (Magun, 1979).

This leaves us with three markers: the execution point of a temperature-sensitive mutant, *Ma 67* (Laffler et al., 1979), the results of recent elegant fusion experiments (Loidl, 1979), and the latest inhibition of mitosis by cycloheximide, all of which strongly support the idea of a distinct transition point in the mitotic cycle at prophase (about 15 min before metaphase). However, inhibition by cycloheximide does not tell us whether a general shock to the ribosome system as a whole, or the lack of a short-lived protein – a true regulatory molecule – has led to the observed defect (Bernstam, 1978).

No experiment in *Physarum* plasmodia has yet achieved the separation of mitosis from the S phase. To the contrary, even if mitosis is blocked at the metaphase and reversed to the interphase, a complete S phase takes place, yielding tetraploid nuclei (Wolf et al., 1979). As for initiation of the S phase, a very drastic change in an unusual phosphonucleotide (AP_4A), a putative regulatory molecule for DNA polymerase α, has recently been demonstrated to change 30-fold between the prophase and the beginning of the S phase (Grummt F., Würzburg, 1981, personal communication).

The main questions, which concern the *Physarum* plasmodium, still await an answer: Why do the nuclei divide, and why do they do so in synchrony?

There are two classes of hypotheses (see Tyson and Sachsenmaier, 1978, for a review). In the popular one, the hourglass model, some stimulatory factor accumulates to a threshold, thus triggering mitosis. Mitosis equals the turning of the hourglass and therefore is part of the mitotic cycle.

According to the other hypothesis, the limit cycle, at least two factors interact continuously with each other by auto- and cross-catalysis, and a certain concentration quotient triggers mitosis. Here mitosis is not part of the mitotic cycle; the clock runs on continuously and every 6 h at the shortest and up to every 48 h at the longest (according to the metabolic state of the mold), the nuclei must divide in synchrony because the quotient is right. Generally, it is believed that whatever signals trigger mitosis, they must come from the cytoplasm. The question is, Do they?

This leads us to the control of the other cyclic event of the plasmodium, the synchronous oscillatory contractions. No doubt, the actomyosin fibrils contract the ectoplasm tube, even if the endoplasm is replaced by buffer, and this requires ATP and calcium. The oscillator may be a self-sustained calcium-binding and calcium-releasing mechanism (Rapp and Berridge, 1977). Synchrony could result from electrical, physical, or physiological signals.

Why is the contraction synchronous? However, there is no large action potential which precedes the contraction. On the other hand, the hydraulic pressure of the endoplasm against the ectoplasm tube may trigger contraction by stretch entrainment. Furthermore, the contractions are not that synchronous, and they could be the result of the averaging of many individual oscillators (little vesicles with calcium pumps) that get shuttled along in the endoplasm.

Thus the following question arises: Could it also be, as so many fusion experiments can be explained by the averaging of the mixed plasmodia, that nuclei synchronize themselves? We can learn something if we look at the mitotic and contraction cycles at the same time. When small plasmodia undergo mitosis, shuttle streaming stops. Maybe all the calcium is needed to dismantle the mitotic spindle? However, we must not conclude, as a touch with a blunt needle stops the shuttle too, that mitosis can be brought about with a gentle touch. As of today, mitosis still requires a magic touch.

IV. THE DEVELOPMENTAL REPERTOIRE OF *PHYSARUM*

The previous section has revealed how little we understand of mitosis and motility and intracellular regulations in *Physarum*. This slime mold is in good company (Sauer, 1980). However, with the limited options of its life cycle, *Physarum* is open to clear-cut analyses as few other developmental systems are.

We have observed the following: (1) that the chromatin is organized, replicated, and expressed as in higher organisms, yet in part eliminated during sporulation; (2) the variably timed sets of reactions that result in mitosis, meiosis, and cellularization; (3) how signals are generated, received, and transduced; (4) the many uses of membranes, the cytoskeleton, and the cellular motile apparatus in response to (5) patterning and scaling processes, all of which are integrated in (6) morphogenetic movements and the creation of distinct cellular shapes, as well as (7) cytodifferentiation; and, finally, even if we do not know how it happens, *Physarum* tells us a little about (8) irreversible commitment (i.e., the determination process) and offers a unique look at how a regulatory gene works as it creates (9) a kind of multicellularity. Recently, an attempt has been made to integrate this developmental repertoire into the life cycle of *Physarum* (Sauer, 1982).

Why *Physarum* can become an even better model for development will be summarized in the following section.

V. OPPORTUNITIES IN *PHYSARUM*

A. Encystment

Cysts result from a developmental sequence in response to many different stimuli. Many changes in metabolism and structure have been described during macrocyst formation. However, the process is not highly synchronous, nor has any parameter been clearly associated with differentiation, because starvation and encystment occur in concert.

Two methods are now available to synchronize the process. A treatment of synchronous macroplasmodia with mannitol produces macrocysts within 24 h (Jalouzot and Toublan, 1981) and colcemide has been claimed to synchronize young microplasmodia in the metaphase (Prior et al., 1980).

Furthermore, three different ways of reversibly blocking cyst formation can be utilized to search for and dissect the developmental sequence: a low dose of alcohol (Kleinig, 1974) or butyrate (Jalouzot and Toublan, 1981), or illumination (Schreckenbach et al., 1981). In addition, the action of the germination inhibitor may provide information on how the differentiated state is maintained and reversed.

The aging process can be separately studied. It is claimed to be a random event (Lott and Clark, 1980), and it can be elegantly reversed by making a plasmodium migrate through a filter with holes small enough to filter out the big "senescent" nuclei (Clark and Hakim, 1980).

Finally, it has been asked whether encystment, as it occurs in nature, is really a sociobiological affair in which the plasmodia produce three kinds of cells, namely, one which is sacrificed to allow the other type to go on growing and maybe help the third kind to differentiate into macrocysts (Brown and Hardman, 1980).

B. Sporulation

The sporulation sequence is more complex; Its end result, the spores, cannot be predicted to be made with certainty, nor can one say which of the many nuclei of a plasmodium will survive nuclear elimination. Moreover, sporulation, in contrast to encystment, is highly temperature dependent, with an optimum temperature of 26°C (Afanasieva et al., 1980).

Usually morphogenesis, presporangial mitosis, cellularization, and meiosis follow in temporal sequence. However, cellularization before morphogenesis leads to cysts, and meiosis in these premature cells results in spores inside the plasmodium (Laane and Haugli, 1976). These observations demonstrate that sporulation is not a sequence of events, but a network of developmental changes, most of which also occur in unicellular organisms. The terms *spacing*, *timing*, and *coupling* can be used to describe this phenomenon. Therefore one can ask whether sporulation is just a different ordering of the encystment program, and a mutant has been obtained and named *alc* (for amebaless cycle) in which the spore is really a uninucleate cyst that turns right away into a little plasmodium after germination (Truitt and Holt, 1979). These variations may permit an understanding of how the different classes of Myxomycetes have evolved.

On the other hand, starvation and the other common elements of sporulation and encystment may not be the essential events in sporulation.

Illumination, which reversibly blocks spherulation, may not activate the sporulation program by inhibiting spherulation. In contrast, illumination seems to act like an "inducer" in embryonic development, in evoking sporulation like the inductor of primary determination in an amphibian gastrula. Illumination needs to act only for a while, after which the mold is committed to sporulate, as the gastrula is committed to neurulate.

It has been shown that the effect of light can be mimicked by injection of an illuminated yellow pigment fraction (even if obtained from a growing plasmodium) (Wormington and Weaver, 1976); but the true blue light receptor is probably a flavoprotein, the still elusive cryptochrome of many lower organisms (Schreckenbach et al., 1981). Conceptionally, the light action has been compared to a hormone signal, as changes in cAMP, calcium, and pH may all be involved in inducing sporulation, including selective gene expression (Daniel, 1980).

Illumination is only effective when the mold is in the right metabolic state, that is, competent. The term is well known from embryology but is little understood. It has been speculated that competence reflects the generation of "metabolic information" (Daniel, 1966). As another suggestion, plasmodia are only competent during the S phase of a quantal cell cycle (Sauer et al., 1969), because DAN synthesis inhibitors added during illumination block sporulation. However, the inhibitors were supplied in fresh sporulation medium lacking the sporulation factor, present in the "conditioned" medium (Wilkins and Reynolds, 1979), and this fact and not DNA replication might explain the earlier result.

At any rate, competence is not the earliest prerequisite for sporulation. The incidence of sporulation depends critically on the time when a growing culture is transferred to sporulation medium (Daniel, 1966). A 24-h growth period as a synchronous macroplasmodium has been found useful. As this time span happens to be the time of the third postfusion mitosis, a distinct point in the S phase or some sensitive state of the cell cycle may be responsible for the "initiation" of the sporulation program.

Initiation, competence, induction, and commitment do not seem to be involved in the encystment process, which makes sporulation quite different and more similar to irreversible developmental changes in higher organisms, yet perhaps more amenable to experimentation.

One approach to understand sporulation would come from a specimen that has lost the capability to sporulate but which can regain it under controlled conditions. In the absence of any TS mutants Hosoda (1980) has performed an ingenious experiment. When plasmodia have been kept in vegetative growth in an axenic medium for some years, the frequency of sporulation decreases and some strains do not sporulate at all. It is conceivable that the genes required for sporulation have somehow been eliminated. However, one feeding of oat flakes to such spore-deficient plasmodia is sufficient to allow sporulation to reoccur.

At first sight this observation indicates that the genome has not been altered over the years of culture. However, in view of the results on DNA methylation, nuclear elimination, the flexible yeast mating type, the DNA rearrangements in the immune system, and the general notion of the "fluid genome" (Bonner, 1981), it could be useful to check whether the rejuvenating diet has an effect on DNA composition.

As an unusual view in comparing the life cycle of *Physarum* to higher organisms like insects or amphibia, the transformation of a plasmodium into sporangia might be likened to the metamorphosis of pupae or tadpoles. Extending this analogy, one might predict that the gametes in *Physarum*, once formed, are as terminally differentiated as eggs and sperm cells, which would make the ameba-gamete transition the third irreversible differentiation step in the life cycle of *Physarum*.

This brings us to the last point, the reversal of the plasmodium-spore transition, and although we have no idea of the genetic changes in plasmodium initiation, we have seen that two slightly different copies of a single gene locus enact a major developmental switch, the transformation of a gamete into a plasmodium.

C. Initiation of Plasmodium Formation

Amebas and plasmodia are of quite distinct phenotypes derived from a common genome and, at first, many mutants were only expressed in one or the other form. However, when the same culture medium was used, fewer mutants were restricted to one developmental stage (Burland and Dee, 1979).

Many mutants have been described that cannot make a plasmodium. Also, starting from a selfing strain, mutants – and backmutants – have been obtained, that need to mate to initiate a plasmodium. Furthermore, starting with a heterothallic strain, mutants – and backmutants – were found that became selfers. The amazing result of this work is that almost every mutant can be assigned to the same gene, *mat A* (Youngman et al., 1981). This genetic locus has previously been described as a mating-type locus and has been identified as another multiallelic system, separate from *mat B*. *Physarum* strains obtained from different geographical locations each have a different pair of alleles of *mat A*.

Like the gene *mat B* already described, two different alleles of this gene must be active in one cell for it to become a plasmodium, and thus this gene is also prone to frequent "mutations."

However, unlike *mat B*, whose function may be restricted to sexual fusion, gene *mat A* is essential for the plasmodial state. This can be concluded from the following matings. When two gametes express different alleles of *mat B*, but the same alleles of *mat A*, they can fuse. Their nuclei will also fuse (in prophase) and then mitosis occurs, yielding two amebas, but not a plasmodium.

However, gametes differing at *mat A* and *mat B* turn into a plasmodium after cell fusion and nuclear fusion (in interphase). In theory, the selfer can now be explained as expressing two different alleles of *mat A* constantly, whereas illegitimate plasmodia, which very rarely arise in any clone of an ameba, may have arisen after transient expression of a second allele of *mat A*. We seem to be approaching a rather unifying picture.

But what is the function of *mat A*? While we could point to analogies for *mat B* with the yeast mating type and fusion of eggs and sperm generally, we have no idea how *mat A* may work: It is a novelty! At the same time it may be a fossil trait of how single cells became true animals. While this interpretation is favored by one group (Youngman et al., 1981), the two multiallelic genes *mat A* and *mat B* can also be viewed as a means to prevent inbreeding among slime molds (Kirouac-Brunet et al., 1981). Furthermore, it has been questioned whether *mat A* governs plasmodium formation via a positive or negative control mechanism (Honey et al., 1979). These observations provide challenging opportunities into studying the development of molds and, perhaps, man.

VI. THE STATE OF THE MITOTIC CYCLE IN *PHYSARUM*

As so many researchers are concerned with the synchronous mitotic cycle of *Physarum* plasmodia, and as a more than coincidental correlation may exist between differential (quantal) cell cycles and differentiation (Holtzer et al., 1975), the cell cycle concept may provide a systemic approach to the life cycle in general. Indeed, although an independent confirmation of the various possible routes of differentiation in *Physarum* is required, the following remarkable correlations between mitotic cycle parameters and the development of *Physarum* can be drawn.

The growing and the starved plasmodium and the growing amebas have no G_1 phase; they display a G_1^- phenotype. It has been argued that this situation reflects a prototype of proliferative cell cycles, which is also observed in early embryonic development, while a G_1^+

phenotype of proliferative tissue culture cells may be an artifact (Liskay and Prescott, 1978).

On the other hand, the macrocysts and microcysts of *Physarum* seem to be arrested in the G_1 phase, as they have the quantity of DNA (2C or 1C) expected when a diploid or haploid nucleus is arrested before DNA replication. As the yeast mating-type factor specifically arrests the cell cycle of yeast at the start position, thus creating a G_1 phase (Hartwell, 1974), it will be interesting to see whether the gamete-producing factor of *Physarum* does likewise during the ameba-gamete transition. As a general rule, differentiated cells of higher organisms are postmitotic (Prescott, 1976) and may thus have been (specifically?) arrested in the G_1 phase.

Finally, the mature spore of *Physarum* seems to be arrested in the G_2 phase (Mohberg and Rusch, 1971; Mohberg et al., 1973), the cell cycle phase that is first prolonged after ending the proliferative cell cycles in *Drosophila* embryos (McKnight and Miller, 1979).

Although mere correlations at this time, the arrest of the mitotic cycle either before (or during?) or after the S phase may indicate that the cell cycle must be broken to open up the life cycle, and this concept can be probed with new experiments to help us understand biological development.

ACKNOWLEDGMENTS

We wish to thank Dorle Wolf for typing, and R. Wolf, Würzburg, for reading the manuscript. Most of the work in our laboratory was supported by the Deutsche Forschungsgemeinschaft, through grants awarded to H. W. S. during his tenure at the universities of Heidelberg, Marburg, Konstant, and Würzburg (Germany). G. P. appreciates a grant from the Volkswagen Stiftung.

REFERENCES

Achenbach, F., Achenbach, U., and Wohlfarth-Bottermann, K. E. (1979). Plasmalemma invaginations, contraction and locomotion in normal and caffeine-treated protoplasmic drops of *Physarum*. *Eur. J. Cell Biol.* *20*:12-23.

Adams, D. S., Nooman, D., and Jeffery, W. R. (1981). Cytoplasmic polyadenylate processing events accompany the transfer of mRNA from the free mRNP particles to the polysomes in *Physarum*. *Proc. Nat. Acad. Sci. USA* *78*:83-87.

Afanasieva, T. P., Belozerskaya, T. A., Filippovich, S., Chernyshova, E. K., and Kritsky, M. S. (1980). Influence of cultivation temperature on the development of functionally active photoregulatory systems in cells of *Neurospora crassa* and *Physarum polycephalum*. *App. Biochem. Microbiol. USSR* *16*:156-161.

Aldrich, H. C., and Daniel. J. W. (Eds). (1982). *Cell Biology of Physarum and Didymium*, Monograph in Cell Biology Series, Academic, New York.

Alexopoulos, C. J. (1960). Gross morphology of the plasmodium and its possible significance in the relationships among Myxomycetes. *Mycologia 52*:1-20.

Baeckmann, A. (1980). Bestimmung der Sequenzkomplexität der poly(A)+RNA von *Physarum polycephalum.* Ph.D. thesis, University of Würzburg.

Bernstam, V. A. (1978). Heat effects on protein biosynthesis. *Annu. Rev. Plant Physiol. 29*:25-46.

Bernstam, V. A., Gray, R. H., and Bernstein, I. A. (1981). Effect of microtubule disrupting drugs on protein and RNA synthesis in *Physarum polycephalum* amoeba. *Arch. Microbiol. 128*:34-40.

Bonner, J. T. (1967). *The Cellular Slime Molds*, 2nd ed. Princeton University, Princeton, NJ.

Bonner, J. T. (1981). *Evolution and Development*. Springer-Verlag, Heidelberg.

Brewer, E. N., and Oleinick, N. L. (1980). Histone phosphorylation during radiation included mitotic delay in synchronous plasmodia of *Physarum polycephalum. Int. J. Radiat. Biol. 38*:697-702.

Brown, A. J. P., and Hardman, N. (1980). Utilization of polyadenylated mRNA during growth and starvation in *Physarum polycephalum. Eur. J. Biochem. 110*:413-420.

Burland, T. G., and Dee, J. (1979). Temperature-sensitive mutants of *Physarum polycephalum*. Expression of mutations in ameba and plasmodia. *Genet. Res. 34*:33-40.

Carlile, M. J. (1970). Nutrition and chemotaxis in the Myxomycete *Physarum polycephalum*: The effect of carbohydrates on the plasmodium. *J. Gen. Microbiol. 63*:221-226.

Carlile, M. J. (1973). Cell fusion and somatic incompatibility in Myxomycetes. *Ber. Dtsch. Bot. Ges. 86*:123-139.

Carlile, M. J., and Gooday, G. W. (1978). Cell fusion in Myxomycetes and fungi. In *Cell Surface Reviews*, Vol. 5. Poste and G. L. Nicolson (Eds.). Elsevier/North-Holland, New York, pp. 219-265.

Chahal, S. S., Matthews, H. R., and Bradbury, E. M. (1980). Acetylation of histone H4 and its role in chromatin structure and function. *Nature 287*:76.

Clark, J. (1980). Competition between plasmodial-incompatibility phenotypes of the Myxomycete *Didymium iridis*. II. Multiple-clone crosses. *Mycologia 72*:512-522.

Clark, J., and Hakim, R. (1980). Nuclear sieving of *Didymium iridis* plasmodia. *Exp. Mycol. 4*:17-22.

Collins, O. R. (1979). Myxomycete biosystematics: Some recent developments and future research opportunities. *Bot. Rev. 45*:145-201.

Daniel, J. W. (1966). Light-induced synchronous sporulation of a Myxomycete. In *Cell Synchrony, Studies in Biosynthetic Regulation*, I. L. Cameron and G. M. Padilla (Eds.). Academic, New York, pp. 117-152.

Daniel, J. W. (1980). The role of light-induced Ca and cAMP metabolism in *Physarum polycephalum*. *J. Cell Biol.* *87*:19a.

Daniel, J. W., and Rusch, H. P. (1961). The pure culture of *Physarum polycephalum* on a partially defined soluble medium. *J. Gen. Microbiol.* *25*:47-59.

Davidson, E. H. (1976). *Gene Activity in Early Development*. Academic, New York.

Dee, J. (1973). Aims and techniques of genetic analysis in *Physarum polycephalum*. *Ber. Dtsch. Bot. Ges.* *86*:93-121.

Dee, J. (1978). A gene unlinked to mating type affecting crossing between strains of *Physarum polycephalum*. *Genet. Res. Cambridge* *31*:85-92.

Dove, W. F., and Rusch, H. P. (Eds.). (1980). *Growth and Differentiation in Physarum polycephalum*. Princeton University, Princeton, NJ.

Dove, W. F., Laffler, T. G., Chang, M. T., Gross, C. A., Schedl, T., Warren, N., and Gorman, J. A. (1980). Periodic activities in the mitotic cycle of *Physarum polycephalum*. In *Abstracts of the Second International Congress on Cell Biology*, Berlin, *Eur. J. Cell Biol.* *22*:492.

Fischer, S. G., and Laemmli, U. K. (1980). Cell cycle changes in *Physarum polycephalum* histone H1 phosphate: Relationship to deoxyribonucleic acid binding and chromosome condensation. *Biochemistry* *19*:2240.

Freed, J. C., and Zimmerman, A. M. (1980). Patterns of DNA-dependent RNA polymerase activity in *Physarum polycephalum*. *J. Cell Biol.* *87*:7a.

Funderud, S., Andreassen, R., and Haugli, F. (1979). DNA replication in *Physarum polycephalum*: Electron microscopic and autoradiographic analysis of replicating DNA from defined stages of the S-period. *Nucl. Acid Res.* *6*:1417-1431.

Gawlitta, W., Hinssen, H., and Stockem, W. (1980). The influence of an actin-modulating protein (AM-protein) from *Physarum polycephalum* on the cell motility of *Amoeba proteus*. *Eur. J. Cell Biol.* *23*:43-52.

Goodman, E. M. (1980). *Physarum polycephalum*: A review of a model system using a structure-function approach. *Int. Rev. Cytol.* *63*:1-57.

Goodman, E. M., and Rusch, H. P. (1970). Ultrastructural changes during spherule formation in *Physarum polycephalum*. *J. Ultrastruct. Res.* *30*:172-183.

Grainger, R. M., and Ogle, R. C. (1978). Chromatin structure of the ribosomal RNA genes in *Physarum polycephalum*. *Chromosoma* *65*:115-126.

Gröbner, P. (1979). Thymidine kinase enzyme variants in the life cycle of *Physarum polycephalum*. *J. Biochem.* *86*:1595-1605.

Gröbner, P., and Mohberg, J. (1980). Thymidine kinase enzyme variants in the life cycle of *Physarum polycephalum*. *Exp. Cell Res.* *126*:137-142.

Gubler, U., Wyler, T., Seebeck, T., and Braun, R. (1980). Processing of ribosomal precursor RNAs in *Physarum polycephalum*. *Nucl. Acid Res.* *8*:2647-2664.

Guttes, E., Guttes, S., and Rusch, H. P. (1961). Morphological observations on growth and differentiation of *Physarum polycephalum* grown in pure culture. *Dev. Biol.* *3*:588-614.

Hardman, N., and Gillespie, D. A. F. (1980). DNA replication in *Physarum polycephalum*. Analysis of replicating nuclear DNA using the electron microscope. *Eur. J. Biochem.* *106*:161-167.

Hartwell, L. H. (1974). *Saccharomyces cerevisiae* cell cycle. *Bacteriol. Rev.* *38*:167-198.

Hatano, S., Ishikawa, H., and Sato, H. (1979). *Cell Motility: Molecules and Organization*. University Park, Baltimore, Md.

Hejnowicz, Z., and Wohlfarth-Bottermann, K. E. (1980). Propagated waves induced by gradients of physiological factors within plasmodia of *Physarum polycephalum*. *Planta* *150*:144-152.

Henney, H. R., and Asgari, M. (1975). Growth of the haploid phase of the Myxomycete *Physarum flavicomum* in defined minimal medium. *Arch. Microbiol.* *102*:175-178.

Holt, C. E., and Hüttermann, A. (1980). Genetic determination of plasmodium formation in *Physarum polycephalum* (Myxomycetes). *Institut für den Wissenschaftlichen Film*, Göttingen. Film No. B 1337.

Holtzer, H., Rubinstein, N., Fellini, S., Yeoh, G., Chi, J., Birnbaum, J., and Okayama, M. (1975). Lineages, quantal cell cycles, and the generation of diversity. *Q. Rev. Biophys.* *8*:523-557.

Honey, N. K., Poulter, R. T. M., and Teale, D. M. (1979). Genetic regulation of differentiation in *Physarum polycephalum*. *Genet. Res. Cambridge* *34*:131-142.

Hosoda, E. (1980). Culture methods and sporulation of *Physarum polycephalum*. *Mycologia* *72*(3):500-504.

Hüttermann, A. (1973). Biochemical events during spherule formation of *Physarum polycephalum*. *Ber. Dtsch. Bot. Ges.* *86*:55-76.

Ishida, N., Kurihara, K., and Kobatake, Y. (1979). Selective modification of positive chemotaxis in the true slime mold *Physarum polycephalum* by ethylenediaminetetraacetic acid treatment. *Biochim. Biophys. Acta* *587*:89-98.

Jacobson, D. N. (1980). Locomotion of *Physarum polycephalum* amoebae is guided by a short range interaction with *E. coli*. *Exp. Cell Res.* *125*:441-452.

Jalouzot, R., and Toublan, B. (1981). Multiple effects of 5 mM sodium butyrate on *Physarum polycephalum* macroplasmodia. *Cell Tissue Res.* *214*:195-200.

Jeffery, W. R. (1979). Translational regulation of polysome formation during dormancy of *Physarum polycephalum*. *J. Bacteriol.* *140*:490-497.

Jerzmanowski, A., and Staron, K. (1980). Mg^{2+} as a trigger of condensation-decondensation transition of chromatin during mitosis. *J. Theor. Biol.* *82*:41-46.

Johnson, E. M., Campbell, G. R., and Allfrey, V. G. (1979). Different nucleosome structures on transcribing and non-transcribing ribosomal gene sequences. *Science* *206*:1192.

Kirouac-Brunet, J., Masson, S., and Palotta, D. J. (1981). Multiple allelism at the *mat B* locus in *Physarum polycephalum*. *Can. J. Genet. Cytol.* *23*9-16.

Kleinig, H. (1974). Differentiation of *Physarum polycephalum*: Inhibition by alcohols. *Cytobiology* *9*:240-243.

Laane, M. M., and Haugli, F. B. (1974). Division centers in mitotic nuclei of *Physarum polycephalum* plasmodia. *Norw. J. Bot.* *21*:309-318.

Laane, M. M., and Haugli, F. B. (1976). Nuclear behavior during meiosis in the Myxomycete *Physarum polycephalum*. *Norw. J. Bot.* *23*:7-21.

Laffler, T. G., Wilkins, A. S., Selvig, S., Warren, N., Kleinschmidt, A., and Dove, W. F. (1979). Temperature sensitive mutants of *Physarum polycephalum*: Viability, growth and nuclear replication. *J. Bacteriol.* *138*:499-504.

Lewis, W. H. (1942). The relation of the viscosity changes of protoplasm to ameboid locomotion and cell division. In *The Structure of Protoplasm*, W. Seifriz (Ed.). Iowa State College, Ames, Ia., pp. 163-197.

Liskay, R. M., and Prescott, D. M. (1978). Genetic analysis of the G1-period: Isolation of mutants (or variants) with a G1-period from a Chinese hamster cell line lacking G1. *Proc. Nat. Acad. Sci. USA* *75*:2873-2877.

Loidl, P. (1979). Regulation der synchronen Mitose und DNA-Synthese in *Physarum polycephalum*. Ph.D. thesis, Innsbruck University, Austria.

Loomis, W. F. (1975). *Dictyostelium discoideum*: *A Developmental System*. Academic, New York.

Lord, A., Nicole, L., and La Fontaine, J. G. (1977). Ultrastructure and radioautographic investigation of the nucleolar cycle in *Physarum polycephalum*. Characterization of DNA-containing subunits. *J. Cell Sci.* *23*:25-42.

Lott, J., and Clark, J. (1980). Plasmodial senescence in the acellular slime mold *Didymium iridis*. *Exp. Cell Res.* *128*:455.

McCormick, J., Blomquist, J. C., and Rusch, H. P. (1970). Isolation and characterization of a galactosamine wall from spores and spherules of *Physarum polycephalum*. *J. Bacteriol.* *104*:1119-1125.

McKnight, S. L., and Miller, O. L., Jr. (1979). Post replicative nonribosomal transcription units in *Drosophila melanogaster* embryos. *Cell* *17*:551-563.

Magun, B. E. (1979). Changes in cytoplasmic DNA-binding phosphoproteins during the cell cycle of *Physarum polycephalum*. *Cell Differ.* *8*:157-172.

Martin, G. W. (1940). The Myxomycetes. *Bot. Rev.* *6*:356-388.

Matthews, H. R. (1980). Chromosome condensation in mitosis. *J. Theor. Biol.* *83*:367.

Melera, P. W., Davide, J. P., and Hession, C. (1979). Identification of mRNA in the slime mold *Physarum polycephalum*. *Eur. J. Biochem.* *96*:373-378.

Mitchison, J. M. (1977). Enzyme synthesis during the cell cycle. In *Cell Differentiation in Microorganisms, Plants and Animals*. L. Nover and K. Mothes (Eds.). VEB Fischer Verlag, Jena, pp. 377-401.

Mitchelson, K. R., Chambers, T., Bradbury, E. M., and Matthews, H. R. (1978). Activation of histone kinase in G2-phase of the cell cycle in *Physarum polycelphalum*. *FEBS Lett.* *92*:339-342.

Mitchelson, K. R., Bekers, A. G. M., and Wanka, F. (1979). Isolation of a residual protein structure from nuclei of the Myxomycete *Physarum polycephalum*. *J. Cell Sci.* *39*:247-256.

Mohberg, J., and Rusch, H. P. (1971). Isolation and DNA content of nuclei of *Physarum polycephalum*. *Exp. Cell Res.* *66*:305-316.

Mohberg, J., Babcock, K. L., Haugli, F. B., and Rusch, H. P. (1973). Nuclear and DNA contents and chromosome numbers in the Myxomycete *Physarum polycephalum*. *J. Bacteriol.* *97*:1411-1418.

Olive, L. S. (1975). *The Mycetozoans*. Academic, New York.

Ozaki, K., and Maruyama, K. (1980). Connectin, an elastic protein of muscle. A connectin-like protein from the plasmodium of *Physarum polycephalum*. *J. Biochem.* *88*:883-888.

Palotta, D. J., Youngman, P. J., Shinnick, T. M., and Holt, C. E. (1979). Kinetics of mating in *Physarum polycephalum*. *Mycologia* *71*:68-84.

Pepe, F. E., Sanger, J. W., and Nachmias, V. T. (1979). *Motility in Cell Function*. Academic, New York.

Pierron, G., and Sauer, H. W. (1980). More evidence for replication-transcription-coupling in *Physarum polycephalum*. *J. Cell Sci.* *41*:105-113.

Prescott, D. M. (1976). *Reproduction of Eukaryotic Cells*. Academic, New York.

Prior, C., Cantor, C. R., Johnson, E. M., and Allfrey, V. G. (1980). Incorporation of exogenous pyrene labeled histone into *Physarum* chromatin, a system for studying changes in nucleosomes assembled in vivo. *Cell 20*:597-608.

Rakoczy, L. (1980). Effect of blue light on metabolic processes, development and movement in true slime molds. In *The Blue Light Syndrome*, H. Senger (Ed.). Springer, Berlin.

Rakoczy, L. (1981). *Biology of Physarum: Current Problems*. Jagiellonian University, Kracow, Poland.

Rapp, P. E., and Berridge, M. J. (1977). Oscillations in calcium-cyclic AMP control loops from the basis of pacemaker activity and other high frequence biological rhythms. *J. Theor. Biol. 66*:497-526.

Roobol, N., Pogson, C. J., and Gull, K. (1980). Identification and characterization of microtubule proteins from myxamoebae of *Physarum polycephalum*. *Biochem. J. 189*:305-312.

Rusch, H. P. (1970). Some biochemical events in the life cycle of *Physarum polycephalum*. In *Advances in Cell Biology*, Vol. 1. D. M. Prescott, L. Goldstein, and E. McConkey, (Eds.). Appleton-Century-Crofts, New York, pp. 297-327.

Satoh, H., Mito, Y., Ueda, T., Kurihara, A., and Kobatake, Y. (1981). Conversion of positive to negative chemotaxis by calcium ionophore treatment in plasmodium of *Physarum polycephalum*. *Biochim. Biophys. Acta 633*:436-443.

Sauer, H. W. (1973). Differentiation in *Physarum polycephalum*. *Symp. Soc. Gen. Microbiol. 23*:373-406.

Sauer, H. W. (1978). Regulation of gene expression in the cell cycle of *Physarum*. In *Cell Cycle Regulation, Cell Biology Series of Monograph*. J. R. Jeter, I. L. Cameron, G. M. Padilla, and A. M. Zimmermann, (Eds.), Academic, New York, pp. 149-156.

Sauer, H. W. (1980). *Entwicklungsbiologie – Ansätze zu einer Synthese*. Springer-Verlag, Heidelberg.

Sauer, H. W. (1982). *Developmental Biology of Physarum*. Cambridge University, Cambridge.

Sauer, H. W., Babcock, K. L., and Rusch, H. P. (1969). Sporulation in *Physarum polycephalum*. A model system for studies on differentiation. *Exp. Cell. Res. 57*: 319-327.

Scheer, U., Zentgraf, H. W., and Sauer, H. W. (1981). Different chromatin structure in *Physarum*. A special form of transcriptionally active chromatin devoid of nucleosomal particles. *Chromosoma* (in press).

Schreckenbach, T., Walckhoff, B., and Verfuerth, C. (1981). Blue light receptor in a white mutant of *Physarum polycephalum* mediates inhibition of spherulation and regulation of glucose metabolism. *Proc. Nat. Acad. Sci. USA 78*:1009-1013.

Seebeck, T., and Braun, R. (1980). Transcription in acellular slime molds. In *Advances in Microbial. Physiology*, Vol. 21. A. H. Rose and J. G. Morris (Eds.). Academic, New York, pp. 1-45.

Smith, S. S., and Braun, R. (1981). New polypeptide chains associated with highly purified RNA polymerase II or B from *Physarum polycephalum*. *FEBS Lett.* *125*:107-110.

Strathern, J. N., Spatola, E. McGill, C., and Hicks, J. B. (1980). Structure and organization of transposable mating type cassettes in *Saccharomyces* yeasts. *Proc. Nat. Acad. Sci. USA* *77*:2839-2843.

Stosch, von, H. A., (1965). Wachstums- und Entwicklungsphysiologie der Myxomyceten. In *Handbuch der Pflanzenphysiologie*, Vol. 15, W. Ruhland (Ed.). Springer, Berlin, pp. 641-679.

Truitt, C. L., and Holt, C. E. (1979). Analysis of an *alc* mutation. In *Current Research on Physarum*, W. Sachsenmaier (Ed). University of Innsbruck, Innsbruck, pp. 26-30.

Tso, W., and Wong, M. (1978). Chemotaxis not a prerequisite for plasmodia coalescence in *Physarum polycephalum*. *Biochem. Biophys. Res. Commun.* *84*:993-997.

Turnock, G. (1980). Patterns of nucleic acid synthesis in *Physarum polycephalum*. *Prog. Nucl. Acid Res. Mol. Biol.* *23*:53-103.

Tyson, J., and Sachsenmaier, W. (1978). Is nuclear division in *Physarum* controlled by a continuous limit cycle oscillator? *J. Theor. Biol.* *73*:723-738.

Ueda, T., and Kobatake, Y. (1979). Spectral analysis of fluorescence of 8-anilino-1-naphthalenesulfonate in chemoreception with a white plasmodium of *Physarum polycephalum*: Evidence for conformational change in chemoreceptive membrane. *Biochim. Biophys. Acta* *557*: 199-207.

Vandekerckhove, J., and Weber, K. (1978). The amino acid sequence of *Physarum polycephalum* actin. *Nature* *276*:720-721.

Wegener, G., and Sauer, H. W. (1981). No growth cycle in *Physarum polycephalum*? *Exp. Cell Res.* *135*:399-406.

Weiss, P. A. (1973). Differentiation in retrospect. *Differentiation* *1*: 3-10.

Whittaker, P. A., McLachlan, A., and Hardman, N. (1981). Sequence organization in nuclear DNA from *Physarum polycephalum*: Methylation of repetitive sequences. *Nucl. Acids Res.* *9*:801-814.

Wilkins, A. S., and Reynolds, G. (1979). The development of sporulation competence in *Physarum polycephalum*. *Dev. Biol.* *72*:175-181.

Wohlfarth-Bottermann, K. E., and Block, I. (1981). The pathway of photosensory transduction in *Physarum polycephalum*. *Cell Biol. Int. Rep.* *5*:365-373.

Wolf, R., Wick, R., and Sauer, H. W. (1979). Mitosis in *Physarum polycephalum*: Analysis of time lapse films and DNA replication of normal and heat-shocked macroplasmodia. *Eur. J. Cell Biol.* *19*: 49-59.

Wormington, W. M., and Weaver, R. F. (1976). Photoreceptor pigment that induces differentiation in the slime mold *Physarum polycephalum*. *Proc. Nat. Acad. Sci. USA 73*:3896-3899.

Wright, M., Moisand, A., and Mir, L. (1981). Centriole maturation in the amoebae of *Physarum polycephalum*. *Protoplasma 105*:149-160.

Wright, M., and Tollon, Y. (1979). *Physarum* thymidine kinase: A step or peak enzyme depending upon temperature of growth. *Eur. J. Biochem. 96*:177-181.

Wright, M., Moisand, A., and Mir, L. (1979). The structure of the flagellar apparatus of the swarm cell of *Physarum polycephalum*. *Protoplasma 100*:231-250.

Youngman, P. J., Adler, P. N., Shinnick, T. M., and Holt, C. E. (1977). An extracellular inducer of asexual plasmodium formation in *Physarum polycephalum*. *Proc. Nat. Acad. Sci. USA 74*:1120-1124.

Youngman, P. J., Anderson, W. R., and Holt, C. E. (1981). Two multiallelic mating compatibility loci separately regulate zygote formation and zygote differentiation in the Myxomycete *Physarum polycephalum*. *Genetics 97*:513-530.

Zaar, K., and Kleinig, H. (1975). Spherulation of *Physarum polycephalum*. I. Ultrastructure. *Cytobiologie 10*:306-328.

Zubrzycka-Gaarn, E., Korczak, B., and Osinska, H. E. (1979). Identification of sarcoplasmic reticulum-like system in *Physarum polycephalum*. *FEBS Lett. 107*:335-339.

5
The Cell Division Cycles of Yeasts

DAVID LLOYD *University College, Cardiff, Wales*

I. INTRODUCTION

A. Cellular Organization in Space and Time

The cell division cycle consists of series of events that mark the progress of the development of a cell from the time when it becomes an autonomous individual until it gives rise to daughter cells. Thus the processes to be studied include the biochemistry of growth, the development of extended surfaces and of new organelles, and the mechanics of segregation of components at cell division. If these processes occurred continuously, there would be no observable biochemical or morphological events; that this is manifestly not the case points to the necessity for the close examination of the cell division cycle. The classical microbiological studies of growth neglect entirely the concept of discontinuity; distinction between the growth of individual cells and increases in cell numbers is lost in references to "increasing biomass." The other oversimplification which pervades the microbiology of yeast growth is the concept of the "homogeneity" of a culture, especially in a chemostat under steady state conditions. Measurements made on such cultures result from contributions of individual cells at different stages of their cell division cycles, and thus represent a time average over the cell cycle time (Edwards, 1981; Lloyd et al., 1982).

In order to study the rich complexity of the sequentially ordered processes that characterize the cell division cycle, we must examine events either in single cells or in a population composed of individuals all of a defined age in terms of cell cycle time. The former approach is limited by the restricted availability of adequately sensitive noninvasive analytical methods, but must ultimately be pursued in order to appreciate the extent of variation permissible during the normal development of individual cells. The latter approach, for which several practical solutions are available, has yielded sufficient amplification to allow conventional biochemical analyses. However, many of the published observations are controversial, because investigators have often failed to distinguish between phenomena which are real cell cycle events from those resulting from the manipulative procedures necessary to obtain a temporally homogeneous population of cells or to sample from that population. This problem is unfortunately not entirely confined to the older methods of induced synchrony, but also seriously hampers useful interpretation of much recent work based on cell selection techniques. This review makes only scant references to information obtained from induced synchrony, except in the cases where deliberate perturbation has yielded insights into the control mechanisms which regulate cell cycle traverse. It emphasizes recent methods of cell cycle research and outlines data relating to the time course of biosynthesis and accumulation of molecular components and their assembly into structures. Most important and fundamental is the breakthrough in the understanding of the genetic control of the expression of cell cycle stages which must underlie future elucidation of key steps in cell cycle control processes.

Recent advances in understanding all these aspects of the cell division cycle have relied to a great extent on yeasts as systems for study. Most of the available information has been obtained with experiments with *Schizosaccharomyces pombe* and *Saccharomyces cerevisiae*; cell cycle maps for these two species indicate the major morphological events observable (Fig. 1). Apart from the intrinsic interest in discovering cell cycle-dependent events and the relevant control systems operating in these organisms, the wider and indeed far-reaching implications of this research rely on the hope that, in essence, the principles of cell cycle control mechanisms will not be too different in yeasts from those in mammalian cells.

II. METHODS FOR STUDY OF THE CELL DIVISION CYCLE

A. Preparation of Synchronous Cultures

1. Selection of Synchrony by Velocity Sedimentation

These methods rely upon the selection from a growing culture of a population of cells which are all at a similar stage of the cell division cycle. The selected population is then grown up separately as the synchronous culture. In order to minimize potential perturbing influences, the selection procedure should be as rapid as possible and not involve the removal of organisms from their growth medium, as even transient nutrient and oxygen depletion can seriously affect the progress of subsequent metabolic events (Lloyd and Ball, 1979). Changes of temperature should be avoided and exposure to extreme centrifugal forces, liquid shear, and hydrostatic pressure may not be without effect. As even the most satisfactory of the recently developed methods do not eliminate these physical hazards, parallel studies on identically treated control (asynchronous) cultures are mandatory. At present there is little information on the susceptibility of different metabolic systems to these various perturbing influences. All selection methods employed for yeasts depend either on (a) the velocity sedimentation principle, in which separation of an exponentially growing cell population into different age classes is based on size and density differences, or (b) equilibrium and sedimentation based entirely on density differences. The behavior of a yeast approximates that of an ideal spherical particle for which the rate of sedimentation is proportional to the square of the radius and to the density difference between the particle and the suspending fluid.

Centrifugation Through Gradients: Velocity sedimentation through density gradients of aqueous sucrose, the method introduced by Mitchison and Vincent (1965) for both fission and budding yeasts,

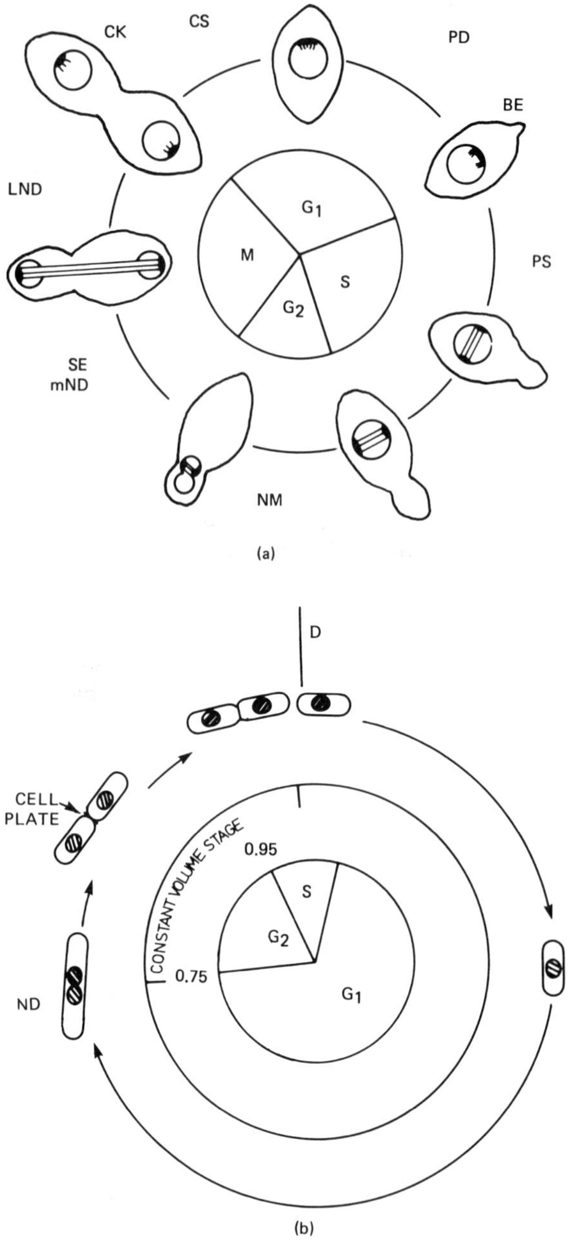

FIG. 1 The cell division cycles of (a) *S. cerevisiae* and (b) *Schiz. pombe* (PD, plaque duplication; BE, bud emergence; PS, plaque separation; NM, nuclear migration; SE, spindle elongation; mND, medial nuclear division; LND, late nuclear division; CK, cytokinesis; CS, cell separation).

provided a popular method for selection; about 10% of the original population (usually the most slowly sedimenting size class) is used as the inoculum for synchronous growth. Gradients of lactose, glucose, and glycerol have also been employed; Mitchison and Carter (1975) give many practical hints. Higher-molecular-weight media used include dextran (Wiemken et al., 1970), Renografin (Hartwell, 1970), Urografin (Salmon, 1980), Percoll (Dwek et al., 1980), and colloidal silica (Ludox) (Shulman et al., 1973; Shulman, 1978). These osmotically less active compounds are often used where the changes in density which occur during the cell cycle of *S. cerevisiae* are exploited for cell selection. Scaling up from gradients in centrifuge tubes (up to 50 ml) to those in zonal rotors (up to 2 liter) provides larger gradient capacity and hence larger synchronous culture volumes, but also leads to the inevitable protraction of handling times (Poole et al., 1973; Poole and Lloyd, 1974). Although centrifugation through gradients can provide excellent cell selection and a high degree of synchrony, undoubtedly some disturbance of cellular physiology occurs; even in the most rapid and "careful" procedures preliminary harvesting and concentration of the cells (either by filtration or centrifugation) leads to anaerobiosis and transient nutrient depletion. For this reason, gradient centrifugation is less useful for selection synchrony than for cell cycle fractionation.

Continuous Flow Centrifugal Selection: Passage of an exponentially growing culture through a continuous flow centrifuge under predetermined conditions of rotor speed and flow rate permits selection of the most slowly sedimenting subpopulation, which escapes harvesting and emerges in the rotor effluent (Fig. 2). Both the "continuous action rotor" of MSE (Lloyd et al., 1975) and the Sharples "Laboratory Super-centrifuge" (Edwards and Jones, 1977) may be employed at flow rates of up to 2 liter/min. Extensively used for *Schiz. pombe* (Lloyd et al., 1975; Edwards and Lloyd, 1977; Poole, 1977a; Bullock and Coakley, 1979; Walker and Duffus, 1980), the method has also been used for *Candida utilis* (Lloyd et al., 1975; Kader and Lloyd, 1979). The attractive features of this procedure are scale-up (we have prepared synchronous cultures of volumes up to 20 liters), high resolution of selection and hence good synchrony indices, and speed of handling. The organisms are never removed from the culture medium, are not exposed to excessive centrifugal forces, and at high flow rates have a short mean residence time in the rotor (e.g., 0.2 min at 2 liter/min). Thus several potential sources of perturbation (anaerobiosis, nutrient deprivation, and temperature change) are minimized. If required, it is not difficult to carry out all procedures aseptically and/or anaerobically. Some disadvantages are the following: (1) it is impossible to obtain control asynchronous cultures which have received *identical* flow and centrifugal treatment, and (2) lack of selection in those cases where the inverse relationships between size and

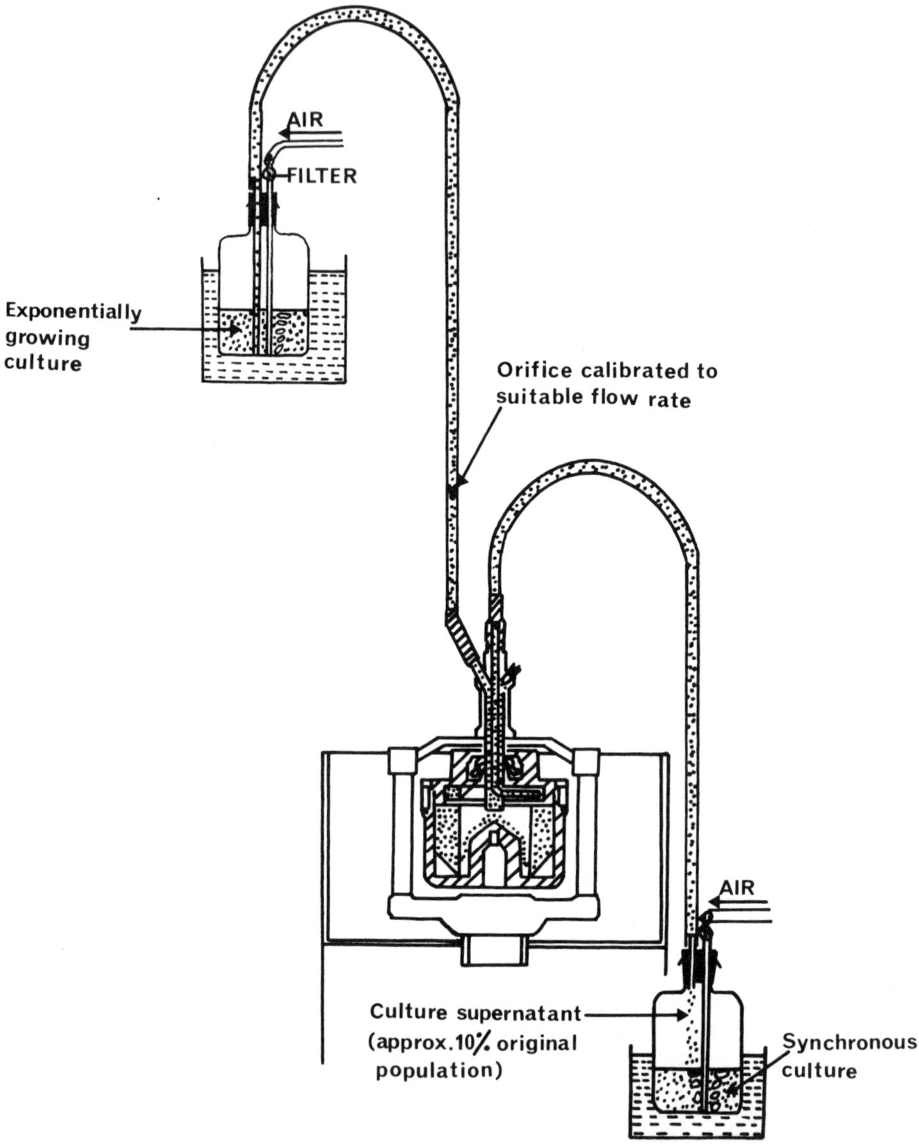

FIG. 2 Diagram of apparatus for the preparation of synchronous cultures by continuous flow size selection using the MSE "continuous action rotor." (From Lloyd et al., 1975, with permission.)

density compensate for one another (R. I. Scott and R. K. Poole, 1980, unpublished data for *Escherichia coli*). That changes in density accompany increasing size during cell cycle traverse is well established for fission and budding yeasts, but it has not been reported that this hinders selection in these instances. A recent claim that continuous flow centrifugation subjects organisms to metabolic shock (Walker et al., 1980) is difficult to evaluate on the basis of the evidence presented.

Centrifugal Elutriation (Counterflow Centrifugation): In the Beckman elutriator rotor the force tending to sediment particles in a centrifugal field is counterbalanced by flowing the suspending fluid in the opposite direction. Fractionation of a cell population into discrete size classes is achieved by incrementally increasing the flow rate while maintaining the centrifuge rotor at constant speed. Gordon and Elliott (1977) first demonstrated the ability of this method to give good separations of small unbudded cells of *S. cerevisiae*, and Creanor and Mitchison (1979) used it to obtain synchronous cultures of *Schiz. pombe*. Maximum rotor loading occurred at 7×10^9 cells and took between 10 and 40 min. It was shown that control asynchronous cultures exposed to centrifugal elutriation showed less perturbation (measured by the time course of [^{3}H]leucine incorporation) than those for velocity sedimentation separations on sucrose gradients in tubes. The disadvantages of this procedure stem from the small volume of the flow compartment of the rotor (4.5 ml) and the protracted handling times. During the centrifugation, organisms are accumulated at such high concentrations that they must become nutrient and oxygen depleted, although perfused by fresh growth medium. Only small cultures are available and the device is costly and has to be sterilized chemically.

Differential Sedimentation in the Culture Medium: A rapid and simple method of selection synchrony has been used for *Acanthamoeba* (Chagla and Griffiths, 1978; Edwards and Lloyd, 1978; Edwards et al., 1981). This promises to provide an alternative to continuous flow centrifugation as a selection method likely to cause only minimal perturbation of yeast cultures. (J. L. Williams and D. Lloyd, 1981, unpublished results.) Aseptic transfer of exponentially growing yeast cultures to sterile 1-liter centrifuge pots immediately preceded centrifugation at room temperature in the 6×1-liter rotor of a MSE "Mistral" centrifuge. For glycerol-grown *C. utilis*, acceleration to 2500 rev min^{-1} took 50 s; the centrifuge was held at this speed for 30 s and then decelerated with the brake on. Total centrifugation time was 2 min 30 s and the supernatant (consisting of the most slowly sedimenting cells and containing about 10% of the initial population) was carefully decanted (avoiding disturbance to the loosely pelleted sediment) and grown as a synchronous culture. Conditions for glucose-grown

Schiz. pombe selection were identical, except that a rotor speed of 2500 rev min^{-1} was maintained for 2 min. Asynchronous control cultures were obtained by mixing the contents of centrifugal pots after identical centrifugal treatment and growing the entire populations as shaken cultures.

2. *Induction Procedures*

A number of procedures have been employed to induce synchronous growth and division in yeasts. For *S. cerevisiae*, Williamson and Scopes (1960, 1962) used cycles of starvation and refeeding. A similar approach involving dilution of stationary phase cultures was developed by Sando (1963) for *Schiz. pombe* and by Tauro and Halvorson (1966) for *S. cerevisiae*. A similar philosophy has led to the continuous phased growth and periodic refeeding method for chemostat cultures of *C. utilis* (Dawson, 1965) and of *S. cerevisiae* (von Meyenburg, 1969). Some divalent cation ionophores or chelators (e.g., A23187, EDTA, Na_4 pyrophosphate) inhibit cell division in *Schiz. pombe*, and release from inhibition (by inoculation into fresh growth medium) gives a degree of synchrony (Duffus and Paterson, 1974; Ahluwalia et al., 1978; Walker and Duffus, 1979, 1980). A similar method has been applied to *Kluyveromyces fragilis* (Penman and Duffus, 1975). Inhibitor-induced synchrony of *Schiz. pombe* has relied upon the DNA synthesis inhibitors 2'-deoxyadenosine (Mitchison and Creanor, 1971b; Poole, 1977b) or hydroxyurea (Miyata et al., 1978b), or the mitochondrial protein synthesis inhibitor chloramphenicol (Quinton and Poole, 1977). The availability of purified *S. cerevisiae* α factor provides a method of G_1 arrest in the *a* strain of this organism; subsequent release provides synchronous growth and division (Hartwell, 1973; Chan, 1977; Shulman, 1978). Cultures of *Schiz. pombe* can be division synchronized after treatment with five to six heat shocks, during each of which the temperature is increased to 41°C for 30 min (Kramhøft and Zeuthen, 1975). Between heat shocks, the optimal growth temperature (32°C) is maintained for 110 min; an automated device provides these conditions.

B. Cell Cycle Fractionation

The methods used for obtaining slowly sedimenting size class by density gradient centrifugation can be extended and modified to sort all the cells of an exponentially growing culture. Thus under appropriate conditions successive fractions from a gradient represent subpopulations of successive stages of the cell cycle. Cell cycle fractionation of yeasts has been achieved both on the basis of size (velocity sedimentation) and density (equilibrium sedimentation). Almost linear increases in mean cell volume with fraction number indicate satisfactory separation in linear sucrose gradients primarily by size (and thus by age in the cell cycle) for *S. cerevisiae* and *S. lactis* (Sebastian et al., 1973; Halvorson et al., 1971; Carter and Halvorson, 1973). Rate sedimentation of *Schiz. pombe* in a reorientating gradient zonal rotor (Wells and James, 1973) and in a high speed (HS) zonal rotor by both velocity and equilibrium sedimentation methods (Poole and Lloyd, 1973; Poole et al., 1974; Lloyd and Edwards, 1977) gives a valid analysis of the first 0.75 of the cell cycle. Cell cycle fractionation of *Candida albicans* has also been reported (Chaffin and Sogin, 1976).

The techniques of cell cycle fractionation permit identical exposure of all the cells of an exponentially growing culture to some treatment before instantaneous arrest (by chemical fixation or rapid cooling). The range of applications reported includes analysis of enzyme activities (Poole and Lloyd, 1973), or chemical analysis (Poole et al., 1974) of organisms at successive cell cycle stages, and measurement of enzyme inducibility (Sebastian et al., 1973) and susceptibility to mutagenesis through the cell cycle (Dawes and Carter, 1974).

The correlation of age in the cell division cycle with cell size (necessary for interpretation of results as timings on cycle maps) depends upon determination of mean cell volumes using a resistive particle counter with multichannel particle size analyzer. Comparison of size distributions of subpopulations obtained in cell cycle fractionation experiments with those of successive stages in the cell cycle of a synchronous culture is necessary. It should be considered that alterations in cell volume may occur on cooling cultures, or during centrifugation. Despite these complications, cell cycle fractionation provides a high-resolution procedure for obtaining large quantities of sorted cells for analysis. A rapid method using the continuous flow principle has been described (Lloyd et al., 1977).

III. MACROMOLECULAR BIOSYNTHESIS DURING THE CELL DIVISION CYCLE

A. Synthesis of DNA

1. *Nuclear DNA*

Recent determinations of the timing of nuclear DNA synthesis in yeasts by biochemical analysis of synchronous cultures confirm cytochemical results for several yeasts. Thus the initiation of DNA synthesis in *S. cerevisiae* occurs at about the same time as bud emergence (Matur and Berry, 1978; Sena et al., 1975). Cell cycle analysis by centrifugal elutriation confirms that the peak of DNA synthesis occurs at between 0.3 and 0.4 of the cell cycle (Elliott and McLaughlin, 1978). In *C. utilis* the midpoint of DNA replication occurs at 0.94 of the cell cycle (Wain et al., 1976).

Protein synthesis is required for the initiation of DNA synthesis, but not for its continuation in *S. cerevisiae* (Hereford and Hartwell,

1973). Thus protein synthesis inhibitors are without effect when presented in mid S phase (Williamson, 1973). Chromatin assembly may proceed from pools of presynthesized histones (Moll and Wintersberger, 1976), as ongoing DNA synthesis is insensitive to protein synthesis inhibitors. Growth of *Schiz. pombe* with 2-phenylethanol shifts the S phase forward from its usual timing just before cell division (Bostock, 1970; Mitchison and Creanor, 1971a) but leaves the most heat-sensitive stage unaltered (Bullock and Coakley, 1979); DNA damage is thus not the dominant cause of cell killing by heat. Using a similar approach, two ultraviolet-sensitive cell cycle phases have been resolved (Williams, 1980).

2. *Mitochondrial DNA and 2-μm Plasmid DNA*

A haploid strain of *S. cerevisiae* contains about 50 copies of mitochondrial (mt DNA) per cell (Williamson, 1970), usually distributed in aggregates or "chondrolites" (Williamson et al., 1977). That mitochondria have their own distinctive DNA polymerase has posed the question of the extent of independence of replication of the organelle and nuclear genomes. Continuous replication of mt DNA throughout the cell cycle, first reported by Williamson and Moustacchi (1971), has been confirmed by several different analytical approaches (Sena et al., 1975; Cottrell, 1979). On the other hand, in cell cycle fractionation experiments, a discontinuous mode of mt DNA replication occurring just prior to or at the same time as the nuclear S period has also been observed (Wells, 1974). The rate of nitrosoguanidine mutagenesis yielding cytoplasmically inherited mutations is highest at a time in the cell cycle different from the nuclear S period (Dawes and Carter, 1974), and the simplest interpretation (that mutagenesis occurs at replication forks) favors discontinuous replication of mt DNA. In the most recent study, Cottrell (1981) concluded from three independent methods of analysis (during three consequentive cell division cycles in starvation-synchronized cultures) that mt DNA replication occurs continuously. Results from isotope dilution analysis and measurements of the ratio of mt DNA to total cellular DNA clearly showed this, whereas pulse label incorporation indicated maximal [^{3}H] adenine uptake into mt DNA at the time of nuclear DNA synthesis. Possible complications of the pulse labeling patterns make this latter result inconclusive. A similar critical analysis of the time course of the mt DNA replication in synchronous culture established by minimally perturbing selection methods would further clarify the issue.

Possible coordination between nulcear DNA synthesis and mt DNA synthesis has also been explored by using *cdc* mutants. In DNA initiation mutants (e.g., *cdc 4* and *cdc 7*), mt DNA synthesis continued for a long time after shifting the organisms to the restrictive temperature (Cottrell et al., 1973; Cryer et al., 1973; Newlon and Fangman, 1975). The product of these genes are not therefore required for mt

DNA replication. However, in the DNA initiation mutants (*cdc 8* and *cdc 21*), mt DNA replication virtually ceased at the restrictive temperature and rapid accumulation of mitochondrial *petites* occurred (Wintersberger et al., 1974; Newlon et al., 1979). Mitochondrial marker transmission has been studied in mutant *cdc 8* (Kruszewska et al., 1976).

Replication of the 2-μm plasmid DNA of *S. cerevisiae* requires a functional *cdc 8* gene product for a normal rate of replication (Petes and Williamson, 1975). Similar results were also obtained for *cdc 28*, *cdc 4*, and *cdc 7* mutants (Livingston and Kupfer, 1977); extracts from mutants *cdc 7* or *cdc 8* show diminished ability to stimulate DNA synthesis in vitro on supercoiled 2-μm DNA (Jazwinski and Edelman, 1979). Thus synthesis of nuclear DNA, mt DNA, and 2-μm plasmid DNA is dependent on the *cdc 8* gene product (probably involved in DNA elongation) and the *cdc 21* product (involved in thymidine metabolism). A correlation between mt DNA synthesis and the double-stranded RNA replication of viruslike particles of killer strains has been suggested (Weiser and Shalitin, 1978).

B. Synthesis of RNA

The constant rate of ribosomal and polyadenylated (messenger) RNA synthesis in the G_1 phase of *S. cerevisiae* was abruptly doubled during the S phase (Fraser and Carter, 1976), although under different growth conditions an exponential continuous increase was observed (Sogin et al., 1974; Hynes and Phillips, 1976; Elliott and McLaughlin, 1978). The gene dosage effect in *Schiz. pombe* (Wain and Staatz, 1973) has also been observed in mutants (Fraser and Moreno, 1976; Fraser and Nurse, 1978, 1979; Barnes et al., 1979). This suggests a size-related control of RNA synthesis to maintain RNA content at a fixed proportion of cell mass. Information on the possible changes in the proportion of different species of rRNA or mRNA is not yet available.

C. Proteins

1. *Total Cellular Protein*

After refeeding starved (stationary phase) cultures, the resulting synchronously dividing population of *S. cerevisiae* cells accumulated protein continuously (Williamson and Scopes, 1960; Gorman et al., 1964). Similar results were obtained for synchronous cultures of *Schiz. pombe* (Mitchison and Wilbur, 1962; Wain, 1971; El'Khayat, 1980; Toma, 1980). Continuous flow size selection of *C. utilis*, however, gives cultures in which total protein (as assayed by the Folin procedure) does not show a smooth accumulation (Lloyd, Edwards, and Williams, 1981). Instead, an oscillation in total protein was observed; this suggests that in this organism generalized protein turnover involving the majority of individual protein species may be organized into discrete phases of synthesis and degradation which alternate throughout the cell cycle. Control cultures indicate that these discontinuities do not result from perturbation.

Experiments using a dual-label technique (preincubation with [^{3}H]-yeast protein hydrolysate for 3 h followed by 10 min with [^{31}S]methionine) provides a method for estimating the rates of synthesis of individual polypeptides after centrifugal elutriation (Elliott and McLaughlin, 1978, 1979). In *S. cerevisiae* only 2 out of 110 proteins (identified by two-dimensional gel electrophoresis) showed rates of synthesis which were discontinuous through the cell cycle (Elliott et al., 1979). Thus at present it seems that, except in *C. utilis*, it is difficult to detect any fluctuations in the time course of accumulation of total cellular protein; evidence for rapid turnover in many systems suggests that further attempts might be worthwhile.

2. *Enzymes*

The patterns of accumulation of individual enzymes through the cell cycle of yeasts has been intensively investigated as a means of understanding the temporal control of gene expression, and a number of models have been proposed to account for the observations. Accumulation of enzymes as "peak" patterns of activity may provide information of great importance to an understanding of the in vivo stability of enzymes. Extensively reviewed (Mitchison, 1971; Halvorson et al., 1971), much of the earlier work has been criticized by Mitchison (1977), who believed that insufficient attention had been payed to metabolic disturbances produced even by gradient methods for selection synchrony. Thus Mitchison (1977) provided instances of 18 enzymes which show continuous increases through the cell cycle of *Schiz. pombe*, and only one (thymidine monophosphate kinase) which is expressed periodically. Recent reports of enzymes which increase in activity abruptly at some time during the cycle ("step enzymes") include ornithine and aspartate transcarbamylases and alcohol dehydrogenase (Sissons et al., 1973), acid phosphatase (Miyata and Miyata, 1978), and sucrase (Miyata et al., 1980).

Discontinuous accumulation of several organelle-associated enzymes was observed in synchronous cultures of glucose-grown *Schiz. pombe* after size selection on gradients (Poole and Lloyd, 1973) and confirmation was obtained by two independent cell cycle fractionation procedures. Thus mitochondrial enzymes (NADH-cytochrome *c* oxidoreductase, succinate dehydrogenase, NADP-linked isocitrate dehydrogenase, and fumarate hydratase) showed two maxima in activity during the cell cycle. Similar results were obtained for catalase, acid p-nitrophenylphosphatase, NADPH-cytochrome *c* oxidoreductase, and malate dehydrogenase. Cytochrome *c* oxidase showed a single peak. Cell cycle fractionation indicated that cytochromes $a + a_3$ and b_{563} (b_T)

showed two peaks per cycle and measured in low-temperature difference spectra (which assays total amounts of hemoproteins rather than activities); interestingly the ratio of the CO-binding species (cytochrome a_3) to total *a*-type cytochrome showed striking variations (Poole et al., 1974). Continuous flow selection of glucose-grown cells gave synchronous cultures in which the activity of mitochondrial oligomycin-sensitive ATPase oscillated over a considerable range (Edwards and Lloyd, 1977). Sensitivity to oligomycin and 10 other specific inhibitors varied greatly at different stages in the cell cycle (Lloyd and Edwards, 1977). For ATPase, independent immunological assays using antibody raised to purified F_1-ATPase indicated that oscillations in enzyme activity are a consequence of oscillations in enzyme protein (J. B. Evans, G. El'Khayat, S. W. Edwards, and D. Lloyd, 1980, unpublished results). That different enzymes of the inner mitochondrial membrane are not expressed at identical times in the cell cycle suggests that the stoichiometry of individual components in the membrane (e.g., cytochrome *c* oxidase, ATPase, and succinate dehydrogenase) may vary at different stages. Discontinuous expression of the inhibitor-binding sites of mitochondrial ATPase may have several different possible explanations, as discussed by Lloyd and Edwards (1977).

Results with glycerol-grown cells (i.e., catabolite derepressed) indicated double-step patterns of expression of cytochrome *c* oxidase activity, which correlated with the cellular content of cytochrome $a + a_3$ determined in difference spectra and with a double step in respiration rate (or 0.45 and 0.95 cycle) (Poole and Lloyd, 1974). Mitochondrial ATPase showed similarly timed step increases after continuous flow selection (Edwards and Lloyd, 1977).

Discontinuous enzyme expression would appear to be the rule in *S. cerevisiae* (Halvorson, 1977); more than 30 enzymes measured suggest this, and a model based on an ordered transcription of the yeast genome has been elaborated to explain the temporal regulation of enzyme synthesis (Halvorson et al., 1971). Exceptions to this rule include ribosomal proteins (Shulman et al., 1973) and RNA polymerase I (Sebastian et al., 1974). Recent additions to the list of "step" enzymes in synchronous cultures obtained by gradient selection include sucrase and alkaline phosphatase (Matur and Berry, 1978) and exo- and endo-1,3-β-glucanase activities (Delrey et al., 1979). Selection by differential centrifugation indicates continuous accumulation of the large isomer (glycoprotein bound) of invertase, but a single-step pattern for the small (carbohydrate-free) form (Gallili and Lampen, 1977). The technique of synchronization by G_1 arrest indicated low levels of activity of ornithine decarboxylase in G_1 (Kay et al., 1980) and increasing levels of total RNAase (McFarlane, 1980). Induction of synchrony by reinoculation of stationary phase cultures gave a step pattern of increase of cytochrome *c* oxidase activity which did not, however, correlate with the continuously increasing cellular content of cytochrome $a + a_3$ (Cottrell et al., 1975). Ribonucleotide reductase showed a peak of activity at the onset of the S phase coinciding with the time of bud emergence after cycles of heat treatment (Lowdon and Vitols, 1973).

It is still uncertain whether *Schiz. pombe* and *S. cerevisiae* operate fundamentally different control mechanisms for enzyme regulation, or whether the observed discrepancies stem from differences in experimental conditions. It is of key importance to demonstrate the following: (1) absence of discontinuities in control (asynchronous) cultures which have been subjected to experimental procedures identical to those used for the establishment of synchronous cultures, (2) that the periodic event is repeated in two or more successive cycles, and (3) that independent techniques (e.g., cell cycle fractionation and synchronous cultures) give identical results. Where these criteria have not been met, a degree of uncertainty remains regarding the final acceptance of discontinuities in enzyme activities as genuine cell cycle events.

IV. DEVELOPMENT OF SUBCELLULAR STRUCTURES

A. The Nucleus

Fluctuations in the numbers of nuclear pores per unit membrane during the cell cycle are not understood (Jordan et al., 1977). Embedded in the nuclear membrane in unbudded cells of *S. cerevisiae* is a single spindle pole body which duplicates in the S phase and eventually separates to the nuclear poles of a complete intranuclear fiber (Byers and Goetsch, 1975). It is the elongation of this fiber which forces the poles of the nucleus into mother and bud before cytokinesis. Condensation of chromatin into discrete chromatids has been followed (Peterson and Ris, 1976; Gordon, 1977). The nuclear membrane remains intact throughout mitosis (Matile et al., 1969). A nucleolar function (synthesis of ribosomal RNA) has been assigned to a densely staining crescent within the nucleus (Smitt et al., 1973).

B. Mitochondria

Early reports that yeast contains large numbers of small mitochondria (Lloyd, 1974) conflict with more recent, careful studies using reconstructions based on serial sectioning and high-voltage electron microscopy. In some cases *S. cerevisiae* has been described as having a single giant mitochondrion (Hoffman and Avers, 1973), but more commonly a small number of mitochondria (less than 10 per cell) have been identified (Stevens, 1977). This latter definitive study elegantly described the presence of highly branched forms of a giant mitochondrion coexisting alongside smaller individuals within a single cell. In some, but not all, budding cells the giant reticulate organelle extended into the buds. No specific cell cycle-dependent changes in mitochondrial numbers or distributions have been determined. A deeper appreciation of the population dynamics of mitochondria is needed to understand the mechanism of segregation of mt DNA at budding. Extensive mitochondrial reticula have also been observed in *Schiz. pombe* and *C. utilis* (Davison and Garland, 1977). In *Sterigmatomyces halophilus* the activities of inner mitochondrial marker enzymes (cytochrome *c* oxidase and succinate dehydrogenase) oscillated similarly in buds and in mother cells, suggesting close parallel development of mitochondrial activities in the two structures (Salmon and Poole, 1980).

C. The Vacuole

During bud initiation in *S. cerevisiae* the large vacuole shrinks and fragments (Wiemken et al., 1970) prior to segregation between mother and daughter cells. These changes are thought to underlie changes in the buoyant density of organisms during the cell cycle (Wiemken et al., 1970; Hartwell, 1970; Sierra et al., 1973). Further evidence for the regulation of vacuole permeability comes from the observation of release of nitrogen sources from the vacuole upon G_1 arrest (Sumrada and Cooper, 1978).

D. Ribosomes and Polysomes

There has been no recent examination of changes in yeast ribosome numbers or activity during the yeast cell cycle. The sum total of available information suggests that the proteins of both large and small ribosomal subunits in *S. cerevisiae* are continuously synthesized (Shulman et al., 1973) and that ribosomal numbers increase through most of the cell cycle of *Schiz. pombe* (Maclean, 1965).

E. Cell Walls

The patterns of increase of cell wall polymers during the cell cycle of *S. cerevisiae* have been studied extensively. The synthesis of both mannan and glucan is greatly reduced at the time of division (Hayashibe et al., 1977; Biely, 1978). Chitin synthesis begins just after the start of bud formation and ends at bud separation (Cabib and Farkaš, 1971). Confirmation of this result was obtained using polyoxin D, an inhibitor of chitin synthesis (Cabib and Bowers, 1975), and also by chitin analysis after dissolving the β-glucan (Vřsanská et al., 1979). That the enzymes of cell wall biosynthesis are performed in the previous cell cycle has been demonstrated by Kopecká (1977). Thus normal fibrillar wall fraction was produced by protoplasts of cell division cycle mutants, irrespective of whether they were arrested in G_0, G_1, or S.

The developing bud wall grows mostly at the bud tip (Johnson and Gibson, 1966) while newly synthesized (concanavalin A-labeled) mannan is deposited around the bud (Tkacz and Lampen, 1972). Transfer of surface structure from mother to bud does not occur (Skutelsky and Bayer, 1979).

The mode of extension of *Schiz. pombe* cell wall surfaces has also been studied by a number of methods. Fluorescent antibody attachment indicated that elongation occurs exclusively at the poles (May, 1962). The percentage of cells which extend at both ends may be as low as 20% (Johnson, 1965) or as high as 80% (Streiblová, 1977). The division scars which persist for several generations aid these studies (Calleja et al., 1980). Cessation of pole growth occurs as the cell plate appears (Mitchison, 1970) to divide the cell into equal volume compartments, which may, however, be of unequal lengths (Johnson et al., 1979).

V. BIOENERGETICS OF THE CELL DIVISION CYCLE

A. Oxygen Uptake, Carbon Dioxide Evolution, and Adenylate Pools

Early studies on the respiration of *S. cerevisiae* in synchronous cultures have been summarized by Lloyd (1974). A uniform interpretation of the results is not possible, as different investigators used different yeast strains, different growth conditions, and different procedures for obtaining synchronous cultures. It is not surprising, therefore, that a variety of patterns for oxygen uptake rates are recorded, including steps (Nejedlý and Greksák, 1977), continuous increases, and oscillations. Dharmalingam and Jayaraman (1973) pointed out that the respiratory oscillations observed when glucose was used as carbon source were not detected with maltose.

Respiratory oscillations (two major maxima per cell cycle) in glucose-grown synchronous cultures of *Schiz. pombe* (Poole et al., 1973) are replaced by stepwise increases (two per cycle) in glycerol-grown cultures (Poole and Lloyd, 1974). The oscillations were attributed to a Pasteur effect control mechanism on the basis of maximal uncoupler (m-Cl-carbonylcyanide phenylhydrazone) action at respiratory maxima. Heat production proceeded uniformly through the cell cycle of glucose-grown cells. It has not yet been determined whether the rate limitation on overall respiratory activity depends on the timing of synthesis of a functional enzyme or redox component, or whether it is exerted by regulatory metabolites. The relative contributions of these two kinds of control may well alter at different times in the cell cycle. Pool sizes of adenine nucleotides show complex fluctuations in synchronous cultures of this organism (El'Khayat, 1980), with adenylate charge values in the range 0.73-0.90 and ATP/ADP ratios varying between 1.2 and 4.9.

Oscillatory respiration has also been observed in cultures of *Schiz. pombe* synchronized by the addition and subsequent removal of 2 mM 2'-deoxyadenosine; respiratory maxima corresponded to minimal values for ATP/ADP ratios (Poole and Salmon, 1978). Evidence has been discussed that the mechanisms responsible appear quite different in these cultures from those obtained by selection.

Linear increases in overall glycolytic activity have been observed in cultures of *Schiz. pombe* grown synchronously on complex medium after gradient size selection or induced by cycle heat treatment (Hamberger et al., 1977; Kramhøft et al., 1978). Manometric measurements of CO_2 evolution by synchronous cultures, growing on defined medium after gradient selection, showed a sharp doubling in rate at about the time of nuclear division (0.75 cycle) and then remained constant, that is, increased in a step (Creanor, 1978a). Inhibitors of DNA synthesis (2'-deoxyadenosine) or nuclear division (mitomycin C) did not affect this pattern: O_2 uptake rates increased as two steps per cycle in the absence or presence of these inhibitors (Creanor, 1978b). Synchronous growth with yeast extract gave a linear increase in glycolytic CO_2 production with a rate change. Synchronization by the 2'-deoxyadenosine method gave a continuously increasing pattern for both O_2 uptake and CO_2 evolution, rather than the stepwise changes seen in the cell cycle of normal synchronous cultures.

Synchronous cultures of *C. utilis*, prepared by continuous flow selection, showed respiratory oscillations when the carbon source was either glucose, acetate, or glycerol (Kader and Lloyd, 1979). The periodicity of the oscillations (about one-third of a cell cycle) was unaltered by the nature of the carbon source, but the effects of an uncoupler, KCN, and DCCD (an ATP synthetase inhibitor) indicated that the metabolic mechanisms responsible for the oscillations are different for glucose from those produced with acetate or glycerol. Interactions between glycolytic reactions and oxidative metabolism (possibly in a Pasteur effect as seen in *Schiz. pombe*) lead to the oscillations with glucose, whereas those with the nonfermentable substrates reflect in vivo respiratory control (i.e., the control of respiration rates by ADP levels). Fluctuations in ADP levels may, in turn, be a consequence of the oscillatory protein accumulation observed in this organism (Lloyd, Edwards, and Williams, 1981), that is, be a direct consequence of periodic turnover, as in *Acanthamoeba castellanii* (Edwards and Lloyd, 1978; Lloyd and Edwards, 1980). Thus it is the energy demand of biosynthesis which dictates the rate of energy generation. Variations in glycolytic and hexosemonophosphate pathway-mediated glucose utilization as detected by radiorespirometry have been reported for phased cultures of *C. utilis* (Dawson and Westlake, 1975; Dawson, 1977; Dawson and Steinhauer, 1980); changes in adenylate charge in the range 0.78-0.94 were also noted (Thomas and Dawson, 1977).

B. Transport of Nutrients

Increased rates of glucose uptake in *S. cerevisiae* occurred at budding and at the midpoint of the cell cycle (Golombek and Wintersberger, 1974). Rates of adenine and serine uptake were constant initially, but doubled during the S phase (Kubitschek and Edvenson, 1977). In *Schiz. pombe*, transport capacities for glucose and glycine were constant for most of the cycle (implying a constant rate of mass increase: see Kubitschek and Clayman, 1976); rates of glycine uptake doubled at 0.9 cycle.

VI. GENETICS OF THE CELL DIVISION CYCLE

Cell cycle traverse necessitates a multiplicity of gene function at any one time combined with differential gene activation at different times to give a complex network of genetic expression in strict temporal sequences. Pioneering studies which have contributed to our understanding of the genetic controls involved have been carried out with *S. cerevisiae* (Hartwell et al., 1974; Hartwell, 1978). Most of this work has employed heat-sensitive mutants that show normal growth and division at a permissive temperature, but which arrest at a specific stage of the cell cycle at the restrictive temperature (usually 36°C). Morphological examination of more than 2000 temperature-sensitive clones obtained after mutagenesis at a haploid strain (A364A) with nitrosoguanidine or ethyl methane sulfate gave 148 cell division cycle (*cdc*) mutants which showed uniform phenotypes after incubation at 36°C. Some of these phenotypes were normal (e.g., unbudded cells), whereas some were abnormal (e.g., multinucleate cells), arising as a consequence of arrest and/or defect in an event of the normal cell cycle. Tetrad analysis indicated that all the mutations were nuclear. Complementation established 32 groups. Mapping of 14 genes gave no evidence for clustering. In most of the mutants, elevation of temperature produced arrest either in the ongoing cycle or in the following one. These "first-cycle arrest" alleles are probably temperature sensitive for function (Hartwell, 1974; Hartwell et al., 1973). On the other hand, in some mutants, arrest occurred only after several cycles ("multicycle arrest"). The conclusion here is that the gene product is present in excess of the immediate requirements for cell division.

A list of cell cycle mutations of *S. cerevisiae* has been presented by Simchen (1978). In *Schiz. pombe*, *cdc* mutants analogous to those in *S. cerevisiae* have been described (Bonatti et al., 1972; Nurse et al., 1976; Minet et al., 1979; Nasmyth, 1977, 1979).

The transition point (also known as the block point or execution point) is defined as the last point in the cycle at which a shift from a permissive to a restrictive temperature brings about arrest of the ongoing cycle (Hartwell et al., 1970). A transition point may occur long before the point of arrest or it may closely precede attainment of the terminal phenotype (Hartwell, 1971). The eventual goal of studies with cell cycle mutants is the identification of the primary gene products and the understanding of the functions of each of these products in the normal cell cycle. Examples where the defective gene products of *cdc* mutants have been identified include thymidylate synthetase in *cdc 21* (Bisson and Thorner, 1977) and DNA ligase involved in both replication and repair in *cdc 9* (Johnston and Nasmyth, 1978).

The isolation of cell cycle mutants has provided a powerful means of investigation of cell cycle progression and its control. Like a biochemical pathway, the cell cycle consists of a multistep process, each step being regulated by a specific gene product. The ordered sequence in time consists of both consecutive dependent events and parallel nondependent events, perhaps controlled by a cellular clock (Edmunds and Adams, 1981). Genetic analysis of the cell cycle reveals cycle control by those genes giving products which are required only once during a specific part of the cycle. Many cell cycle mutations which specifically block an early event also block other events which occur later. A model for the cell cycle of *S. cerevisiae* as two dependent sequences of events (Hartwell et al., 1974) and a similar set of dependent sequences for *Schiz. pombe* (Nurse et al., 1976) are proposed on these observations (Fig. 3). So the dependent sequences of the cell cycle are made up of a large number of events, each of which is dependent upon the completion of a previous event. The timing of completion of gene function corresponds to the timing of the transition point and is determined by measuring the proportion of cells that go on to complete division after shifting to the restrictive temperature.

In the case of *S. cerevisiae* six clearly defined landmarks were assigned to the sequence: initiation of DNA synthesis, DNA synthesis, medial nuclear division, late nuclear division, cytokinesis, and cell separation. Any mutant with a defect in one process failed to complete later events in this sequence. Thus mutants defective in the initiation of DNA synthesis (*cdc 4* and *cdc 7*) were unable to complete any subsequent event, but were able to undergo bud emergence and nuclear migration. Mutants defective in late nuclear division (*cdc 14* and *cdc 15*) could complete all events prior to this (and undergo bud emergence and nuclear migration), but could not complete cytokinesis or cell separation. A "reciprocal shift" procedure (Jarvick and Botstein, 1973; Hereford and Hartwell, 1974) enables further dissection of dependent sequences, particularly with respect to the relationship between the action of inhibitors and key cell cycle events.

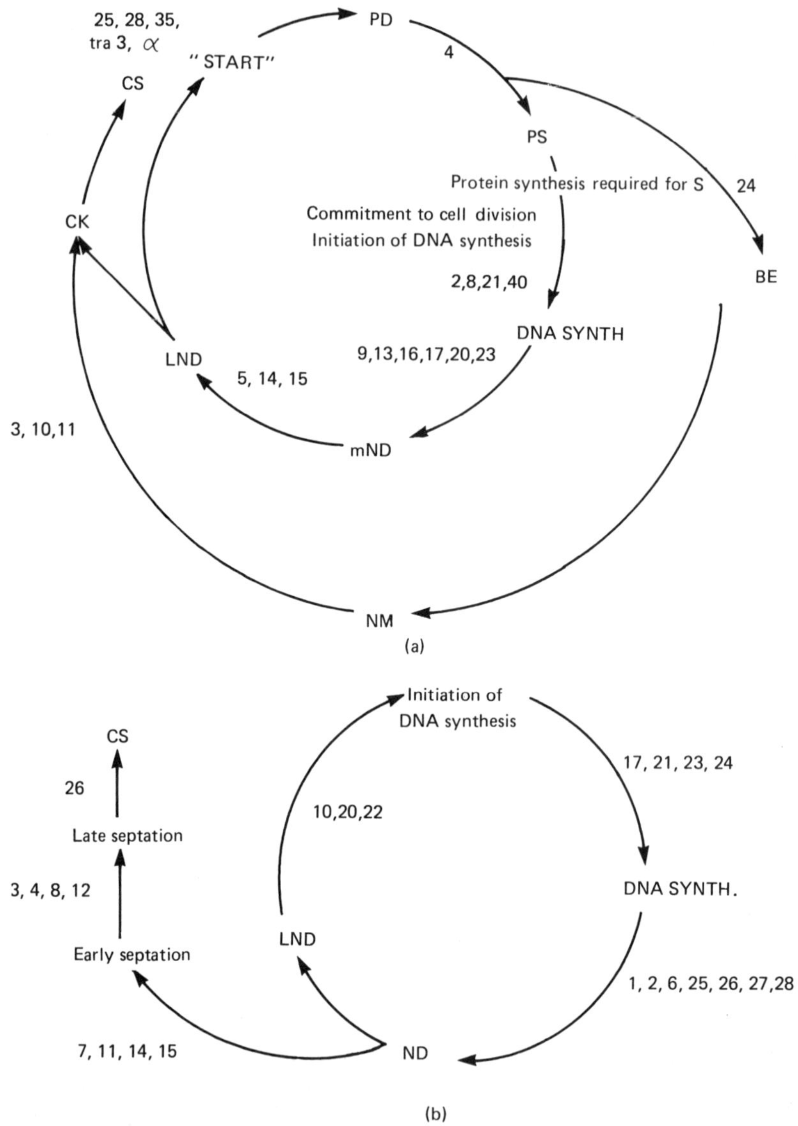

FIG. 3 Dependent sequences of cell cycle gene-controlled events in (a) *S. cerevisiae* and (b) *Schiz. pombe* (abbreviations as in Fig. 1; numbers refer to *cdc* mutations). (Redrawn from Hartwell, 1978, and Nurse, 1981, with permission.)

VII. CONTROL OF GROWTH AND CELL DIVISION

A. Introduction

Some important insights into the mechanisms by which cells control their rates of growth and timings of their divisions have come from cell cycle studies with yeasts. In general, models of the timing of cell division fall into three classes. In the transition probability model, variability of cell cycle times is attributed to a random step from an indeterminate portion of the cell cycle to a second stage in which a strictly programmed sequence of events occurs (Smith and Martin, 1973). The second class of models invokes timekeeping by a mitotic oscillator (mitogen or division protein) (Zeuthen, 1974; Klevecz et al., 1978; Edmunds and Adams, 1981). Finally, we have the dependent pathways model, in which one or more causal sequences finally join to give a closed loop of states. Evidence for this third model is provided by the *cdc* mutants of *S. cerevisiae* (Hartwell, 1974), and also by two parallel sequences of events constituting a "DNA division cycle," which runs alongside the "growth cycle" of *Schiz. pombe* (Mitchison, 1971; Mitchison and Creanor, 1971a; Sissons et al., 1973). Although it is possible that cell cycle controls may be different for mammalian and microbial cells, it seems likely that various combinations of all three models apply universally (Nurse, 1980).

B. Growth and Division Control in *S. cerevisiae*

A key regulatory cell cycle step has been identified in the G_1 phase of *S. cerevisiae*. Thus the product of gene *cdc 28* mediates an event called "start" which initiates the cell cycle (Hartwell, 1974, 1978), provided that carbon and nitrogen sources, sulfur, phosphorus, and required amino acids are present (Hartwell, 1974; Johnston et al., 1977; Shilo et al., 1978) and the mating pheromone is absent (Bücking-Throm et al., 1973). It is at this critical stage that cells usually arrest when growth media become exhausted; cells can remain viable in this stage for long periods of time in stationary phase cultures as single, unbudded G_1 cells containing a single spindle plaque (Byers and Goetsch, 1975). Defects in cytokinesis, cell separation, or bud emergence do not prevent a cell from embarking on a new sequence initiated by "start" (Hartwell et al., 1974). Attempts to identify the *cdc 28* gene product by molecular cloning of the gene have recently been reported (Nasmyth and Reed, 1980). On the basis of the sedimentation behavior of chromosomes from *cdc 28* mutants arrested in G_1, it has been suggested that this gene product is essential for the functional integrity of folded chromosomes (Piñon, 1979). Two other mutants with defective controls in the "start" events have been isolated (Sudbery et al., 1980). The *whi-1* mutation results in bud initiation when the parent cell is only half the size at which bud initiation takes place in the wild type; *whi-2* has lost the control, leading to G_1 arrest in stationary phase.

Starvation for any one of many nutrients results in arrest in G_1 (Williamson and Scopes, 1962; Unger and Hartwell, 1976). Methionine auxotrophs have been used to show that a signal for SO_4^{2-} starvation involves an intermediate of the sulfate assimilation pathway at or beyond the formation of methionyl tRNA (Unger, 1977). The other requirement for uniform arrest of a population in G_1 is that sufficient capacity for protein synthesis is available so that cells can finish cycles that are in progress. Thus G_1 arrest of nitrogen-starved cells can be prevented by addition of arginine (Cooper, 1979); partial inhibition of protein synthesis by cycloheximide produced a similar result (Shilo et al., 1978). A requirement at "start" for a labile protein with a half-life of about 6 min has also been discovered (Shilo et al., 1979); such a protein may allow rapid modification of the proliferative response to environmental changes. Transcriptional blocks at "start" can be produced by nalidixic acid (Singer and Johnston, 1979), o-phenanthroline (Johnston and Singer, 1978), L-ethionine (Singer et al., 1978), and β-2-DL thienylalanine (Bedard et al., 1980). All these agents decrease the formation of the 35-S ribosomal RNA precursor.

The primary source of variability of division times among individual cells is at "start" (Shilo et al., 1976, 1977). It has been repeatedly stressed that it is the unbudded portion of the cycle which is selectively lengthened at slow growth rates (Williamson, 1965; von Meyenburg, 1968; Slater et al., 1977); the use of metabolic inhibitors leads to the same conclusion (Hartwell and Unger, 1977). Populations growing at different rates on different carbon or nitrogen sources also show G_1 to be the most variable phase in terms of the proportion of the total cell cycle time (Johnston et al., 1980). Furthermore, the expandable time is confined to the phase before "start" in cells growing with generation times less than 6 h (Jagadish and Carter, 1977). In these cells the time from "start" to division did not vary and took about 2 h; in cells growing more slowly the "start" to division period was also lengthened. Alteration of the dilution rate in chemostat cultures indicated that expansion of the cell cycle time is primarily a consequence of increased time between division and the completion of S (Jagadish and Carter, 1978). Thus variability in cell cycle time is attributable to variability in G_1, as the duration of S is constant (Barford and Hall, 1976). Decreased growth rate at a lower temperature extended the cycle phases before and after completion of S. Different observations have been reported by Rivin and Fangman (1980a,b), who have shown that in nitrogen-limited cells, the duration of S remains at about 50% of the total cycle time as a consequence of decreased replication rate.

As in the case with bacteria, faster-growing cells are bigger than slow-growing cells (Mor and Fiechter, 1968; von Meyenburg, 1968; McMurrough and Rose, 1969). The duration of the budded phase B

(= S + G_2 + M), constant over a wide range of growth conditions in many investigations has been given a value $B = 0.17\tau + 87.4$ min (Hartwell and Unger, 1977) which gives similar results to the equation $B = 0.5\tau + 27$ min (Tyson et al., 1979). These data, together with those from nutritional shift-up experiments (Carter et al., 1978), are all consistent with the proposal that initiation of the cell cycle at "start" is followed by a constant time period before attainment of a critical size. This critical size (mandatory for the initiation of budding; see Johnston et al. 1977) is dependent on growth rate (Carter and Jagadish, 1978a,b; Lorincz and Carter, 1979; Johnston et al., 1979a).

This model appears to hold also for *C. utilis* growing in the phased chemostat system; faster growth of daughters, so that mothers and daughters reach division size simultaneously, indicated roles for "sister" and "timer" controls (Thomas et al., 1980). Control of *S. cerevisiae* cell cycle progression at "start" by the α factor pheromone is dealt with in Chapter 6; this control effectively prevents commitment to cell cycle traverse.

C. Growth and Division Control in *Schiz. pombe*

In the fission yeast *Schiz. pombe*, cells of size larger than the mean at birth have, on the average, shorter cell cycle times (James et al., 1975). Fantes (1977) has further demonstrated that deviations in length at division must be homeostatically compensated for, mainly within one division: That was so both for the wild type and the abnormally long mutant *cdc 2-33*. The mechanism by which this is achieved is probably as follows. A cell that undergoes an abnormally long cycle will divide at a larger cell size than the average and will give rise to two large daughters. Each of these will have shorter than average cell cycle times, so that a long interdivision time in one cycle leads to the next cycle being shorter than average. Restoration of normal cell length after the formation of elongated cells in the presence of inhibitors has also been reported (Mitchison and Creanor, 1971b; Quinton and Poole, 1977; Miyata et al., 1978a,b; Poole and Salmon, 1978).

Mutant *wee 1-50* divides at about the same size as the wild type at 25°C, but at half that size at 35°C (Nurse, 1975). Premature advance into cell division on shift to the high temperature indicates that the *wee* gene product plays some role in a major rate-controlling step for completion of the cell cycle (Thuriaux et al., 1978). Nutritional shift-up experiments with wild-type cells also indicated a control acting at the stage of nuclear division (Fantes and Nurse, 1977; Fantes, 1979). Two genes, *wee 1* and *cdc 2*, are possibly involved in the control that monitors cell mass and determines mitotic timing (Nurse, 1981). A second "sizer" control (Fig. 4) is proposed to exert its action on the initiation of DNA synthesis both in *wee* mutants (Nurse, 1975; Nurse and Thuriaux, 1977; Fantes and Nurse, 1978) and in the wild type

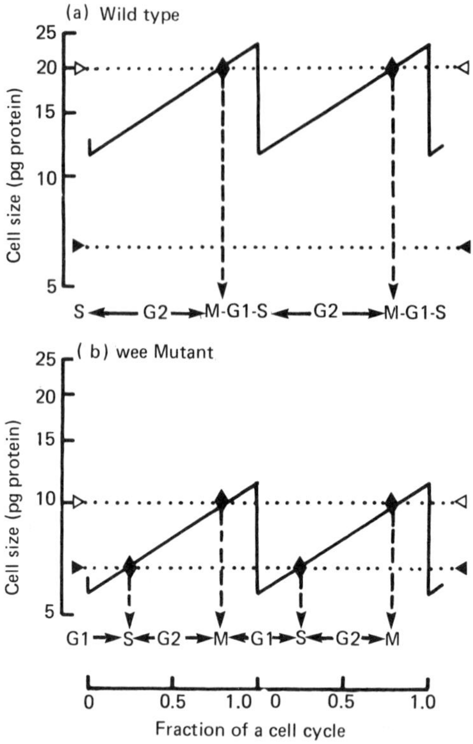

FIG. 4 Model of cell size controls acting over DNA replication and nuclear division in *Schiz. pombe*: (a) Wild type and (b) *wee* mutant. The graphs show the increase in cell mass during two consecutive cell cycles. Nuclear division is initiated when the cell attains the mass shown by the open triangles. This is reduced in the *wee* mutant compared with the wild type. In the wild type, the S phase follows rapidly after mitosis, whereas in the *wee* mutant the cell has to grow for a quarter of a cell cycle to attain the minimum mass required for DNA replication shown by the closed triangles. (Redrawn from Nurse and Thuriaux, 1977, with permission.)

(Nurse and Thuriaux, 1977). The period between S phase and cell division is constant in slow-growing cells of *Schiz. pombe* (Nasmyth, 1979) as is the case for *S. cerevisiae* (Hartwell and Unger, 1977; Jagadish and Carter, 1977). This stage may contain a cell cycle event which takes a long time for completion, and may thus be a major rate-controlling step unaffected by cell mass. Critical cell mass is required for DNA replication (Nasmyth, 1979), as is the product of the *cdc 10* gene (Nurse et al., 1976). It has been suggested that *cdc 10* in *Schiz. pombe* defines a "start" gene by analogy to *cdc 28* in *S. cerevisiae*.

Accumulation of cells at the beginning of the cycle when deprived of nutrients occurs by a different mechanism for *Schiz. pombe* from that previously described for *S. cerevisiae*. Under poor nutritional conditions the reduction in cell mass of the fission yeast at mitosis eventually leads to the production of cells which are below the critical mass for DNA replication (Fantes and Nurse, 1977; Nurse and Thuriaux, 1977). A mechanism which ensures that only one septum is produced per cell cycle has been proposed by Minet et al. (1979).

Magnesium ions are essential for cell division (Duffus and Patterson, 1974; Penman and Duffus, 1975; Ahluwalia et al., 1978; Walker and Duffus, 1979, 1980); cells exposed to a variety of chelating agents or incubated in Mg^{2+}-deficient media continue to elongate, but nuclear division and cell plate formation are prevented. Thus a falling intracellular Mg^{2+} concentration may act as a transducer of cell size, eventually triggering tubulin polymerization and spindle formation and a membrane change which permits influx of Mg^{2+} just before cell division. Division delays in *Schiz. pombe* produced by heat shocks and pulses of cycloheximide (Polanshek, 1977), colcemid (Stetten and Lederberg, 1973), and anaesthetics (Smith and Mitchison, 1976) provide information on the importance of microtubules in mitotic and cell division-related events.

REFERENCES

Ahluwalia, B., Duffus, J. H., Paterson, L. J., and Walker, G. M. (1978). Synchronization of cell division in the fission yeast *Schizosaccharomyces pombe* by ethylenediaminetetra-acetic acid. *J. Gen. Microbiol. 106*:261-264.

Barford, J. P., and Hall, R. J. (1976). Estimation of the length of cell cycle phases from asynchronous cultures of *Saccharomyces cerevisiae*. *Exp. Cell Res. 102*:276-284.

Barnes, A., Nurse, P., and Fraser, R. S. S. (1979). Analysis of the significance of a periodic cell size-controlled doubling in rates of macromolecular synthesis for the control of balanced exponential growth of fission yeast cells. *J. Cell Sci. 35*:41-51.

Bedard, D. P., Singer, R. A., and Johnston, G. C. (1980). Transient cell cycle arrest of *Saccharomyces cerevisiae* by amino acid analog β-2-DL-thienyl alanine. *J. Bacteriol. 141*:100-105.

Biely, P. (1978). Changes in the rate of wall polysaccharides during the cell cycle of yeast. *Arch. Microbiol. 119*:213-214.

Bisson, L., and Thorner, J. (1977). Thymidine 5'-monophosphate-requiring mutants of *Saccharomyces cerevisiae* are deficient in thymidylate synthetase. *J. Bacteriol. 132*:44-50.

Bonatti, S., Similt, M., and Abbondandolo, A. (1972). Isolation of temperature-sensitive mutants of *Schizosaccharomyces pombe*. *J. Bacteriol. 109*:484-491.

Bostock, C. J. (1970). The effect of 2-phenyl ethanol on the DNA synthesis cycle of *Schizosaccharomyces pombe*. *J. Cell Sci. 7*:523-530.

Bücking-Throm, E., Duntze, W., Hartwell, L. H., and Manney, T. R. (1973). Reversible arrest of haploid yeast cells at the initiation of DNA synthesis by a diffusible sex factor. *Exp. Cell Res. 76*:99-110.

Bullock, J. G., and Coakley, W. T. (1979). The cell cycle thermal-inactivation sensitive stage of *Schizosaccharomyces pombe* is independent of 2-phenylethanol-induced changes in S-phase location. *Exp. Cell Res. 121*:441-445.

Byers, B., and Goetsch, L. (1975). Behaviour of spindles and spindle plaques in the cell cycle and conjugation of *Saccharomyces cerevisiae*. *J. Bacteriol. 124*:511-523.

Cabib, E., and Bowers, B. (1975). Timing and function of chitin synthesis in yeast. *J. Bacteriol. 124*:1586-1593.

Cabib, E., and Farkaš, V. (1971). The control of morphogenesis: An enzymatic mechanism for the initiation of septum formation in yeast. *Proc. Nat. Acad. Sci. USA. 68*:2052-2056.

Calleja, G. B., Zuker, M., Johnson, B. F., and Yoo, B. Y. (1980). Analysis of fission scars as permanent records of cell division in *Schizosaccharomyces pombe*. *J. Theor. Biol. 84*:523-545.

Carter, B. L. A., and Halvorson, H. O. (1973). An evaluation of the oscillatory repression model of periodic enzyme synthesis in yeast. *Exp. Cell Res. 76*:152-158.

Carter, B. L. A., and Jagadish, M. N. (1978a). The relationship between cell size and cell division in the yeast *Saccharomyces cerevisiae*. *Exp. Cell Res. 112*:15-24.

Carter, B. L. A., and Jagadish, M. N. (1978b). Control of cell division in the yeast *Saccharomyces cerevisiae* cultured at different growth rates. *Exp. Cell Res. 112*:373-383.

Carter, B. L. A., Lorincz, A., and Johnston, G. C. (1978). Protein synthesis, cell division and the cell cycle in *Saccharomyces cerevisiae* following a shift to a richer medium. *J. Gen. Microbiol. 106*:221-225.

Chaffin, W. L., and Sogin, S. J. (1976). Germ tube formation from zonal rotor fractions of *Candida albicans*. *J. Bacteriol. 126*:771-776.

Chagla, A. H., and Griffiths, A. J. (1978). Synchronous cultures of *Acanthamoeba castellanii* and their use in the study of encystation. *J. Gen. Microbiol. 108*:39-43.

Chan, R. K. (1977). Recovery of *Saccharomyces cerevisiae* mating-type a cells from G_1 arrest by α-factor. *J. Bacteriol. 130*:766-774.

Cooper, S. (1979). A unifying model for the G_1 period in prokaryotes and eukaryotes. *Nature 280*:17-19.

Cottrell, S. F. (1979). Efficiency of ethidium-bromide mutagenesis as a function of different stages in the cell cycle of *Saccharomyces cerevisiae. Exp. Cell Res. 118*:398-401.

Cottrell, S. F. (1981). Mitochondrial DNA synthesis in synchronous cultures of the yeast *Saccharomyces cerevisiae. Exp. Cell Res. 132*:89-98.

Cottrell, S., Rabinowitz, M., and Getz, G. S. (1973). Mitochondrial deoxyribonucleic acid synthesis in a temperature-sensitive mutant of deoxyribonucleic acid replication of *Saccharomyces cerevisiae. Biochemistry 12*:4374-4378.

Cottrell, S. F., Rabinowitz, M., and Getz, G. S. (1975). Cytochrome synthesis in synchronous cultures of *Saccharomyces cerevisiae. J. Biol. Chem. 250*:4087-4094.

Creanor, J. (1978a). Carbon dioxide evolution during cell-cycle of fission yeast *Schizosaccharomyces pombe. J. Cell Sci. 33*:385-397.

Creanor, J. (1978b). Oxygen-uptake during cell-cycle of fission yeast *Schizosaccharomyces pombe. J. Cell Sci. 33*:399-411.

Creanor, J., and Mitchison, J. M. (1979). Reductions of perturbations in leucine incorporation in synchronous cultures of *Schizosaccharomyces pombe* made by elutriation. *J. Gen. Microbiol. 112*: 385-388.

Cryer, D. R., Goldthwaite, C. D., Zinker, S., Lam, K. B., Storm, E., Hirschberg, R., Blamire, J., Finkelstein, D. B., and Marmur, J. (1973). Studies on nuclear and mitochondrial DNA of *S. cerevisiae. Cold Spring Harbor Symp. Quant. Biol. 38*:17-29.

Davison, M. T., and Garland, P. B. (1977). Structure of mitochondria and vacuoles of *Candida utilis* and *Schizosaccharomyces pombe* studies by electron microscopy of serial thin sections and model building. *J. Gen. Microbiol. 98*:147-153.

Dawes, I. W., and Carter, B. L. A. (1974). Nitrosoguanidine mutagenesis during nuclear and mitochondrial gene replication. *Nature 250*:709-712.

Dawson, P. S. S. (1965). Continuous phased growth with a modified chemostat. *Can. J. Microbiol. 11*:893-903.

Dawson, P. S. S. (1977). Radio-respirometry of *Candida utilis* during cell cycle. *FEMS Microbiol. Lett. 1*:21-24.

Dawson, P. S. S., and Steinhauer, L. P. (1980). Dynamic aspects of radiorespirometry during the cell cycle of *Candida utilis*: Some observations in phased culture under carbon-nitrogen-and-phosphorous-limited growth. *Biotechnol. Bioeng. 22*:137-156.

Dawson, P. S. S., and Westlake, D. W. S. (1975). Changes in patterns of respiration and glucose utilization during the cell cycle: Some variations with growth rate. *Can. J. Microbiol. 21*:210-219.

Delrey, F., Santos, T., Garcia-Acha, I., and Nombeta, C. (1979). Synthesis of 1,3-β-glucanases in *Saccharomyces cerevisiae* during the mitotic cycle, mating and sporulation. *J. Bacteriol. 139*:924-931.

Dharmalingam, K., and Jayaraman, J. (1973). Mitochondriogenesis in synchronous cultures of yeast. 1. Oscillatory pattern of respiration. *Arch. Biochem. Biophys. 157*:197-202.

Duffus, J. H., and Paterson, L. J. (1974). Control of cell division in yeast using the ionophore A23187 with calcium and magnesium. *Nature 251*:626-627.

Dwek, R. D., Kobrin, L. H., Grossman, N., and Ron, E. Z. (1980). Synchronization of cell division in microorganisms by Percoll gradients. *J. Bacteriol. 144*:17-21.

Edmunds, L. N., and Adams, K. J. (1981). Clocked cell cycle clocks. *Science 211*:1002-1013.

Edwards, C. (1981). *The Microbial Cell Cycle.* Nelson, Surrey, p. 81.

Edwards, C., and Jones, C. W. (1977). Respiratory properties of synchronous cultures of *Alcaligenes eutrophus* H16 prepared by a continuous-flow size selection method. *J. Gen. Microbiol. 99*:383-388.

Edwards, S. W., and Lloyd, D. (1977). Mitochondrial adenosine triphosphatase of the fission yeast *Schizosaccharomyces pombe* 972 h^-. Changes in activity and oligomycin-sensitivity during the cell cycle of catabolite-repressed and -derepressed cells. *Biochem. J. 162*:39-46.

Edwards, S. W., and Lloyd, D. (1978). Oscillations of respiration and adenine nucleotides in synchronous cultures of *Acanthamoeba castellanii*: Mitochondrial respiratory control *in vivo*. *J. Gen. Microbiol. 108*:197-204.

Edwards, S. W., Evans, J. B., and Lloyd, D. (1981). Oscillatory accumulation of catalase during the cell cycle of *Acanthamoeba castellanii*. *J. Gen. Microbiol. 125*:459-462.

El'Khayat, G. H. (1980). Changes in energy conservation systems during the growth and starvation of *Schizosaccharomyces pombe*. Ph.D. thesis, University of Wales, Cardiff.

Elliott, S. G., and McLaughlin, C. S. (1978). Rate of macromolecular synthesis through cell cycle of yeast *Saccharomyces cerevisiae*. *Proc. Nat. Acad. Sci. USA 75*:4384-4388.

Elliott, S. G., and McLaughlin, C. S. (1979). Synthesis and modification of proteins during the cell cycle of the yeast *Saccharomyces cerevisiae*. *J. Bacteriol. 137*:1185-1190.

Elliott, S. G., Warner, J. R., and McLaughlin, C. S. (1979). Synthesis of ribosomal proteins during the cell cycle of the yeast *Saccharomyces cerevisiae*. *J. Bacteriol. 137*:1048-1050.

Fantes, P. (1977). Control of cell size and cycle time in *Schizosaccharomyces pombe*. *J. Cell Sci.* *24*:51-67.

Fantes, P. (1979). Epistatic gene interactions in the control of division in fission yeast. *Nature 279*:428-430.

Fantes, P., and Nurse, P. (1977). Control of cell size at division in fission yeast by a growth modulated size control over nuclear division. *Exp. Cell Res. 107*:377-386.

Fantes, P., and Nurse, P. (1978). Control of the timing of cell division in fission yeast. *Exp. Cell Res. 115*:317-329.

Fraser, R. S. S., and Carter, B. L. A. (1976). Synthesis of polyadenylated messenger RNA during the cell cycle of *S. cerevisiae*. *J. Mol. Biol. 164*:223-242.

Fraser, R. S. S., and Moreno, F. (1976). Rates of synthesis of polyadenylated messenger RNA during the cell cycle of *Schizosaccharomyces pombe*. *J. Cell Sci. 21*:497-521.

Fraser, R. S. S., and Nurse, P. (1978). Novel cell cycle control of RNA synthesis in yeast. *Nature 271*:726-730.

Fraser, R. S. S., and Nurse, P. (1979). Altered patterns of ribonucleic acid synthesis during the cell cycle: A mechanism compensating for variation in gene concentration. *J. Cell Sci. 35*:25-40.

Gallili, G., and Lampen, J. O. (1977). Large and small invertase and the yeast cell cycle. *Biochim. Biophys. Acta 475*:113-122.

Golombek, J., and Wintersberger, E. (1974). Glucose uptake in the cell cycle of *Saccharomyces cerevisiae*. *Exp. Cell Res. 86*:199-202.

Gordon, C. N. (1977). Chromatin behavior during the mitotic cell cycle of *Saccharomyces cerevisiae*. *J. Cell Sci. 24*:81-93.

Gordon, C. N., and Elliott, S. G. (1977). Fractionation of *Saccharomyces cerevisiae* cell populations by centrifugal elutriation. *J. Bacteriol. 129*:97-100.

Gorman, J., Tauro, P., LaBerg, M., and Halvorson, H. O. (1964). Timing of enzyme synthesis during synchronous division in yeast. *Biochem. Biophys. Res. Commun. 15*:43-49.

Halvorson, H. O. (1977). The cell division cycle. A review of current models on temporal expression in *Saccharomyces cerevisiae*. In *Cell Differentiation in Microorganisms, Plants and Animals*. L. Nover and K. Mottres. (Eds.). Gustav Fischer Verlag, Jena, pp. 361-376.

Halvorson, H. O., Carter, B. L. A., and Tauro, P. (1971). Synthesis of enzymes during the cell cycle. *Adv. Microbiol. Physiol. 6*: 47-106.

Hamberger, K., Kramhøft, B., Nissen, S. B., and Zeuthen, E. (1977). Linear increase in glycolytic activity through the cell cycle of *Schizosaccharomyces pombe*. *J. Cell Sci. 24*:69-79.

Hartwell, L. H. (1970). Periodic density fluctuation during the yeast cell cycle and the selection of synchronous cultures. *J. Bacteriol. 104*:1280-1285.

Hartwell, L. H. (1971). Genetic control of the cell division cycle in yeast. 11. Genes controlling DNA replication and its initiation. *J. Mol. Biol. 59*:183-194.

Hartwell, L. H. (1973). The additional genes required for DNA synthesis in *S. cerevisiae*. *J. Bacteriol. 115*:966-974.

Hartwell, L. H. (1974). *Saccharomyces cerevisiae* cell cycle. *Bacteriol. Rev. 38*:164-198.

Hartwell, L. H. (1978). Cell division from a genetic perspective. *J. Cell Biol. 77*:627-637.

Hartwell, L. H., Culotti, J., and Reid, B. J. (1970). Genetic control of the division cycle in yeast. 1. Detection of mutants. *Proc. Nat. Acad. Sci. USA 66*:352-359.

Hartwell, L. H., Mortimer, R. K., Culotti, J., and Culotti, M. (1973). Genetic control of the cell division cycle in yeast. V. Genetic analysis of *cdc* mutants. *Genetics 74*:267-286.

Hartwell, L. H., Culotti, J., Pringle, J. R., and Reid, B. J. (1974). Genetic control of the cell division cycle in yeast: A model. *Science 183*:46-51.

Hartwell, L. H., and Unger, M. W. (1977). Unequal division in *Saccharomyces cerevisiae* and its implications for the control of cell division. *J. Cell Biol. 75*:422-435.

Hayashibe, M., Abe, N., and Matsui, M. (1977). Model of increase in cell wall polysaccharides in synchronously growing *Saccharomyces cerevisiae*. *Arch. Microbiol. 114*:91-92.

Hereford, L. M., and Hartwell, L. H. (1973). Role of protein synthesis in the replication of yeast DNA. *Nature London New Biol. 244*:129-131.

Hereford, L. M., and Hartwell, L. H. (1974). Sequential gene function in the initiation of *Saccharomyces cerevisiae* DNA synthesis. *Mol. Biol. 84*:445-461.

Hoffman, H. P., and Avers, C. J. (1973). Mitochondrion of yeast; ultrastructural evidence for one giant branched organelle per cell. *Science 181*:749-751.

Hynes, N. E., and Phillips, S. L. (1976). Rate of synthesis of polyadenylate containing ribonucleic acid during yeast cell cycle. *J. Bacteriol. 128*:502-505.

Jagadish, M. N., and Carter, B. L. A. (1977). Genetic control of cell division in yeast cultured at different growth rates. *Nature 269*:145-147.

Jagadish, M. N., and Carter, B. L. A. (1978). Effects of temperature and nutritional conditions on the mitotic cell cycle of *Saccharomyces cerevisiae*. *J. Cell Sci. 31*:71-78.

James, T. W., Hemond, P., Czer, G., and Bohman, R. (1975). Parametric analysis of volume distributions of *Schizosaccharomyces pombe* and other cells. *Exp. Cell Res. 94*:267-276.

Jarvick, J., and Botstein, D. (1973). A genetic method for determining the order of events in a biological pathway. *Proc. Nat. Acad. Sci. USA 70*:2046-2050.

Jazwinski, S. M., and Edelman, S. M. (1979). Replication *in vitro* of the 2 micrometer DNA plasmid of yeast. *Proc. Nat. Acad. Sci. USA 76*:1223-1227.

Johnson, B. F. (1965). Autoradiographic analysis of regional cell wall growth of yeast: *Schizosaccharomyces pombe*. *Exp. Cell Res. 39*:613-624.

Johnson, B. F., and Gibson, E. J. (1966). Autoradiographic analysis of regular cell wall growth of yeasts. III. *Saccharomyces cerevisiae*. *Exp. Cell Res. 41*:580-591.

Johnson, B. F., Calleja, G. B., Boisclair, I., and Yoo, B. Y. (1979). Cell division in yeasts. II. The biased asymmetric location of the septum in the fission yeast, *Schizosaccharomyces pombe*. *Exp. Cell Res. 123*:253-259.

Johnston, G. C., and Singer, R. A. (1978). RNA synthesis and the control of cell division in the yeast *S. cerevisiae*. *Cell 14*:951-959.

Johnston, G. C., Ehrhardt, C. W., Lorincz, A., and Carter, B. L. A. (1979a). Regulation of cell size in the yeast *Saccharomyces cerevisiae*. *J. Bacteriol. 137*:1-5.

Johnston, G. C., Pringle, J. R., and Hartwell, L. H. (1977). Coordination of growth with cell division in the yeast *Saccharomyces cerevisiae*. *Exp. Cell Res. 105*:79-98.

Johnston, G. C., Singer, R. A., Sharrow, S. O., and Slater, M. L. (1980). Cell division in the yeast *S. cerevisiae* growing at different rates. *J. Gen. Microbiol. 118*:479-484.

Johnston, L. H., and Nasmyth, K. A. (1978). *Saccharomyces cerevisiae* cell cycle mutant *cdc* 9 is defective in DNA ligase. *Nature 274*:891-893.

Jordan, E. G., Severs, N. J., and Williamson, D. H. (1977). Nuclear pore formation and the cell cycle in *Saccharomyces cerevisiae*. *Exp. Cell Res. 104*:446-449.

Kader, J., and Lloyd, D. (1979). Respiratory oscillations and heat evolution in synchronous cultures of *Candida utilis* NCYC 193. *J. Gen. Microbiol. 114*:455-461.

Kay, D. G., Singer, R. A., and Johnston, G. C. (1980). Ornithine decarboxylase activity and cell cycle regulation in *Saccharomyces cerevisiae*. *J. Bacteriol. 141*:1041-1046.

Klevecz, R. R., Kros, J., and Gross, S. D. (1978). Phase response versus positive and negative division delay in animal cells. *Exp. Cell Res. 116*:285-290.

Kopecká, M. (1977). Biogenesis of the wall fibrillar component in the cell cycle of *Saccharomyces cerevisiae*. *Folia Microbiol. Prague 22*:440-441.

Kramhøft, B., and Zeuthen, E. (1975). Synchronization of the fission yeast *Schizosaccharomyces pombe* using heat shocks. In *Methods in Cell Biology*, Vol. 12. D. Prescott (Ed.). Academic, New York, pp. 373-380.

Kramhøft, B., Hamberger, K., Nissen, S. B., and Zeuthen, E. (1978). The cell cycle and glycolytic activity of *Schizosaccharomyces pombe* synchronized in defined medium. *Carlsberg Res. Commun. 43*:227-239.

Kruszewska, A., Szczesniak, B., and Gajewski, W. (1976). Effect of auxotrophic starvation on mitochondrial marker transmission in the *cdc* mutant of *S. cerevisiae. Mol. Gen. Genet. 148*:65-77.

Kubitschek, H. E., and Clayman, R. V. (1976). Transport of glucose and glycerine in *Schizosaccharomyces pombe* during the cell cycle. *J. Bacteriol. 127*:109-113.

Kubitschek, H. E., and Edvenson, R. W. (1977). Midcycle doubling of uptake rates of adenine and serine in *Saccharomyces cerevisiae. Biophys. J. 20*:15-21.

Livingston, D. M., and Kupfer, D. M. (1977). Control of *S. cerevisiae* 2 μm DNA replication by cell division cycle genes that control DNA replication. *J. Mol. Biol. 116*:249-260.

Lloyd, D. (1974). *The Mitochondria of Micro-organisms*. Academic, London.

Lloyd, D., and Ball, J. (1979). Perturbation of respiration in *Candida utilis*: Induction of metabolic oscillations. *J. Gen. Microbiol. 114*:463-473.

Lloyd, D., and Edwards, S. W. (1977). Mitochondrial adenosine triphosphatase of the fission yeast *Schizosaccharomyces pombe* 972 h^-. Changes in inhibitor sensitivities during the cell cycle indicate similarities and differences in binding sites. *Biochem. J. 162*: 581-590.

Lloyd, D., John, L., Edwards, C., and Ghagla, A. H. (1975). Synchronous cultures of micro-organisms: Large-scale preparation by continuous-flow size selection. *J. Gen. Microbiol. 88*:153-158.

Lloyd, D., and Edwards, S. W. (1980). Oscillatory expression of mitochondrial activities and components during the cell cycle of *Acanthamoeba castellanii* is coupled to a clock. *Soc. Gen. Microbiol. Q. 8*:25.

Lloyd, D., John, L., Hamill, M., Phillips, C., Kader, J., and Edwards, S. W. (1977). Continuous-flow cell cycle fractionation of eukaryotic micro-organisms. *J. Gen. Microbiol. 99*:223-227.

Lloyd, D., Edwards, S. W., and Williams, J. L. (1981). Oscillatory accumulation of total cellular protein in synchronous cultures of *Candida utilis. FEMS Microbiol. Lett., 12*:295-298.

Lloyd, D., Poole, R. K., and Edwards, S. W. (1982). *The Cell Division Cycle: Temporal Organization and Control of Cellular Growth and Reproduction*. Academic, London.

Lorincz, A., and Carter, B. L. A. (1979). Control of cell size at bud initiation in *Saccharomyces cerevisiae*. *J. Gen. Microbiol. 113*: 287-295.

Lowdon, M., and Vitols, V. (1973). Ribonucleotide reductase activity during the cell cycle of *S. cerevisiae*. *Arch. Biochem. Biophys. 158*:177-184.

McFarlane, E. S. (1980). Ribonuclease activity during G_1 arrest of the yeast *S. cerevisiae*. *Arch. Microbiol. 124*:243-247.

Maclean, N. (1965). Ribosome numbers in a fission yeast. *Nature 207*:322-323.

McMurrough, I., and Rose, A. H. (1969). Effect of growth rate and substrate limitation on the compositions and structure of the cell wall of *Saccharomyces cerevisiae*. *Biochem. J. 105*:189-203.

Matile, P. H., Moor, H., and Robinow, C. F. (1969). Yeast cytology. In *The Yeasts*, Vol. 1. A. H. Rose and J. S. Harrison (Eds.). Academic, London, pp. 219-302.

Matur, A., and Berry, D. R. (1978). Use of step enzymes as markers during meiosis and ascospore formation in *Saccharomyces cerevisiae*. *J. Gen. Microbiol. 109*:205-213.

May, J. W. (1962). Sites of cell wall extension demonstrated by the use of fluorescent antibody. *Exp. Cell Res. 27*:170-172.

Minet, M., Nurse, P., Thuriaux, P., and Mitchison, J. M. (1979). Uncontrolled septation in a cell division cycle mutant of the fission yeast *Schizosaccharomyces pombe*. *J. Bacteriol. 137*:440-446.

Mitchison, J. M. (1971). *The Biology of the Cell Cycle*. Cambridge University, Cambridge.

Mitchison, J. M. (1977). Enzyme synthesis during the cell cycle in *Cell Differentiation in Micro-organisms, Plants and Animals*, L. Nover and K. Mothes (Eds.). Gustav Fischer Verlag, Jena, pp. 377-401.

Mitchison, J. M., and Carter, B. L. A. (1975). Cell cycle analysis. In *Methods in Cell Biology*, Vol. 11. D. M. Prescott (Ed.). Academic, New York, pp. 201-219.

Mitchison, J. M., and Creanor, J. (1971a). Further measurements of DNA synthesis and enzyme potential during cell cycle of fission yeast *Schiz. pombe*. *Exp. Cell Res. 69*:244-247.

Mitchison, J. M., and Creanor, J. (1971b). Induction synchrony in the fission yeast *Schizosaccharomyces pombe*. *Exp. Cell Res. 67*: 368-374.

Mitchison, J. M., and Vincent, W. S. (1965). Preparation of synchronous cultures by sedimentation. *Nature 205*:987-989.

Mitchison, J. M., and Wilbur, K. M. (1962). The incorporation of protein and carbohydrate precursors during the cell cycle of a fission yeast. *Exp. Cell Res. 26*:144-157.

Miyata, M., and Miyata, H. (1978). Relationship between extracellular enzymes and cell growth during cell cycle of fission yeast. *Schizosaccharomyces pombe. J. Bacteriol. 136*:558-564.

Miyata, H., Miyata, M., and Ito, M. (1978a). The cell cycle in the fission yeast *Schizosaccharomyces pombe. 1.* Relationship between cell size and cycle time. *Cell Struct. Function 3*:39-46.

Miyata, H., Miyata, M., and Ito, M. (1978b). The cell cycle in the fission yeast *Schizosaccharomyces pombe.* II. Recovery mode of the period of a cell extended in pulse treatment with hydroxyurea. *Cell Struct. Function 3*:153-159.

Miyata, M., Imura, K., and Miyata, H. (1980). Relationship between extracellular enzymes and cell growth during the cell cycle of the fission yeast *Schizosaccharomyces pombe. J. Gen. Appl. Microbiol. 26*:109-118.

Moll, R., and Wintersberger, E. (1976). Synthesis of yeast histone in the cell cycle. *Proc. Nat. Acad. Sci. USA 73*:1863.

Mor, J. R. and Fiechter, A. (1968). Continuous cultivation of *Saccharomyces cerevisiae. Biotech. Bioeng. 10*: 159-176.

Nasmyth, K. A. (1977). Temperature-sensitive lethal mutants in the structural gene for DNA ligase in the yeast *Schizosaccharomyces pombe. Cell 12*:1109-1120.

Nasmyth, K. A. (1979). A control acting over the initiation of DNA replication in the yeast *Schizosaccharomyces pombe. J. Cell Sci. 36*: 155-168.

Nasmyth, K. A., and Reed, S. I. (1980). Isolation of genes by complementation in yeast molecular cloning of a cell cycle gene. *Proc. Nat. Acad. Sci. USA 77*:2119-2123.

Nejedlý, K., and Greksák, M. (1977). Mitochondriogenase a buněcný cyklus. *Chem. listy. 71*:394-411.

Newlon, C. S., and Fangman, W. L. (1975). Mitochondrial DNA synthesis in cell cycle mutants of *S. cerevisiae. Cell 5*:423-428.

Newlon, C. S., Luoesche, R. D., and Walter, S. K. (1979). *Mol. Gen. Genet. 169*:189-194.

Nurse, P. (1975). Genetic control of cell size at cell division in yeast. *Nature 256*:547-555.

Nurse, P. (1980). Cell cycle control, both deterministic and probabilistic? *Nature 286*:9-10.

Nurse, P. (1981). Genetic analysis of the cell cycle. *Symp. Soc. Gen. Microbiol. 31*:291-315.

Nurse, P., and Thuriaux, P. (1977). Controls over the timing of DNA replication during the cell cycle of fission yeast. *Exp. Cell Res. 107*:365-375.

Nurse, P., Thuriaux, P., and Nasmyth, K. (1976). Genetic control of the cell division cycle in the fission yeast *Schizosaccharomyces pombe. Mol. Gen. Genet. 146*:167-178.

Penman, C. S., and Duffus, J. H. (1975). 2'-Deoxyadenosine and A23187 as agents for inducing synchrony in the budding yeast *Kluyveromyces fragilis*. *J. Gen. Microbiol.* *90*:76-80.

Peterson, J. B., and Ris, H. (1976). Electron microscopic study of the spindle and chromosome movement in the yeast *Saccharomyces cerevisiae*. *J. Cell Sci.* *22*:219-242.

Petes, T. D., and Williamson, D. H. (1975). Replicating circular DNA molecules in yeast. *Cell* *4*:249-253.

Piñon, R. (1979). A probe into the nuclear events during the cell cycle of *Saccharomyces cerevisiae*. *Chromosoma* *70*:337-352.

Polanshek, M. M. (1977). Effects of heat shock and cycloheximide on growth and division of the fission yeast *Schizosaccharomyces pombe*. *J. Cell Sci.* *23*:1-23.

Poole, R. K. (1977a). Development of respiration activity during the cell cycle of *Schizosaccharomyces pombe* 972 h^-: Respiratory oscillations and heat dissipation in cultures synchronized with 2'-deoxyadenosine. *J. Gen. Microbiol.* *103*:19-27.

Poole, R. K. (1977b). Preparation of synchronous cultures of microorganisms by continuous-flow selection; which cells are selected. *FEMS Microbiol. Lett.* *1*:305-307.

Poole, R. K., and Lloyd, D. (1973). Oscillations of enzyme activities during the cell cycle of a glucose-repressed fission-yeast *Schizosaccharomyces pombe* 972 h^-. *Biochem. J.* *136*:195-207.

Poole, R. K., and Lloyd, D. (1974). Changes in respiratory activities during the cell cycle of the fission yeast *Schizosaccharomyces pombe* 972 h^- growing in the presence of glycerol. *Biochem. J.* *144*:141-148.

Poole, R. K., and Salmon, I. (1978). The pool sizes of adenine nucleotides in exponentially growing, stationary phase and 2'-deoxyadenosine synchronized cultures of *Schizosaccharomyces pombe* 972 h^-. *J. Gen. Microbiol.* *106*:153-164.

Poole, R. K., Lloyd, D., and Kemp, R. B. (1973). Respiratory oscillations and heat evolution in synchronously dividing cultures of the fission yeast *Schizosaccharomyces pombe* 972 h^-. *J. Gen. Microbiol.* *77*:209-220.

Poole, R. K., Lloyd, D., and Chance, B. (1974). The development of cytochromes during the cell cycle of a glucose-repressed fission yeast *Schizosaccharomyces pombe* 972 h^-. *Biochem. J.* *138*:201-210.

Quinton, A., and Poole, R. K. (1977). Induction of a shortened synchronous cell cycle by removal of chloramphenicol from an inhibited culture of a fission yeast *Schizosaccharomyces pombe*. *J. Gen. Microbiol.* *103*:271-275.

Rivin, C. J., and Fangman, W. L. (1980a). Cell cycle phase expansion in nitrogen-limited cultures of *Saccharomyces cerevisiae*. *J. Cell Biol.* *85*:96-107.

Rivin, C. J., and Fangman, W. L. (1980b). Replication fork rate and origin activation during the S phase of *Saccharomyces cerevisiae*. *J. Cell Biol.* *85*:108-115.

Salmon, I. (1980). Spatial and temporal development in *Sterigmatomyces halophilus*. Ph.D. thesis, University of London, London.

Salmon, I., and Poole, R. K. (1980). Activities of some mitochondrial enzymes in the cell cycle of *Sterigmatomyces halophilus*. *Proc. Soc. Gen. Microbiol.* *7*:85.

Sando, N. (1963). Biochemical studies on the synchronized culture of *Schizosaccharomyces pombe*. *J. Gen. Appl. Microbiol.* *9*:233-241.

Sebastian, J., Carter, B. L. A., and Halvorson, H. O. (1973). Induction capacity of enzyme synthesis during the cell cycle of *Saccharomyces cerevisiae*. *Eur. J. Biochem.* *37*:516-522.

Sebastian, J., Takano, I., and Halvorson, H. O. (1974). Independent regulation of RNA polymerases I and II in the cell cycle of *Saccharomyces cerevisiae*. *Proc. Nat. Acad. Sci. USA* *71*:769-773.

Sena, E. P., Welch, J. W., Halvorson, H. O., and Fogel, S. (1975). Nuclear and mitochondrial DNA replication during mitosis in *Saccharomyces cerevisiae*. *J. Bacteriol.* *123*:497-504.

Shilo, B., Shilo, V., and Simchen, G. (1976). Cell cycle initiation in yeast follows first-order kinetics. *Nature* *244*:767-770.

Shilo, B., Shilo, V., and Simchen, G. (1977). Transition probability and cell cycle integration in yeast. *Nature* *267*:648-649.

Shilo, B., Simchen, G., and Paradee, A. B. (1978). Regulation of cell cycle initiation in yeast by nutrients and protein synthesis. *J. Cell Physiol.* *97*:177-188.

Shilo, B., Riddle, V. G. H., and Paradee, A. B. (1979). Protein turnover and cell cycle initiation in yeast. *Exp. Cell Res.* *123*:221-227.

Shulman, R. W. (1978). Yeast cell selection and synchrony in density gradients and mating factor block. In *Methods in Cell Biology*, Vol. 20, D. M. Prescott (Ed.). Academic, New York, pp. 35-43.

Shulman, R. W., Hartwell, L. H., and Warner, J. R. (1973). Synthesis of ribosomal proteins during the yeast cell cycle. *J. Mol. Biol.* *73*:513-525.

Sierra, J. M., Sentandreu, R., and Villaneuva, J. R. (1973). Regulation of wall synthesis during *Saccharomyces cerevisiae* cell cycle. *FEBS Lett.* *34*:285-290.

Simchen, G. (1978). Cell cycle mutants. *Annu. Rev. Genet.* *12*:161-191.

Singer, R. A., and Johnston, G. C. (1979). Nalidixic acid causes a transient G_1 arrest in the yeast *Saccharomyces cerevisiae*. *Mol. Gen. Genet.* *176*:37-39.

Singer, R. A., Johnston, G. C., and Bedard, D. P. (1978). Methionine analogs and cell division regulation in the yeast *Saccharomyces cerevisiae*. *Proc. Nat. Acad. Sci. USA* *75*:6083-6087.

Sissons, C. H., Mitchison, J. M., and Creanor, J. (1973). Enzyme synthesis and potential during induction synchrony in the fission yeast *Schizosaccharomyces pombe*. *Exp. Cell Res.* *82*:63-72.

Skutelsky, E., and Bayer, E. A. (1979). The ultrastructural delineation of cell growth and division processes using the avidin-biotin complex. *Exp. Cell Res.* *121*:331-336.

Slater, M. L., Sharrow, S. O., and Gart, J. J. (1977). Cell cycle of *Saccharomyces cerevisiae* in populations growing at different rates. *Proc. Nat. Acad. Sci. USA* *74*:3850-3854.

Smith, H. T. B., and Mitchison, J. M. (1976). Anaesthetics delay and accelerate division in the fission yeast *Schiz. pombe*. *Exp. Cell Res.* *99*:432-435.

Smith, J. A., and Martin, L. (1973). Do cells cycle? *Proc. Nat. Acad. Sci. USA* *70*:1263-1267.

Smitt, W. W. S., Vlak, J. M., Molenaar, I., and Rozijn, Th. H. (1973). Nuclear function of the dense crescent in the yeast nucleus. *Exp. Cell Res.* *80*:313-321.

Sogin, S. J., Carter, B. L. A., and Halvorson, H. O. (1974). Changes in the rate of ribosomal RNA synthesis during the cell cycle in *Saccharomyces cerevisiae*. *Exp. Cell Res.* *89*:127-138.

Stetten, G., and Lederberg, S. (1973). Colecemid sensitivity of fission yeast. II. Sensitivity of stages of the cell cycle. *J. Cell Biol.* *56*:259-262.

Stevens, B. J. (1977). Variation in number and volume of the mitochondria in yeast according to growth conditions: A study based on serial sectioning and computer graphics reconstruction. *Biol. Cell.* *28*:37-56.

Streiblová, E. (1977). Dependence of cell wall growth and conjugation on cycle of development of *Schizosaccharomyces pombe*. *Microbiologiya* *46*:589-595.

Sudbery, P. C., Goodey, A. R., and Carter, B. L. A. (1980). Genes which control cell proliferation in the yeast *Saccharomyces cerevisiae*. *Nature* *288*:401-444.

Sumrada, R., and Cooper, T. G. (1978). Control of vacuole permeability and protein degradation by the cell cycle arrest signal in *Saccharomyces cerevisiae*. *J. Bacteriol.* *136*:234-246.

Tauro, P., and Halvorson, H. O. (1966). Effect of gene position on the timing of enzyme synthesis in synchronous cultures of yeast. *J. Bacteriol.* *92*:652-661.

Thomas, K. C., and Dawson, P. S. S. (1977). Variations in the adenylate charge during phased growth (cell cycle) of *Candida utilis* under energy excess and energy-limited growth conditions. *J. Bacteriol.* *132*:36-43.

Thomas, K. C., Dawson, P. S. S., and Gamborg, B. L. (1980). Differential growth rates of *Candida utilis* mother and daughter cells under phased cultivation. *J. Bacteriol. 141*:1-9.

Thuriaux, P., Nurse, P., and Carter, B. L. A. (1978). Mutants altered in the control co-ordinating cell division with cell growth in the fission yeast *Schizosaccharomyces pombe*. *Mol. Gen. Genet. 161*:215-220.

Tkacz, J., and Lampen, J. O. (1972). Wall replication in *Saccharomyces* sp: Use of fluorescein-conjugated concanavalin A to reveal the site of mannan insertion. *J. Gen. Microbiol. 72*:243-247.

Toma, Z. F. H. (1980). Mechanism of enzyme instability in yeasts. Ph.D. thesis, University of Wales, Cardiff.

Tyson, C. B., Lord, P. G., and Wheals, A. E. (1979). Dependency of size of *Saccharomyces cerevisiae* cells on growth rate. *J. Bacteriol. 138*:92-98.

Unger, M. W. (1977). Methionyl-transfer RNA deficiency during G_1 arrest of *S. cerevisiae*. *J. Bacteriol. 130*:11-19.

Unger, M. W., and Hartwell, L. H. (1976). Control of cell division in *Saccharomyces cerevisiae* by methionyl-tRNA. *Proc. Nat. Acad. Sci. USA 73*:1664-1668.

von Meyenburg, H. K. (1968). Der sprossungszykenes van *Saccharomyces cerevisiae*. *Pathol. Microbiol. 31*:117-127.

von Meyenburg, H. K. (1969). Energetics of the budding cycle of *Saccharomyces cerevisiae* during glucose limited aerobic growth. *Arch. Microbiol. 66*:289-303.

Vȓsanská, M., Krátky, Z., Biely, P., and Machala, S. (1979). Chitin structures of the cell walls of synchronously grown virgin cells of *Saccharomyces cerevisiae*. *Z. Allg. Mikrobiol. 19*:357-362.

Wain, W. H. (1971). Synthesis of soluble protein during the cell cycle of the fission yeast *Schizosaccharomyces pombe*. *Exp. Cell Res. 69*:49-56.

Wain, W. H., Price, M. F., Brayton, A. R., and Cawson, R. A. (1976). Macromolecular synthesis during cell cycles of yeasts and hyphal phases of *Candida albicans*. *J. Gen. Microbiol. 97*:211-217.

Wain, W. H., and Staatz, W. D. (1973). Rates of synthesis of ribosomal protein and total RNA through the cell cycle of the fission yeast *Schizosaccharomyces pombe*. *Exp. Cell Res. 81*:269-278.

Walker, G. M., and Duffus, J. H. (1979). An investigation into the potential use of chelating agents and antibiotics as synchrony inducers in the fission yeast *Schizosaccharomyces pombe*. *J. Gen. Microbiol. 114*:391-400.

Walker, G. M., and Duffus, J. H. (1980). Magnesium ions and the control of the cell cycle in yeast. *J. Cell Sci. 42*:329-356.

Walker, G. M., Thompson, J. C., Slaughter, J. C., and Duffus, J. H. (1980). Perturbation of enzyme activity in cells of the fission yeast *Schizosaccharomyces pombe* subjected to continuous-flow centrifugation. *J. Gen. Microbiol. 119*:543-546.

Weiser, I., and Shalitin, C. (1978). Induction of virus-like particles in cell cycle mutants of *S. cerevisiae*. *Int. J. Med. Sci.* *14*:821.

Wells, J. R. (1974). Mitochondrial DNA synthesis during the cell cycle of *Saccharomyces cerevisiae*. *Exp. Cell Res.* *85*:278-286.

Wells, J. R., and James T. W. (1973). Cell cycle analysis by culture fractionation. *Exp. Cell Res.* *75*:465-474.

Wiemken, A., Matile, P. H., and Moor, H. (1970). Vacuolar dynamics in synchronously budding yeast. *Arch. Microbiol.* *70*:89-103.

Williams, E. (1980). The ultraviolet irradiation sensitivity of asynchronous and synchronous cultures of *Schizosaccharomyces pombe* grown in the presence and absence of 2-phenylethanol. M. Sc. thesis, University of Wales, Cardiff.

Williamson, D. H. (1965). The timing of deoxyribonucleic acid synthesis in the cell cycle of *Saccharomyces cerevisiae*. *J. Cell Biol.* *25*: 517-528.

Williamson, D. H. (1970). The effect of environmental and genetic factors on the replication of mitochondrial DNA in yeast. In *Control of Organelle Development*. Vol. 24. Symposia of the Society for Experimental Biology, P. L. Miller, (Ed.), Cambridge University, Cambridge, pp. 247-276.

Williamson, D. H. (1973). Replication of the nuclear genome in yeast does not require concomitant protein synthesis. *Biochem. Biophys. Res. Commun.* *52*:731-740.

Williamson, D. H., and Moustacchi, E. (1971). The synthesis of mitochondrial DNA during the cell cycle in the yeast *Saccharomyces cerevisiae*. *Biochem. Biophys. Res. Commun.* *42*:195-201.

Williamson, D. H., and Scopes, A. W. (1960). The behavior of nucleic acids in synchronously-dividing cultures of *Saccharomyces cerevisiae*. *Exp. Cell Res.* *20*:338-349.

Williamson, D. H., and Scopes, A. W. (1962). A rapid method for synchronizing division in the yeast *Saccharomyces cerevisiae*. *Nature* *193*:256-257.

Williamson, D. H., Johnston, L. H., Richmond, K. M. V., and Game, J. C. (1977). Mitochondrial DNA and the heritable unit of the yeast mitochondrial genome: A review. In *Mitochondria 1977, Genetics and Biogenesis of Mitochondria*, W. Bandlow, R. J. Schweyen, K. Wolf, and F. Kaudewitz (Eds.). Berlin, Walter de Gruyter, pp. 1-24.

Wintersberger, U., Hirsch, J., and Fink, A. M. (1974). Studies on nuclear and mitochondrial DNA replication in a temperature-sensitive mutant of *S. cerevisiae*. *Mol. Gen. Genet.* *131*:291-299.

Zeuthen, E. (1974). Synchrony of *Tetrahymena* and *Schizosaccharomyces*. In *Cell Cycle Controls*, G. M. Padilla, I. L. Cameron and A. Zimmerman (Eds.). Academic, New York, pp. 1-30.

6
Ascospore Formation in Yeast

DAVID R. BERRY *University of Strathclyde, Glasgow, Scotland*

I. INTRODUCTION

Ascospores are spores produced during sexual reproduction in ascomycetous fungi. Most of the commonly recognized yeasts, including bakers' yeast, *Saccharomyces cerevisiae*, are Ascomycetes and, when they exhibit sexual reproduction, produce ascospores in asci. Any discussion of ascospore formation should concern itself with the whole sexual process, including plasmogamy as well as meiosis and spore formation. However, the mechanism and control of plasmogamy is a fascinating study in itself (Crandall et al., 1977) which cannot be adequately covered in this review.

Since sporulation has been most intensively studied in *S. cerevisiae*, this review concentrates almost exclusively on this species. This is not intended to minimize the contribution made by studies on other species such as *Schizosaccharomyces pombe* (Calleja et al., 1980), *Hansenula wingei* (Crandall and Lawrence, 1980), and *Wickerhamia fluorescens* (Rooney and Moens, 1973).

Much of the early work concerning the physiology, biochemistry, and genetics of yeast sporulation has previously been reviewed (Fowell, 1969; Tingle et al., 1973), and the techniques for studying yeast sporulation have been reviewed by Haber and Halvorson (1975). The parameters shown in Fig. 1 give an outline of the major events of ascospore formation in *S. cerevisiae* and provide a suitable background against which more recent developments can be discussed.

II. CYTOLOGY

A. Nuclear Division

At the cytological level, the processes of meiosis and ascospore formation are intimately associated. The stages of meiosis and ascospore formation can be followed by staining the nuclei with Giemsa's stain. Using this stain, the uninucleate, binucleate, and tetranucleate stages can be detected before refractile ascospores appear toward the end of sporulation. If further detail is required however, electron microscopy must be used. The following description is taken from the work of Moens and Rapport (1971a) and Esposito and Esposito (1978).

In *S. cerevisiae*, the nuclear membrane remains intact throughout meiosis and the four haploid ascospore nuclei arise as lobes on the parent nucleus. In the early stages of ascus formation, the nucleus contains a single spindle pole body which is associated with the nuclear membrane. Microtubules radiate from the spindle pole body into the nucleus (Fig. 2a). A structure known as the polycomplex body and synaptonemal complexes appear after 6 h (Fig. 2b). After 8 h, the spindle pole body duplicates (Fig. 2c). Initially, the two newly formed spindle bodies remain side by side, connected by an electron dense layer. Later the spindle pole bodies separate and move to opposite sides of the nucleus. At this stage the two spindle pole bodies are joined by microtubules. Synaptonemal complexes are not visible (Fig. 2d). The end of first meiotic division is indicated by the elongation of spindle and nucleus (Fig. 2e). The spindle pole bodies divide a second time, then separate and move apart (Fig. 2f) to form the spindles for the second meiotic division. As the spindles elongate, the nucleus

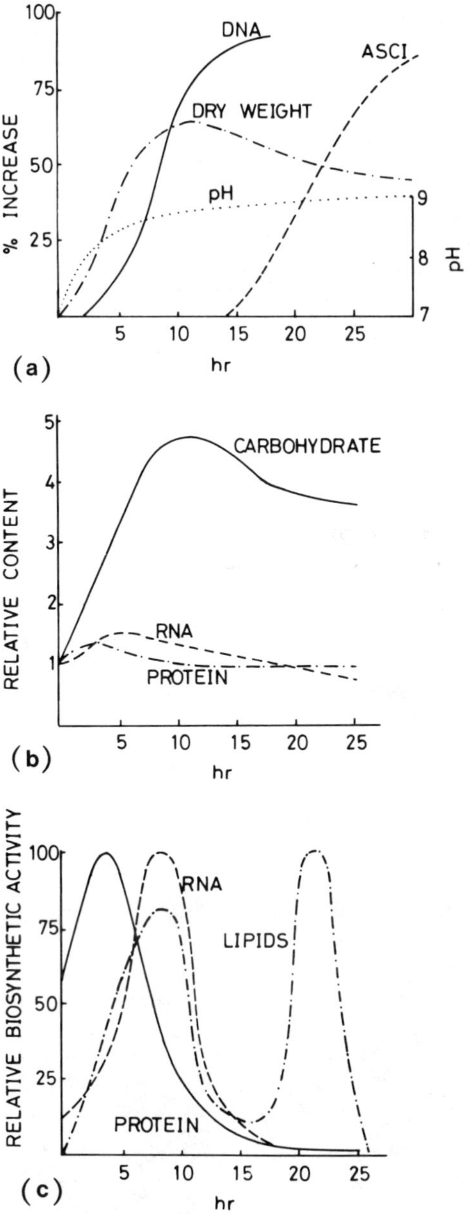

FIG. 1 Major biochemical events during sporulation. (From Haber and Halvorson, 1975.) (a) Changes in dry weight, DNA, medium pH, and appearance of mature asci. (b) Changes in level of RNA, protein, and storage carbohydrate. (c) Periodicity of biosynthesis of protein, RNA, and lipids.

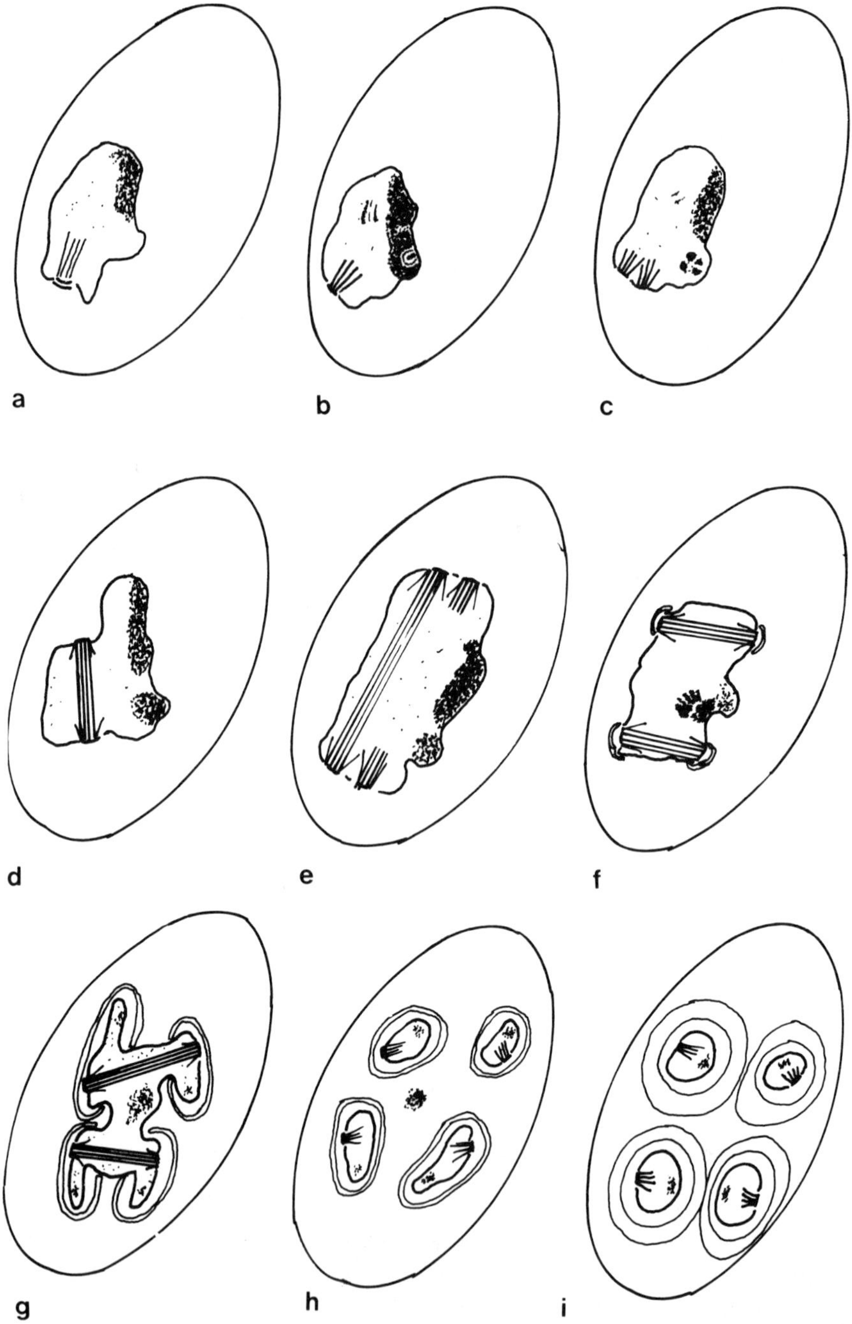

FIG. 2 Cytological events of meiosis and sporulation (for explanation see text). (From Esposito and Esposito, 1978.)

becomes lobed and one spindle pole body moves into each lobe (Fig. 2g). At this stage two electron-dense membranes separated by an electron-transparent zone develop adjacent to each of the four spindle pole bodies. These two membranes, which constitute the prospore wall, extend outwards and enclose some of the cytoplasm and the lobes of the nucleus (Fig. 2h). The nucleolus from the original parent nucleus is excluded from the nuclear material which is enclosed in the new ascospore nuclei. Finally, ascospore wall material is laid down between the two prospore wall membranes. The inner spore wall becomes the plasmalemma and the outer membrane contributes to the protein layer of the mature ascospore (Fig. 2i).

B. Synaptonemal Complex Formation

During the prophase of meiosis, a spherical body appears in the nucleus, often associated with the nucleolus (Moens and Rapport, 1971b). Structures which resemble small segments of synaptonemal complex have been found to associate with this so called polycomplex body. It has been reported that in the first 2 h of sporulation, only the lateral components of the synaptonemal complex are visible, but between 3 and 6 h these lateral components become associated with the central region to form the characteristic tripartite structure of a synaptonemal complex (Zickler and Olson, 1975). Synaptonemal complexes can occur in isolation; however, diplotene nuclei often contain structures composed of several layers of material which resemble synaptonemal complex material in a polycomplex body. Since synaptonemal complexes appear during meiotic prophase in many organisms, they have been associated with recombination. However the evidence for this is not strong. Olson and Zimmerman (1978) found that gene conversion and a low level of crossing over occurred before the appearance of synaptonemal complexes and concluded that meiotic recombination did not require the formation of synaptonemal complexes in *S. cerevisiae*. In *Schiz. pombe*, synaptonemal complexes were not detected during sporulation, although unpaired lateral elements of synaptonemal complexes were observed (Olson et al., 1978).

C. Formation of Aberrant Asci

During studies on yeast sporulation, it is not uncommon to observe asci containing one, two, or three spores in addition to normal four-spored asci. In most strains, these aberrant asci arise because not all four nuclei produced during meiosis become enclosed by an ascospore wall. However, some strains which consistently produce two spores have been isolated (Grewal and Miller, 1972). In these apomictic strains, the nuclei remain diploid throughout the life cycle, and meiosis is replaced by a single nuclear division. By adjusting the level of zinc in the presporulation medium, Bilinski and Miller (1980) have produced four-spored asci from such strains in which the four spores are haploid and of meiotic origin. It appears that an apomictic strain can be stimulated to undergo normal meiosis by adjusting the level of zinc in the growth medium.

A technique using thermal arrest to reduce the number of ascospores per ascus was used by Davidow et al., (1980). Analysis of two-spored asci produced by this technique indicated that the spores were not produced at random, but that each ascus contained nonsister products of the reductional division. The loss of ascospores appeared to be caused by failure of the outer layer of the spindle plaque to develop outside one spindle pole body.

A technique of inducing the formation of two-spored asci by transfer to water at T_4-T_8 has been reported (Promod et al., 1981).

III. PHYSIOLOGY

A. Conditions Favoring Sporulation

Sporulation in *S. cerevisiae* can be induced by the transfer of diploid vegetative cells from a rich presporulation medium to a sporulation medium containing acetate. The importance of the composition of the presporulation and sporulation media have been extensively discussed elsewhere (Fowell, 1969, 1975; Tingle et al., 1973). Traditionally, a rich medium containing high levels of complex nitrogen and glucose has been preferred as the presporulation medium. However, since transfer from such a medium to acetate medium involves transfer from catabolite-repressed conditions to nonfermentative conditions, many of the changes observed can be attributed to adaptation to acetate rather than the induction of sporulation, and many workers have preferred to use acetate as the carbon source in the presporulation medium. Improved rates of sporulation and a higher degree of synchrony have been reported using acetate (Roth and Halvorson, 1969). Other nonfermentable carbon sources such as galactose, pyruvate, lactate, and ethanol have also been used as the carbon source in the presporulation medium (Patel and Miller, 1972; Fast, 1973).

B. Induction of Synchronous Sporulation

Studies on differentiation can be facilitated by using synchronously differentiating systems. Several attempts have been made to obtain a high degree of synchrony in yeast sporulation. Improved synchrony was reported by Roth and Halvorson (1969) when acetate was used in the presporulation medium. In contrast, a technique which involves the addition of 0.1% Bacto yeast extract and 0.05% glucose to the presporulation medium was used by Petersen et al., (1978) to obtain a rapid and synchronous sporulation in glucose-grown cells at a cell density of 2.5×10^8 cells per milliliter.

Watson and Berry (1979) developed a procedure that permitted growth and sporulation to occur in fermenter-controlled conditions and which gave a high degree of synchrony. In this procedure, a rich growth medium containing glucose was replaced by acetate sporulation medium by a process of continuous filtration so that the cells were never removed from the fermenter or packed on a membrane or in a centrifuge tube. Studies using this system indicated that significant metabolic changes occurred during the period of medium exchange. In view of the results obtained using selection synchrony techniques on vegetative yeast cells, it was suggested that the centrifugation and washing step adopted by most workers for medium exchange could be responsible for some of the physiological effects observed during sporulation. Acetate presporulation medium was not used in these studies because it had been observed that when nitrogen became depleted, low levels of sporulation could occur in the acetate presporulation medium before transfer to the sporulation medium had been made.

A method of synchrony which was based on the sporulation of synchronously mated cells was described by Hartig and Breitenbach (1977); however, the rate of sporulation and degree of synchrony were not as good as have been obtained with other techniques. An alternative technique to using synchronous sporulation is to separate out homogeneous fractions of cells at different stages of sporulation. This has been achieved by centrifugation on Urografin gradients (Dawes et al., 1980b). With this technique, eight fractions were identified from a culture in which asci containing ascospores were just visible. A correlation was found to exist between the density of the cells and the stage of sporulation. The greater the cell density, the later the stage of sporulation reached. This technique of separating cells into classes after the experimental period, for example, after a period of isotope labeling, has been used with considerable success in studies on the vegetative cell cycle, since it eliminates any perturbation to the system caused by the induction or selection of a synchronous population at the beginning of the experiment (Elliott and McLaughlin, 1978). The use of such a technique in sporulation studies, however, does not preclude the need to induce sporulation, but eliminates the need for a high degree of synchrony.

The ability to obtain sporulation in protoplasts (Kopecká, 1974) may be valuable in certain types of study.

C. Carbon Metabolism

Yeast sporulation is an aerobic process. When acetate is provided as an exogenous carbon source, there is an initial increase in oxygen consumption in the first 2 h, after which the rate of oxygen consumption declines steadily throughout sporulation (Croes, 1967; Hopper et al., 1974). Studies using labeled acetate have indicated that 62% of the acetate was metabolized to carbon dioxide, 16% incorporated into cellular material, and 22% remained in the soluble pool of intermediates (Esposito et al., 1969).

However, acetate is not essential as a carbon source during sporulation and is effective as an inducer when supplied for only a short period of time (Tingle et al., 1973). In the absence of acetate, endogenous respiration has been reported to increase during the first 14 h of sporulation (Vezinhet et al., 1971).

A high level of tricarboxylic acid cycle and glyoxalate cycle activity has been reported during yeast sporulation (Miyake et al., 1971). However, Croes (1967) has proposed that sporulation was controlled by an insufficiency of the glyoxalate cycle and Betz and Weiser (1976b) have reported that isocitrate lyase and malate dehydrogenase actually decreased during sporulation.

It has been estimated that 67% of the dry weight increase during the early stages of sporulation can be attributed to glycogen accumulation. Glycogen accumulates until the stage at which ascospore formation can be detected, and then decreases. Trehalose increases throughout sporulation and appears to be the storage carbohydrate of the ascospores (Tingle et al., 1973). The glycogen accumulation observed in sporulating cells also occurs when $\alpha\alpha$ or *aa* cells are transferred to sporulation medium (Hopper et al., 1974; Kane and Roth, 1974), so it appears that glycogen accumulation is a response to the transfer of cells into sporulation medium rather than a sporulation-specific event. On the other hand, the decrease in glycogen observed during the later stages of sporulation after 8 h only occurs in $a\alpha$ cells and not in $\alpha\alpha$ controls, so this appears to be sporulation specific (Hopper et al., 1974).

The breakdown of glycogen during sporulation has been studied by Calonna and Magee (1978), who found that the decrease in glycogen content was associated with a 30- to 40-fold increase in the level of intracellular glucose. This increase in glucose level has also been shown to be sporulation specific. The levels of several glycolytic intermediates have been studied in sporulating and in control cultures. The levels of glucose-6-phosphate (G6P), fructose-6-phosphate (F6P), and fructose-1,6-diphosphate ($F16P_2$) were low and the level of phosphonenol pyruvate (PEP) was high both in $a\alpha$ and $\alpha\alpha$ strains growing on acetate presporulation medium. These results would be consistent with a cell grown on acetate and exhibiting gluconeogenesis (Fonzi et al., 1979).

D. Lipid Metabolism

Biochemical and cytological observations have indicated that the lipid content of developing asci increases and that most of the lipid is enclosed in the developing ascospores (Chassang et al., 1972; Illingworth et al., 1973). Most of the increase in the lipid content of asci can be attributed to the synthesis of triglycerides and sterol esters

(Illingworth et al., 1973). Two periods of lipid biosynthesis have been reported during yeast sporulation; however, only the second period appears to be sporulation specific (Tingle et al., 1973).

An inositol-requiring mutant has been studied by Schroeder and Breitenbach (1981), who found that 90% of exogenously supplied inositol was incorporated into phospholipid during two periods of lipid synthesis. Eight phospholipids which incorporated inositol were identified, of which one was sporulation specific. Upon transfer to sporulation medium the levels of G6P, F6P, and PEP remained at a similar level, while the level of $F16P_2$ increased by 500% in sporulating cells and by a somewhat smaller amount in αα controls.

The enzymes involved in glycolysis have been studied by Betz and Weiser (1976b), who reported that phosphofructokinase, G6P dehydrogenase, and pyruvate carboxylase did not change during sporulation, but that malate dehydrogenase, isocitrate lyase, and $F16P_2$ decreased.

E. Nitrogen Metabolism

The inhibitory effect of ammonium ions on sporulation is well documented, although the mechanism of this inhibition is not understood. Piñon (1977) reported that ammonium sulfate at a 2-4 mM concentration inhibited sporulation but had no effect on the respiratory activity of the cell. Ammonium did not appear to inhibit commitment to recombination or the initiation of DNA replication. However, DNA replication was arrested after initiation. In these experiments the addition of ammonium to sporulating cells did not stimulate a reversal to vegetative growth. Although ammonia has been implicated in the control of glycogen biosynthesis in vegetative cells (Suomalainen and Oura, 1971), it has been reported that the addition of ammonia has no effect on glycogen biosynthesis but that it inhibits glycogen breakdown during sporulation (Fonzi et al., 1979). The same authors also failed to observe any effect of ammonium on the level of several glycolytic enzymes.

Both ammonium and glutamine have been reported to inhibit RNA and protein synthesis when added 4 h after transfer to sporulation medium (Durieu-Trautmann and Delavier-Klutchko, 1977). However, glutamine and glutamine synthetase appear to be necessary for sporulation and a strain lacking glutamine synthetase has been referred to which had lost the ability to sporulate (Delavier-Klutchko et al., 1980).

F. Catabolite Repression and cAMP

Sporulation can be repressed by the addition of glucose or other readily metabolized sugars. This has been considered as a form of catabolite repression and it was reported that it could be reversed by the addition of cAMP (Tsuboi et al., 1972). However, we have not been able to repeat this when glucose was present in concentrations higher than 1% (Watson, 1978). Stimulation of sporulation by cAMP has also been reported in *Schiz. pombe* (Calleja et al., 1980).

One of the problems of this type of experiment is the difficulty of ensuring that the exogenously supplied cAMP actually enters the cell. Studies on the level of cellular cAMP during sporulation have been carried out by several workers. Watson and Berry (1977) found that the level of cAMP during sporulation was influenced by the level of glucose in the presporulation medium, in that a high level of glucose suppressed the level of cAMP in the subsequent sporulation phase. The level of cAMP was found to fluctuate during sporulation, with peaks occurring at 4, 8, and 12 h. It was proposed that these fluctuations could be related to cell cycle-dependent fluctuations, which has been reported in mammalian cells. Similar fluctuations were reported by Hartig and Breitenbach (1977); however, Fonzi has more recently reported that no significant changes in the level of cAMP occur during sporulation (Fonzi et al., 1979).

G. pH and Cell Permeability

During sporulation on acetate sporulation medium the pH rises to 9.6. This high value favors sporulation; however, it has presented problems for the study of sporulation, since the uptake of labeled precursors and inhibitors is inhibited at these pH values. It has been reported that reducing the pH to 7 or less inhibits sporulation (Mills, 1972; Arnaud and Galzy, 1973). One solution to this problem is to bring the pH of the sporulation culture down to pH 6 for a period not exceeding 1 h, during which precursors can be fed to the cells (Mills, 1972). An alternative approach, the use of a mutant strain *SK1*, which has improved permeability, did not prove to be satisfactory, the strain giving good uptake of precursors but poor sporulation (Mills et al., 1977). Sporulation has now been obtained at pH 6 in cultures in which the pH was controlled using the following buffers: succinate:2-(N-morpholine ethane sulfonic acid) (MES); piperizine-N, N-bis(2-ethane sulfonic acid) (PIPES); or morpholinosulfonic acid (MOPS) (McCusker and Haber, 1977).

H. Enzymes of the Ascospore Wall

In view of the different properties of the ascospores in vegetative cells (Tingle et al., 1973), it would not be surprising to find differences in the activity of biosynthetic and degradative enzymes associated with cell wall formation. β1-3-Glucanase has been observed to increase slowly from T_0 to T_7, then more rapidly after T_7. This period of rapid increase was reported to be sporulation specific (del Rey et al., 1980).

Evidence has also been presented that N-acetylglucosamine synthesis is involved in ascospore formation. A mutant strain GCN1 has been isolated which lacks L-glutamine-D-fructose-6-phosphate aminotransferase, an enzyme which is essential for glucosamine synthesis. Diploids homozygous for this gene were able to progress to the four nuclear stage during sporulation, but not to produce normal ascospores. Studies on the ascospores produced indicated that these lacked an outer electron-dense layer, which is normally present in developing ascospores (Whelan and Ballou, 1975). The N-acetylglucosamine is probably involved in linking proteins to mannan in the cell wall, and not primarily in chitin formation.

IV. GENETICS

A. Introduction

Elucidation of the genetic control of sporulation is a very complex problem; however, some progress has been made in isolating mutations which affect sporulation and identifying their site of action and time of action during sporulation (Haber et al., 1975). Since the process of sporulation involves several coordinated events such as DNA replication, meiotic nuclear division, recombination, and ascospore delimitation, it is possible to distinguish different categories of mutation. These would include mutations which affect the production of mature spores (*spo*), the recovery of intragenic prototrophic recombinants (*mei, con*), the control of sporulation by mating type (*csp, sca, rme*), and the initiation of meiosis (*spd*). However, sporulation is also affected by several types of mutation which were originally selected for their effect on other functions, for example, cell cycle (*cdc*) and recombination (*rec*) (Baker et al., 1976).

B. Control by Mating-Type Locus

The process of sporulation appears to be subject to overall control by the genes which control mating, *mat a* and *mat* α, and only diploids which have the *a*α genotype normally sporulate. Recombination and premeiotic RNA replication are also repressed when αα or *aa* diploids are transferred to sporulation medium. However, mutations which permit homozygous diploids to sporulate have been isolated. One such strain which produces some asci containing 4α ascospores contains the dominant gene *CSP1* (Hopper and Hall, 1975). In *CSP1* strains, all the cells are capable of premeiotic DNA replication, but only 20% of the asci develop mature ascospores. Intragenic recombination occurs in the sporulating cells but not in the others, so *CSP1* is capable of releasing premeiotic DNA replication, but not necessarily recombination from control by the mating-type locus. Other mutations which affect mating-type control include *RME*/*rme*, in which the diploid recessive is able to sporulate irrespective of the mating type (Kassir and Simchen, 1976), and the *SAD1* gene, which allows MATα/MATα strains to sporulate without affecting their α mating properties (Hopper and MacKay, 1980).

An interesting technique was adopted by Klar (1980), who produced a heterokaryon between αα or *aa* cells and a haploid nucleus, obtaining a gene *kar* (karyogamy defective) and either an *a* or α mating type.

The asci produced from this strain contained six ascospores, 4α or 4*a*, from the meiotically dividing diploid nucleus, and 2α or 2*a* from the mitotically dividing haploid nucleus.

C. Mutations Affecting the Sporulation Process

Several types of mutation which block sporulation at different stages of the process of meiosis and ascospore development have been investigated in an attempt to establish the site of action of different genes and the nature of their products. The site of action of some of these mutations is indicated in Fig. 3. The most extensively studied series of mutations are the *SPO* strains isolated by Esposito and Esposito (1974). Most of these are temperature-sensitive mutations that are incapable of producing mature ascospores but which are capable of vegetative growth. The recessive *spo* mutations isolated have been classified into 11 loci, *spo 1-11*, by complementation tests. However, it has been estimated that on the order of 50 loci are probably required for the formation of mature ascospores (Esposito and Esposito, 1978; Baker et al., 1976). Three dominant sporulation-defective mutations have also been isolated.

Strains *spo 7, 8, 9,* and *11* are blocked before premeiotic S phase, as are certain *cdc* mutations, *cdc 2, 7, 8,* and *21* (Schild and Byers, 1978), *cdc 28* (Reid and Hartwell, 1977), and *spd 1* (Vezinhet et al., 1979). This period before the onset of the DNA replication is a critical phase in the control of sporulation. The transition from mitosis to meiosis has been shown to occur before the site of the *cdc 28* block in the G1 phase (Reid and Hartwell, 1977). As indicated previously, *mat* genes also block in G1. The *spd 1* gene is involved with the release of sporulation from nitrogen repression. Since it acts at the same time as *cdc 28*, it would appear that nitrogen repression also occurs in the G1 phase. Strains which are blocked in G1 do not exhibit genetic recombination (Esposito and Esposito, 1978). DNA synthesis may be essential for intragenic recombination, since inhibition of RNA synthesis by 80H quinoline inhibited recombination at the *leu 2* locus by over 70% (Mills, 1978). However, a pulse of hydroxyurea between T_2 and AT_4 increased the frequency of recombination and also delayed the time of commitment to recombination (Simchen et al., 1978).

DNA replication does not, however, appear to be essential for several events in ascospore development. Strain *spo 11*, which is blocked

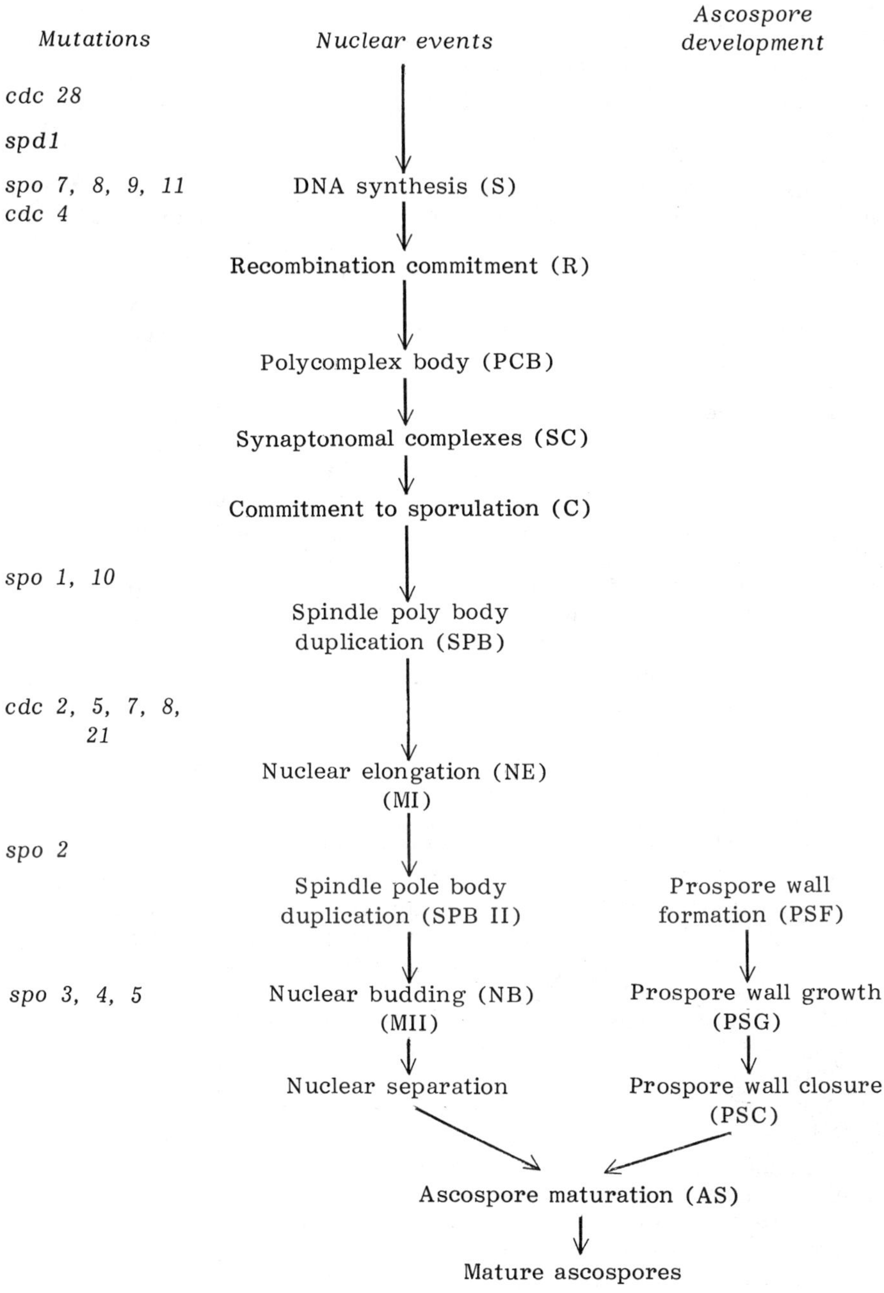

Fig. 3 Possible sequence of events during meiosis and sporulation as indicated by the site of action of different mutations which block sporulation (adapted from Esposito & Esposito, 1974).

in DNA replication, is capable of forming synaptonemal complexes and the kinetic apparatus of meioses I and II. It is clear that DNA replication is not essential for these functions (Moens et al., 1977). Studies on *cdc 4*, which block DNA replication, also indicate that commitment to recombination is not dependent on the completion of DNA replication (Simchen and Hirschberg, 1977).

Whereas commitment to sporulation appears to occur at T_7, just before spindle pole body duplication (SPB), commitment to recombination occurs at the time of the initiation of DNA synthesis. If cells are removed from sporulation medium after commitment to recombination but before commitment to sporulation, it is possible to obtain recombination in cells which return to mitosis. It is apparent that commitment to recombination does not inevitably lead to commitment to ascospore formation.

Strains *spo 1* and *spo 10* are blocked after DNA replication, but before spindle pole body duplication. This demonstrates that spindle pole body duplication is not essential for DNA replication, commitment to recombination, or synaptonemal complex formation (Esposito and Esposito, 1978). Several *cdc* mutations, *cdc 2, 5, 7, 8, 14,* and *21*, all appear to act after spindle pole body duplication but before spindle elongation (Schild and Byers, 1978; Schild and Byers, 1980). Studies on *spo 2* have indicated that meioses I and II are under common control and that prospore wall formation and closure are independent of the entry of a nucleus into the prospore, but that a normal ascospore wall is only established around a prospore which contains a nucleus. Strains *spo 3, 4,* and *5*, which are blocked at meiosis II, also develop immature ascospores, usually lacking nuclei (Esposito and Esposito, 1978).

Esposito and Esposito (1978) have proposed a model for the genetic control of sporulation. In this, two genes, *spo 11* and *APO* (apomixis), play a key role in the control of meiosis. Mutant *spo 11* prevents cells from undergoing reductional division until DNA synthesis and the prophase of meiosis I have been completed. Mutant *APO* is thought to prevent cells from entering second meiotic division until they have completed the first division.

The relationship between meiosis and the normal mitotic cell division has been investigated by studying the effect of *cdc* mutations on both the vegetative cell cycle and on meiosis and sporulation (Simchen, 1974). Some 32 different loci have been identified which affect cell division. Simchen (1974) examined the effect of 20 of these *cdc* mutations on meiosis and ascospore formation. The results obtained indicated that most of the genes required for DNA replication and mitotic nuclear division are also essential for meiosis. Only genes involved with cytokinesis and bud emergence appeared not to be essential for sporulation (*cdc 3, 10,* and *11*).

Recombination is, of course, an essential feature of sporulation in yeast. Mutations exhibiting a reduced rate of recombination have been isolated by Roth and Fogel (1971). These can be divided into two groups; *mei* strains, which lacked premeiotic DNA replication, and *con* strains, which exhibited premeiotic DNA replication but not gene conversion events.

Evidence has recently been presented that the product of a gene *rad 52* plays a role in recombination in yeast, as well as in DNA repair during vegetative growth (Resnick et al., 1981).

V. CELL BIOLOGY

A. DNA Replication

Many studies have indicated that the level of DNA per ascus doubles during the period starting around 2-4 h after transfer to sporulation medium. This increase in DNA is a sporulation-specific event which does not occur when $\alpha\alpha$ cells are transferred to sporulation medium (Hopper et al., 1974). The period occupied by the synthesis of DNA is variable, depending upon the degree of synchrony of the culture. The shortest period reported indicated a premeiotic S phase of approximately 2 h (Williamson et al., 1980). DNA synthesis in yeast sporulation is of interest not just because it appears to be under separate control from vegetative DNA replication, but also because it coincides with the period during which genetic recombination occurs (Hopper et al., 1974). The S phase is completed before the binucleate stage of ascus development, which represents the end of meiosis I.

Yeast DNA is replicated in short segments called replicons. The size of the replicon and the rate of DNA replication have been determined for premeiotic DNA replication by autoradiographically measuring the size of loops and the rate of fork movement (Williamson et al., 1980). The average replicon was found to be 30 μm, and the rate of movement 1.0 μm/min. If all the replicons were initiated at the same time, then it should be possible to complete DNA replication within 30 min according to this data. The observation that the premeiotic S phase takes 2 h suggests that the initiation of different replicons does not occur at the same time.

An analysis of histone biosynthesis has shown that the same histones which occur in vegetative cells, H2a, H2b, K3, and H4, also occur in sporulating cells and are produced in similar amounts. However, whereas histone synthesis coincides with DNA replication and is coupled to it in the vegetative cell cycle, it occurs ahead of DNA replication during sporulation. This may indicate a different structural requirement during the period of recombination (Marian and Wintersberger, 1980).

Transition from mitotic G1 to premeiotic DNA replication is probably a key point in the initiation of sporulation. A mutant strain *spd 1* has been isolated which blocks the cell cycle in the G1 phase but promotes sporulation under certain conditions (Vezinhet et al., 1979). A 20,000-dalton polypeptide has been identified which is constitutively synthesized in *spd 1* mutants, produced during sporulation conditions in wild-type strains, and absent from *spo* strains (Dawes et al., 1980a).

B. RNA Metabolism

Several groups have studied the RNA content of developing ascospores, the rate of RNA formation, and the timing of RNA synthesis with a view to elucidating the mechanism which controls gene expression during the transition from vegetative growth to sporulation (Tingle et al., 1973; Haber, 1975; Curiale and Mills, 1975; Sogin et al., 1975).

The level of RNA increases during the early stages of sporulation and decreases to 50% of the original level during the remainder of the period of sporulation. Early studies using isotopic labeling techniques indicated that there were two major periods of RNA synthesis at T_{10} and T_{25}. However, these results may have been influenced by the effect of pH on the uptake of precursors. Studies in which the pH was controlled at pH 6 indicated that a single peak of incorporation of tritiated adenine occurred between T_2 and T_{10} (Hopper et al., 1974). This period of incorporation was observed in $a\alpha$ and $\alpha\alpha$ diploids in sporulation medium, so does not appear to be sporulation specific. Since no exogenous nitrogen is present in sporulation medium, new RNA synthesis probably involves reutilization of bases released by the breakdown of existing RNAs. The breakdown of RNA has been investigated in sporulating $a\alpha$ cells and control $\alpha\alpha$ cells. RNA breakdown occurs in sporulating cells, but not in control cells (Hopper et al., 1974). This breakdown of RNA is dependent upon protein synthesis and is inhibited by the addition of cycloheximide (Magee and Hopper, 1974). Magee and Hopper (1974) have suggested that this inhibition of RNA degradation by cycloheximide could be explained if only free ribosomal units are available for degradation. The addition of cycloheximide would block translation and the release of free ribosomes.

One 20-S RNA particle which was reported in sporulating cells has been identified as a precursor of the methylated 18-S ribosome of RNA (Tingle et al., 1973). A different 20-S RNA particle has been reported by Wejksnora and Haber (1978) in sporulating cells. This particle is derived from a 32-S spherical ribonucleoprotein particle. The RNA appears to be resistant to RNAase when present in the 32-S particle, although it is single stranded and sensitive to RNAase when extracted.

The protein fraction associated with this 32-S particle consists of 18-20 copies of a 23,000-dalton protein subunit. Production of the 20-S RNA only occurs during sporulation and is resistant to inhibition by actinomycin D. Garvik and Haber (1978) presented evidence that the formation of this 20-S RNA particle is controlled by a new cytoplasmic element which is independent of phage translation suppressor, URE 3, Killer, 2-μm DNA, and mitochondrial DNA.

One of the prime objectives of studies on sporulation is to establish which genes are expressed during sporulation, and especially which are unique to sporulation. Since gene expression involves mRNA formation, several studies on mRNA metabolism during sporulation have been carried out.

Oligo (dT10) cellulose columns have been used to isolate RNA which contains polyA fragments from sporulating yeast (Harper et al., 1980). The RNA isolated was polydisperse, had a mean molecular weight of 4×10^5, and a base composition similar to vegetative polyA RNA. The polyA segment varied from 20 to 90 bases, with a peak at 60. This RNA fraction, which is presumably mRNA, has either a 7-methylguanosine-5-triphosphoryl-5-adenosine (^{7}GPPPA) or a 7-methylguanosine-triphosphoryl-5-guanosine (^{7}GPPPG) cap at the 5' end. Similar techniques have been used to demonstrate that sporulating cells produce approximately 7% sporulation-specific mRNA (Mills, 1980). The half-life of polyA RNA has been determined at 20 min in both sporulating and vegetative cells (Kraig and Haber, 1980). The polyA RNA fraction comprised 1.3-1.4% of the total RNA in both sporulating and vegetative cells; however, the average polysome size was smaller in sporulating cells and the template activity of mRNA as indicated by a wheat germ protein-synthesizing system decreased during sporulation and was only 50% of the vegetative level 8 h after the induction of sporulation.

C. Protein Metabolism

1. *Protein Degradation*

Since nitrogen is absent from sporulation medium, it is apparent that any protein biosynthesis occurring in the developing ascus is dependent upon endogenous reserves of nitrogen. It is not surprising, therefore, that sporulation is associated with a high proteolytic activity. Protein degradation is initiated immediately after diploid cells are placed in sporulation medium. In *a*α sporulating cells 30% of the protein is broken down, whereas in αα controls only a 0-10% breakdown has been reported (Hopper et al., 1974; Klar and Halvorson, 1975). Klar and Halvorson (1975) observed a 20-25% increase in the activity in several proteases during sporulation and reported that these increases were blocked by the inhibition of protein synthesis. The increase in protease activity did not occur in nonsporogenic strains. Three proteases, A, B, and C, were separated on diethylaminoethyl sephadex columns and evidence was obtained for a heat-stable inhibitor for proteases A and B. No changes in the type of protease present were observed during sporulation.

Different results, however, have been obtained by Betz and Weiser (1976a,b), who observed a 60-70% breakdown of preexisting proteins in sporulating cells and in haploid controls. G. Zubenko and E. Jones (personal communication) have also reported a 50% breakdown of protein in both *a*α and αα diploids. The importance of protease action for sporulation has been indicated by certain protease mutants. Strain *prb 1* is altered in the structural gene for protease B, the enzyme which is considered to be involved in the activation of chitin synthetase. However, *prb 1* strains were able to grow and bud normally but had a reduced ability to sporulate when the *prb 1* mutation was present in the homozygous state. Such strains showed only 50% of the normal wild-type protein degradation during sporulation and gave rise to abnormal asci. Another strain which was homozygous for the gene *prc 1* and which lacks proteinase C (carboxypeptidase Y) shows 80% of the wild-type level of protein degradation and sporulates normally (Zubenko et al., 1980). Another strain, *pep 4-3*, has been isolated which has a pleiotropic effect, reducing the level of proteases A, B, and C, ribonuclease, and alkaline phosphatase. This strain is unable to undergo meiosis or sporulation (G. Zubenko and E. Jones, personal communication).

2. *Protein Biosynthesis*

The level of protein in asci increases to a limited extent during the early stages of sporulation (T_0-T_6), then decreases slowly. However, since during this period there is a heavy breakdown of protein, there must be a corresponding level of biosynthesis of new proteins. There are two difficulties in studying protein synthesis during yeast sporulation. One, which has been referred to previously, is the effect of pH on the uptake of labeled precursors, the other is that protein synthesis in asci utilizes amino acids, generated endogenously through protein breakdown, which are not easy to label and monitor. Initial results obtained indicated two main periods of protein synthesis: one between T_4 and T_7, when the level of protein is increasing, and the other between T_{23} and T_{27}, when the ascospores are maturing (Tingle et al., 1973). Subsequent work in which the pH of the medium was controlled showed a single peak of protein synthesis between T_1 and T_6 that was found to occur in both *a*α and αα cells when transferred to sporulation medium (Hopper et al., 1974).

Acrylamide gel electrophoresis of the labeled protein showed no difference between *a*α and αα cells, although a difference between sporulation and vegetative cells was observed. The changes observed are therefore probably a response to the transfer of growth to the sporulation medium rather than sporulation-specific events.

Polysomes have been isolated from asci by several workers (Frank and Mills, 1978; Kraig and Haber, 1980). Kraig and Haber reported that the level of polysomes was lower in sporulating cells than in vegetative cells, but that monosomes were still active, since the addition of yeast nitrogen base, which stimulated growth, also stimulated a rapid increase in polysome size. Again, these changes occurred in both *a*α and αα diploids.

3. *Biosynthesis of Specific Proteins*

Several groups have used a combination of isotope labeling techniques and acrylamide gel electrophoresis in an attempt to identify sporulation-specific proteins; however, few sporulation specific proteins have as yet been identified. Out of 700 recognizable spots obtained by two-dimensional electrophoresis, 11 were found to occur in cells grown in sporulation medium but not in vegetative cells. However, all 11 occurred in both *a*α and αα cells (Trew et al., 1979). The use of a combination of isoelectric focusing and acrylamide gel electrophoresis led to the detection of 400 polypeptides in which 45 changes were observed after transfer of growth to sporulation medium. Of these, 22 involved the appearance or disappearance of a polypeptide, whereas 23 involved changes in level. The appearance of 13 polypeptides and changes in the level of 7 polypeptides were found to be sporulation specific (Wright and Dawes, 1979). In these experiments an attempt was made to ensure that all the proteins were labeled by supplying $^{35}SO_4^{2-}$ to the presporulation medium. Since no S was added in the sporulation medium, all proteins produced during sporulation must be labeled with ^{35}S present in the vegetative cells prior to transfer. Artifacts were also minimized by adding PMSF, a protease inhibitor. Three sporulation-specific proteins have also been identified by labeling with ^{32}P by Wright and Dawes (1980), who proposed that phosphorylation of peptides may be the mechanism of protein changes during sporulation.

The effect of nitrogen depletion and acetate on sporulating cells has been investigated by Kraig and Haber (1980). These authors have identified 44 changes in protein when vegetative cells were transferred to sporulation medium. Of these, some appeared to be a response to the lack of nitrogen, whereas others were stimulated by the presence of acetate. The presence of several temperature-sensitive *SPO* mutants was also investigated. Each of them was found to have a pleiotropic effect, causing changes in both sporulation-specific and non-sporulation-specific proteins. The results did not give any support to the concept of an ordered sequence of changes in sporulation-specific proteins during sporulation.

The number of genes which are specific to sporulation has been estimated at 48 ± 27 (Esposito et al., 1972), compared to a total number of around 12,000 genes in yeast. In view of this, it is not surprising that difficulty has been experienced in identifying sporulation-specific proteins and ascribing specific functions to them.

An alternative approach is to study known proteins or enzymes. The rate of formation of ribosomal proteins has been studied by Pearson and Haber (1980), who found that their rate of synthesis was depressed during sporulation to 20% of the rate observed in vegetative cells. The effect of a mutation *rna 2*, which depresses the biosynthesis of ribosomal proteins on sporulation, was also investigated. The results obtained indicated that the mechanism of control of ribosomal proteins in sporulating cells is different from that which operates in vegetative cells, and independent of the gene product of *RNA2*.

Evidence for differences in regulatory systems between sporulating and vegetative cells has also been obtained from studies on alcohol dehydrogenase induction (Eltayeb, 1977). Whereas acetate and ethanol induce ADH II in vegetative cells, in sporulating cells acetate has no stimulatory effect on alcohol dehydrogenase activity, and the addition of alcohol, besides inhibiting sporulation, stimulates ADH I production.

A different approach to the regulation of enzyme formation during sporulation was adopted by Matur and Berry (1978a,b,c), who compared the pattern of appearance of selected enzymes with that observed in synchronized vegetative cells. The step increases in the activity of enzymes have been used as markers of the vegetative cell cycle in yeast (Mitchison, 1973). The observation that the four enzymes studied, ornithine aminotransferase, sucrase, and acid and alkaline phosphotase, exhibited the same pattern of increase in sporulating and in vegetative cells led the authors to postulate that the physiological control of the vegetative cell cycle and the meiotic cell cycle during sporulation were under common control. More recently the appearance of step enzymes in the vegetative cell cycle has been demonstrated to be influenced by the cell selection used to obtain synchronous vegetative cells. These results must cast doubt on the significance of the results obtained in sporulating cells; however, it should be noted that conditions used to obtain sporulation were designed to minimize the physiological shock of transfer to sporulation medium. Whereas vegetative cell synchrony can be considered as a laboratory artifact, the transition from growth conditions to sporulation conditions is a natural phenomenon.

An interesting question which arises when considering the regulation of the formation of spore components is, When do the genes in the haploid ascospores begin to be expressed? Studies on a cycloheximide-resistant mutant which alters the ribosomal protein have indicated that this may be as early as meiosis II (Emanuel and Magee, 1981).

VI. CONCLUSIONS

Considerable progress has been made in recent years in our understanding of sporulation in *S. cerevisiae*. This process provides an excellent model system for studying sexual differentiation, including the mechanisms of recombination and meiotic cell division in eukaryotic cells. Some of the advantages of studying sporulation in yeast include the following:

1. A high percentage (80%) of cells can be induced to sporulate.
2. A high degree of synchrony of sporulation can be induced.
3. Cytological events of sporulation are well described.
4. Several stages of sporulation can be easily monitored.
5. The genetic system of *S. cerevisiae* is well studied and techniques of genetic analysis are available.
6. The biochemistry of *S. cerevisiae* is well studied.
7. It is possible to isolate sporulation-defective mutations.
8. Mutations of other cellular functions, for example, *cdc*, which affect sporulation are available.

REFERENCES

Arnaud, A., and Galzy, P. (1973). On the role of pH in the sporulation of *Saccharomyces cerevisiae*. *Folia Microbiol.* *18*:281-285.

Baker, B. S., Carpenter, A. T. C., Esposito, M. S., Esposito, R. E., and Sandler, L. (1976). Genetic control of meiosis. *Annu. Rev. Genet.* *10*:53-134.

Betz, H., and Weiser, U. (1976a). Protein degradation and proteinases during yeast sporulation. *Eur. J. Biochem.* *62*:65-76.

Betz, H., and Weiser, U. (1976b). Protein degradation during yeast sporulation. *Eur. J. Biochem.* *70*:385-392.

Bilinski, C. A., and Miller, J. J. (1980). Induction of normal ascosporogenesis in two spored *S. cerevisiae* by glucose, acetate and zinc. *J. Bacteriol.* *143*:343-348.

Calleja, G. B., Johnson, B. F., and Yoo, B. Y. (1980). Macromolecular changes and commitment to sporulation in the fission yeast. *Schizosaccharomyces pombe*. *Plant Cell Physiol.* *21*:613-625.

Calonna, W. J., and Magee, P. T. (1978). Glycogenolytic enzymes in sporulating yeast. *J. Bacteriol.* *134*:844-853.

Chassang, A., Roger, M., Vezinhet, F., and Galzy, P. (1972). Variation of the lipid content of yeast cells during sporulation. *Folia Microbiol.* *17*:241-247.

Crandall, M., and Lawrence, L. J. (1980). Sporulation in *Hansenula wingei* is induced by nitrogen starvation in maltose-containing media. *J. Bacteriol.* *142*:276-284.

Crandall, M., Egel, R., and MacKay, V. L. (1977). Physiology of mating in three yeasts. *Adv. Microbiol. Physiol.* *15*:307-398.

Croes, A. F. (1967). Induction of meiosis in yeast. *Planta Berl.* *76*:209-226.

Curiale, M. S., and Mills, D. (1975). Characterization of pulse-labeled RNA in polysomes of sporulating yeast cells. In *Spores*, Vol. 6. P. Gerhart, R. Castilow, and H. Sadoff (Eds.). American Society for Microbiology, Washington, D.C., pp. 165-171.

Davidow, L. S., Goetsch, L., and Byers, B. (1980). Preferential occurrence of non-sister spores in two spored asci of *S. cerevisiae*. *Genetics* *94*:581-595.

Dawes, I. W., Calvert, G. R., and Martin, T. A. (1980a). Initiating of meiosis and sporulation in *S. cerevisiae*. In *Tenth International Conference on Yeast Genetics and Molecular Biology*. Louvaine La Neuve, Belgium, pp. 152.

Dawes, I. W., Wright, J. F., Vezinhet, F., and Ajam, N. (1980b). Separation on urografin gradients of subpopulations from sporulating *Saccharomyces cerevisiae* cultures. *J. Gen. Microbiol.* *119*:165-171.

Delavier-Klutchko, C., Durieu-Trautmann, O., Allemand, P., and Tarlitzki, J. (1980). Assimilation of ammonia during sporogenesis of *S. cerevisiae*. *J. Gen. Microbiol.* *116*:143-148.

Durieu-Trautmann, O., and Delavier-Klutchko, C. (1977). Effect of ammonia and glutamine on macromolecular synthesis and breakdown during sporulation of *S. cerevisiae*. *Biochem. Biophys. Res. Commun.* *79*:438-442.

Elliott, S. G., and McLaughlin, C. S. (1978). Rate of macromolecular synthesis through the cell cycle of the yeast *Saccharomyces cerevisiae*. *Proc. Nat. Acad. Sci. USA* *75*:4384-4388.

Eltayeb, Y. (1977). Ph.D. thesis, Strathclyde University, Glasgow, Scotland.

Emanuel, J. R., and Magee, P. T. (1981). Timing of ribosome synthesis during ascosporogenesis of yeast cells. *J. Bacteriol.* *145*:1342-1350.

Esposito, M. S., and Esposito, R. E. (1974). Genes controlling meiosis and spore formation in yeast. *Genetics* *78*:215-225.

Esposito, M. S., and Esposito, R. E. (1978). Aspects of the genetic control of meiosis and ascospore development inferred from the study of *spo* (sporulation-deficient) mutants of *Saccharomyces cerevisiae*. *Biol. Cell.* *33*:93-102.

Esposito, M. S., Esposito, R. E., Arnaud, A., and Halvorson, H. O. (1969). Acetate utilization and macromolecular synthesis during sporulation of yeast. *J. Bacteriol.* *100*:180-186.

Esposito, R. E., Frink, N., Bernstein, P., and Esposito, M. S. (1972). The genetic control of sporulation in *Saccharomyces*. *II*. Dominance and complementation mutants of meiosis and spore formation. *Mol. Gen. Genet.* *114*:241-248.

Fast, D. (1973). Sporulation synchrony of *S. cerevisiae* grown in various carbon sources. *J. Bacteriol. 116*:925-930.

Fonzi, W. A., Shanley, M., and Opheim, O. J. (1979). Relationship of glycolytic intermediates, glycolytic enzymes and ammonia to glycogen metabolism during sporulation in the yeast *S. cerevisiae. J. Bacteriol. 137*:285-294.

Fowell, R. R. (1969). Sporulation and hybridization of yeasts. In *The Yeasts, Biology of yeasts*, Vol. 1. A. H. Rose and J. S. Harrison (Eds.). Academic, London, pp. 303-383

Fowell, R. R. (1975). Ascospores in yeast. In *Spores VI.* P. Gerhart, R. Castilow, and A. Sadoff (Eds.). American Society for Microbiology, Washington, D.C., pp. 124-131.

Frank, K. R., and Mills, D. (1978). Ribosome activity and degradation in meiotic cells of *S. cerevisiae. Mol. Gen. Genet. 160*:59-65.

Garvik, B., and Haber, J. E. (1978). New cytoplasmic genetic element that controls 20-S RNA synthesis during sporulation in yeast. *J. Bacteriol. 134*:261-269.

Grewal, N. S., and Miller, J. J. (1972). Formation of asci with two diploid spores by diploid cells of *Saccharomyces. Can. J. Microbiol. 18*:1897-1905.

Haber, J. E. (1975). Requirement for expression of genes controlling synthesis of ribosomal RNA during sporulation of *Saccharomyces cerevisiae* cells. In *Spores*, Vol. 6. P. Gerhart, R. Castilow, and H. Sadoff (Eds.). American Society for Microbiology, Washington, D.C., pp. 158-164.

Haber, J. E., and Halvorson, H. O. (1975). Methods in sporulation and germination of yeasts. In *Methods in Cell Biology*, Vol. 11. D. M. Prescott (Ed.). Academic, New York, pp. 46-69.

Haber, J. E., Esposito, M. S., Magee, P. T., and Esposito, R. E. (1975). Current trends in genetic and biochemical study of yeast sporulation. In *Spores*, Vol. 6. P. Gerhart, R. Castilow, and H. Sadoff (Eds.). American Society of Microbiology, Washington, D.C., pp. 132-137.

Harper, J. F., Clancy, M. J., and Magee, P. T. (1980). Properties of polyadenylate associated RNA from *Saccharomyces cerevisiae* ascospores. *J. Bacteriol. 143*:958-965.

Hartig, A., and Breitenbach, M. (1977). Intracellular cAMP in synchronous sporulation in *S. cerevisiae. FEMS Lett. 1*:77-82.

Hopper, A. K., and Hall, B. D. (1975). Control of yeast sporulation by the mating type locus. In *Spores*, Vol. 6. P. Gerhart, R. Castilow, and H. Sadoff (Eds.). American Society of Microbiology, Washington, D.C., pp. 138-146.

Hopper, A. K., and MacKay, V. L. (1980). Control of sporulation in yeast. *SADI* a mating type specific unstable alteration that uncouples sporulation from mating type control. *Mol. Gen. Genet. 180*:301-314.

Hopper, A. K., Magee, P. T., Welsch, S. K., Friedman, M., and Hall, B. D. (1974). Macromolecular synthesis and breakdown in relation to sporulation and meiosis in yeast. *J. Bacteriol. 119*:619-628.

Illingworth, R. F., Rose, A. H., and Beckett, A. (1973). Changes in lipid composition and fine structure of *S. cerevisiae* during ascus formation. *J. Bacteriol. 113*:373-386.

Kane, S. M., and Roth, R. (1974). Carbohydrate metabolism during ascospore development in yeast. *J. Bacteriol. 118*:8-14.

Kassir, Y., and Simchen, G. (1976). Regulation of mating and meiosis in yeast by the mating type region. *Genetics 82*:187-206.

Klar, A. J. S. (1980). Mating type functions for meiosis and sporulation in yeast act through cytoplasm. *Genetics 94*:597-605.

Klar, A. J. S., and Halvorson, H. O. (1975). Proteinase activities of *S. cerevisiae*. *J. Bacteriol. 124*:863-869.

Kopecká, M. (1974). Sporulation in protoplasts of the yeast *Saccharomyces cerevisiae*. *J. Gen. Microbiol. 83*:171-178.

Kraig, E., and Haber, J. E. (1980). mRNA and protein metabolism during sporulation of *S. cerevisiae*. *J. Bacteriol. 144*:1089-1112.

McCusker, J. H., and Haber, J. E. (1977). Efficient sporulation of yeast in media buffered near pH 6.0. *J. Bacteriol. 132*:180-185.

Magee, P. T., and Hopper, A. K. (1974). Protein synthesis in relation to sporulation and meiosis in yeast. *J. Bacteriol. 119*:952-960.

Marian, B., and Wintersberger, U. (1980). Histone synthesis during sporulation in yeast. *FEBS Lett. 117*:63-67.

Matur, A., and Berry, D. R. (1978a). The use of step enzymes as markers during meiosis and ascospore formation in *Saccharomyces cerevisiae*. *J. Gen. Microbiol. 109*:205-213.

Matur, A., and Berry, D. R. (1978b). Step enzyme formation during yeast sporulation in controlled fermenter conditions. *FEMS Lett. 4*:91-95.

Matur, A., and Berry, D. R. (1978c). Effects of hydroxyurea and cycloheximide on step enzyme formation during sporulation of *Saccharomyces cerevisiae*. *FEMS Lett. 4*:195-198.

Mills, D. (1972). Effect of pH on adenine and amino acid uptake during sporulation in *S. cerevisiae*. *J. Bacteriol. 112*:519-526.

Mills, D. (1978). 8-Hydroxy quinoline inhibition of DNA synthesis and intragenic recombination during yeast meiosis. *Mol. Gen. Genet. 162*:221-228.

Mills, D. (1980). Quantitative and qualitative analysis of polyadenylated RNA sequences during meiosis of *S. cerevisiae*. In *Tenth International Conference on Yeast Genetics and Molecular Biology*. Louvaine La Neuve, Belgium, pp. 55.

Mills, D., Schiller, M., and Petryna, M. M. (1977). Characteristics of pulse labeled RNA in relation to medium pH and acetate concentration during meiosis of yeast. *Mol. Gen. Genet. 156*:71-78.

Mitchison, J. M. (1973). Cell cycle of a eukaryote. *Soc. Gen. Microbiol. Symp. 23*:189-208.

Miyake, S., Sando, N., and Sato, S. (1971). Biochemical changes in yeast during sporulation. II. Acetate metabolism. *Dev. Growth Differ. 12*:285-295.

Moens, P. B., and Rapport, E. (1971a). Spindles, spindle plaques, and meiosis in the yeast *Saccharomyces cerevisiae* (Hansen). *J. Cell Biol. 50*:344-361.

Moens, P. B., and Rapport, E. (1971b). Synaptic structures in the nuclei of sporulating yeast, *Saccharomyces cerevisiae* (Hansen). *J. Cell Sci. 9*:665-677.

Moens, P. B., Mowat, M., Esposito, M. S., and Esposito, R. E. (1977). Meiosis in a temperature sensitive DNA synthesis mutant and in an apomictic yeast strain *(Saccharomyces cerevisiae). Philos. Trans. R. Soc. Ser. B. 277*:351-358.

Olson, L. W., and Zimmerman, F. K. (1978). Mitotic recombination in the absence of synaptonemal complexes in *S. cerevisiae. Mol. Gen. Genet. 166*:161-165.

Olson, L. W., Eden, U., Egel-Matani, M., and Egel. R. (1978). Asynaptic meiosis in fission yeast. *Hereditas 89*:189-200.

Patel, P. V., and Miller, J. J. (1972). Stimulation of yeast sporulation by glycerol. *J. Appl. Bacteriol. 35*:63-69.

Pearson, M. J., and Haber, J. E. (1980). Changes in regulation of ribosomal protein synthesis during vegetative growth and sporulation of *S. cerevisiae. J. Bacteriol. 143*:1411-1419.

Petersen, J. G. L., Olson, L. W., and Zickler, D. (1978). Synchronous sporulation of *Saccharomyces cerevisiae* at high cell concentrations. *Carlsberg Res. Commun. 43*:241-254.

Piñon, R. (1977). Effect of ammonium ions on sporulation of *S. cerevisiae. Exp. Cell Res. 105*:367-378.

Promod, K., Scrivastava, P. K., Harashima, S., and Oshima, Y. (1981). Formation of two spored asci by interrupted sporulation in *S. cerevisiae. J. Gen. Microbiol. 123*:39-48.

Reid, B. J., and Hartwell, L. H. (1977). Regulation of mating in cell cycle of *S. cerevisiae. J. Cell Biol. 75*:335-365.

Resnick, M. A., Kasimos, J. M., Game, J. C., Braun, R. J., and Roth, R. M. (1981). Changes in DNA during meiosis in a repair deficient mutant (*rad 52*) of yeast. *Science 212*:543-545.

del Rey, F., Santos, T., Garcia-Acha, I., and Norribela, C. (1980). Synthesis of β glucanases during sporulation in *S. cerevisiae. J. Bacteriol. 143*:621-627.

Rooney, L., and Moens, P. B. (1973). The fine structure of ascospore delimitation in the yeast *Wickerhamia fluorescens*. *Can. J. Microbiol.* *19*:1389-1392.

Roth, R., and Fogel, S. (1971). A system selective for yeast mutants deficient in meiotic recombination. *Mol. Gen. Genet.* *112*:295-305.

Roth, R., and Halvorson, H. O. (1969). Sporulation of yeast harvested during logarythmic growth. *J. Bacteriol.* *98*:831-832.

Schild, D., and Byers, B. (1978). Meiotic effects of DNA-defective cell division cycle mutations of *Saccharomyces cerevisiae*. *Chromosoma* *70*:109-130.

Schild, D., and Byers, B. (1980). Diploid spore formation and other meiotic effects of two cell division cycle mutations of *S. cerevisiae*. *Genetics* *96*:859-876.

Schroeder, R., and Breitenbach, M. (1981). Metabolism of *myo* inositol during sporulation of *myo* inositol-requiring *S. cerevisiae*. *J. Bacteriol.* *146*:775-783.

Simchen, G. (1974). Are mitotic functions required in meiosis? *Genetics* *76*:745-753.

Simchen, G., and Hirschberg, J. (1977). Effect of the mitotic cell cycle mutation *cdc 4* on yeast meiosis. *Genetics* *86*:57-72.

Simchen, G., Idar, D., and Kassir, Y. (1978). Recombination and hydroxy urea inhibition of DNA synthesis in yeast meiosis. *Mol. Gen. Genet.* *144*:21-27.

Sogin, S. J., Chaffin, W. L., and Halvorson, H. O. (1975). Phosphate uptake and RNA synthesis during sporulation in *Saccharomyces cerevisiae* cells. In *Spores*, Vol. 6. P. Gerhart, R. Castilow, and H. Sadoff (Eds.). American Society for Microbiology, Washington, D.C., pp. 241-258.

Suomalainen, H., and Oura, E. (1971). Yeast nutrition and solute uptake. In *The Yeasts*, Vol. 2. A. H. Rose and J. S. Harrison (Eds.). Academic, New York, pp. 3-74.

Tingle, M., Klar, A. J. S., Henry, S. A., and Halvorson, H. O. (1973). Ascospore formation in yeast. *Soc. Gen. Microbiol. Symp.* *23*:209-243.

Trew, B. J., Friesen, J. D., and Moens, P. B. (1979). Two dimensional protein patterns during growth and sporulation in *S. cerevisiae*. *J. Bacteriol.* *138*:60-69.

Tsuboi, M., Komisaka, S., and Yanagishima, N. (1972). Effect of cyclic 3'5' adenosine monophosphate on sporulation in *S. cerevisiae*. *Plant Cell Physiol.* *13*:585-588.

Vezinhet, F., Arnaud, A., and Galzy, P. (1971). Étude de la respiration de la levure au conis de la sporulation. *Can. J. Microbiol.* *17*:1179-1184.

Vezinhet, F., Kinnaird, J. H., and Dawes, I. W. (1979). The physiology of mutants derepressed for sporulation in *S. cerevisiae*. *J. Gen. Microbiol.* *115*:391-402.

Watson, D.C. (1978). Ph.D. thesis, University of Strathclyde, Glasgow, Scotland.

Watson, D. C., and Berry, D. R. (1977). Intracellular and extracellular cAMP in *S. cerevisiae*. *FEMS Lett.* *1*:83-86.

Watson, D. C., and Berry, D. R. (1979). Use of an exchange filtration technique to obtain synchronous sporulation in an extended batch fermentation. *Biotechnol. Bioeng.* *21*:213-220.

Wejksnora, P. J., and Haber, J. E. (1978). Ribonucleoprotein particle appearing during sporulation in yeast. *J. Bacteriol.* *134*:246-260.

Whelan, W. L., and Ballou, C. E. (1975). Sporulation in D glucosamine auxotrophs of *Saccharomyces cerevisiae* meiosis with defective ascospore wall formation. *J. Bacteriol.* *124*:1545-1557.

Williamson, D. H., Johnson, L. H., Fennell, D. J., and Simchen, G. (1980). Premeiotic DNA synthesis in *S. cerevisiae* replicon: size and rate of replicon fork movement. In *Tenth International Conference on Yeast Genetics and Molecular Biology*. Louvaine La Neuve, Belgium, pp. 181.

Wright, J. F., and Dawes, I. W. (1979). Sporulation specific protein changes in yeast. *FEBS Lett.* *104*:183-186.

Wright, J. F., and Dawes, I. W. (1980). Phosphorylation as a possible mechanism for protein changes during yeast sporulation. In *Tenth International Conference on Yeast Genetics and Molecular Biology*. Louvaine La Neuve, Belgium, pp. 155.

Zickler, D., and Olson, L. W. (1975). The synaptonemal complex and the spindle plaque during meiosis in yeast. *Chromosoma* *50*:1-23.

Zubenko, G., Mitchell, A., and Jones, E. (1980). Septum formation, cell division and sporulation in mutants of yeast deficient in proteinase B. *Proc. Nat. Acad. Sci. USA* *76*:2395-2399.

7

The Blastocladialean γ Particle: Once Viral Endosymbiont (?), Now "Chitosome" Progenitor

EDWARD C. CANTINO and GARY L. MILLS *Michigan State University, East Lansing, Michigan*

I. INTRODUCTION

The aquatic fungus *Blastocladiella emersonii* Cantino and Hyatt exhibits a life cycle seemingly devoid of conventional gametic copulation. Its several developmental pathways have been described (Cantino, 1966; Hennessy and Cantino, 1972; Lovett, 1975), illustrated (Lovett, 1975; Cantino and Mills, 1978), and filmed (for source and film outline, see Cantino, 1975; Cantino and Myers, 1974).

The nongrowing, motile stage in the life history of *B. emersonii* is represented by a single-celled zoospore (Fig. 1, top) containing a number of structures and organelles (Fig. 2). Some of them (e.g., lipid globules, ribosomes, nucleus, etc.) appear to be identical to similar structures present in the growing thallus from which the zoospores are formed; others arise by fusion (mitochondrion, symphyomicrobody) or by reorganization (nuclear cap) of entities also present in the previously growing thallus. One group of membrane-bound organelles, however – the γ particles – does *not* preexist in the thallus; instead, these particles are formed de novo during zoosporogenesis, and they disappear again when a zoospore encysts and then develops a germ tube. Although γ particles are not transmitted by sexual fusions in *B. emersonii*, they may be transferred from cell to cell, albeit at low frequency, via temporary cytoplasmic bridges (Cantino and Horenstein, 1954; Fig. 1, bottom).

The γ particle was discovered (Cantino and Horenstein, 1954) and first described (Cantino and Horenstein, 1956) 25 years ago, but meaningful studies of it had to await the application of electron microscopy and developments in other kinds of instrumentation and labora-

---

FIG. 1 Photomicrographs of zoospores of *B. emersonii* taken by scanning electron microscopy. Plates of PYG (peptone, yeast extract, glucose) agar containing mature thalli were flooded with water, fixed with 1.5% glutaraldehyde, and prepared for examination by critical point drying (Courtesy of Dr. Marie Cantino and the Electron Microscope Facility, Center for Bioengineering, University of Washington, Seattle). Top: Single zoospore. The bulge at the arrow is probably formed by the symphyomicrobody-lipid complex and its adjacent mitochondrion (× 6400). Bottom: Two zoospores connected by a cytoplasmic bridge (× ca. 6400).

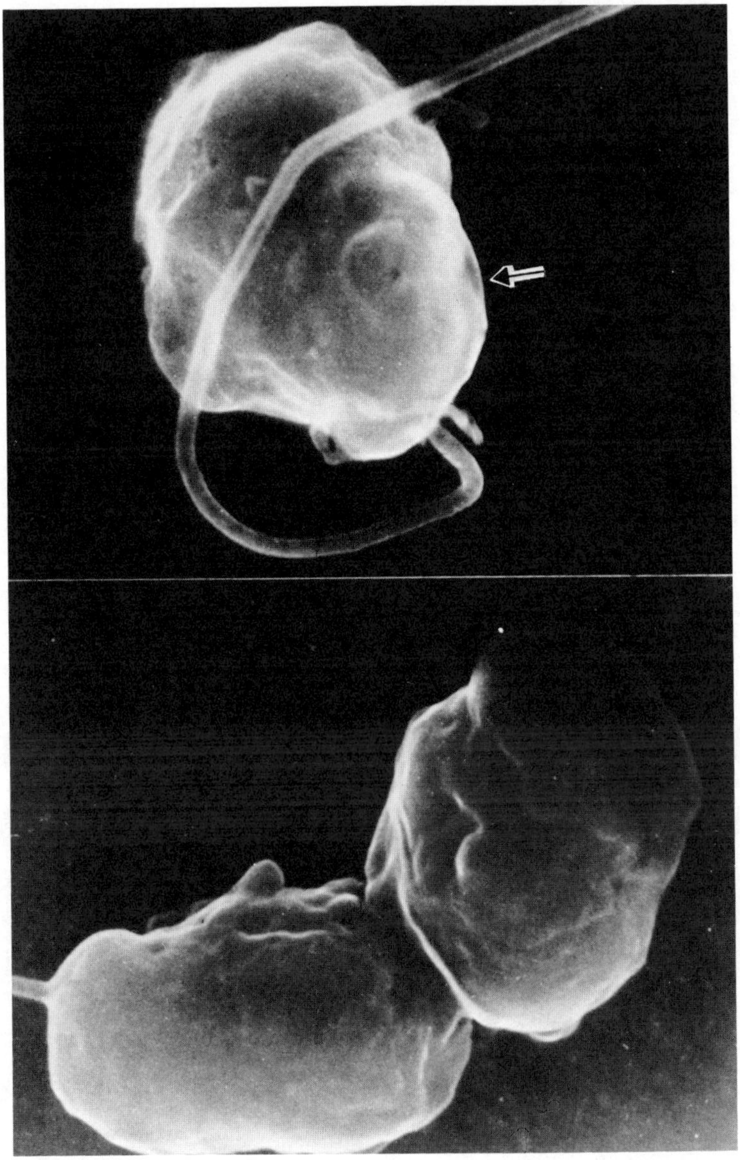

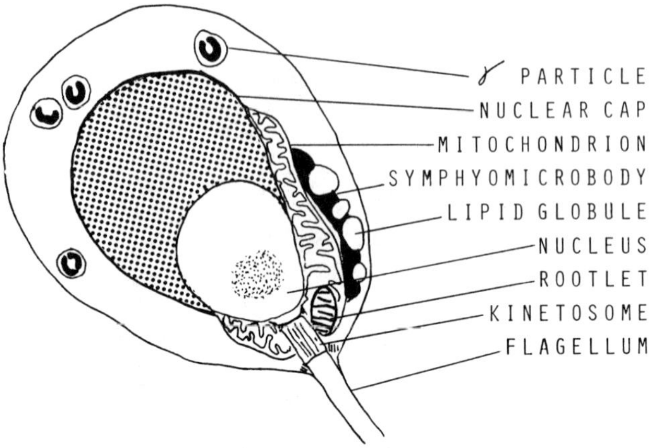

FIG. 2 The *B. emersonii* zoospore. The spore contains 12 (±7) γ particles, 8.2 (±1.5) × 10^5 ribosomes in the nuclear cap, 9 (±2) lipid globules, and one each of the other labeled structures. The developmental origin of all of them has been established, and all but three of them (nucleus, rootlet, kinetosome) have been isolated and studied biochemically in vitro (for references, see Cantino and Mills, 1979).

tory techniques. The biology of the *B. emersonii* γ particle has been summarized from various points of view by Truesdell and Cantino (1971), Cantino and Myers (1974), Myers and Cantino (1974), Lovett (1975), and Cantino and Mills (1976, 1979).

The purpose of this essay is to consolidate results of recent studies of γ particles with some of the older information and then to bring about some speculative synthesis at three levels of integration: ontogeny and function, the potential practical utility of the γ particle's chitin-synthesizing mechanism, and phylogeny. The first of these levels will emphasize the now clearly established presence of γ particles in three different families of aquatic fungi, some saprophytic and some parasitic, summarize what is known of their genesis and decay, and point to the probability that γ particles are progenitors of "chitosomes" (Bracker et al., 1976) in the zoospores of this group of fungi. The second level, based upon the presence of a distinctive glycolipid present in the γ particle, will aim at the possibly practical significance of glycosyldiacylglycerols as potential targets for control of opportunistic fungal pathogens. The third will point to the novel possibility that the γ particle was once a viral endosymbiont, perhaps a sort of pox virus.

II. THE γ PARTICLE IS WIDELY DISTRIBUTED

There is now very good evidence that organelles similar to the *B. emersonii* γ particle occur in at least two other genera of fungi (*Allomyces* and *Catenaria*) and almost certainly in a third (*Coelomomyces*); hence γ particles are to be found in three families of aquatic fungi, two of them obligately or facultatively parasitic. For the first two of these genera, the evidence was derived from ultrastructural and developmental studies employing synchronous and semisynchronous cultures; for the third genus, the evidence is based only (but convincingly) on ultrastructural considerations.

A. The *Blastocladiella emersonii* γ Particle as a Point of Reference: Some Selected Aspects

1. *Physical Form*

The γ particle is approximately 500 nm in diameter and consists of a cup-shaped core surrounded by at least one membrane (Cantino and Mack, 1969), and possibly two (Cantino and Mills, 1979). Random thin sections through γ particles in situ most frequently consist of a matrix shaped like a horseshoe (Fig. 3) or crescent, less frequently two opposing crescents or a cylinder, with or without discontinuities, enclosed by a membrane. The matrix is usually differentiated into two regions of differing electron transparency, the outer region being the most electron opaque (Cantino and Mack, 1969; Truesdell and Cantino, 1970).

2. *Method of Development*

The γ particle is formed (Fig. 4, top) during zoosporogenesis (Barstow and Lovett, 1975). Electron-opaque granules 40 nm in diameter with electron-transparent centers appear in cisternae of rough endoplasmic reticulum, increase in number, and seemingly fuse to form doublets and triplets. Further coalescence yields aggregates 100 nm in diameter (the endoplasmic reticulum by this time is smooth surfaced), and these, in turn, aggregate to yield the characteristic γ-particle matrix surrounded by a membrane. The latter is presumed to be the cisternal membrane within which the original 40-nm "subunits" appeared.

3. *Method of Decay*

The γ particle decays at the time of zoospore encystment (Truesdell and Cantino, 1970). Vesicles about 80 nm in diameter appear in the inner (i.e., electron-transparent) portion of the γ-particle matrix (Fig. 4, bottom). Simultaneously, the γ particle's surrounding membrane buds off vesicles about 50-100 nm in diameter which migrate to the cell surface, where they appear to fuse with the plasma membrane

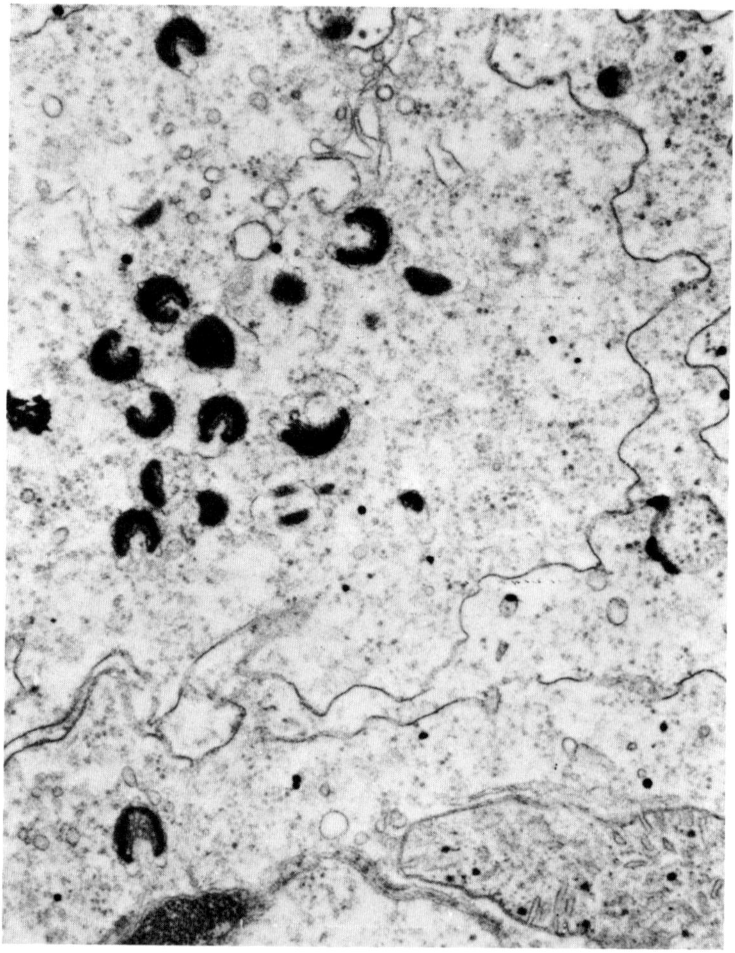

FIG. 3 Profiles of γ particles in a sectioned *B. emersonii* zoospore (before release from sporangium), fixed ("fix II") and processed according to Truesdell and Cantino (1970) (× 16,000).

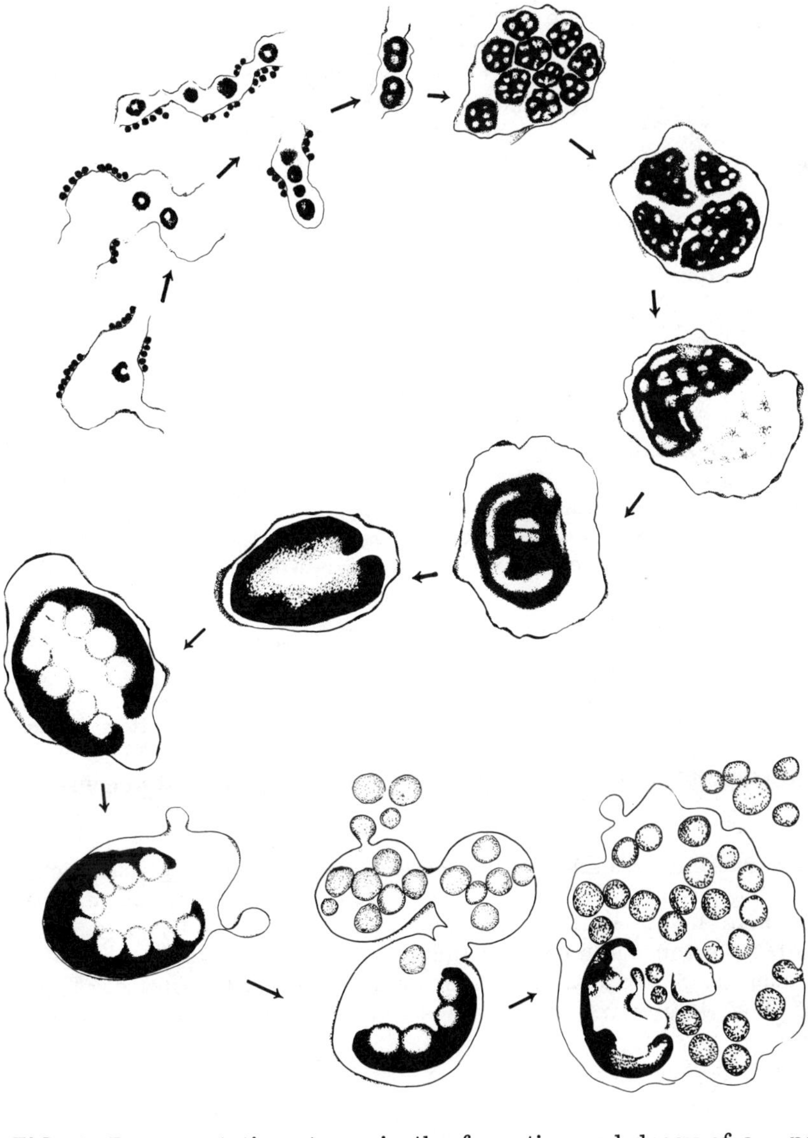

FIG. 4 Representative stages in the formation and decay of a γ particle (see text for explanation). Drawings are based on Figs. 1-10 in Barstow and Lovett (1975) and Figs. 4-11 in Truesdell and Cantino (1970), which should be consulted for details.

at the time that deposition of the first cell wall is detected (Truesdell and Cantino, 1970, 1971). Meanwhile, some of the 80-nm vesicles generated by the γ particle's matrix fuse with the particle's surrounding membrane while the latter is budding off vesicles. The γ-particle matrices continue to decay by releasing 80-nm vesicles, and as the γ particle's surrounding membrane continues to bud off 50 to 100-nm vesicles, the γ-particle matrices are consumed and eventually disappear. In the final stages of the process, the surrounding membranes of decaying γ particles fuse with one another, thus generating increasingly large multivesicular bodies.

B. The *Allomyces* γ Particle

1. *Physical Form*

Although profiles that somewhat resemble those of *B. emersonii* γ particles have been pictured in several publications on the motile cells of *Allomyces* (Barstow and Pommerville, 1980, and references therein), the best and most recent evidence for the existence of γ particles in *Allomyces* is to be found in the work of Barstow (1979). Although the three-dimensional form of the *Allomyces* γ particle has not yet been reconstructed from serial sections, it is now quite clear that *Allomyces* zoospores *do* contain cytoplasmic organelles which, when sectioned, display an ultrastructure unquestionably like that of sectioned γ particles of *B. emersonii*.

2. *Method of Development*

The sequence of intracellular transitions leading to γ-particle formation in *Allomyces macrogynous* has been traced electron microscopically (Barstow, 1979), using cultures with a definable level of synchrony, thus permitting a reasonable reconstruction of the events involved. The first identifiable progenitors of the γ-particle matrix are about 60-nm-diameter granules within cisternae of rough endoplasmic reticulum. The granules then increase in number and generate larger granules about 130 nm in diameter, the latter fuse to form larger spheres about 215 nm in diameter, and these eventually fuse to form a $\sim$500-nm-diameter γ particle. In all essential respects, this process is like that (Fig. 4) involved in the genesis of the *B. emersonii* γ particle.

3. *Method of Decay*

Synchronously germinating *A. macrogynous* mitospores have also been used to follow the decay of γ particles (Barstow and Pommerville, 1980). Ultrastructural studies showed that during spore encystment, $\sim$35- to 70-nm-diameter vesicles became concentrated at the spore surface and appeared to be fusing with the plasma membrane during the formation of the cyst wall, the vesicles apparently having arisen from γ particles. These results suggest that the response of *Allomyces* to induction of encystment, involving the formation and release of vesicles from γ particles, is very similar to that (Fig. 4) of the *B. emersonii* zoospore.

Observations by scanning electron microscopy have also revealed that both *Allomyces arbuscula* (Seale et al., 1974) and *Allomyces macrogynous* (Rønne and Bøye, 1978) form vesicles, interpreted as probably arising from γ particles, which accumulate at the surface of the zoospore plasma membrane at the time that cell wall material is being deposited.

C. The *Catenaria* γ Particle

1. *Physical Form*

Electron microscopic evidence suggesting the existence of structures resembling *B. emersonii* γ particles in the motile cells of *Catenaria* was provided by Manier (1977) and Olson et al. (1978). The most recent documentation (Barstow, 1979) not only supports this conclusion, but also demonstrates that the profiles of *Catenaria* γ particles, while only about half the size of the *B. emersonii* γ particles, are otherwise very similar to them. Their three-dimensional form, however, has not yet been reconstructed from serial sections.

2. *Method of Development*

Barstow (1979) has traced the ultrastructural intracellular transitions leading to the formation of γ particles in *Catenaria anguillulae*. Although the fungus could not be induced to differentiate zoosporangia and zoospores synchronously, it was grown and cells were selected for study in such a fashion that plausible correlations could be made between the steps in γ-particle formation and the stages in the development of the fungus. The first identifiable progenitors of γ particles are electron-opaque granules, 45 nm in diameter, that appear within cisternae of *smooth* endoplasmic reticulum. As is the case for *A. macrogynous* and *B. emersonii*, these granules then form larger aggregates which eventually fuse to form the differentiated γ-particle matrix, the final γ particles of *C. anguillulae*, however, being only 250 nm in diameter, that is, smaller than the ∿500-nm-diameter γ particles of *A. macrogynous* and *B. emersonii*.

The method of decay for the γ particles of *Catenaria* has not yet been established.

D. The *Coelomomyces* γ Particle

1. *Physical Form*

Travland (1979) has documented the existence of γ-like bodies in the motile cells of *Coelomomyces psorophorae*. While none of the pictured profiles look exactly like the profiles of *B. emersonii*, *A. macrogynous*, or *C. anguillulae*, their dense matrix (with its inner electron-transparent region) as well as their apparent capacity to vesiculate during encystment, taken together with the profiles published by Loubes and Manier (1974), strongly suggest that these organelles resemble γ particles and that in all likelihood they are essentially equivalent to the γ particles in the three aforementioned species.

E. Consolidation

Barstow (1979) made the reasonable suggestion that use of the terms γ *particle* and γ*-like particle* ought to be reserved for cytoplasmic organelles in the motile cells of aquatic fungi that have a morphology, method of development, or pattern of decay similar to that of the γ particles in *B. emersonii*. We sympathize with this point of view and, in fact, have even suggested one additional requirement (Cantino and Mills, 1976): that biochemical and other features of the isolated organelles should also be used as criteria for purposes of organelle identification. Indeed, such information would be *essential* for establishing whether or not an organelle in question could be called an "encystosome" (Cantino and Mills, 1976), for which the *B. emersonii* γ particle serves as the "type specimen."

However, since the γ particles from *B. emersonii* are the only ones that have been isolated thus far and studied in vitro, comparative considerations must be limited for the time being to nonphysiological aspects. In this respect, an important point must be made here. While the presence of certain morphological features may logically permit tentative identification of a particular organelle as a γ particle, it does not follow that the absence of the same morphological features can always lead just as logically to the opposite conclusion, that is, serve to rule out such an identification. As a case in point, consider the "γ-like particles" (Cantino and Truesdell, 1971) which occur in the zoospores of another species of *Blastocladiella, Blastocladiella britannica* Horenstein and Cantino. These organelles clearly resemble *B. emersonii* γ particles in size, average number per spore, strongly osmiophillic nature, and apparent capacity to produce and bud off vesicles. However, the *B. britannica* particles differ ultrastructurally (Cantino and Truesdell, 1971) from the typical *B. emersonii* γ particles in that they did not exhibit horseshoelike profiles when sectioned.

On the other hand, under certain conditions, the γ particles of *B. emersonii* can take on an appearance greatly resembling that of the γ-like particles of *B. britannica*. This becomes immediately evident when the profiles of the *B. emersonii* γ particles, as found in sections through zoospores that have rapidly lysed, are compared with the profiles of the *B. britannica* γ-like particles, for example, as seen in Fig. 4 of Cantino and Truesdell (1971). This may simply mean that γ-like particles in *B. britannica* zoospores tend to decay rather quickly on "provocation" (e.g., when they are manipulated, fixed for microscopy, etc.), whereas the γ particles of *B. emersonii* are more resistant to such perturbations, and extremely rapid "decay" is induced only rarely, such as under conditions potent enough to cause the spores to lyse. Other explanations for the apparent presence of vesicles in the *B. britannica* γ-like particles are possible, but irrespective of the explanation, the fact is that the zoospores of two different species of *Blastocladiella* contain γ particles that can appear to be quite dissimilar structurally under certain conditions of fixation and staining, yet they look structurally similar when spores are fixed under a different set of circumstances.

The most recent investigations (W. E. Barstow, 1980, personal communication of unpublished observations) of *B. britannica*, using the same strain employed by Cantino and Truesdell (1971), demonstrate that the γ particles in *B. britannica* zoospores can, indeed, look very similar to the γ particles of *Allomyces*!

Clearly, some variation in the apparent structure of γ particles in the Chytridiomycetes should be expected. The "identification" of γ particles will have to depend not only on their apparent structure, but also, in good measure, upon studies of their formation during synchronous sporogenesis and their decay during synchronous zoospore encystment, as well as upon in vitro biochemical studies of the isolated γ particles.

In conclusion, the *B. emersonii* particle occurs not only in at least two different species of *Blastocladiella*, but, unquestionably, also in at least three genera of Chytridiaceous fungi, and almost certainly in a fourth. These fungi are distributed among three families of the order Blastocladiales, parasitic as well as saprophytic in habitat and worldwide in distribution. Recognizing that even within two species of the same genus (*B. emersonii* and *B. britannica*), apparent variation in γ-particle morphology can occur (at least as seen *after* cells are processed for microscopy), it can reasonably be expected that γ particles may be even more abundant than is presently envisioned.

III. CHITIN SYNTHESIS: THE γ-PARTICLE-CHITOSOME CONNECTION

While the zoospores of *B. emersonii* do not contain chitin, they do possess the capacity for synthesizing it (Camargo et al., 1967; Myers and Cantino, 1974). All of the enzymes involved in the biosynthetic path for chitin have been found in the zoospore. All of them are "soluble", except the last enzyme of the path, chitin synthetase, which is localized in γ particles (Myers and Cantino, 1974; Norrman et al., 1975; Giddings and Cantino, 1974; Selitrennikoff et al., 1976; Selitrennikoff and Sonneborn, 1976); hence γ particles are biochemically involved in chitin synthesis.

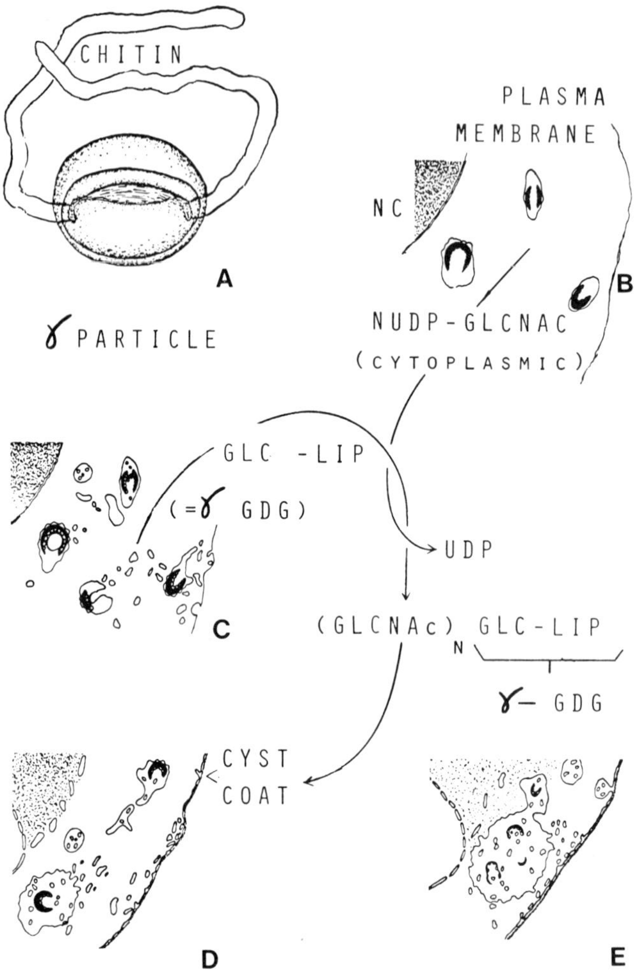

FIG. 5 Relationship between γ-particle decay and initiation of chitin synthesis during zoospore encystment in *B. emersonii* (see text for explanation). This figure was originally used at a symposium lecture, "The role of the γ particle in cyst wall synthesis in *Blastocladiella emersonii*", Second International Mycological Congress, Tampa, Florida, 1977 (NC = nuclear cap; GLC = glucose, GLC-LIP = monoglucosyldiglyceride).

A. Mechanism of the Interactions Between *B. emersonii* γ Particles, Chitin Synthesis, and Wall Formation During Zoospore Encystment

Ultrastructural evidence has shown that γ particles are involved in cyst wall formation (Truesdell and Cantino, 1970, 1971). As summarized in Fig. 4, γ particles undergo a number of distinct ultrastructural changes when nongrowing motile zoospores are induced to convert synchronously into nonmotile growing cells. The γ particles generate and release vesicles which migrate to the plasma membrane. As the vesicles fuse with the plasma membrane, a cyst wall is formed.

Biochemical evidence has shown that a sugar-linked lipid, a monoglucosyldiglyceride present in the γ particle, is involved in chitin synthesis (Mills and Cantino, 1978, 1980). This fact points to the probable role of the γ particle monoglucosyldiglyceride in cell wall formation during encystment.

These ultrastructural and biochemical observations can be integrated as follows (Fig. 5). In the zoospore, γ particles are intact (Fig. 5b) and contain chitin synthetase and the lipid intermediate. If such γ particles are isolated, they exhibit chemically detectable chitin synthetase activity. Additionally, chitin macrofibrils emanating from them (Fig. 5A) are demonstrable in shadow-cast preparations derived from the above reaction mixtures (Mills and Cantino, 1978).

However, the events described above undoubtedly do not happen in vivo. The γ-particle surrounding membrane presumably prevents contact between the chitin synthetase, which is associated with the γ particle (Myers and Cantino, 1974) and its substrate, UDP-N acetylglucosamine (UDP-GlcNAc), which is in the cytosol (Selitrennikoff et al., 1976).

When zoospores are triggered to encyst and γ particles start to "unravel" (Fig. 5C) and release their core material in the form of vesicles containing chitin synthetase and the glycosyldiglyceride (γ-GDG), the enzyme and substrate become mutually accessible. The γ-GDG acts as a primer molecule (Mills and Cantino, 1978), that is, as the initial acceptor of the GlcNAc units in the chitin synthetase reaction. Increasing numbers of GlcNAc molecules become attached to the lipid, forming a chitin molecule (Fig. 5). The lipid-chitin complex associated with the vesicles is carried to the plasma membrane where the cyst wall is constructed (Fig. 5D,E).

The foregoing model is based upon the results of (1) biochemical and ultrastructural analyses in experiments which involved isolated γ particles, and (2) ultrastructural analyses of zoospores that had been induced to encyst synchronously. The former observations showed that isolated, *intact* γ particles exhibit chitin synthetase activity and can produce chitin fibrils. The latter observations showed that γ particles do not remain intact during zoospore encystment but, instead, produce vesicles which migrate to the cell's surface and fuse

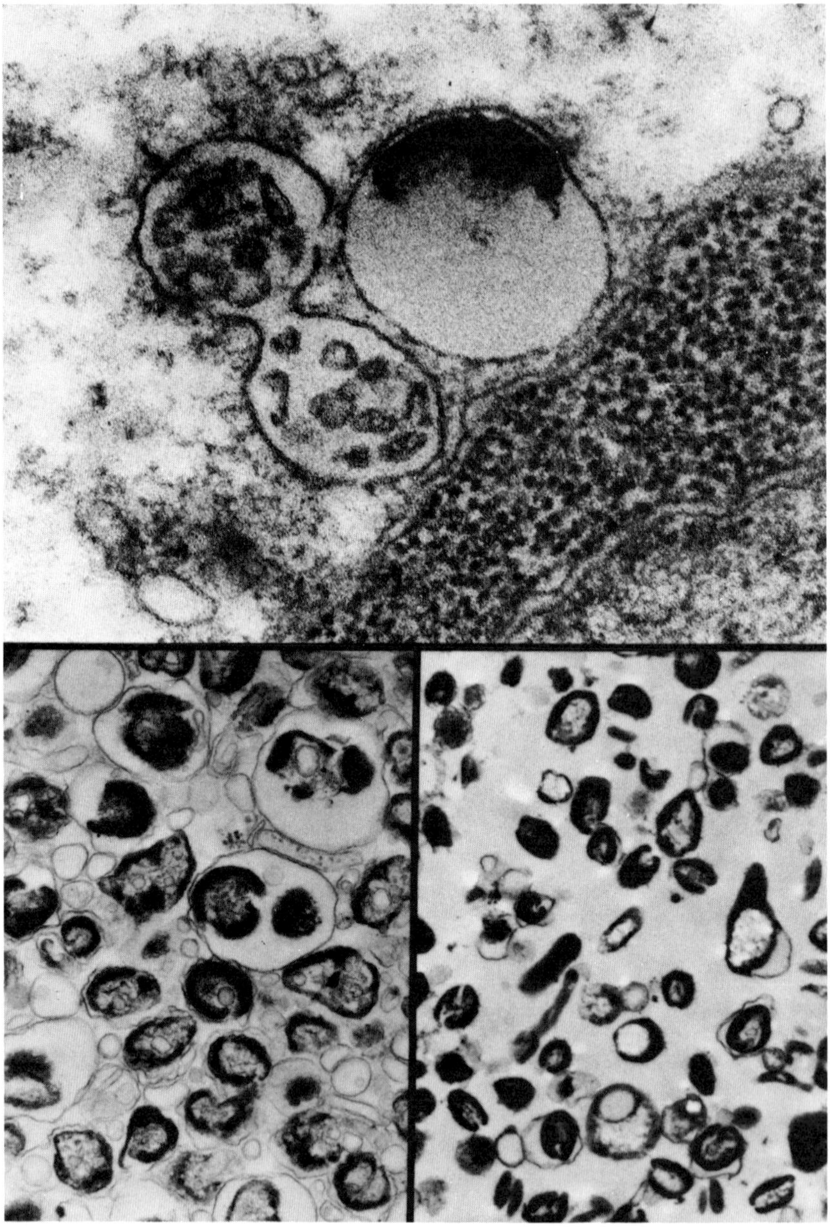

with the plasma membrane. It may seem reasonable to suppose, therefore, that the vesicles which come from the γ particles also contain chitin synthetase and are capable of forming chitin fibrils. However, until recently there was no experimental evidence to support such an assumption. In the following section, however, we will show that these vesicles are, in fact, capable of performing these functions and that γ-particle vesicles resemble another class of chitin-synthesizing particles.

B. The Vesicles Generated and Released by γ Particles Synthesize Chitin and Resemble Chitosomes

The results described here are based upon recent work (Mills and Cantino, 1981). Isolated γ particles will vesiculate when they are incubated in a cell-free zoospore homogenate obtained from spores that have previously been induced to encyst with KCl. Both the way in which vesiculation occurs in vitro and the nature of the vesicles produced closely resemble what happens during γ-particle vesiculation as observed in vivo (Fig. 6). Incubation of γ particles in the vesiculation medium results in a threefold increase in chitin synthetase activity, as well as a shift in the buoyant density of the particulates in the suspension: from 1.32 g/cm^3 for γ particles to 1.18 g/cm^3 for γ-particle vesicles. Under optimal conditions, more than 90% of the isolated γ particles can be induced to vesiculate completely. Optimal conditions are defined in terms of the nature of the vesiculation medium. For best results, the vesiculation medium must be prepared from a large number of zoospores (about 5×10^9 spores) in which less than 5% of them have encysted. The identity of the "active" component in the vesiculation medium is unknown.

Isopycnic centrifugation of the γ-particle vesicles through a linear sucrose gradient (0.88-1.5 M) shows that chitin synthetase activity is associated with particles with buoyant densities between 1.17 and 1.20 g/cm^3. Negatively stained samples of the peak fraction contain small spheroidal bodies with diameters of 70-120 nm. Each particle is surrounded by a delimiting shell 8-16 nm in thickness. When such particles are incubated with UDP-GlcNAc, the substrate for chitin synthetase, chitin microfibrils, are formed.

These results illustrate that γ particles are capable of producing vesicles in vitro, and that these vesicles contain chitin synthetase and can produce chitin microfibrils. They also support and strengthen the

FIG. 6 Top: In situ appearance of γ-particle vesicles during zoospore encystment (× 80,000). (From Truesdell and Cantino, 1970.) Bottom left: Isolated γ particles incubated in vesiculation medium for 30 min (× 23,000). Bottom right: Isolated γ particles not incubated in vesiculation medium (× 20,500). (From Myers and Cantino, 1974.)

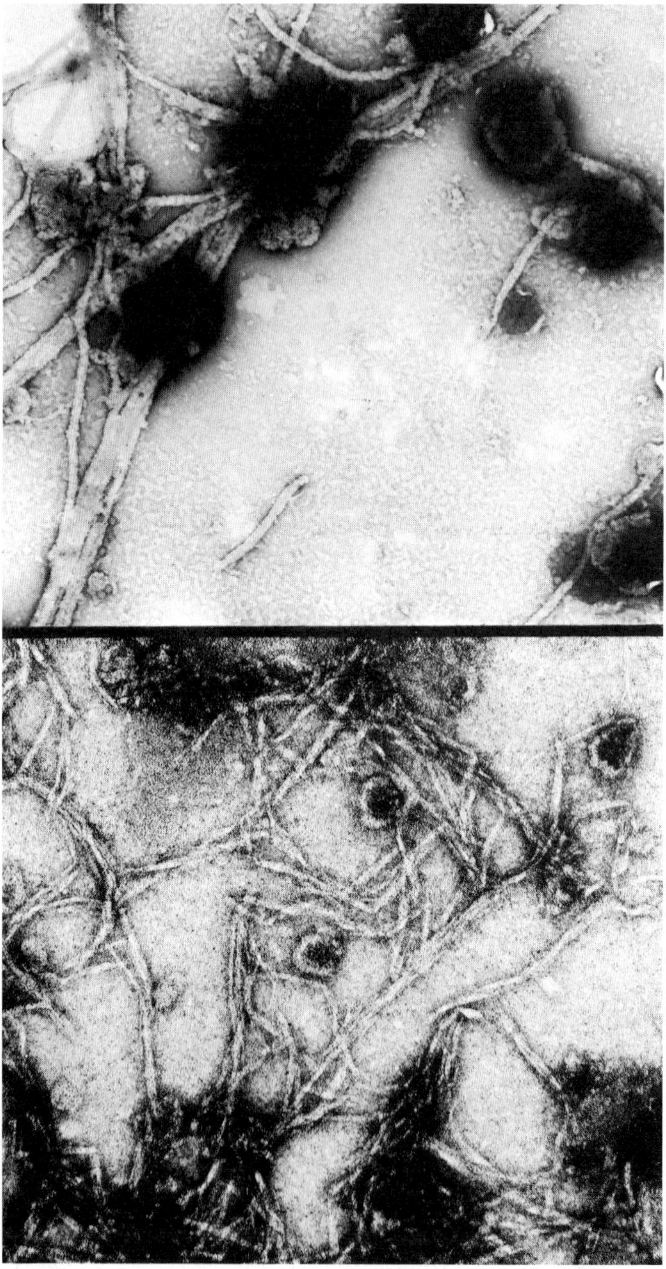

TABLE 1 A Comparison of γ Particles and γ-Particle Vesicles

	γ Particle	γ-Particle vesicle
Organelle diameter[a]	450-750 nm	70-120 nm
Thickness of fibril	40-100 nm	8-17 nm
Thickness of shell	Not applicable	8-16 nm
Chitin synthetase specific activity[b]	14	104
Buoyant density	1.32 g/cm^3	1.18 g/cm^3[c]

[a]Size measurements were made on gradient-purified, negatively stained particles and fibers.
[b]Nanomoles of GlcNAc incorporated per milligram of protein per minute.
[c]Based on the position of a tight band formed after isopycnic centrifugation of γ-particle vesicles that had been frozen prior to centrifugation; if not frozen, vesicles formed a broad peak (1.17-1.20 g/cm^3).
Source: Based upon Mills and Cantino (1981).

original interpretation (Truesdell and Cantino, 1970) concerning the relationship between the two types of particles and their function during zoospore encystment. Even though both γ particles and γ-particle vesicles exhibit chitin synthetase activity and can both produce chitin fibrils in vitro, it is undoubtedly the γ-particle vesicles, not the γ particles, which function in vivo in the synthesis of the cyst wall (see Fig. 7 and Table 1 for a comparison of the two).

It has also been emphasized by Mills and Cantino (1981) that there is a very close resemblance between γ-particle vesicles and chitosomes, as follows. The chitin-synthesizing granules (Ruiz-Herrera et al., 1975) of *Mucor rouxii*, named "chitosomes" by Bracker et al. (1976), are spheroidal bodies, 40-70 nm in diameter, with delimiting shells 6-12 nm in thickness. Their "morphological correlates" (Bracker et al., 1976), as seen in situ in thin sections, are microvesicular structures about 30-50 nm in diameter with tripartite shells;

FIG. 7 Top: Negatively stained γ particles and chitin macrofibrils (× 32,000). Bottom: Negatively stained γ-particle vesicles and chitin microfibrils (× 120,000). (From Mills and Cantino, 1981.)

they occur either singly or in the form of multivesicular bodies (MVBs).

The chitin-synthesizing (Cantino and Myers, 1972) γ particles produce microvesicular structures (Truesdell and Cantino, 1970) which, as seen in situ in thin sections, are about 80 nm (± 20 nm) in diameter, with tripartite shells about 6 nm thick; they occur either singly or in the form of MVBs.

The strong similarity between the profiles of both types of microvesicular particles becomes immediately evident when Fig. 4 of Bracker et al. (1976) is compared with Fig. 6 of Truesdell and Cantino (1970). A view of the vesicles produced in vivo by decaying γ particles is illustrated here in Fig. 6 (top). As already indicated in Sec. II.B, the zoospores of another aquatic fungus, *A. macrogynus*, also contain γ particles which, upon encystment, produce microvesicular structures that occur either singly or as MVBs (Barstow and Pommerville, 1980).

Finally, the corresponding negatively stained, isolated particles also have similar morphology. There is a slight size difference, the γ-particle vesicles being larger.

In summary, γ-particle vesicles and chitosomes are morphologically equivalent, possibly even identical, subcellular structures of comparable size (Table 2). They are also functionally equivalent in that they exhibit chitin synthetase activity, produce chitin microfibrils in vitro, and apparently function in transporting chitin synthetase to the surface of a cell.

Chitosomes are seemingly ubiquitous in chitinous fungi (Bartnicki-Garcia et al., 1978); they have been isolated from Zygomycetes, Chytridiomycetes, Hemiascomycetes, Ascomycetes, and Basidiomycetes, including a number of developmental stages – yeast and mycelial phases of *Mucor rouxii*, mycelia of *Allomyces macrogynous* and *Neurospora crassa*, budding yeast cells of *Saccharomyces cerevisiae*, and the basidiocarp of *Agaricus bisporus*. To this assemblage, we can now add one more genus and one more developmental form: the encysting zoospore of *Blastocladiella emersonii*.

C. Chitosomes: Are They Artifacts or Naturally Occurring Subcellular Particles?

The exact in vivo location of chitin synthetase in fungi has been in question. The literature indicates that the enzyme is generally associated with "microsomal" fractions, the composition of which is essentially ill defined. However, it has been reported that chitin synthetase is localized in cell walls (McMurrough et al., 1971; Mishra and Tatum, 1972; Peberdy and Moore, 1975), plasma membranes (Jan, 1974; Duran et al., 1975; Gooday, 1977; Braun and Calderone, 1978), and chitosomes (see Bartnicki-Garcia et al., 1979; Bartnicki-Garcia,

TABLE 2 Comparison of Chitosomes and γ-Particle Vesicles

	Chitosomes[a]	γ-Particle vesicles[b]
From studies of thin sections of particles in situ		
Diameter	35-50 nm[c]	80 nm (range, 60-100 nm)[d]
Thickness of shell	6.5-7.0 nm (tripartite)	4.8-5.7 nm[e] (tripartite)
Origin	Unknown	γ Particles (formed *via* ER; see Sec. II.A)
Fate	Postulated to move to cell surface[f]	Move to and fuse with cell surface during spore encystment[d]
From studies of isolated particles in vitro: structural aspects		
Diameter of shadow-cast particles	35-100 nm	125 nm
Diameter of negatively stained particles	40-70 nm	70-120 nm
Thickness of shell	In thin section, 6.5-7.0 nm; after negative staining, 6-12 nm	8-16 nm
From studies of isolated particles in vitro: functional aspects		
Chitin synthetase activity associated with particles?	Yes	Yes
Chitosomes isolated on sucrose gradient form	Single broad peak	Single broad peak[g]

TABLE 2 (Continued)

	Chitosomes[a]	γ-Particle vesicles[b]
Protease treatment needed to activate synthetase?	Yes	Not known
Chitin microfibrils "arising" from granules demonstrable during chitin synthetase assays?	Yes	Yes
Microfibril thickness	12-18 nm	8-17 nm
Microfibril morphology	Ribbon shaped	Ribbon shaped
Number of microfibrils	1-2 per granule	1 per granule
Microfibrils increase in length during incubation?	Yes	Yes

[a]All tabulated information is for *M. rouxii* and based on reports by Ruiz-Herrera et al. (1975), Bracker et al. (1976), and Bartnicki-Garcia et al. (1978), unless otherwise specified.
[b]All tabulated information is for *B. emersonii* and based on the report by Mills and Cantino (1981).
[c]"Morphological correlates" (Bartnicki-Garcia et al., 1978) considered to be equivalent to the chitosomes isolated and studied in vitro.
[d]Truesdell and Cantino (1970, 1971).
[e]G. L. Mills and E. C. Cantino, unpublished data.
[f]Based on the enzyme content of isolated chitosomes and other observations; see Bartnicki-Garcia et al. (1978).
[g]Produce tight peak if previously frozen; see Table 1.

1980, and references therein). The plasma membrane and cytoplasmic chitosomes have been the most generally accepted sites. On the other hand, it has been postulated (Farkaŝ, 1979) that chitosomes may be artifacts produced during isolation procedures and caused by fragmentation and vesiculation of other cell components, presumably plasma membranes. Therefore, the question is, Are chitosomes real, naturally occurring subcellular particles?

To provide a satisfying answer, it must be conclusively shown that chitosomes do, in fact, occur in vivo; the suspicion that they may be artifacts produced during isolation must be eliminated. Bracker et al. (1976) have published photomicrographs of thin sections of yeast cells of *M. rouxii* containing microvesicular structures that are morphologically similar to thin sections of isolated chitosomes. Furthermore, it has been postulated over the years that various vesicles seen in thin sections of fungi are involved in cell wall synthesis, particularly during the growth of hyphal tips. Even so, the gap between seeing a thin-sectioned vesicle in situ and seeing an isolated, chitin-synthesizing granule has not been convincingly bridged with acceptable experimental evidence. One way to bridge this gap would be to work with a chitin-synthesizing subcellular organelle that also qualifies, simultaneously, as a *marker*, one that can (1) be easily and reliably identified, (2) isolated, and (3) once isolated, retain its identity and perform its presumed function, in this instance the synthesis of chitin. The zoospores of *B. emersonii* provide such a marker: the γ particle.

The γ particles are the only "free" organelles found in the cytosol of the zoospore of *B. emersonii* (Fig. 2). They are easily identified by their characteristic "horseshoe" and double-crescent shapes when viewed in thin sections and their intense staining with osmium (Fig. 3). In a free-swimming zoospore, the cytosol around γ particles is devoid of other organelles. They maintain their characteristic shapes and there is no cell wall synthesis. However, once the zoospores are triggered to encyst, the γ particles undergo their characteristic decay (Figs. 4 and 6) to form MVBs which participate in cell wall formation.

Since the zoospores do not possess a cell wall, γ particles can be released from the cell by a gentle procedure involving a simple osmotic lysis (Myers and Cantino, 1974). They can then be isolated by centrifugation through a linear sucrose gradient. The particles form a tight band and have a high buoyant density of 1.32 g/cm^3, corresponding to about 66% sucrose. Due to their exceptionally high buoyant density, the isolated γ particles are remarkably free of other cellular contaminants; *in particular, they are devoid of plasma membranes*. The isolated particles retain their characteristic shapes and are easily recognized (Fig. 6, bottom right). They can be induced to vesiculate in a normal fashion (Fig. 6, bottom left), and the vesicles produced are capable of forming chitin microfibrils (Fig. 7, top). Such vesicles are morphologically and functionally identical to chitosomes (Table 2). In *B. emersonii, then, intact γ particles are the progenitors of chitosomes, hence our evidence indicates convincingly that chitosomes do, in fact, exist as real subcellular particles, and that they are not artifacts.*

Accordingly, our data provide direct experimental support for the chitosome concept and the interpretation of Bartnicki-Garcia (1980), that "the chitosome appears to be a central piece in the cellular machinery of fungal wall synthesis; its mission is to deliver an organized package of chitin synthetase zymogen to specific sites where cell wall growth takes place." Although we do not know whether the chitin synthetase in the zoospore of *B. emersonii* is a zymogen, clearly it is *not* manufacturing a chitinous cell wall while it is localized in the particle in vivo. Upon encystment, the enzyme is transported through the cytosol by means of γ-particle vesicles, and it is either activated en route or after it reaches its final destination at the plasma membrane. We can conclude, therefore, that chitin synthetase is present at different cellular sites in encysting spores of *B. emersonii*: the γ particles, which contain the inactive enzyme; the γ-particle vesicles, which transport the enzyme; and the plasma membrane, which houses the active enzyme.

Dual locations for active and inactive chitin synthetases have also been reported for two other fungi, *Schizophyllum* (Vermeulen et al., 1979) and *Saccharomyces cerevisiae* (Schekman and Brawley, 1979). In addition, the chitin synthetase in the insect *Tetranychus urticae* is localized in both chitosomes and the plasma membrane (Mothes and Seitz, 1981).

D. A Possible Unified Mechanism for Zoospore Encystment and Cell Wall Synthesis

The zoospores of *B. emersonii* can be induced to encyst synchronously, whereby the motile nonwalled spore becomes a nonmotile walled cell. A variety of morphological and subcellular changes accompany this encystment process. These include rupturing of the backing membrane, flagellar retraction, γ-particle vesiculation, rounding up of the cell, development of cell adhesiveness, fusion of vesicles with the plasma membrane, deposition of a cyst wall, progressive disruption of the nuclear cap membrane, and rearrangement of the mitochondrion and its associated symphyomicrobody and lipid globules. These dramatic changes occur very rapidly, being completed in about a minute, and occur without concomitant RNA or protein synthesis. For a thorough review on this remarkable example of dedifferentiation, see Lovett (1975).

Although a variety of different triggering agents have been used to induce zoospore encystment, the one most commonly employed – and perhaps the one that can be most easily thought of in connection with naturally occurring triggers (see Van Brunt and Harold, 1980) – is the K^+ ion. How do K^+ ions trigger the zoospore to encyst, and what is the sequence of events that leads to cell wall deposition? Obviously, the answers to these questions have not been settled with any certainty, but we take this opportunity to bring together the applicable literature to postulate a possible mechanism for zoospore encystment. The more important observations upon which it is based are listed in Table 3.

TABLE 3 Relevant Observations as a Basis for a Mechanism of Encystment in *B. emersonii*[a]

Observation	Reference
1. Monovalent cations trigger zoospores to encyst.	Cantino et al., 1968; Soll and Sonneborn, 1969; Truesdell and Cantino, 1971; Van Brunt and Harold, 1980
2. Ca^{2+} stabilizes zoospores, that is, prevents zoospore encystment.	Cantino et al., 1968; Soll and Sonneborn, 1969; Truesdell and Cantino, 1971; Suberkropp and Cantino, 1972
3. ^{45}Ca is released from encysting zoospores, but not from nonencysting zoospores.	Soll and Sonneborn, 1972; Gomes et al., 1980a
4. Hypothesis: K^+ may function by displacing Ca^{2+} at some intracellular site.	Soll, 1969; Hutchinson et al., 1977; Van Brunt and Harold, 1980; Gomes et al., 1980a,b
5. Zoospores contain a functional chitin synthetase; Ca^{2+} inhibits chitin synthetase activity.	Camargo et al., 1967
6. γ Particles vesiculate during zoospore encystment.	Truesdell and Cantino, 1970
7. Zoospore chitin synthetase is compartmentalized in γ particles.	Myers and Cantino, 1974
8. All γ particles contain Ca^{2+} in high amounts, but only some of them contain K^+.	Hutchinson et al., 1977
9. γ Particles have a high glycolipid content; the major lipid is a diglucosyldiglyceride, while a monoglucosyldiglyceride functions in chitin synthesis.	Mills and Cantino, 1978, 1980
10. The major lipid component of the zoospore plasma membrane is a diglucosyldiglyceride; ESR analyses show that the middle break point (T_m) of the plasma membrane is Ca^{2+} dependent	Leonards and Haug, 1980a,b,c

TABLE 3 (Continued)

Observation	Reference
10. (Continued) and due to lipid-lipid interactions, *not* lipid-protein interactions.	
11. K^+ functions by depolarizing the plasma membrane.	Van Brunt and Harold, 1980; Jen and Haug, 1981
12. Germination depends on Ca^{2+} mobilization and cyclic AMP.	Gomes et al., 1980a,b
13. ATP generates an "energized membrane" which permits K^+ ions to reverse the stabilizing effect of Ca^{2+} on zoospores.	Leonards and Haug, 1981
14. γ-Particle vesicles (chitosomes) contain chitin synthetase and produce chitin microfibrils.	Mills and Cantino, 1981

[a]The reader is encouraged to consult the recent essay by Gottschalk and Sonneborn (1981) on the counteraction of encystment by the "zoospore maintenance factor."

If K^+ ions are added to a population of zoospores, they encyst after a short lag period. The K^+ ions constitute the first signal for encystment to begin. This involves depolarization of the ATP-energized plasma membrane. The K^+ ions then accumulate in the cytosol, forming a concentration gradient which elicits two secondary signals. One of these is a direct activation of cyclic AMP, which sets into motion the organelle rearrangements that occur during zoospore encystment; the other one causes the γ particles to commence the vesiculation process, this being the first step in the chain of events that leads to cell wall synthesis. This latter signal occurs when K^+ ions enter the γ particles and cause the displacement of Ca^{2+} ions from this particular intracellular location (Ca^{2+} mobilization might also be facilitated by calmodulin, a calcium transport protein in zoospores of *B. emersonii*; Gomes et al., 1979). The ion exchange initiates the unraveling of the γ particles to form γ-particle vesicles. These vesicles, still containing Ca^{2+} ions which inhibit chitin synthetase activity, migrate via electrical forces toward the cell surface and fuse with the plasma membrane. The Ca^{2+} ions are associated with the glucose head groups of the diglucosyldiglyceride in the vesicles (diglucosyldiglyceride accounts for 47% of the lipid in the plasma membrane and 35% of the lipid in the γ-particle vesicles; Leonards and Haug, 1980c, and G. L. Mills and E. C. Cantino, 1981, unpublished results, respectively). It is this Ca^{2+}-glycolipid complex which directly interacts with the plasma membrane diglucosyldiglyceride to bring about vesicle fusion and a suitable insertion of the chitin synthetase into the plasma membrane. After fusion of the vesicles with the plasma membrane, the Ca^{2+} ions are released from the spore; the metabolic block on chitin synthetase activity is thereby removed. The enzyme becomes activated and the chitinous cyst wall is formed via the lipid primer. The release of Ca^{2+} ions provides a more fluid membrane which allows movement of a γ-particle vesicle in a plane along the plasma membrane, the chitin microfibril being layed down as the vesicle moves laterally through the membrane.

IV. THE POSSIBLE RELEVANCE OF γ PARTICLES TO DISEASE: CONTROL OF OPPORTUNISTIC FUNGI

Time and again in the history of science, results of pure research have served sooner or later as cornerstones for applied research. However, as Cohen (1979) points out with respect to drugs that have been used against microbial agents of disease, insight into the way in which such drugs work has usually developed *after* the fact. We wish to call attention, therefore, to the possibility that γ-particle glycolipid may have a potential for useful applications.

The role of the monoglucosyldiacylglycerol primer in the γ particle suggests the possibility of an entirely new approach for the control of medically important opportunistic fungi. Such an approach would be based upon the fact that the γ-particle glycolipid involved in the synthesis of the chitinous cyst wall in *B. emersonii* is structurally different from the dolichylphosphates involved in the glycosyl transfer reactions which lead to the synthesis of glycoproteins in most living organisms. If glycolipids similar to the one in the γ particle mediate synthesis of wall material in other fungi, it may be possible to exploit the distinctive chemical difference between the two types of lipid intermediates to achieve control over the growth of fungal pathogens without significantly altering the growth of their hosts. In the following paragraphs, we have selected medically important fungi as the vehicle for developing our argument, but the basic ideas may also apply to fungi that cause plant disease.

A. Background

Certain infectious diseases caused by microorganisms that occur either in the normal flora of the host or in the immediate environment of the host, but which are, nonetheless, generally nonpathogenic, are designated as opportunistic infections, implying derangement in the normal defense mechanism of the host. Fungi represent one of the major groups of microorganisms that cause opportunistic infections (Emmons et al., 1970). Some sort of uniform therapy – ideally, a single antifungal agent – that could be utilized for *all* opportunistic fungi would be very useful.

B. Rationale for Selecting an Attackable Target for Control of Opportunistic Fungi

The one thing that most, if not all, fungi have in common is chitin. Either it is the major cell wall component or it is a minor constituent that acts as an endoskeleton upon which other wall components are deposited. Therefore this should logically be a primary target area for achieving control over or preventing the growth of opportunistic fungi. More specifically, the mechanism whereby chitin is formed via a lipid intermediate is the site where the attack should be made.

Why specify the lipid intermediate? Because lipid intermediates are involved in the synthesis of cell envelopes, cell wall mannan-protein complexes, higher plant cell walls, and plant and animal glycoproteins (see Hemming, 1977, and Elbein, 1979, for reviews). Usually, the lipid intermediate is a phosphorylated polyisoprenol, often dolichol pyrophosphate, which accepts nucleoside phosphate-linked sugars and attaches the sugars to a growing polymer. The lipid is then released, after which it can accept more nucleoside phosphate-linked sugars, and the cycle starts again.

However, as seen in Sec. III.A, the synthesis of chitin by encysting zoospores of *B. emersonii* involves a different kind of lipid intermediate. Functionally, it does not act as an intermediate in a cyclic process; rather, it serves as a primer upon which the chitin is built. The final product, a lipid-chitin complex, is presumably deposited at the cell surface while the cyst wall is laid down. Structurally, it is not a polyisoprenol and does not contain phosphorus; rather, it is a monoglycosyldiacylglycerol in which palmitic acid accounts for over half of the acyl groups present.

We suggest that it may be possible to exploit this chemical difference between the lipid intermediate involved in the synthesis of a chitinous wall in *B. emersonii* and the lipid intermediates involved in the synthesis of nonfungal cell walls and membrane glycoproteins in other organisms so as to achieve control of the growth of a fungal pathogen without significantly altering the growth of its host.

C. Rationale for Thinking That Other Fungi May Employ a Lipid Primer in Chitin Synthesis Similar or Identical to That in *B. emersonii*

It seems to have been generally accepted that chitin synthesis can proceed by the direct transfer of GlcNAc from UDP-GlcNac to a chitin molecule without the aid of a cyclic lipid intermediate (Farkâs,

1979). This is undoubtedly true for the addition of GlcNac to a preexisting, chitin polymer, but what about the addition of the very first GlcNAc that starts the formation of chitin? To what is the GlcNAc to be attached? Here, the evidence seems to point in different directions. Several substances have been suggested as initiators of chitin synthesis: chitodextrans (Glaser and Brown, 1957), GlcNAc (Camargo et al., 1967), glycolipid (Mills and Cantino, 1978, 1980), and protein (Bartnicki-Garcia, 1980).

In the zoospores of *B. emersonii*, where there are no preformed chitin polymers (Selitrennikoff et al., 1976), chitin synthesis is initiated by a lipid primer. Is there any direct evidence that lipids can act as primers for chitin synthesis in other fungi? The answer is no. However, lipids do stimulate chitin synthetase activity in two fungi known to contain chitosomes, *M. rouxii* (McMurrough and Bartnicki-Garcia, 1971) and *S. cerevisiae* (Duran and Cabib, 1978). In addition, maximal chitin synthetase activity in *Candida albicans* requires the presence of an enzyme-associated lipid (Braun and Calderone, 1978). Are such lipids glycolipids? For the two yeasts, the stimulatory lipid appears to be not a glycolipid, but phosphatidylserine or its lyso form. For *M. rouxii*, however, this is not the case; its chitosomes are devoid of phosphatidylserine (Hernandez et al., 1981). It is noteworthy, too, that the lipids used to test chitin synthetase activity in *S. cerevisiae* and *C. albicans* were purchased commercially, whereas the lipids used to test the chitin synthetase from *M. rouxii* were obtained from extracts of the fungus itself. The polar lipid fraction from *M. rouxii* (phospholipids and glycolipids) stimulated chitin synthetase activity, while the neutral lipid fraction inhibited it. Such evidence, while indirect, is consistent with the possibility that a glycolipid may be involved as a lipid primer for *M. rouxii*.

Other indirect support stems from analyses of chitosomes isolated from the yeast form of *M. rouxii* (Hernandez et al., 1981) which indicate that they contain significant amounts of glycolipids. The data for the *M. rouxii* chitosomes compare quite favorably with data derived from analyses of γ-particle vesicles (G. L. Mills and E. C. Cantino, unpublished results), as follows. Phospholipids contributed about equally to the total lipid content of both kinds of organelles, phosphatidylcholine and phosphatidylethanolamine being the major phospholipids in both of them as well. There is about twice as much neutral lipid in chitosomes than in γ-particle vesicles, but the distribution of neutral lipid components is almost identical in the two particles: In γ-particle vesicles and chitosomes, respectively, diglycerides plus monoglycerides contribute 37.5 and 31.1%; free sterols plus sterol esters, 37.4 and 48.4%; free fatty acids plus their esters, 18.8 and 16.5%; and triglycerides, 6.3 and 4.0%. There is about three times as much glycolipid in γ-particle vesicles as in chitosomes, but there are four different glycolipids in both types of particles. Whether or not one of the four chitosome glycolipids is the same as the γ-particle glycolipid primer is not known.

Therefore the argument for thinking that other fungi may employ a lipid primer for chitin synthesis similar to the one in *B. emersonii* rests primarily upon indirect evidence. However, such evidence may provide a reasonable case if all the facts are considered together: (1) γ Particles occur in several genera of aquatic fungi, not just in *B. emersonii* (Sec. II); (2) the vesicles released by γ particles are either chitosomes or their equivalent (Sec. III); (3) construction of cell walls by taxonomically diverse nonzoosporic as well as zoosporic fungi is probably mediated by chitosomes; (4) both chitosomes and γ-particle vesicles exhibit lipid-stimulated chitin synthetase activity; and (5) chitosomes and γ-particle vesicles have a similar lipid composition and contain an appreciable amount of glycolipid.

It should not be too surprising to find that, as a group, the fungi have evolved a distinctive molecular mechanism for the synthesis of a vital carbohydrate polymer; they also differ in other distinctive and fundamental ways from the rest of the kingdoms of living organisms.

D. Conclusion: Control of Chitin Synthesis at the Lipid Primer Target

We think that it is likely that lipid primers like the one in γ particles, that is, glycosyldiacylglycerols, will be found in the chitosomes of opportunistic and systemic fungi. Since both γ-particle vesicles and chitosomes produce chitin microfibrils in vitro, it should be possible to find or synthesize a compound that will inhibit chitin microfibril synthesis in vitro by blocking the first step in polymer formation at the lipid primer level, that is, by interfering with the initial glycosyl transferase reaction that brings about the addition of N-acetylglucosamine to the lipid.

The potential importance of such an inhibitor is that it could lead to a method for controlling the growth of fungal pathogens. This control would be exerted at a specific biochemical focal point. If the biochemical model for chitin formation in *B. emersonii* is universally applicable to all fungi, chitin synthesis – and, therefore, in effect, growth of fungal pathogens – could be stopped. The fact that the lipid primer involved in the formation of chitin differs from the lipid intermediates involved in the formation of other biologically important polysaccharide polymers means that it should be feasible to apply an inhibitor that will stop chitin synthesis without introducing detrimental effects upon the infected host. If this is achieved, its importance will lie not only in the fact that fungal pathogens, heretofore so hard to treat, might be controllable, but also in the advantages provided by a standardized chemotherapeutic approach for controlling the growth of fungi.

Just as the cell walls of prokaryotic organisms, differing as they do from those of eukaryotes, have provided sites for selective chemotherapy (Cohen, 1979), so the chitinous components of fungal cell walls, having no counterparts among higher animals and plants, may provide similarly appropriate sites for preventive medicine and treatment.

V. THE POSSIBLE RELEVANCE OF γ PARTICLES TO PHYLOGENY: ENDOSYMBIOSIS

Cantino and Mills (1979) posed the question, What *is* the γ particle? They were led to conclude that from certain structural, functional, and biochemical points of view, the γ particles of *B. emersonii* apparently resemble pox viruses, especially entomopoxviruses – at least, more so than they resemble any other extant groups of prokaryotes or any of the well-known and widely distributed organelles present in plants and animals. Although the idea that γ particles may have originated from viruses was thought-provoking, it was not pursued further at that time; we wish to call attention to this idea now, albeit very briefly.

Most biologists accept the existence of two structurally and functionally different kinds of cells: prokaryotic cells and eukaryotic cells. A frequently discussed theory for the origin of the latter holds that today's mitochondria and chloroplasts are descendants of primitive prokaryotes that invaded primitive eukaryotes and established symbiotic relationships; the latter presumably persisted because of their inheritable, selective advantages and thus evolved into present-day cell types. In the most recent critical analysis of the evidence behind the theory, Dodson (1979) concluded that a relatively strong case can be made for the genetic connection between chloroplasts and blue-green algae, although the probability that mitochondria evolved from endosymbiotic, aerobic bacteria "falls short of that for the endosymbiotic origin of chloroplasts" (for other recent discussions of this concept, see Phillips and Carr, 1977; King, 1977; Margulis, 1976, and references therein).

If the inclusion of chloroplasts and mitochondria into primitive eukaryotic cells did not constitute unique events but, rather, just "patent examples of a process commonplace in early evolution" (King, 1977), then it seems reasonable to expect to find examples of present-day cells that harbor organelles descendent from *viruses* as well. If it is true that organisms only remotely related may sometimes have entered "into associations with great consequences for evolution" (Margulis, 1976), perhaps viruses represented the outer limit of such remoteness. From some fascinating points of view, viruses and evolution have not been ignored (Reanney, 1974), and some have wondered

(Cohen, 1970) if and even postulated (Benveniste and Todaro, 1974) that exogenous viruses have become incorporated into inheritable germ lines from primitive ancestors. But, such ideas notwithstanding, viruses still stand out by the inconspicuous role they have played in recent thinking about the origins of eukaryotic *organelles*. The reason is not hard to find: Useful suggestive examples have been all but lacking.

For many years and a variety of reasons (Cantino, 1966), it has been thought that the Blastocladiales possess various primitive characteristics. Recent evidence supports this older point of view. If the length of poly(A) reflects the approximate time of origin of many organisms (Carlin, 1980), then the length of the poly(A) in *B. emersonii* (Jawarski, 1976) places this fungus in a very primitive evolutionary position, close to the presumed time of origin of chloroplasts and mitochondria (Carlin, 1980), about 2.1 billion years ago. This conclusion coincides nicely with the time frame needed to accommodate the idea of an early evolution of a viral-protozoosporic association. The circumstantial evidence that we recently compiled (Cantino and Mills, 1979) documents some evident similarities between present-day pox viruses and γ particles. Indeed, it can be argued that, on the whole, γ particles and entomopoxviruses resemble one another even more closely than mitochondria and chloroplasts resemble present-day prokaryotes such as bacteria. In conclusion, the known facts point to the possibility that γ particles may have originated in the form of invading virions turned symbionts.

REFERENCES

Bartnicki-Garcia, S. (1980). Chitosomes and the origin of chitin microfibrils. In *Advances in Protoplast Research*, Vol. 5. L. Ferenczy and G. L. Farkâs (Eds.). Pergamon, London, pp. 475-484.

Bartnicki-Garcia, S., Bracker, C. E., Reyes, E., and Ruiz-Herrera, J. (1978). Isolation of chitosomes from taxonomically diverse fungi and synthesis of chitin microfibrils in vitro. *Exp. Mycol. 2*:173-192.

Bartnicki-Garcia, S., Ruiz-Herrera, J., and Bracker, C. E. (1979). Chitosomes and chitin synthesis. In *Fungal Walls and Hyphal Growth*, J. H. Burnett and A. P. J. Trinci (Eds.). Cambridge University, Cambridge, pp. 149-168.

Barstow, W. E. (1979). Ultrastructure of the formation of γ particles during zoosporogenesis in *Allomyces macrogynus* and *Catenaria anguillulae*. *Exp. Mycol. 3*:28-41.

Barstow, W. E., and Lovett, J. S. (1975). Formation of gamma particles during zoosporogenesis in *Blastocladiella emersonii*. *Mycologia 67*:518-529.

Barstow, W. E., and Pommerville, J. (1980). The ultrastructure of cell wall formation and of gamma particles during encystment of *Allomyces macrogynus* zoospores. *Arch. Microbiol. 128*:179-189.
Benveniste, R. E., and Todaro, G. J. (1974). Evolution of C-type viral genes: Inheritance of exogenously acquired viral genes. *Nature 252*:456-459.
Bracker, C. E., Ruiz-Herrera, J., and Bartnicki-Garcia, S. (1976). Structure and transformation of chitin synthetase particles (chitosomes) during microfibril synthesis in vitro. *Proc. Nat. Acad. Sci. USA 73*:4570-4574.
Braun, P. C., and Calderone, R. A. (1978). Chitin synthesis in *Candida albicans*: Comparison of yeast and hyphal forms. *J. Bacteriol. 135*:1472-1477.
Camargo, E. P., Dietrich, C. P., Sonneborn, D., and Strominger, J. L. (1967). Biosynthesis of chitin in spores and growing cells of *Blastocladiella emersonii*. *J. Biol. Chem. 242*:3121-3128.
Cantino, E. C. (1966). Morphogenesis in aquatic fungi. In *The Fungi*, Vol. 2. G. C. Ainsworth and A. S. Sussman (Eds.). Academic, New York, pp. 283-337.
Cantino, E. C. (1975). Developmental pathways of the water mold, *Blastocladiella emersonii*. *ASM News 41*:782.
Cantino, E. C., and Horenstein, E. A. (1954). Cytoplasmic exchange without gametic copulation in the water mold *Blastocladiella emersonii*. *Am. Nat. 88*:143-154.
Cantino, E. C., and Horenstein, E. A. (1956). Gamma and the cytoplasmic control of differentiation in *Blastocladiella*. *Mycologia 48*:443-446.
Cantino, E. C., and Mack, J. P. (1969). Form and function in the zoospore of *Blastocladiella emersonii*. I. The γ Particle and satellite ribosome package. *Nova Hedwigia Z. Kryptogamankd. 18*:115-148.
Cantino, E. C., and Mills, G. L. (1976). Form and function in Chytridiomycete spores. In *The Fungal Spore: Form and Function*. D. J. Weber and W. M. Hess (Eds.). Wiley, New York, pp. 501-556.
Cantino, E. C., and Mills, G. L. (1978). *Blastocladiella emersonii*. In *Lower Fungi in the Laboratory*. M. S. Fuller (Ed.). University of Georgia, Athens, Ga., pp. 39-40.
Cantino, E. C., and Mills, G. L. (1979). The gamma particle in *Blastocladiella emersonii: What is it?* In *Viruses and Plasmids in Fungi*, P. A. Lemke (Ed.). Marcel Dekker, New York, pp. 441-484.
Cantino, E. C., and Myers, R. B. (1972). Concurrent effect of visible light on γ-particles, chitin synthetase, and encystment capacity in zoospores of *Blastocladiella emersonii*. *Arch. Mikrobiol. 83*:203-215.

Cantino, E. C., and Myers, R. B. (1974). The γ particle and intracellular interactions in *Blastocladiella emersonii*. *Brookhaven Symp. Biol.* *25*:51-74.

Cantino, E. C., and Truesdell, L. C. (1971). Cytoplasmic γ-like particles and other ultrastructural aspects of zoospores of *Blastocladiella britannica*. *Trans. Br. Mycol. Soc.* *56*:169-179.

Cantino, E. C., Truesdell, L. C., and Shaw, D. S. (1968). Life history of the motile spore of *Blastocladiella emersonii*: A study in cell differentiation. *J. Elisha Mitchell Sci. Soc.* *84*:125-146.

Carlin, R. K. (1980). Poly(A): A new evolutionary probe. *J. Theor. Biol.* *82*:353-362.

Cohen, S. S. (1970). Are/were mitochondria and chloroplasts microorganisms? *Am. Sci.* *58*:281-289.

Cohen, S. S. (1979). Comparative biochemistry and drug design for infectious disease. *Science* *205*:964-971.

Dodson, E. O. (1979). Crossing the procaryote-eucaryote border: Endosymbiosis or continuous development? *Can. J. Microbiol.* *25*: 651-674.

Duran, A., and Cabib, E. (1978). Solubilization and partial purification of yeast chitin synthetase. Confirmation of the zymogenic nature of the enzyme. *J. Biol. Chem.* *253*:4419-4425.

Duran, A., Bowers, B., and Cabib, E. (1975). Chitin synthetase zymogen is attached to the yeast plasma membrane. *Proc. Nat. Acad. Sci. USA* *72*:3952-3955.

Elbein, A. D. (1979). The role of lipid-linked saccharides in the biosynthesis of complex carbohydrates. *Annu. Rev. Plant Physiol.* *30*:239-272.

Emmons, C. W., Binford, C. H., and Utz, J. P. (1970). *Medical Mycology*. Lea and Febiger, Philadelphia, Pa.

Farkâs, V. (1979). Biosynthesis of cell walls of fungi. *Microbiol. Rev.* *43*:117-144.

Giddings, T. H., and Cantino, E. C. (1974). Partial purification and properties of D-glucosamine, 6-phosphate N-acetyltransferase from zoospores of *Blastocladiella emersonii*. *J. Bacteriol.* *120*:976-979.

Glaser, L., and Brown, D. H. (1957). The synthesis of chitin in cell free extracts of *Neurospora crassa*. *J. Biol. Chem.* *228*:729-742.

Gomes, S. L., Mennucci, L., and Maia, J. C. C. (1979). A calcium-dependent protein activator of mammalian cyclic nucleotide phosphodiesterase from *Blastocladiella emersonii*. *FEBS Lett.* *99*:39-42.

Gomes, S. L., Mennucci, L., and Maia, J. C. C. (1980a). Calcium efflux during germination of *Blastocladiella emersonii*. *Dev. Biol.* *77*:157-166.

Gomes, S. L., Mennucci, L., and Maia, J. C. C. (1980b). Induction of *Blastocladiella emersonii* germination by cyclic adenosine-3',5'-monophosphate. *Cell Differ.* *9*:169-179.

Gooday, G. W. (1977). Biosynthesis of the fungal wall – Mechanisms and implications. *J. Gen. Microbiol. 99*:1-11.
Gottschalk, W. K., and Sonneborn, D. R. (1981). *Blastocladiella emersonii* maintenance factor: A quantitative bioassay used to characterize extracellular appearance and maintenance of factor and interactions between factor, zoospores, and salt. *Exp. Mycol. 5*: 1-14.
Hemming, F. W. (1977). Dolichol phosphate, a coenzyme in the glycosylation of animal membrane-bound glycoproteins. *Biochem. Rev.* 5:1223-1231.
Hennessy, S. W., and Cantino, E. C. (1972). Lag-phase sporogenesis in *Blastocladiella emersonii*: Induced formation of unispored plantlets. *Mycologia 64*:1066-1087.
Hernandez, J., Lopez-Romero, E., Cerbon, J., and Ruiz-Herrera, J. (1981). Lipid analysis of chitosomes, chitin synthesizing microvesicles from *Mucor rouxii*. *Exp. Mycol.* 5:349-356.
Hutchinson, T. E., Cantino, M. E., and Cantino, E. C. (1977). Calcium is a prominent constituent of the gamma particle in the zoospore of *Blastocladiella emersonii* as revealed by X-ray microanalysis. *Biochem. Biophys. Res. Commun. 74*:336-342.
Jan, Y. N. (1974). Properties and cellular localization of chitin synthetase in *Phycomyces blakesleanus*. *J. Biol. Chem. 249*:1973-1979.
Jawarski, A. J. (1976). Synthesis of polyadenylic acid RNA during zoospore differentiation and germination in *Blastocladiella emersonii*. *Arch. Biochem. Biophys. 173*:201-209.
Jen, C. J., and Haug, A. (1981). Potassium-induced depolarization of the transmembrane potential in *Blastocladiella emersonii* zoospores precedes encystment. *Exp. Cell Res. 131*:79-87.
King, G. A. M. (1977). Symbiosis and the evolution of prokaryotes. *Biosystems 9*:35-42.
Leonards, K. S., and Haug, A. (1980a). Effects of cations on the plasma membrane of *Blastocladiella emersonii* zoospores. *Biochim. Biophys. Acta 600*:805-816.
Leonards, K. S., and Haug, A. (1980b). Electron spin resonance study of the isolated lipid components from *Blastocladiella emersonii* zoospores. *Biochim. Biophys. Acta 600*:817-830.
Leonards, K. S., and Haug. A. (1980c). Isolation and characterization of the plasma membrane of *Blastocladiella emersonii* zoospores. *Exp. Mycol. 4*:207-220.
Leonards, K. S., and Haug. A. (1981). Potassium- and calcium-induced alterations in lipid interactions of isolated plasma membranes from *Blastocladiella emersonii*. Evidence for an adenosine 5'-triphosphate requirement. *Biochemistry 20*:1784-1789.

Lovett, J. S. (1975). Growth and differentiation of the water mold *Blastocladiella emersonii:* Cytodifferentiation and the role of ribonucleic acid and protein synthesis. *Bacteriol. Rev. 39*:345-404.

Loubes, C., and Manier, J. F. (1974). Etude ultrastructurale de *Coelomycidium simulii* Debaisieux, 1920. Sa position systématique parmi les Chytridiomycètes. *Protistologica 1*:47-57.

McMurrough, I., and Bartnicki-Garcia, S. (1971). Properties of a particulate chitin synthetase from *Mucor rouxii. J. Biol. Chem. 246*:4008-4016.

McMurrough, I., Flores-Carreon, A., and Bartnicki-Garcia, S. (1971). Pathway of chitin synthesis and cellular localization of chitin synthetase in *Mucor rouxii. J. Biol. Chem. 246*:3999-4007.

Manier, J. F. (1977). Cycle, ultrastructure d'une *Caternaria* (Phycomycetes, Blastocladiales) parasite de crustacés Cyclopoides. *Ann. Parasitol. 52*:363-376.

Margulis, L. (1976). Genetic and evolutionary consequences of symbiosis. *Exp. Parasitol. 39*:277-349.

Mills, G. L., and Cantino, E. C. (1978). The lipid composition of the *Blastocladiella emersonii* γ particles and the function of γ-particle lipid in chitin formation. *Exp. Mycol. 2*:99-109.

Mills, G. L., and Cantino, E. C. (1980). The glycolipid involved in chitin synthesis by zoospores of *Blastocladiella emersonii* is a monoglucosyldiacylglycerol. *Exp. Mycol. 4*:175-180.

Mills, G. L., and Cantino, E. C. (1981). Chitosome-like vesicles from gamma particles of *Blastocladiella emersonii* synthesize chitin. *Arch. Microbiol. 130*:72-77.

Mishra, N. C., and Tatum, E. L. (1972). Effects of L-sorbose on polysaccharide synthetase of *Neurospora crassa. Proc. Nat. Acad. Sci. USA 69*:313-317.

Mothes, U., and Seitz, K. A. (1981). A possible pathway of chitin synthesis as revealed by electron microscopy in *Tetranychus urticae* (Acari, Tetranychidae). *Cell Tissue Res. 214*:443-448.

Myers, R. B., and Cantino, E. C. (1974). *The Gamma Particle.* S. Karger, Basel, pp. 1-117.

Norrman, J., Giddings, T. H., and Cantino, E. C. (1975). Partial purification and properties of L-glutamine: D-Fructose 6-Phosphate aminotransferase from zoospores of *Blastocladiella emersonii. Phytochemistry 14*:1271-1274.

Olson, L. W., Lange, L., and Reichle, R. (1978). The zoospore and meiospore of the aquatic Phycomycete *Catenaria anguillulae. Protoplasma 94*:53-71.

Peberdy, J. F., and Moore, P. M. (1975). Chitin synthetase in *Mortierella vinacea,* properties, cellular location and synthesis in growing cultures. *J. Gen. Microbiol. 90*:228-236.

Phillips, D. O., and Carr, N. G. (1977). Nucleic acid analysis and the endosymbiotic hypothesis. *Taxon 26*:3-42.

Reanney, D. C. (1974). Viruses and evolution. In *International Review of Cytology*, Vol. 37. G. H. Bourne and J. F. Danielli (Eds.). Academic, New York, pp. 21-52.

Rønne, M., and Bøye, H. A. (1978). Cell wall synthesis in *Allomyces macrogynus*. *Cytobios*. *19*:159-170.

Ruiz-Herrera, J., Sing, V. O., Van der Woude, W. J., and Bartnicki-Garcia, S. (1975). Microfibril assembly by granules of chitin synthetase. *Proc. Nat. Acad. Sci. USA 72*:2706-2710.

Schekman, R., and Brawley, V. (1979). Localized deposition of chitin on the yeast cell surface in response to mating pheromone. *Proc. Nat. Acad. Sci. USA 76*:645-649.

Seale, T. W., Delehanty, J., and Runyan, R. B. (1974). Liberation and development of *Allomyces arbuscula* mitospores viewed by scanning electron microscopy. *J. Bacteriol. 120*:1417-1426.

Selitrennikoff, C. P., and Sonneborn, D. R. (1976). The last two pathway-specific enzyme activities of hexosamine biosynthesis are present in *Blastocladiella emersonii* zoospores prior to germination. *Biochim. Biophys. Acta 451*:408-416.

Selitrennikoff, C. P., Allin, D., and Sonneborn, D. R. (1976). Chitin biosynthesis during *Blastocladiella* zoospore germination: Evidence that the hexosamine biosynthetic pathway is post-translationally activated during cell differentiation. *Proc. Nat. Acad. Sci. USA 73*:534-538.

Soll, D. R. (1969). Germination in the water mold, *Blastocladiella emersonii*: The ionic basis of control and the involvement of protein synthesis. Ph.D. thesis, University of Wisconsin, Madison, Wisconsin.

Soll, D. R., and Sonneborn, D. R. (1969). Zoospore germination in the water mold, *Blastocladiella emersonii*. II. Influence of cellular and environmental variables on germination. *Dev. Biol. 20*:218-235.

Soll, D. R., and Sonneborn, D. R. (1972). Zoospore germination in *Blastocladiella emersonii*. IV. Ion control over cell differentiation. *J. Cell Sci. 10*:315-333.

Suberkropp, K. F., and Cantino, E. C. (1972). Environmental control of motility and encystment in *Blastocladiella emersonii* zoospores at high population densities. *Trans. Br. Mycol. Soc. 59*:463-475.

Travland, L. B. (1979). Structure of the motile cells of *Coelomomyces psorophorae* and function of the zygote in encystment on a host. *Can. J. Bot.* 57:1021-1035.

Truesdell, L. C., and Cantino, E. C. (1970). Decay of γ-particles in germinating zoospores of *Blastocladiella emersonii*. *Arch. Mikrobiol. 70*:378-392.

Truesdell, L. C., and Cantino, E. C. (1971). The induction and early events of germination in the zoospore of *Blastocladiella emersonii*. In *Current Topics in Developmental Biology*, Vol. 6. A. A. Moscona and A. Monroy (Eds.). Academic, New York, pp. 1-44.

Van Brunt, J., and Harold, F. M. (1980). Ionic control of germination of *Blastocladiella emersonii* zoospores. *J. Bacteriol.* *141*:735-744.

Vermeulen, C. A., Raven, M. B. J. M., and Wessels, J. G. H. (1979). Localization of chitin synthase activity in subcellular fractions of *Schizophyllum commune* protoplasts. *J. Gen. Microbiol.* *114*:87-97.

8

Macromolecular Synthesis in *Blastocladiella*

JAMES S. LOVETT *Purdue University, West Lafayette, Indiana*

I. INTRODUCTION

The saprobic aquatic fungus *Blastocladiella emersonii* has served as a useful experimental organism for developmental study ever since it was first described by Cantino in 1951. Cantino and his associates early on recognized and exploited the characteristic rapid growth and development of the fungus as single coenocytic cells that could be induced to develop along two alternate morphological pathways. This allowed the study of developmentally associated metabolism without the difficulty of distinguishing between the contributions of growing and differentiating portions of a mycelium, as is typical with filamentous fungi. The development of simple conditions for producing synchronous growth and development (see Lovett, 1975, for references) and shortened cycles (Hennessy and Cantino, 1972; Lodi and Sonneborn, 1974; Barstow and Lovett, 1978) further increased the value of the organism for detailed study. As a result, there is now a rather extensive background of information on cytodifferentiation at the light and electron microscopic levels and on the basic changes in macromolecular synthetic patterns that accompany these events. Much of the earlier work relevant to the subject of this chapter has been reviewed and requires no detailed repetition here (Lovett, 1975). However, it may be useful to provide a brief résumé of the key events to serve as background for the discussion of recent work to follow.

II. DEVELOPMENT AND METABOLISM

Two stages in the vegetative cycle of *B. emersonii* involve major differentiative events: sporulation and spore germination. The latter has received the most attention and will also be emphasized to some extent here, but the status of the motile zoospore and its germination clearly depend on events that take place during sporulation and can be more adequately interpreted in that context. Sporulation in *Blastocladiella* really involves two phases. In the first the saclike, coenocytic growing cell is converted into a zoosporangium by formation of a basal septum to separate the thallus from the basal rhizoid system that is presumably no longer needed for nutrition, and in which the cytoplasm subsequently degenerates (Lessie and Lovett, 1968). A discharge papilla (or papillae) is also formed at the apical end of the cell. In the latter process secretory vesicles appear to carry the amorphous material that fills a pore simultaneously dissolved in the chitinous cell wall (Lessie and Lovett, 1968; Barstow and Lovett, 1978). Sporangium differentiation requires about 2-3 h and can be induced at virtually any stage of the growing phase by subjecting the cells to starvation. There is little or no cytological evidence that any microscopically visible events associated with zoospore differentiation occur until sporangium development is completed, though this is not necessarily also true for activities at the biochemical level, as will become evident shortly. The uninucleate zoospores are rapidly differentiated during the last hour of sporulation in a process that includes the appearance of new structures, most notably a flagellum and associated flagellar rootlets, as well as a number of small membrane-enclosed γ particles, and the rearrangement of others, such as ribosome aggregation to form the membrane-enclosed nuclear cap, mitochondrial fusion (Lessie and Lovett, 1968), and microbody fusion (Mills and Cantino, 1979), and the association of the fused microbody and mitochondrion with lipid droplets to form the characteristic "side body" apparatus. Production of membranes via vesicle fusion is a prominent activity throughout, involving both compartmentation of organelles and cytoplasmic cleavage. In cells induced to sporulate in the early growth phase this must require considerable membrane synthesis. In older cells membranes may be derived from the elaborate "macrotubular" system formed during the growth phase, since it disappears during sporulation (Lessie and Lovett, 1968; Barstow and Lovett, 1975).

The most obvious initial metabolic response to the induction of sporulation in *Blastocladiella* is a rapid shutdown of both RNA and protein accumulation immediately after a shift to media lacking carbon and nitrogen sources. After a delay, the cellular content of both RNA and protein as well as the dry weight decrease steadily until zoospore release, while the DNA content remains essentially constant (Murphy and Lovett, 1966). The highly differentiated zoospores may remain motile for many hours in solutions containing low levels of calcium and lacking potassium, but germinate rapidly and synchronously if the K^+ concentration is increased to 50 mM (Soll et al., 1969).

Zoospore germination is an irreversible process and also consists of two phases, encystment and germ tube (initial rhizoid) production. Encysting cells round up, retract their flagellar axoneme, and form a thin cyst wall in a process that requires about 1 min (Truesdell and Cantino, 1971). It occurs in the absence of new RNA or protein synthesis, and with the exception of the cyst wall appears to involve primarily membrane rearrangements (Truesdell and Cantino, 1971; Soll and Sonneborn, 1971a). It is followed by germ tube emergence about 20 min later, which does require new protein synthesis, but not RNA synthesis (Lovett, 1968; Soll and Sonneborn, 1971b). However, continued rhizoid development after 45-50 min will not occur in the presence of actinomycin D and is presumed to require new RNA synthesis. In untreated cultures the germinated cells grow exponentially by expansion of the cell body, with elongation and extensive branching of the basal rhizoid system. Cells can encyst and produce a germ tube and branching rhizoid even in non-growth-supporting inorganic salts media, but arrest before the first nuclear division (Lovett, 1968, 1975). It is assumed that such cells develop at the expense of extensive turnover of endogenous materials, since their RNA and protein content gradually declines.

In a complete synthetic medium the dry weight initially decreases by about 25% and begins to increase again at 60 min; net RNA accumulation begins at 40 min post inoculation, and protein accumulation at about 70 min. The DNA content doubles between 120 and 240 min. The only other major cellular components that have been measured during development are the lipid and glycogen contents of the cells. The total lipid content approximately doubles during sporulation (Smith and Silverman, 1973) and the composition changes; that is, phospholipid decreases while neutral lipid and glycolipid increase. Smith and Silverman (1973) reported all lipids to decrease by 30-40% during the first 15 min of germination, while Mills et al. (1974) reported an increase in phospholipid, constant neutral lipid, and decreased glycolipid levels following encystment. In non-growth-supporting medium the glycogen content of germinating cells (originally around 10% of the dry weight; Suberkropp and Cantino, 1973) decreases by 90% in the 20 min following encystment (Suberkropp and Cantino, 1972). The glycogen content has not been examined during synchronous sporulation.

The preceding paragraphs suffice to show that the two major phases of cellular differentiation in the vegetative life cycle of *B. emersonii*, sporulation and zoospore germination, also represent periods of major change in the rates of macromolecule synthesis. Part of this is attributable to biosynthetic adjustments to the availability of medium nutrients, and part must be related to the developmental response. A good example of the former would be the regulation of the rate of ribosomal RNA synthesis and processing into ribosomes. Control experiments with early-growth-phase cells (J. S. Lovett, 1975, unpublished data) have shown that a variety of step-down conditions designed to alter the rate of protein synthesis through reduced amino acid availability result in an immediate sharp reduction in the synthesis and processing of rRNA precursors. In each case the level of polysomes drops dramatically immediately after the step-down, but stabilizes at 40-50% of the ribosomes, indicating probable protein turnover. After a $\sim$50-min delay the cells again begin to accumulate protein at a lower rate (step-down to minimal medium) without net RNA synthesis, or both the protein and RNA content of the cells begin to decrease (step-down to inorganic salts medium). Similar patterns of lowered polysome levels and blocked rRNA synthesis and processing occur as the first phase of induced sporulation (Adelman and Lovett, 1974b; Lovett, 1975) and in cells germinated in inorganic salts medium (Lovett, 1975); both are situations under which protein accumulation fails to occur. On the other hand, when spores germinate in complete medium, polysomes rapidly form and the rates of rRNA and total RNA synthesis rise coordinately (Leaver and Lovett, 1974). New RNA synthesis does not become essential for development, however, until 45-50 min (Lovett, 1968). Thus though new ribosome synthesis is triggered almost immediately following zoospore encystment, presumably via the resumption of protein synthesis, it is not obligatory for early development (Lovett, 1975).

It seems reasonable to propose that during the lag following the initial responses to an inducing medium step-down, the cells switch to a developmental program leading to sporulation in which the synthesis of protein may be controlled by special mechanisms. The reduced rate of protein synthesis supported by turnover may be sufficient to hold new rRNA synthesis in abeyance through normal growth phase regulatory mechanisms.

During zoospore differentiation a ribosome-associated translation inhibitor is produced (Adelman and Lovett, 1974a) and the ribosomes aggregate via an unknown mechanism to form the membrane-enclosed nuclear cap (Lessie and Lovett, 1968; Barstow and Lovett, 1978). In the motile zoospore neither RNA nor protein synthesis is detectable, although about 10% of the ribosomes still remain in an inactive polyribosomal fraction (Gong and Lovett, 1977). Except for a low level of elongation factors, nearly all of the necessary components for protein synthesis have been found with the ribosomes in the membrane-enclosed nuclear cap organelle; this includes charged aminoacyl-tRNA and poly(A)$^+$RNA, as well as initiation and some elongation activity (Schmoyer and Lovett, 1969; Lovett, 1975; Gong and Lovett, 1977; Johnson et al., 1977). It is not at all clear what effect ribosome aggregation and the enclosing double membrane have on ribosome activity. The cap membrane could in some way protect the ribosomes from degradation, or serve to prevent entry of the ATP and GTP required for protein synthesis. The latter has not been examined, but the ribosomes will function if isolated nuclear caps are lysed and briefly dialyzed against 0.5 M KCl to remove the inhibitory activity (Schmoyer and Lovett, 1969; Lovett, 1975; Gong and Lovett, 1977). The inhibitor activity decreases in parallel with an increase in the in vivo polypeptide chain elongation rate during germination and early development (Adelman and Lovett, 1974a; Leaver and Lovett, 1974). The inhibitor thus appears to be a sufficient, though not necessarily the only control mechanism involved in suppressing protein synthesis in motile zoospores. It is not known how the inhibitor is inactivated or removed during zoospore germination; this is of considerable interest as a developmental control, because protein synthesis is essential for development beyond the encystment stage. Since polysome formation and polypeptide synthesis are initiated even in the absence of a growth-supporting medium, the release of inhibition is not simply a response to amino acids in the medium, indicating a developmental control.

III. DEVELOPMENTALLY SPECIFIC MACROMOLECULAR SYNTHESIS

The biochemical experiments and electron microscopic studies, briefly summarized above, have described the *Blastocladiella* developmental system in considerable detail. They have not really clarified how the shift to starvation medium induces the developmental sequence of sporulation, nor the mechanisms that regulate spore germination. On the other hand, work in several laboratories has disclosed a number of associated intracellular changes that strongly suggest the necessary participation of developmentally specific macromolecule synthesis, that is, differential gene expression. In this section I would like to review this suggestive evidence, and then in the concluding section consider some direct evidence for differential gene expression.

During sporulation the cyclic GMP pool that is barely detectable in growing cells increases from 50- to 100-fold (Silverman and Epstein, 1975), and this is accompanied by a comparable increase in guanylate cyclase activity (Silverman, 1976). Actinomycin D and cycloheximide each separately block the appearance of the cyclase if added up to 30 min before the end of sporangium formation, indicating a requirement for de novo synthesis. Although the function of the high-cGMP pool in zoospores is not known, the appearance of the cyclase clearly qualifies as a developmentally specific event, since the cGMP pool and cyclase activity drop to extremely low levels upon zoospore germination and remain so during the growth phase (Silverman, 1976).

Activities associated with cyclic AMP metabolism also increase rather dramatically in sporulating cells. Adenylcyclase has been reported to increase about 70-fold (Gomes et al., 1978) and cAMP phosphodiesterase about 10- to 20-fold (Maia and Camargo, 1974; Epstein and Silverman, 1978) during sporulation. The pool of cAMP also rises several fold at or immediately prior to the time of zoospore release. The increase in diesterase activity is blocked by cycloheximide and is presumed to require de novo protein synthesis (Epstein and Silverman, 1978). Cyclic AMP-dependent protein kinase and cAMP binding protein activities also increase during sporulation (Silverman, 1978), while both cAMP-dependent and cAMP-independent kinases have been reported in zoospores (Juliani and Maia, 1979). The cyclic AMP pool and associated activities decrease rapidly during germination (Silverman and Epstein, 1975; Maia and Camargo, 1974); however, both a transient increase (Vale et al., 1976) and a sharp decrease (Silverman and Epstein, 1975) in the cAMP pool have been reported during the first 45 min of germination. Since the groups reporting these results used different growth media, the variance in the pool behavior may have been due to metabolic responses to specific environments. The striking changes in cyclic nucleotide metabolism are clearly associated with the two phases of active developmental change, and some obviously require new protein synthesis. It is not yet known whether the changes are both developmentally associated and regulatory in differentiation or result from the concomitant drastic changes in the medium used to induce sporulation and germination. In the latter case they could be simply part of a metabolic response to C/N starvation and its reversal and have little developmental significance. Control experiments using different media to induce sporulation and germination will be needed to clarify this issue.

A second clearly defined developmental event is the synthesis of two proteases during starvation-induced sporulation (Lodi and Sonneborn, 1974). These caseinolytic proteases are apparently absent in growth phase cells and both increase dramatically during sporulation. This is correlated with a greatly increased rate of in vivo proteolysis (12% per hour versus 1% per hour in growing cells) during starvation-induced sporulation. One protease with a pH optimum between 7 and 10 is released from the cells at the end of sporangium differentiation. The second, pH 5 enzyme is retained in zoospores and may contribute to the 5% per hour rate of proteolysis (of uniformly labeled zoospore proteins) measured during the first 2 h of zoospore germination. Although it was not directly demonstrated by Lodi and Sonneborn (1974) that the appearance of the proteases represented de novo synthesis, it appears reasonable, since they were undetectable in growing cells prior to induction. The pH 7-10 protease can be inhibited by antipain and if cells are treated with this inhibitor during the first hour of sporulation, subsequent development is completely arrested (Correa et al., 1978). Later additions inhibit proteolysis to the same extent but do not block the completion of sporulation. It seems evident that this protease may have an important role in protein turnover, and that such turnover is an essential event in the conversion of growth phase cells to sporangia.

Another structure produced during sporangium formation is the amorphous discharge papilla. This is formed via localized deposition of material carried to the cell wall in secretory vesicles (Lessie and Lovett, 1968). The vesicles may carry cell wall-degrading enzymes (e.g., chitinase and glucanases) to make the hole in the cell wall, as well as the plug material deposited in its place. The papillae bind concanavalin A (Lemos and Lodi, 1975) and can be readily labeled with radioactive mannose (W. D. Sikkema and J. S. Lovett, unpublished data); their synthesis is also blocked by inhibitors of RNA and protein synthesis and the inhibitors cytochalasin B and tunicamycin (Murphy and Lovett, 1966; W. D. Sikkema and J. S. Lovett, unpublished data). Preliminary studies of partially purified papillae also indicate that they contain glycoprotein as a major component (T. M. Hohn and J. S. Lovett, unpublished data), as do the discharge papillae of the related genus *Allomyces* (Youatt, 1976). Production of both the cell wall lytic enzymes and papilla glycoprotein could require de novo protein synthesis during sporulation. However, owing to the complicated nature of glycoprotein synthesis and secretory vesicle function, one cannot in the absence of other evidence assume that these processes are the direct targets of general inhibitors of protein synthesis that could, and doubtless do, interfere with many activities.

Several structures formed during the phase of zoospore cleavage and differentiation may represent cases of de novo mRNA and protein synthesis, but there is little or no direct evidence available, pro or con. The most evident candidates are flagella and flagellar "rootlet" formation, cleavage and possible plasma membrane modification for the osmotic stability of the free-swimming wall-less zoospores, ribosome aggregation to form the membrane-enclosed nuclear caps, and assembly of the several small membrane-enclosed γ particles found in each zoospore. Experiments on microtubule-tubulin turnover (Abe and Lovett, 1982) suggest that the tubulins of the cytoskeletal system and flagella of *Blastocladiella* zoospores may be different, as they are in *Naegleria* (Kowit and Fulton, 1974). If confirmed, the synthesis of flagella microtubules and associated factors could require proteins not needed during the growth phase.

The progenitor subunits of the γ particles appear in cisternae of the rough endoplasmic reticulum at an early stage of zoospore differentiation (Barstow and Lovett, 1975) and their appearance can be blocked by addition of actinomycin D or cycloheximide prior to their appearance (W. E. Barstow and J. S. Lovett, unpublished data). The mature γ particles have been reported by Cantino and coworkers to contain most (or all) of the zoospore chitin synthetase activity in zymogen form (Cantino and Myers, 1972; Myers and Cantino, 1974; Cantino and Mills, 1979), as well as some lipid (Mills and Cantino, (1978). The latter would be consistent with the presence of a surrounding cytoplasmic membrane. However, Mills and Cantino (1978) have also proposed that the glycolipid fraction has a role in chitin polymerization and that a monoglucosyladiacylglycerol may serve as a direct lipid primer (Mills and Cantino, 1980). The γ particles give rise to numerous vesicles during early zoospore germination, and these appear to migrate to and fuse with the cell surface (Soll et al., 1969; Truesdell and Cantino, 1970). It is assumed by Myers and Cantino (1974) that these vesicles deliver the chitin synthetase to the surface, where it is activated to produce the chitinous cyst wall. Although this intriguing hypothesis needs further documentation, it is consistent with the repeated observation that the cyst cell wall can be formed in the absence of new protein synthesis (Lovett, 1968; Soll and Sonneborn, 1971a,b; Silverman et al., 1974; Leaver and Lovett, 1974) and the existence of the complete but feedback-inhibited biosynthetic pathway for UDP-N-acetylglucosamine in zoospores (Norrman et al., 1973; Giddings and Cantino, 1974; Selitrennikoff and Sonneborn, 1976; Selitrennikoff et al., 1976; Selitrennikoff et al., 1980). The γ particles contain several polypeptides based on SDS-polyacrylamide electrophoresis, but two proteins of about 40,000 and 37,000 daltons make up about 60-70% of the total protein (Cantino and Mills, 1979). The functional identity is not known for any of the γ particle-associated proteins observed on the gels. Chitin synthetase is a normal growth phase enzyme and the γ particle may thus represent a special storage organelle at this developmental stage lacking a cell wall. Its apparent complexity does, however, suggest that its assembly could require proteins that are not needed during growth when the enzyme would be continuously secreted more or less directly to the cell surface. It is hoped that present attempts to purify and characterize the two major proteins will help to further clarify the role of the γ particles and their fate during zoospore germination.

An alternative approach that can be taken to detect stage-specific protein synthesis is to pulse label developing cells and examine the patterns of labeled proteins by one- or two-dimensional gel electrophoresis. No systematic study of this kind has been made of the proteins synthesized during *Blastocladiella* sporulation. However, Jaworski and Stumhofer (1981) have examined proteins produced during sporulation that are found to be associated with zoospore ribosomes with interesting results. Several polypeptides that can only be labeled with ^{35}S-methionine during late sporulation (T = 180-210 min postinduction, i.e., during the zoospore differentiation phase) are found to be associated with ribonucleoprotein particles that sediment mainly between 20 and 80 S on gradients. These polypeptides with estimated molecular weights of 120,000, 105,000, 64,000, 56,000, and 42,000 disappear from the ribosome fraction during the first 45 min of zoospore germination as protein synthesis is reactivated. At the same time a new 178,000-dalton protein, also synthesized during sporulation, becomes associated with the 80-S ribosomes. The functions of these proteins are not known, but Jaworski and Stumhofer (1981) suggested that the RNP particles could well be stored mRNP, since the synthetic period of the proteins corresponds to the period when stable poly(A)$^+$RNA also accumulates (Jaworski and Thompson, 1980; see below). They also proposed that the 178,000-dalton protein that appears in the 80-S fraction at early stages of germination may have a role in polypeptide initiation, or some event required to activate the previously inactive, inhibited ribosomes. Although it has not yet been demonstrated that these proteins are synthesized only during sporulation, the results are nevertheless intriguing, since they may lead to a better understanding of how protein synthesis is suppressed during the motile zoospore stage and then rapidly reactivated.

As part of a study of the possible turnover of microtubular proteins during zoospore germination (Abe and Lovett, 1982), cells were pulse labeled with [^{14}C] amino acids during the first 75 min of germination and the pattern of total soluble labeled proteins on two-dimensional gels was examined. The results suggested that the tubulins of the retracted flagellar axoneme do not reenter the soluble pool of tubulin dimers detected by colchicine binding, while those released by disassembly of the intracellular microtubules do. We were unable to detect new tubulin synthesis before 30 min and this synthesis failed to occur in the absence of new RNA synthesis. This suggests that the pool is already adequate to provide at least the initial microtubules found in the newly formed primary rhizoid (Barstow and Lovett, 1974), a conclusion consistent with a rapid 50% increase in the soluble tubulin pool size observed at 30 min.

FIG. 1 Proteins synthesized during *Blastocladiella* germination. Zoospores at a final density of 2×10^6 per milliliter were germinated at 20°C in 20 ml of minimal medium containing glutamic acid and methionine at 10^{-6} M, 0.1% glucose, and in the presence or absence of 10 μg/ml of actinomycin D. Cultures were labeled either from 0 to 30 min or from 30 to 40 min by addition of 25 μCi of uniformly labeled [^{14}C]amino acid mixture (sp. act., 54 mCi,/m atom). After chilling and washing, the cells were broken and the soluble proteins separated by the two-dimensional gel electrophoresis procedure of O'Farrell (1975); 200,000 cpm were applied to each gel. Fluorographs of the gels are presented. The arrows numbered 1 and 2 indicate polypeptides whose labeling was blocked by actinomycin D; those numbered 3 and 4 mark proteins that were only detectable after 30 min with or without actinomycin D. The unlabeled arrow indicates the position of β-tubulin. The fluorographs from untreated cells are shown at the top, and those exposed to actinomycin D to block RNA synthesis at the bottom. (See pages 220 and 221.)

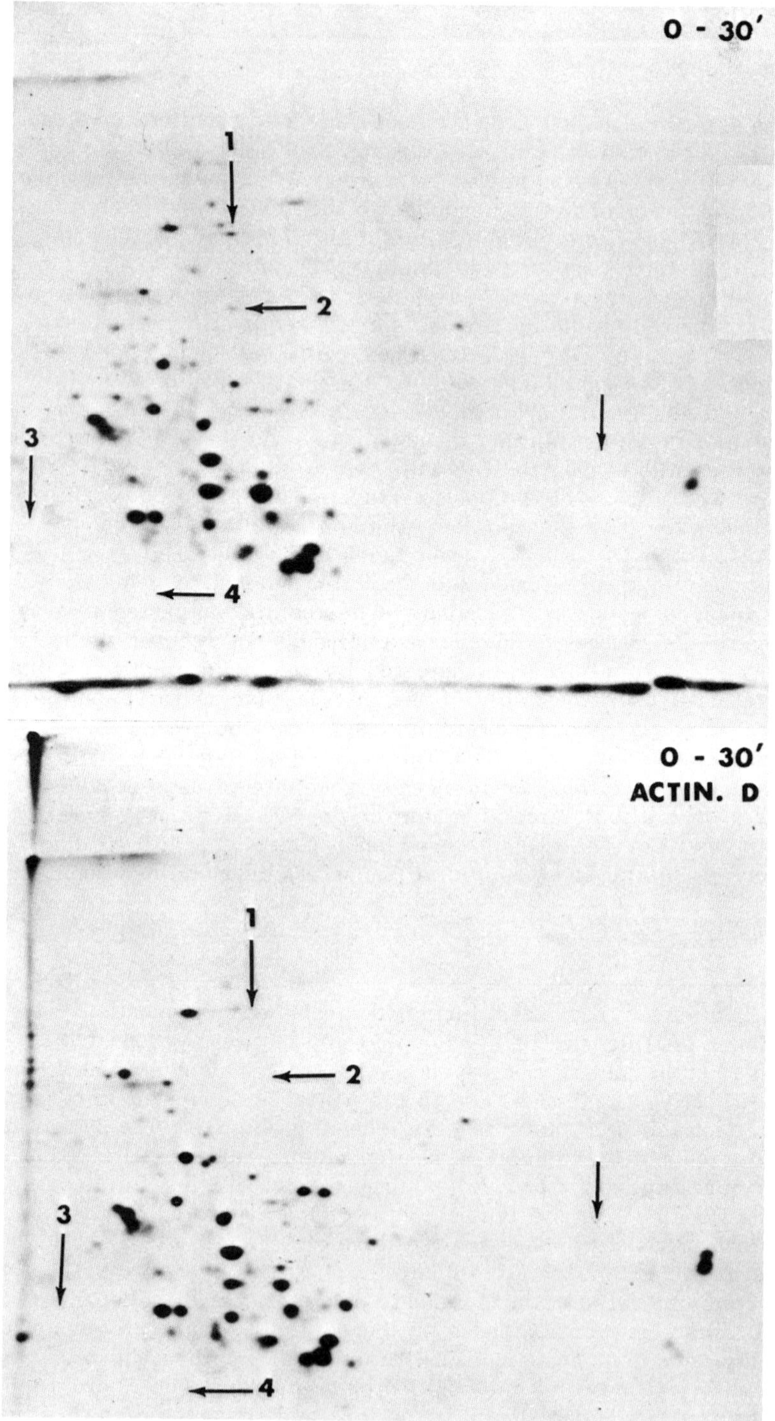
0 - 30′
1
2
3
4
0 - 30′
ACTIN. D
1
2
3
4

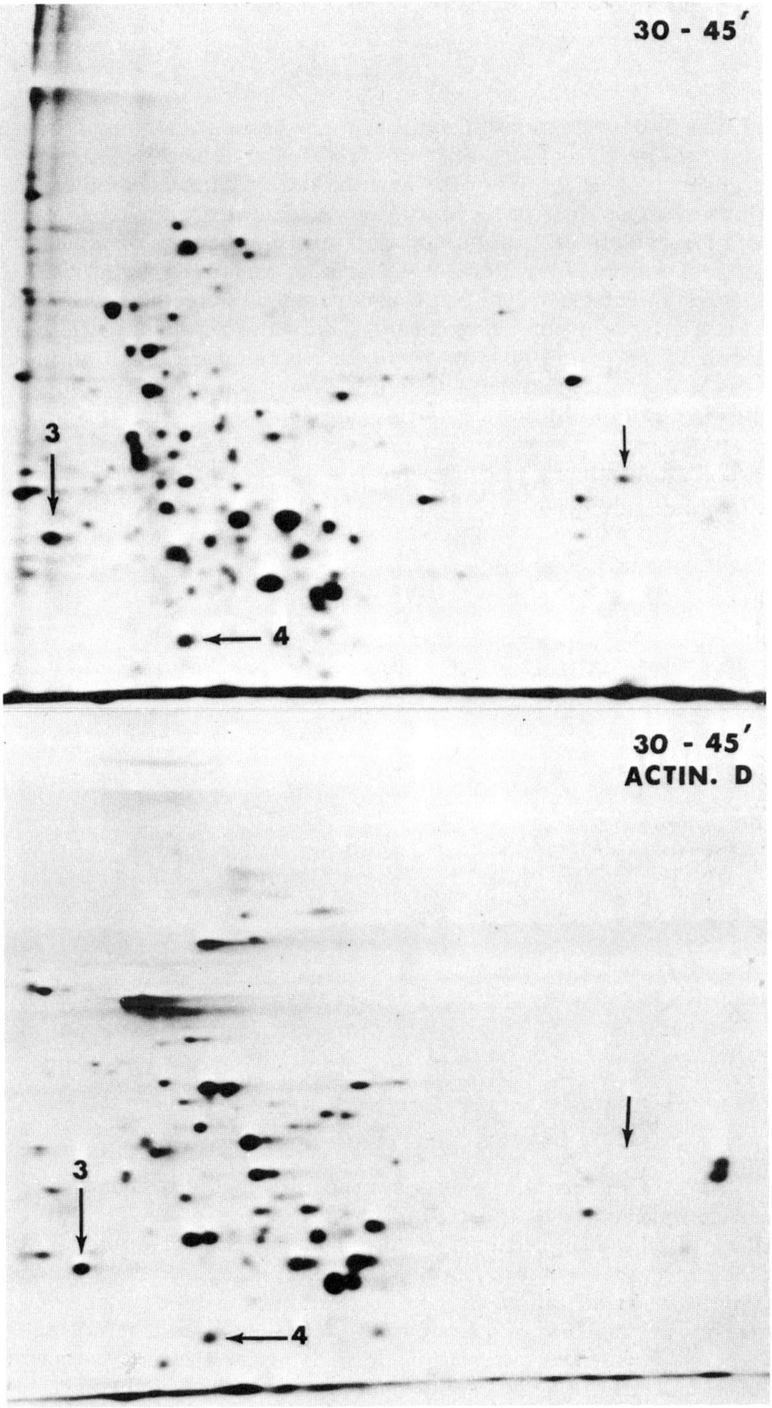
30 - 45′
3
4
30 - 45′
ACTIN. D
3
4

In addition to data on tubulin synthesis, radiofluorographs of two-dimensional gels prepared with extracts from cells labeled in the presence and absence of actinomycin D provided data on the number of proteins produced using stored versus newly synthesized mRNA. Representative radiofluorographs and data on the numbers of labeled proteins are shown in Figs. 1 and 2. The most striking result was that of some 132 proteins whose synthesis could be detected by 30 min (about 20 min after zoospore encystment), only two minor ones (labeled 1 and 2 in Fig. 1) were absent from cells pulse labeled with actinomycin D present throughout germination. Two other proteins (labeled 3 and 4 in Fig. 1) did not appear until after 30 min in either the presence or absence of actinomycin D. Since actinomycin D is an effective inhibitor of RNA synthesis in *Blastocladiella* (Lovett, 1968; Soll and Sonneborn, 1971a; Jaworski, 1976), these results show that nearly all

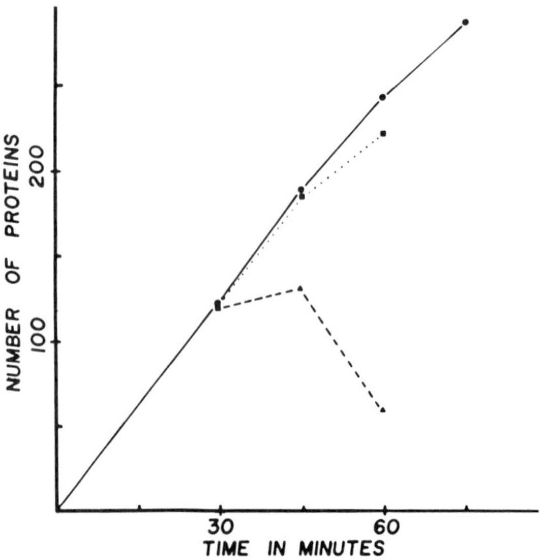

FIG. 2 Number of different proteins synthesized in the presence or absence of actinomycin D. Cultures of cells were pulse labeled and the labeled proteins analyzed by two-dimensional gel electrophoresis, as described in Fig. 1. The detectable labeled proteins were counted on each fluorogram and the results are plotted as the total number of detected spots versus time of germination (● —— ● , untreated culture; ▲ - - - ▲ , cultures with actinomycin D present from time zero; ■ · · · ■ , cultures in which actinomycin D was added 5 min prior to each pulse period). Pulse intervals were 0-30 (Fig. 1), 30-45 (Fig. 1), 45-60, and 60-75 min (not shown).

the major proteins made in the first 30 min could be translated from stored mRNAs already present in zoospores. The latter conclusion is drawn not only because the same proteins are made in the presence or absence of new RNA synthesis, but because the fraction of ribosomes assembled into polysomes is also unaffected by actinomycin D prior to 30 min (Leaver and Lovett, 1974). The normally observed increase in the number of proteins labeled during a 15-min pulse (Fig. 2), and in the ribosomal loading of polysomes, are both severely inhibited by actinomycin D after 30 min and are presumed to depend upon new mRNA synthesis. The delayed appearance of two polypeptides until after 30 min in the presence of the RNA inhibitor may indicate that the translation of the stored messenger RNA is selectively programmed.

IV. EVIDENCE FOR DIFFERENTIAL GENE EXPRESSION

Two direct, though still general, approaches have been taken to assess the role of gene expression in *Blastocladiella* development. The first has involved examination of the poly(A)$^+$RNA content of cells and patterns of poly(A)$^+$RNA synthesis during development. The second has used the approach of molecular hybridization to estimate the information content of messenger RNA.

A. Poly(A)$^+$RNA Studies

The initial report that *B. emersonii* zoospores could begin protein synthesis and develop a small rhizoid in the absence of new RNA synthesis suggested the presence in zoospores of a cryptic mRNA fraction that could be translated in early germination (Lovett, 1968). This was subsequently confirmed by the demonstration that zoospores contained a fraction of poly(A)$^+$RNA (Jaworski, 1976; Schimmelpfeng et al., 1976; Johnson et al., 1977). In order to determine whether the translation of this stored putative mRNA was suppressed by either modification or compartmentation to separate it from the nuclear cap ribosomes, the zoospore poly(A)$^+$RNA was fractionated and characterized (Johnson et al., 1977). The zoospore poly(A)$^+$RNA was found to represent 2.5% of the total cellular RNA and to have a modal molecular weight of 425,000 daltons and an average 3'-poly(A) isostich size of 32 nucleotides. Early-growth-phase cells also contain 2.5% poly(A)$^+$-RNA, but the poly(A)$^+$RNA differs slightly, having a modal molecular weight of 360,000 and an average poly(A) isostich size of 38 nucleotides. The range of poly(A)$^+$RNA molecular weights did not differ, however, and there appeared to be no difference in the 5'-capping groups between the two RNA populations, although these were not separately isolated and characterized. Thus no obviously significant difference was found between the cryptic zoospore poly(A)$^+$RNA population and that of growth phase cells. Although Jaworski (1976) reported the zoospore poly(A) fragments to be 50 nucleotides long as part of a germination study, the difference is probably due to differences in labeling protocols. Jaworski labeled with [H^3]-adenosine only during late sporulation, while Johnson et al. uniformly labeled all the RNA prior to and throughout sporulation. The zoospore fractionation experiments to localize the site of storage took advantage of a procedure to isolate the intact ribosomal nuclear caps (Lovett, 1963). Surprisingly, a minimum of at least 80% of the poly(A)$^+$RNA was found in the nuclear cap fraction and, furthermore, about 85% was associated with the major 80-S monoribosome and minor 10% fraction of inactive polysomes (Johnson et al., 1977). This result agreed with earlier reports that zoospore polysomes could synthesize polypeptides if dialyzed to remove the ribosome-associated translation inhibitor. The ribosomes of isolated nuclear caps have also been fractionated on a Biogel A-15 column in high salt buffer to separate the minor, highly enriched polyribosome fraction and the 80-S monosomes for assay; both the ribosome fractions and RNA purified from them were then translated in a cell-free wheat germ translation system. In both cases 70% of the total translation activity was found in the polysome fraction (65%) and first (heavy?) 80-S ribosome fraction (5%), while the remaining 30% was spread over several fractions of 80-S monosomes (J. S. Lovett and A. Evers, unpublished data). This suggests that a significant fraction of the stored mRNA is in the form of arrested polysomes, with the remainder present either as inactive 80-S ribosome-mRNA complexes or as mRNP. This will be discussed further below.

Despite the evidence for a significant level of stored mRNA [poly(A)$^+$RNA] in zoospores and its apparent close association with the stored nuclear cap ribosomes, new poly(A)$^+$RNA synthesis begins very early in germination (Jaworski, 1976; Jaworski et al., 1976; Johnson et al., 1977). Indeed, although the data are not extensive, it seems clear that immediately after encystment the rate of poly(A)$^+$RNA synthesis exceeds that of rRNA (Jaworski et al., 1976; Johnson et al., 1977) and at least some of it enters the polysome fraction almost immediately (Johnson et al., 1977). In addition to new poly(A)$^+$RNA synthesis, Jaworski et al. (1976) have presented evidence for both polyadenylation of preformed, stored messenger RNA and turnover (shortening) of the 3'-poly(A) sequences of zoospore poly(A)$^+$RNA during the first 45 min of spore germination. Inhibitor and double-label experiments gave results consistent with the addition of poly(A) tails to the 3' end of stored poly(A)$^-$RNA in the absence of new RNA synthesis. They further showed that the turnover of the poly(A) segments of the stored poly(A)$^+$RNA did not require new protein synthesis. These experiments indicated the presence of nonadenylated messenger RNA in zoospores, but did not determine the amount present or the actual extent of adenylation during germination. The poly(A)$^+$RNA analyzed had also been separated after the total RNA was isolated from cells, so it is not yet known whether the changes occurred with molecules that subsequently entered polysomes for translation or had other fates, such as turnover without translation. Since the function of the 3'-poly(A) segment of eukaryotic mRNA is not known in any case, the significance of the early germination poly(A) metabolism is at present hard to evaluate. In fact, none of the work to date on poly(A)$^+$RNA allows a quantitative estimate of the relative contributions of preformed stored poly(A)$^+$RNA and poly(A)$^-$RNA and newly synthesized poly(A)$^+$RNA to polypeptide synthesis in the first 30-40 min of development; although some contribution by all three fractions appears probable, the experiments necessary to measure the amount of each have not been undertaken.

Jaworski and Thompson (1980) have also examined the synthesis of the cryptic zoospore poly(A)$^+$RNA during sporulation. These results along with the data on putative mRNP particles cited earlier (Jaworski and Stumhofer, 1981) have clarified the period during which the stored mRNA is produced and at the same time raised interesting questions about the differential metabolism of poly(A)$^+$RNA during sporulation. First, these results showed that both the poly(A)$^+$RNA and the ribosomal RNA present prior to induction of sporulation were degraded more or less randomly during this stage, a period when the total RNA content decreases by 25-30% (Murphy and Lovett, 1966). Pulse-chase experiments, on the other hand, indicated that new poly-(A)$^+$RNA made during early stages of sporulation was subjected to extensive turnover (about 90%), while that made late in the process (T = 180-240 min postinduction, during late zoospore differentiation) was preferentially conserved in the resulting zoospores (Jaworski and Thompson, 1980). These authors estimated that as much as one-third of the zoospore poly(A)$^+$RNA is accumulated during the final 15-20 min of spore differentiation. Furthermore, a significant fraction of the "late mRNA" appears to end up in the 20- to 80-S subribosomal region when postmitochondrial supernatants are run on sucrose gradients. In contrast, the preinduction poly(A)$^+$RNA of the growth phase that is conserved through the sporulation process is almost entirely associated with the 80-S ribosomes and polysomes (Jaworski and Stumhofer, 1981), where Johnson et al. (1977) found the bulk of the total nuclear cap poly(A)$^+$RNA to be stored and the cell-free translation activity to reside.

The difference in the stabilities of the poly(A)$^+$RNA of the growth phase and of the newly synthesized poly(A)$^+$RNA during sporulation presents an obvious paradox. Why should the preformed poly(A)$^+$RNA be relatively stable when 90% of the newly made poly(A)$^+$RNA is subjected to rapid turnover? The difference between the stability of the new mRNA made early and late could perhaps be explained by changes in nuclease activity during sporulation (Adelman and Lovett, 1974b; Jaworski and Stumhofer, 1981). This rapid turnover may be primarily associated with the first sporangial development phase, as also appears to be the case with proteolysis and protein turnover (Lodi and Sonneborn, 1974; Correa et al., 1978). This period represents about 70% of total time from induction to spore release (Murphy and Lovett, 1966; Lessie and Lovett, 1968; Barstow and Lovett, 1978). The interval when stable poly(A)$^+$RNA sequences accumulate corresponds to the last half of the second or zoospore differentiation phase and only represents the terminal 15% of the total sporulation period, a rather short interval when ribosome aggregation occurs and protein synthesis ceases (Murphy and Lovett, 1966; Lessie and Lovett, 1968; Adelman and Lovett, 1974a,b). Differential nuclease activity does not explain the apparent differences in turnover of the poly(A)$^+$RNA already present before induction and that synthesized during sporulation. This result suggests that the conserved growth phase messages must in some way be protected relative to newly made ones. Although one could propose that they are protected by entering a special nuclease-resistant free-mRNP pool, the different distribution of the preinduction and newly synthesized poly(A)$^+$RNA in the ribosomal versus subribosomal mRNP fractions of zoospores suggests an alternative explanation. This is that the mRNAs associated with the high level of polysomes present before the inducing step-down of the medium becomes resistant to turnover by reason of their tight association with the released inactive ribosomes as 80-S ribosome-mRNA complexes. Reversible formation of such complexes with a different density from free 80-S ribosomes during a temperature-induced shift down in the rate of protein synthesis has been demonstrated by Schochetman and Perry (1972) in mouse L cells. There are additional reasons for suggesting this mechanism in *Blastocladiella*. The first is that the fraction of polysomes decreases by more than half immediately following induction, after which it recovers, but only to a level about 75-80% that present in the growth phase (Adelman and Lovett, 1974b). Thus a significant number of ribosomes could be released in this form. At the same time a ribosome-associated inhibitor activity rapidly appears (Adelman and Lovett, 1974a). The second reason for suggesting 80-S mRNA complexes is based on the extremely rapid increase in the fraction of polysomes during the first few minutes following zoospore encystment (Leaver and Lovett, 1974). The hybridization results to be discussed below support the idea that the data in Fig. 1 mainly represent the products of these polysomes and that they carry mRNAs for abundant proteins. Finally, the polysomes increase much more rapidly than the rate of decrease in the proteins associated with the late sporulation mRNPs (Jaworski and Stumhofer, 1981). If the changes in the latter are associated with the recruitment of their mRNA into the polysome fraction for translation, as proposed by Jaworski and Stumhofer (1981), it seems possible that they augment or replace the messages already associated with the ribosomes as the rate of translation increases, rather than contributing to the formation of the earliest polysomes. It would also be in accord with this model to suggest that the partial postinduction rebound in the size of the polysome fraction would occur by initiations between ribosomal subunits and the newly synthesized messages that turn over rapidly in order to permit the synthesis of newly required proteins for sporulation. The model would also require that the ribosome inhibitor preferentially associate with inactive monoribosome-mRNA complexes to stabilize them and prevent turnover.

Obviously, although the above model provides a reasonable explanation of events, it can only serve as a working hypothesis to guide the design of further experiments. It does indicate that a comparison of the proteins made before, during, and after sporulation could be useful and, furthermore, that we need to know what kind of information is represented by the messenger RNAs in different fractions and at different stages.

B. Molecular Hybridization

In order to obtain some understanding of the extent of gene expression during development in *Blastocladiella* we have examined the hybridization characteristics of RNA from different stages (S. A. Johnson and J. S. Lovett, submitted for publication). The unique sequence complexity of *Blastocladiella* DNA is 1.86×10^7 nucleotide pairs. To evaluate the amount of this genetic complexity that was expressed, Johnson and Lovett isolated the total RNA and poly(A)$^+$RNA from zoospores and the total polysomal RNA from cells during sporulation and germination. These RNAs were then hybridized in RNA excess/single-copy DNA tracer reactions to measure the base sequence complexity of each RNA preparation. It is assumed that the polysomal RNA that hybridizes in such reactions represents the sequence complexity of the gene products being translated at the stage from which it was purified.

The total zoospore RNA was found to have a complexity of 8×10^6 nucleotides, indicating the presence of about 6400 unique average-sized mRNAs [the average-sized zoospore poly(A)$^+$RNA is 1250 NT; Johnson et al., 1977] and representing some 43% of the total genomic complexity. Even more surprising than the extent of the base sequence complexity in the total zoospore stored mRNA was the discovery that only 59% of this complexity [4.7×10^6 NT (equivalent to 3700 unique average-sized mRNAs)] was represented by the purified poly(A)$^+$RNA fraction. A total of 50% of the total RNA complexity (4×10^6 NT) was represented by a nonpolyadenylated fraction and this only hybridized to the extent of 25% with cDNA prepared from the purified poly(A)$^+$RNA fraction indicating only partial overlap. This poly(A)$^-$RNA contained 1.2×10^6 NT of complexity in common with the poly(A)$^+$RNA fraction and 2.8×10^6 NT of complexity not found in the poly(A)$^+$RNA fraction. Thus zoospores store putative mRNA representing some 43% of the organism's genetic sequence complexity, with about 45% of this as poly(A)$^+$RNA, 36% as distinct poly(A)$^-$RNA sequences, and 15% percent in both fractions. This is believed to be the first direct demonstration of a significant nonpoly(A)-containing mRNA fraction in fungi. The intracellular location of the poly(A)$^-$RNA fractions is not yet known, but the major portion of the poly(A)$^+$RNA has already been shown to be in the ribosomal nuclear cap (Johnson et al., 1977).

When polysomal RNAs from cells germinated for 20 min in synthetic medium or an inorganic medium (BSM; Lovett, 1968) were examined, it was found that the complexity present increased 70 and 29%, respectively (equivalent to 4400 and 1900 new average-sized messages). This means that despite the high level of complexity present in the zoospore mRNA population, many new sequences were expressed in the first 30 min of development. Mixing experiments further indicated that in the case of cells germinated in complete synthetic medium, the changes included losses of "old" spore sequences, as well as synthesis of new unique sequences. Changes in the middle- and high-abundance classes were also detected. The high level of expression (about 70% of the genomic complexity in synthetic medium at 30 min) decreased to about 47% as the cells approached the start of the early growth phase at 60 min. The polysomal RNA was also examined at two stages in sporulation, 90 and 210 min postinduction. The sequence complexity expressed at both times was very high, averaging 80% of the total genomic complexity.

The above hybridization data, though not yet extensive, have nevertheless revealed that *Blastocladiella* expresses a significant fraction of its genome during vegetative growth and asexual development. Also, the maximal levels of expression occur during the two periods of rapid developmental and metabolic change represented by sporulation and spore germination, respectively. The highest levels occur in sporulation at periods when Jaworski and Thompson (1980) found very rapid poly(A)$^+$RNA turnover. Furthermore, nearly half of the complexity found in the RNA of sporulation phase polysomes is not found in the resulting zoospores and this loss must occur rather late in zoospore differentiation, that is, during the time when active poly(A)$^+$-RNA turnover ceases and one-third of the zoospore poly(A)$^+$RNA accumulates (Jaworski and Thompson, 1980). An obvious interpretation is that many specific gene sequences are transcribed and expressed during sporulation that are not needed in zoospores. This transcription must then be rapidly terminated and the excess messages must turn over during zoospore differentiation. Perhaps the completion of the continuous nuclear cap membrane and its fusion with the nuclear envelope could have some role in the retention of the "late" mRNAs that are conserved (there are no nuclear pores exposed to the cytosol outside the cap region; Lovett, 1975).

The significance of the early burst of new message synthesis and translation during germination is not known. It could mean that zoospores are preprogrammed to initiate new transcription immediately upon encystment to make an increased amount of information available for translation. Further events could then result in the selection of a specific set of old and new sequences for translation during development under specific circumstances. Since spores germinated in BSM starvation medium undergo normal early development at the expense of endogenous materials (Lovett, 1975), the smaller increase in polysomal RNA sequence complexity expressed by such cells may more nearly reflect a basic set of sequences required for a fixed "developmental program." Certainly the large numbers of sequences expressed in both germination media indicate that the pulse-labeling experiments in Figs. 1 and 2 almost certainly only detected proteins synthesized on abundant-class mRNAs that may or may not be those important in development.

V. EPILOGUE

The experimental results described in the preceding paragraphs, though far from providing a complete description of macromolecule synthesis in *Blastocladiella*, do provide strong evidence that differential gene expression is important in the regulation of development by this simple fungus. The data on poly(A)$^+$RNA metabolism and on the sequence complexity of the messenger RNA expressed during development, and stored as cryptic mRNA in zoospores, is in generally good agreement. Neither approach, unfortunately, identifies the specific products of gene transcription, though both are useful in evaluating the amount of information transcribed and translated. It appears to the author that with this information as a base, more real progress will require detailed investigations of developmentally specific macromolecules and macromolecular structures. The ribonuclear proteins recently described by Jaworski and Stumhofer (1981) represent a good example; other good prospects are the zoospore γ particles being studied by Cantino and Mills (Chapter 7, this volume) and Lovett et al., and the discharge papilla presently under examination in our laboratory and by Youatt (1976) with the related genus *Allomyces*. Characterization of the individual proteins should allow one to identify and isolate the specific mRNAs, as well as the appropriate segments of the genome, for analysis via presently available techniques of molecular biology. *Blastocladiella emersonii* is a favorable organism for this kind of analysis because of its nonmycelial growth form and aquatic habit. The first characteristic makes synchronization of any stage a simple process, and the second means that the cells can be readily labeled with precursors at any stage without perturbing the developmental process of interest.

Finally, no attempt has been made to review the progress of similar work in other fungi, but the results are generally comparable with regard to macromolecule synthesis, particularly during spore germination (see Van Etten et al., Chapter 9, this volume). The expression of genetic information (polysomal RNA complexity) as contrasted to transcription during the life cycle has not yet been reported for other fungi. However, comparably high levels of transcription have been reported for *Neurospora*, *Aspergillus*, and *Saccharomyces*, all Ascomycetes. Dutta and Chaudhuri (1975) found 37% of the total single-copy DNA complexity of *N. crassa* to be represented in conidial RNA, and this increased to 70% in the log phase of growth. Hereford and Rosbash (1977) have also reported that 40% of the single-copy sequence complexity was expressed at a single stage in *S. cerevisiae*. The most comprehensive developmental study has been that of Timberlake (1980) with *Aspergillus nidulans*. He found about 27% of the single-copy sequence complexity represented in the total poly(A)$^{+}$RNA of vegetative mycelia, a value equivalent to 5600-6000 diverse, average-sized mRNAs. Of particular interest was his finding that from 700 to 1100 additional sequences are transcribed during sporulation and a further 300 appear in the resulting conidia. Timberlake and Barnard (1981) have since identified a 13.3 kb region of the *Aspergillus* genome that codes for six poly(A)$^{+}$RNAs that appear to be coordinately expressed only at the time of conidial maturation. This work clearly has the prospect of progressing rapidly toward the identification of developmentally specific products and their regulation.

ACKNOWLEDGMENTS

The author gratefully acknowledges the important contributions of the following individuals to the unpublished work reported herein from his own laboratory: Susan S. Abe, William E. Barstow, Abigail Evers, Thomas M. Hohn, Steven A. Johnson, and William D. Sikkema. This work was supported by Grants from the National Science Foundation and the National Institutes of Health.

REFERENCES

Abe, S. E., and Lovett, J. S. (1982). Microtubular proteins and tubulin pool changes during zoospore germination in the fungus *Blastocladiella emersonii*. *Arch. Microbiol.* *131*:323-329.

Adelman, T. G., and Lovett, J. S. (1974a). Evidence for a ribosome-associated translation inhibitor during differentiation of *Blastocladiella emersonii*. *Biochim. Biophys. Acta* 335:236-245.

Adelman, T. G., and Lovett, J. S. (1974b). Ribosome function in vitro and in vivo during the life cycle of *Blastocladiella emersonii*. *Biochim. Biophys. Acta 349*:240-249.

Barstow, W. E., and Lovett, J. S. (1974). Apical vesicles and microtubules in rhizoids of *Blastocladiella emersonii*: Effects of actinomycin D and cycloheximide on development during germination. *Protoplasma 82*:103-117.

Barstow, W. E., and Lovett, J. S. (1975). Formation of gamma particles during zoosporogenesis in *Blastocladiella emersonii*. *Mycologia 67*:518-529.

Barstow, W. E., and Lovett, J. S. (1978). Ultrastructure of a reduced developmental cycle (minicycle) in *Blastocladiella emersonii*. *Exp. Mycol. 2*:145-155.

Cantino, E. C., and Mills, G. L. (1979). The gamma particle in *Blastocladiella emersonii*. What is it? In *Viruses and Plasmids in Fungi*, P. Lemke (Ed.). Marcel Dekker, New York, pp. 441-484.

Cantino, E. C., and Myers, R. B. (1972). Concurrent effect of visible light on γ-particles, chitin synthetase, and encystment capacity in zoospores of *Blastocladiella emersonii*. *Arch. Microbiol. 83*:203-215.

Correa, J. U., Lemos, E. M., and Lodi, W. R. (1978). Inhibition of sporulation in the water mold *Blastocladiella emersonii* by antipain. *Dev. Biol. 66*:470-479.

Dutta, S. K., and Chaudhuri, R. K. (1975). Differential transcription of nonrepeated DNA during development of *Neurospora crassa*. *Dev. Biol. 43*:35-41.

Epstein, P. M., and Silverman, P. M. (1978). Induction of cyclic AMP phosphodiesterase in *Blastocladiella emersonii* and its relation to cyclic AMP metabolism. *J. Bacteriol. 135*:968-975.

Giddings, T. H., and Cantino, E. C. (1974). Partial purification and properties of D-glucosamine-6-phosphate N-acetyltransferase from zoospores of *Blastocladiella emersonii*. *J. Bacteriol. 120*:976-979.

Gomes, S. L., Mennucci, L., and Maia, J. C. C. (1978). Adenylate cyclase activity and cyclic AMP metabolism during cytodifferentiation of *Blastocladiella emersonii*. *Biochim. Biophys. Acta 541*:190-198.

Gong, C. S., and Lovett, J. S. (1977). Regulation of protein synthesis in *Blastocladiella* zoospores: Factors for synthesis in nonsynthetic spores. *Exp. Mycol. 1*:138-151.

Hennessy, S. W., and Cantino, E. C. (1972). Lag-phase sporogenesis in *Blastocladiella emersonii*: Induced formation of unispored plantlets. *Mycologia 64*:1066-1087.

Hereford, L. M., and Rosbash, M. (1977). Number and distribution of polyadenylated sequences in yeast. *Cell 10*:453-462.

Jaworski, A. J. (1976). Synthesis of polyadenylic acid RNA during zoospore differentiation and germination in *Blastocladiella emersonii*. *Arch. Biochem. Biophys. 173*:201-209.

Jaworski, A. J., and Stumhofer, P. (1981). Stage-specific synthesis of proteins complexed to ribonucleoprotein particles and ribosomes in zoospores of *Blastocladiella emersonii*. *Mol. Cell. Biol. 1*:310-320.

Jaworski, A. J., and Thompson, K. (1980). A temporal analysis of the synthesis of the mRNA sequestered in zoospores of *Blastocladiella emersonii*. *Dev. Biol. 75*:343-357.

Jaworski, A. J., Lu, W. B. L., and Bateman, O. M. (1976). Metabolism of polyadenylic acid sequences during germination of *Blastocladiella emersonii* zoospores. *Arch. Biochem. Biophys. 177*:516-524.

Johnson, S. A., Lovett, J. S., and Wilt, F. H. (1977). The polyadenylated RNA of zoospores and growth phase cells of the aquatic fungus, *Blastocladiella*. *Dev. Biol. 56*:329-342.

Juliani, M. H., and Maia, J. C. C. (1979). Cyclic AMP-dependent and independent protein kinases of the water mold, *Blastocladiella emersonii*. *Biochim. Biophys. Acta 567*:347-356.

Kowit, J. D., and Fulton, C. (1974). Programmed synthesis of tubulin for the flagella that develop during cell differentiation in *Naegleria gruberi*. *Proc. Nat. Acad. Sci. USA 65*:2877-2881.

Leaver, C. J., and Lovett, J. S. (1974). An analysis of protein and RNA synthesis during encystment and outgrowth (germination) of *Blastocladiella* zoospores. *Cell Differ. 3*:165-192.

Lemos, E. M., and Lodi, W. R. (1975). Effect of concanavalin A on sporulation of *Blastocladiella emersonii*. *Exp. Cell Res. 94*:437-440.

Lessie, P. E., and Lovett, J. S. (1968). Ultrastructural changes during sporangium formation and zoospore differentiation in *Blastocladiella emersonii*. *Am. J. Bot. 55*:220-236.

Lodi, W. R., and Sonneborn, D. R. (1974). Protein degradation and protease activity during the life cycle of *Blastocladiella emersonii*. *J. Bacteriol. 117*:1035-1042.

Lovett, J. S. (1963). Chemical and physical characterization of "nuclear caps" isolated from *Blastocladiella* zoospores. *J. Bacteriol. 85*:1235-1246.

Lovett, J. S. (1968). Reactivation of ribonucleic acid and protein synthesis during germination of *Blastocladiella* zoospores and the role of the ribosomal nuclear cap. *J. Bacteriol. 96*:962-969.

Lovett, J. S. (1975). Growth and differentiation of the water mold *Blastocladiella emersonii*: Cytodifferentiation and the role of ribonucleic acid and protein synthesis. *Bacteriol. Rev. 39*:345-404.

Maia, J. C. C., and Camargo, E. P. (1974). c-AMP phosphodiesterase activity during growth and cytodifferentiation in *Blastocladiella emersonii*. *Cell Differ.* *3*:147-155.

Mills, G. L., and Cantino, E. C. (1978). The lipid composition of the *Blastocladiella emersonii* γ-particle and the function of the γ-particle lipid in chitin formation. *Exp. Mycol.* *2*:99-109.

Mills, G. L., and Cantino, E. C. (1979). Trimodal formation of microbodies and associated biochemical and cytochemical changes during development in *Blastocladiella emersonii*. *Exp. Mycol.* *3*:53-69.

Mills, G. L., and Cantino, E. C. (1980). The glycolipid involved in chitin synthesis by zoospores of *Blastocladiella emersonii* is a monoglucosyldiacylglycerol. *Exp. Mycol.* *4*:175-180.

Mills, G. L., Myers, R. B., and Cantino, E. C. (1974). Lipid changes during development of *Blastocladiella emersonii*. *Phytochemistry* *13*:2653-2657.

Murphy, M. N., Sr., and Lovett, J. S. (1966). RNA and protein synthesis during zoospore differentiation in synchronized cultures of *Blastocladiella*. *Dev. Biol.* *14*:68-95.

Myers, R. B., and Cantino, E. C. (1974). The gamma particle. *Monogr. Dev. Biol.* *8*:1-117.

Norrman, J., Myers, R. B., Giddings, T. H., and Cantino, E. C. (1973). Partial purification of L-glutamine: D-Fructose-6-phosphate aminotransferase from zoospores of *Blastocladiella emersonii*. *Biochim. Biophys. Acta* *302*:173-177.

O'Farrell, P. H. (1975). High resolution two-dimensional electrophoresis of proteins. *J. Biol. Chem.* *250*:4007-4021.

Schimmelpfeng, L., Grierson, D., and Zetsche, K. (1976). Preformed poly(A)-containing RNA in zoospores of *Blastocladiella emersonii*. *Biochim. Biophys. Acta* *435*:340-348.

Schmoyer, I. R., and Lovett, J. S. (1969). Regulation of protein synthesis in zoospores of *Blastocladiella*. *J. Bacteriol.* *100*:854-864.

Schochetman, G., and Perry, R. P. (1972). Characterization of the messenger RNA released from L cell polysomes as a result of temperature shock. *J. Mol. Biol.* *63*:577-590.

Selitrennikoff, C. P., and Sonneborn, D. R. (1976). Post-translational control of *de novo* cell wall formation during *Blastocladiella emersonii* zoospore germination. *Dev. Biol.* *54*:37-51.

Selitrennikoff, C. P., Allin, D., and Sonneborn, D. R. (1976). Chitin biosynthesis during *Blastocladiella* zoospore germination: Evidence that the hexosamine biosynthetic pathway is post-translationally activated during cell differentiation. *Proc. Nat. Acad. Sci. USA* *73*:534-538.

Selitrennikoff, C. P., Dalley, N. E., and Sonneborn, D. R. (1980). Regulation of the hexosamine biosynthetic pathway in the water mold *Blastocladiella emersonii*: Sensitivity to end product inhibition is dependent on the life cycle phase. *Proc. Nat. Acad. Sci. USA* *77*:5998-6002.

Silverman, P. M. (1976). Regulation of guanylate cyclase activity during cytodifferentiation of *Blastocladiella emersonii*. *Biochem. Biophys. Res. Commun.* *70*:381-388.

Silverman, P. M. (1978). Cyclic AMP binding proteins and cyclic AMP-dependent protein kinase from *Blastocladiella emersonii*. *J. Bacteriol.* *135*:976-980.

Silverman, P. M., and Epstein, P. M. (1975). Cyclic nucleotide metabolism coupled to cytodifferentiation of *Blastocladiella emersonii*. *Proc. Nat. Acad. Sci. USA* *72*:442-446.

Silverman, P. M., Huh, M. O. H., and Sun, L. (1974). Protein synthesis during zoospore germination in the aquatic Phycomycete *Blastocladiella emersonii*. *Dev. Biol.* *40*:59-70.

Smith, J. D., and Silverman, P. M. (1973). Lipid turnover during morphogenesis in the water mold *Blastocladeilla emersonii*. *Biochim. Biophys. Acta* *54*:1191-1197.

Soll, D. R., and Sonneborn, D. R. (1971a). Zoospore germination in *Blastocladiella emersonii*. III. Structural changes in relation to protein and RNA synthesis. *J. Cell Sci.* *9*:679-699.

Soll, D. R., and Sonneborn, D. R. (1971b). Zoospore germination in *Blastocladiella emersonii*: Cell differentiation without protein synthesis? *Proc. Nat. Acad. Sci. USA* *68*:459-463.

Soll, D. R., Bromberg, R., and Sonneborn, D. R. (1969). Zoospore germination in the water mold *Blastocladiella emersonii*. I. Measurement of germination and sequence of subcellular morphological changes. *Dev. Biol.* *20*:183-217.

Suberkropp, K. F., and Cantino, E. C. (1972). Environmental control of motility and encystment in *Blastocladiella emersonii* zoospores at high population densities. *Trans. Br. Mycol. Soc.* *59*:463-475.

Suberkropp, K. F., and Cantino, E. C. (1973). Utilization of endogenous reserves by swimming zoospores of *Blastocladiella emersonii*. *Arch. Mikrobiol.* *89*:205-221.

Timberlake, W. E. (1980). Developmental gene regulation in *Aspergillus nidulans*. *Dev. Biol.* *78*:497-510.

Timberlake, W. E., and Barnard, E. C. (1981). Organization of a gene cluster expressed specifically in the asexual spores of *A. nidulans*. *Cell* *26*:29-37.

Truesdell, L. C., and Cantino, E. C. (1970). Decay of γ-particles in germinating zoospores of *Blastocladiella emersonii*. *Arch. Microbiol.* *70*:378-392.

Truesdell, L. C., and Cantino, E. C. (1971). The induction and early events of germination in the zoospore of *Blastocladiella emersonii*. *Curr. Top. Dev. Biol.* *6*:1-44.

Vale, V. L., Gomes, S. L., Maia, J. C. C., and Mennucci, L. (1976). Transient cyclic AMP accumulation in germinating zoospores of *Blastocladiella emersonii*. *FEBS Lett.* *67*:189-192.

Youatt, J. (1976). Chemical nature of discharge papillae of *Allomyces*. *Trans. Br. Mycol. Soc.* *66*:113-122.

9
Fungal Spore Germination

JAMES L. VAN ETTEN, KURT R. DAHLBERG,* and GERALDINE M. RUSSO *University of Nebraska-Lincoln, Lincoln, Nebraska*

**Present affiliation*: Ralston Purina Company Central Research, St. Louis, Missouri

I. INTRODUCTION

The subject of fungal spore germination is important for both practical and basic reasons. From a practical standpoint, most fungi causing mycotic diseases (both plant and animal) are disseminated by spores and the disease cycle is initiated when the spores germinate. From a basic standpoint, studies on dormancy and spore germination may provide new insights into fundamental processes regulating cellular development. The germination of fungal spores is a morphologically simple process in which nongrowing, usually metabolically quiescent cells respond to changes in environmental conditions by reactivating physiological processes and genetic functions to produce rapidly growing cells.

Fungi comprise an extremely diverse group of organisms and it is difficult to generalize about any aspect of fungal physiology, including spore germination. General characteristics reported for one fungal spore type may not apply to other spores. Fungal spores may be motile or nonmotile, small (a few microns in diameter) or large (e.g., 350 μm for azygospores of *Gigaspora*), single celled or multicelled, haploid or diploid, uninucleate or multinucleate, thin walled or thick walled, and short-lived (a few hours or days) or long-lived (many years). Many spores are formed as the result of the fusion of two compatible nuclei (sexual spores), whereas others are derived from vegetative cells (asexual spores). Furthermore, many spores are formed in some type of specialized structure and these structures can assume many shapes and sizes.

Terms such as dormancy, activation, and germination were defined by Sussman (1966) and Sussman and Douthit (1973). These definitions are widely accepted in the mycological literature and we have subdivided this review accordingly.

In the past few years, many excellent review articles and several books (e.g., Weber and Hess, 1976; Turian and Hohl, 1981) have appeared that deal, either exclusively or in large part, with fungal spores and spore germination. Also, several reviews on the germination of a single spore or spore type are available [e.g., sporangiospores of *Phycomyces blakesleeanus* (Furch, 1981) and *Rhizopus stolonifer* (Van Etten et al., 1977), conidia of *Neurospora crassa* (Schmidt and Brody, 1976; Brody, 1981), zoospores of *Blastocladiella emersonii* (Lovett, 1975; Lovett et al., 1977), and uredospores of the rust fungi (Staples and Huang, 1981)]. Because of space limitations, this review will focus on more recent information. Furthermore, we have cited recent review articles extensively; in many cases the reader will have to refer to these articles for references to original papers.

II. SPORE DORMANCY

A. General Considerations

Sussman and Douthit (1973) define dormancy as "any rest period or reversible interruption of the phenotypic development of an organism." They classify dormancy into three types: constitutional, exogenous, and induced. Constitutional (constitutive) dormancy is defined as "a condition wherein development is delayed due to an innate property of the dormant stage, such as a barrier to the penetration of nutrients, a metabolic block, or the production of a self-inhibitor." In exogenous dormancy "development is delayed because of unfavorable chemical or physical conditions of the environment." One critical distinction that is frequently made between exogenous and constitutive dormancy is that spores in the latter category require an activation treatment in order to germinate. However, in some instances, the distinction between constitutive and exogenous dormancy may not be clear-cut, as is discussed in Sec. III on activation. When a "nondormant or exogenously dormant spore is made constitutionally dormant," dormancy is said to be induced.

In this section we will briefly refer to some of the unique properties of dormant spores. At the outset it should be noted that two characteristics of almost all spores are a decreased rate of respiration and the absence of detectable protein and nucleic acid synthesis.

B. Ultrastructural Properties

Ultrastructural studies on dormant spores have often been hampered by the difficulty of getting embedding resins into the spore. Nonetheless, ultrastructural correlations between spores and vegetative cells of various fungi have revealed differences in cell wall structure, the abundance and form of endoplasmic reticulum, the presence of lipid bodies, the number, shape, and/or size of mitochondria, and the contour of the plasmalemma. The walls of vegetative cells are generally thinner and less complex than the walls of their corresponding spores, although spores vary greatly with regard to the number of wall layers (Fletcher, 1971; Sussman and Douthit, 1973). In a few fungi, the germ tube wall appears to be ultrastructurally similar to the spore wall [e.g., arthrospores of *Geotrichum candidum* (Steele and Fraser, 1973)].

Many fungal spores have a paucity of endoplasmic reticulum (Sussman and Douthit, 1973; Weber and Hess, 1975; Hawker and Madelin, 1976). In spores where endoplasmic reticulum is abundant, as in the myxomycete *Arcyria cinerea* (Mims, 1971), it is composed of short segments. Characteristic longer segments appear only after germination. Other spores such as zoospores of *B. emersonii* contain specialized organelles such as a nuclear cap and gamma particles, which disappear during germination (Lovett, 1975).

The presence of membrane-bound storage vacuoles containing phospholipids (lipid bodies), polysaccharides (glycogen granules), or other materials in the spore is characteristic of many fungi (Sussman and Douthit, 1973; Hawker and Madelin, 1976). The fact that many of these storage bodies decrease in size or disappear upon spore germination suggests their utilization as a carbon source during this process.

C. Chemical Composition

Although many of the biochemical changes which accompany spore germination will be dealt with in Sec. IV on germination, a few of the properties of dormant spores will be mentioned here. Probably all dormant spores contain carbohydrates and/or lipids, presumably for use as reserve energy sources during dormancy and germination. The types of compounds and levels within spores are quite variable. The disaccharide trehalose is frequently found in dormant spores and may account for 10-30% of the dry weight [e.g., ascospores of *Neurospora tetrasperma* (Lingappa and Sussman, 1959) and sporangiospores of *P. blakesleeanus* (Rudolph and Ochsen, 1969)]. Many spores also contain high concentrations of polyols such as glycerol, erythritol, arabitol, and mannitol. These compounds comprise up to 10-15% of the dry weight of some spores [e.g., *Penicillium* species (Ballio et al., 1964; Eckert et al., 1968)]. The β-glucan polymers in the walls of oospores of many Oomycetes may also serve as an energy reserve, since the inner spore wall is digested during long-term storage and during germination (Bartnicki-Garcia and Hemmes, 1976; Beakes, 1980b).

The lipid content in many fungal spores is high, comprising as much as 70% of the dry weight of *Glomus caledonius* spores (Beilby and Kidby, 1980). A wide range of lipid compounds have been reported in spores, including hydrocarbons, fatty acids, phospholipids, sterols, terpenes, and polyketides (Reisener, 1976; Weete, 1974). Some of these components are unusual, such as the cis-9, 10-epoxyoctadecanoic fatty acid, which comprises up to 78% of the total fatty acids present in some rust uredospores (see the review by Reisener, 1976). In *N. crassa* conidia phospholipids account for 90% of the total lipid, which comprises about 17% of the dry weight (Bianchi and Turian, 1967). In contrast, the mycelium only contains 3-4% phospholipid (Hubbard and Brody, 1975; Scarborough, 1975). The total phospholipid content of the conidia decreases slowly with aging of the spores, suggesting that phospholipids may be used as an energy source during dormancy (Bianchi and Turian, 1967).

Some spores contain large pools of free amino acids; for example, 2.5% of the dry weight of *N. crassa* conidia is glutamic acid (Schmidt and Brody, 1975). The role ascribed to glutamic acid during germination will be discussed in Sec. IV on germination. The first example of a major protein in fungal spores was recently reported in conidia of *Botryodiplodia theobromae* (Van Etten et al., 1979). This protein comprises 20-25% of the total protein present in *B. theobromae* spores and may be actively degraded during germination; the protein has not been detected in vegetative cells. The phenomenon of "storage" protein(s) in spores may be limited among fungi, however, since protein extracted from six other diverse fungal spores did not reveal any major protein(s).

D. Respiration

Dormant spores usually have a reduced rate of oxygen uptake as compared to vegetative cells. This reduced oxygen consumption is generally not due to a deficiency in respiratory enzymes or other components of the mitochondrial electron transport chain, for example, the cytochromes (see Gottlieb, 1976). Exceptions to this are the conidia of *B. theobromae*, which lack a functional cytochrome oxidase and an oligomycin-sensitive ATPase (see the review by Brambl, 1981). A deficiency in respiratory substrates and the oxidized form of the cofactor nicotinamide adenine dinucleotide is responsible for the low levels of oxygen consumption in *Dictyostelium discoideum* spores (Kobilinsky and Beattie, 1977).

It has been reported that conidia of *N. crassa* respire by an alternate respiratory pathway (the so-called SHAM-sensitive pathway) and that during germination, the spores switch over to the normal cyanide-sensitive cytochrome electron transport pathway (see references in Schmidt and Brody, 1976). However, these results could not be repeated by Brambl and his colleagues and he concluded that *N. crassa* conidia respire by the normal cyanide-sensitive cytochrome pathway (Brambl, 1981).

E. Water Content

Most fungal spores have a decreased water content as compared to vegetative cells. Variability in the water content of spores exists, however, and reflects the intrinsic characteristics of different species and the conditions under which the spores are formed, matured, aged, and harvested. Most of the water lost from spores is attributed to natural processes such as dehydration or syneresis, rather than simple drying (Hawker and Madelin, 1976). The functional significance of spore dehydration has interested investigators over the years and several studies have correlated the degree of hydration of both fungal and bacterial spores with resistance to environmental stresses (see Hawker and Madelin, 1976; Keynan, 1978).

The manner in which spores are harvested can have a marked effect on some of their biochemical properties. Exposure of fungal spores to water for brief periods (e.g., during harvesting) can produce metabolic alterations in the spores, even though the spores do not form germ tubes in water. A short exposure of *N. crassa* conidia (Mirkes, 1974) and *R. stolonifer* sporangiospores (Nickerson et al., 1981) to water leads to increased levels of polysomes as compared to dry-harvested spores. Water-harvested spores of *R. stolonifer* also differ from dry-harvested spores in that they incorporate $^{14}CO_2$ into cellular components, lose viability upon storage at room temperature, and exhibit altered nutritional requirements for germination.

Fahey et al. (1978) studied the effect of relative humidity on the water content and other properties of ungerminated *N. crassa* conidia. Conidia stored at 0, 50, and 100% relative humidity were considered to be in dry (5% water content, water absent or tightly "bound"), quasi-dry (5-15% water content, only "bound" water), and wet (40-70% water content, "bound" plus "bulk" water accumulated in the regions between hydrated molecules) hydration states, respectively. Wet conidia exhibited substantial enzymatic activity, whereas the enzymatic activity of quasi-dry and dry conidia was limited and negligible, respectively. The decreased metabolic activity in quasi-dry and dry conidia may result from the effects of dehydration on enzyme-catalyzed reactions, possibly via (1) a reduction in the rate or a reversal in the direction of a reaction in which water is a reactant; (2) a conformational or other structural change in an enzyme, leading to its inactivation; or (3) decreased diffusion of substrates and coenzymes to an enzyme. Dry conidia were also more resistant to heat (Fahey et al., 1978) and freezing (Leef and Mazur, 1978), possibly owing to the stabilizing influence of dehydration on macromolecules, and survived for longer periods of time.

The intracellular concentrations of storage carbohydrates may also influence the water content of spores and hence their survivability. Emyanitoff and Wright (1979) altered the intracellular concentrations of glycogen and trehalose in constitutively dormant spores of *D. discoideum* by growing them under various conditions. Their data indicate that trehalose is at least partly responsible for the wet heat resistance of these spores. They suggested that trehalose might be localized in vesicles in the spore and, if so, might serve an osmoregulatory function by aiding to dehydrate the cytoplasm.

F. Mechanisms to Establish and Maintain Dormancy

Several mechanisms have been proposed to explain dormancy in fungal spores, including permeability barriers, anhydrobiosis, inhibition of de novo synthesis of enzymes or conformational changes in preexisting enzymes, spatial separation of substrates and enzymes by compartmentalization, and the production of self-inhibitors of germination. These mechanisms need not be mutually exclusive.

1. *Permeability*

Early work in the search for a mechanism to establish and maintain constitutive dormancy in fungi focused on the permeability of spores to oxygen, water, and nutrients. Sussman and Lowry (cited in Sussman, 1969) showed that *N. tetrasperma* ascospores rapidly plasmolyzed when suspended in concentrated glycerol. Such spores quickly regained their turgor when suspended in water. Studies using heavy water ($H_2{}^{18}O$), heavy oxygen ($^{18}O_2$), and [^{14}C]glucose indicated free exchange of these materials, thereby ruling out a permeability barrier within the spore as the mechanism for establishing and maintaining dormancy (reviewed by Hawker and Madelin, 1976). Although dormant spores are not impermeable, differences in permeability between dormant and germinating spores exist. About 150 min after *N. tetrasperma* ascospores are activated, Ca^{2+} and Na^+ ions are released from the spore and the spores become sensitive to EDTA. Other substances, including pyruvate, cysteine, and toxic cations, begin to penetrate the spore only at this time (Sussman and Halvorson, 1966). Thus the permeability differences between dormant and germinating ascospores of *Neurospora* appear to be secondary, in that they follow, rather than induce, the initial steps in the activation process.

Similar results have been reported for sporangiospores of *P. blakesleeanus* (Furch, 1981); dormant spores are resistant to inhibitors that have an effect on germinated spores. However, dormant spores readily take up glucose and actively metabolize it. Thus dormant spores are permeable to at least certain substances, so that total lack of permeability cannot explain dormancy in these spores.

2. *Compartmentalization*

A spore could be maintained in the dormant state by physical separation of components (e.g., the separation of an enzyme from its substrate or the separation of subunits of a multicomponent enzyme). A few examples of this type of regulation have been reported in fungi. The enzyme trehalase is stored in the ascospore wall of *N. tetrasperma*, while its substrate is stored in the cytoplasm (Hecker and Sussman, 1973). An alteration in cell membrane permeability of the spore occurs during heat activation, bringing the enzyme and substrate into close proximity. Brambl (1980) has shown that conidia of *Botryodiplodia theobromae* lack a functional cytochrome oxidase because the nuclear encoded enzyme subunits are stored in a precursor form outside the mitochondria and only transported into the mitochondria to yield a functional enzyme when the spores are placed under germinating conditions.

Finally, compartmentalization of ribosomes within the nuclear cap of *B. emersonii* zoospores was considered to be the mechanism whereby protein synthesis is inhibited in these spores (discussed by Lovett,

1975). However, subsequent studies have shown the presence of all necessary components of the translational machinery within the nuclear cap of *B. emersonii* zoospores, although some components are present in lesser amounts in the nuclear cap as compared to the cytoplasm (Lovett et al., 1977). Current evidence suggests that zoospores contain an inhibitor which blocks mRNA translation (Adelman and Lovett, 1974). Concomitant with zoospore encystment, the inhibitor is released into the medium and protein synthesis resumes.

3. *Self-Inhibitors*

A number of fungal spores, encompassing a wide range of taxonomic groups, contain compounds (called self-inhibitors) which prevent their germination; that is, germination occurs only after the compounds are leached from the spore (see the reviews by Allen, 1976; Macko et al., 1976; Macko, 1981). In some cases, the inhibitory substances have been isolated and identified. Examples include methyl-cis-3,4-dimethoxycinnamate (MDC) in uredospores of *Uromyces phaseoli*, methyl-cis-ferulate (MF) in uredospores of *Puccinia graminis tritici*, 5-isobutyroxy-β-ionone (trivial name, "quiesone") in conidia of *Peronospora tabacina*, and 3-(3-amino-3-carboxylpropyl)-6-(3-methyl-2-butenyl amino)purine (trivial name, "discadenine") in spores of *D. discoideum*. MDC and MF also function as self-inhibitors in other species of the *Pucciniaceae*. Self-inhibitors, which are species specific, are active at very low concentrations (a few picograms to a few nanograms per milliliter) and usually only have an affect on early stages of germination.

In rust fungi, self-inhibitors probably function by preventing dissolution of the germ pore plug. This is the only region of the spore wall composed of mannoprotein (Hess et al., 1981) and specific hydrolytic enzymes are involved in its digestion; MDC and MF probably function by inhibiting these enzymes. However, studies on wheat rust uredospores indicate that in addition to inhibiting germ pore plug dissolution, MF prevents the initiation and maintenance of growth of the germ tube prior to emergence. Since germ tube initiation and wall digestion could not be experimentally separated, it is not known which is the primary target of the self-inhibitor.

Less is known about the mode of action of other self-inhibitors. Quiesone is postulated to inhibit protein synthesis in spores of *P. tabacina* by preventing the reassociation of the large ribosomal subunit with the protein synthesis initiation complex (Hollomon, 1973).

Self-inhibitors do not necessarily have to be "unique" chemical compounds. The inhibition of germination at high spore densities of *Phomopsis viticola* alpha spores is attributed to the presence of elemental sulfur (Pezet and Pont, 1975). Likewise, high iron concentrations may be involved in the inhibition of germination of *Colletotrichum musae* conidia (Harper et al., 1980).

III. SPORE ACTIVATION

A. General Considerations

Activation was originally defined by Sussman (1966) as "the application of environmental stimuli which induce germination." It is generally considered that a constitutively dormant spore requires a specialized stimulus such as heat or a chemical compound in order to germinate, while an exogenously dormant spore will germinate if provided with moisture, nutrients, oxygen, and so on. However, the distinction between these two types of dormancy can occasionally create difficulties. For example, the addition of a specific "nutrient" might relieve a metabolic block of a constitutively dormant spore or it might serve as a substrate for an exogenously dormant spore. Thus the fact that a "nutrient" induces germination does not, in itself, allow one to classify spores as constitutively or exogenously dormant. One must know if the "nutrient" relieves a metabolic block or exerts a definite nutrient effect in order to assign a spore to the categories of dormancy.

Additional considerations may further blur the distinction between exogenous and constitutive dormancy. Some fungal spores germinate in response to multiple activation treatments, some of which appear to be relief from exogenous dormancy, and some from constitutive dormancy (see below). Also, some spores which are unequivocably constitutively dormant are capable of germinating in the apparent absence of an activation treatment (autoactivation; see below). In view of this occasional lack of a clear distinction between exogenous and constitutive dormancy, we will consider spore activation according to the broad definition of Sussman and Douthit (1973): "any treatment which induces or stimulates spore germination." This definition does not attempt to account for differences in spore activation arising from different types of dormancy.

B. Spore Activation Treatments

A range of activation treatments induce fungal spore germination (Sussman and Halvorson, 1966; Sussman, 1976; Cotter, 1981). These methods may be divided into several categories which need not be mutually exclusive.

1. *Moisture/Rehydration*

Many fungal spores require no other activation treatment than to be suspended in water or incubated in high water activity (a_W) under moderate environmental conditions. For instance, more than 90% of *Peronophythora litchii* sporangiospores germinate within 6 h after being suspended in distilled water (Kao and Leu, 1980). One might presume that all spores that germinate in distilled water are unequivocably exogenously dormant, but this is not a firm rule. A few fungal spores germinate in water or weak buffers by an "autoactivation" mechanism, as discussed below.

2. *Nutrient Effects*

Many fungal spores will germinate when provided with compounds normally considered as "nutrients," such as ones capable of being used as carbon or nitrogen sources. This is by far the most common method of inducing germination. For instance, *Phyllosticta lycopersici* spores germinate when provided with glucose, NH_4NO_3, PO_4^{-3}, and $MgSO_4$ (Onesirosan, 1976). All compounds in this mixture are required for germination, and the process ceases if the nutrients are removed from the spore suspension. In this case, the effect is unquestionably nutritive.

Other fungal spores are activated by what is ostensibly a nutrient effect, however, this may not be the case. *Dictyostelium discoideum* spores are activated by a mixture of tryptophan, phenylalanine, and methionine (Cotter and Raper, 1966). These amino acids are not, however, required for the duration of spore germination, since they may be removed from the suspension, and spores still germinate normally. It is not known whether they exert a nutrient effect by loading cellular pools to supply amino acids throughout germination or, alternatively, whether they activate a metabolic pathway which allows spores to germinate without additional exogenous sources of the amino acids.

3. *Chemical Activation*

Some fungal spores are activated by compounds not normally considered to be nutrients. Examples of this include activation of *Uromyces phaseoli* spores by β-ionone and various ketones (French et al., 1977) and *D. discoideum* spores by such diverse compounds as urea, guanidine, and dimethylsulfoxide (DMSO) (Cotter, 1981). Chemical activators are generally not metabolized and are not required for the duration of the germination process. *Dictyostelium discoideum* spores are activated by exposure to 20% DMSO for 30-60 min (Cotter et al., 1976). Following this exposure, the presence of DMSO is not required for spore germination; indeed, spores will not germinate unless DMSO is removed.

In some cases, it is impossible to conclude from existing data whether an activation treatment is nutritional or chemical in nature. For instance, exudates of lettuce seeds induce germination of *Bremia lactucae* spores (Morgan, 1978). This activation could be mediated by nutrients or by chemical activators (phenolics, etc.) released from the seeds. Likewise, a number of spores are activated by substances released by the yeast *Rhodotorula glutinis* (Fries, 1977; 1978).

4. *Temperature Activation*

Many fungal spores are activated by the application of heat, including spores of *Byssochlamys fulva*, *Coprinus radiatus*, *Mucor miehi*, *Neurospora tetrasperma*, *Puccinia graminis* (see Sussman, 1976), *Dictyostelium discoideum* (Cotter, 1981), and *Phycomyces blakesleeanus* (Furch, 1981). A variety of thermal regimens have been described for various spore types. Some mild treatments, such as activation of myxomycete spores at 32°C followed by germination at 27°C (Jahn, 1940), may reflect differences in temperature optima for spore germination and outgrowth rather than true activation. Low-temperature activation is less common than high-temperature activation. Nonetheless, some spores respond to cold treatments; ascospores of *Hypoxylon diathrauston* germinate at 12°C only if they are pretreated at -3°C (Ouellette and Ward, 1970).

5. *Autoactivation*

A few fungal spores undergo autoactivation of spore germination. These constitutively dormant spores are activated by substances associated with or produced by the spores. Suspending spores in water results in germination, but this process is different from simple moisture effects, since the spores are constitutively dormant. Autoactivation of spore germination has not been described very often. Whether this results from the comparative rarity of the phenomenon or whether it is simply unrecognized among spores germinating in the absence of exogenous nutrients is unknown.

Dictyostelium discoideum spores are known to autoactivate after a period of aging or as a result of mutations (Cotter and Dahlberg, 1977; Dahlberg and Cotter, 1978a). The process shows a positive spore density dependence, resulting from the production of substances which stimulate spore germination. Filtrates in which spores have previously autoactivated stimulate the germination of other spores. This is in constrast to the situation where self-inhibitors of spore germination are present. In the case of *D. discoideum* spores, both self-inhibitors (Abe et al., 1976) and autoactivators (Cotter and Dahlberg, 1977) are present. The possibility that self-inhibitors are metabolized to autoactivators should not be overlooked.

Other examples of autoactivation include spores of several Myxomycetes (Smart, 1937; Dahlberg and Franke, 1977) and the cellular slime mold *Dictyostelium purpureum* (Kornfeld, 1968; Coco and Kornfeld, 1978). The siderophore-mediated density-dependent germination of *Penicillium*, *Aspergillus*, and *Neurospora* spores stressed by low a_w may be similar to autoactivation (Charlang et al., 1981). *Candida albicans* spores also have an "intrinsic factor" which induces germ tube formation (Pugh and Cawson, 1979) in a manner analogous to autoactivation.

6. *Multiple Methods of Spore Activation*

Several of the better-studied spores are activated by more than one type of treatment. *Neurospora tetrasperma* spores are activated by heat shock or furfurol treatment. Both *P. blakesleeanus* and *D. discoideum* spores are activated by several methods. It is reasonable to propose that some of these treatments merely mimic others. For instance, heat shock, gamma irradiation, DMSO, urea, and guanidine treatments may activate *D. discoideum* spores by a common mechanism. All of these treatments are capable of denaturing proteins (Cotter, 1981) and may affect a theoretical site of dormancy in the same manner. However, the largely uncharacterized "spore germination promoter" (Hashimoto et al., 1976), "autoactivator substance" (Dahlberg and Cotter, 1978a), and "substances released from bacteria" (Dahlberg and Cotter, 1978b) may not mimic the protein denaturation mechanism. Existing evidence suggests that responses to these substances induce *D. discoideum* spore germination by a distinctly different mechanism. For instance, spore germination by autoactivation requires protein synthesis, whereas protein synthesis is not required to initiate spore germination after heat or DMSO treatments (Dahlberg and Cotter, 1978a,b). Cotter (1981) perceives the entire process of *D. discoideum* spore activation as a cascade phenomenon which can be regulated by various activators, autoactivators, self-inhibitors, and perhaps cAMP. Thus there may be a single site of dormancy, but multiple methods of affecting that site. One can envision the conformation of a critical protein as the primary mechanism of maintaining dormancy. Dormancy could then be broken by altering the conformation of that protein, such as by a heat shock or by de novo synthesis of the protein in the "active" conformational state.

C. Mechanism of Spore Activation

Several possibilities have been proposed to explain the mechanism of fungal spore activation. Based upon his studies of heat and chemical activation of *D. discoideum* spores, Cotter (1973) suggested that a regulatory protein in the inner mitochondrial membrane was in a "restricted" conformation in the dormant spore. This restricted conformation blocked the electron transport system and hence prevented oxidative phosphorylation. Heat activation of *D. discoideum* spores induced a variety of conformational changes in this protein, ultimately resulting in its assuming a "relaxed" configuration in which oxidative phosphorylation and spore germination occurred. Cotter's (1973) hypothesis was based on thermodynamic considerations: The magnitude of entropy changes ruled out the involvement of DNA, RNA, and lipids in the activation process. Further support for this hypothesis came from observations that a variety of protein denaturants also activated *D. discoideum* spores and that the spore mitochondria were damaged by supraoptimal activation treatments (Cotter and George, 1975).

Sussman (1976) argued that while Cotter's hypothesis was plausible, it was not the only possible result of the various activation treatments. Citing a considerable amount of data on *Neurospora* ascospore activation, Sussman presented a cogent argument for the involvement of lipids in fungal spore activation. He cited numerous examples in which lipids were required for enzyme activity and showed that at least in the case of furfurol activation of *Neurospora* ascospores, the activating compound acted on membranes. Thus at least in some systems activation treatments appear to be mediated by lipids or membranes.

In a recent review, Cotter (1981) reiterated that lipid transitions in plasma membranes are not sharp enough to account for *D. discoideum* spore activation, nor would the required transitions occur at 42-45°C, the temperatures at which spores are activated. However, Cotter now believes that proteins within the plasma membrane are the primary site of spore activation. The mitochondria have a critical, but secondary role in the regulation of dormant and activated states. Evidence presented by Cotter strongly implies that the hydrophilic regions of membrane proteins account for spore activation, since the high entropy changes observed can occur in these regions. The site of the critical "activation protein" must be on the inner face of the plasma membrane rather than on the external face or spore wall, since penetration of the spores by a variety of protein denaturants is required for spore activation. Additionally, the particle distribution on the inner rather than outer membrane face changes dramatically in response to *D. discoideum* spore activation (Hohl et al., 1978).

Heat activation of *P. blakesleeanus* spore germination has also been extensively studied (see the review by Furch, 1981). Thevelein et al. (1979, 1980) showed that heat activation of *Phycomyces* spores at high pressures was more consistent with a protein-conformational change than a phospholipid phase transition. Protein-conformational transition temperatures were shifted upward by 2-6°K per 1000 atm, while phospholipid phase changes were shifted by 17-22°K per 1000 atm. Thus activation of spores at high pressures allowed these workers to distinguish between the effects on proteins and lipids. Such activation of spores at high pressures suggested that the primary effect was on a protein.

Although the available evidence suggests that the site of fungal spore activation lies in protein conformation, caution must be taken in accepting and broadly applying this conclusion. Sussman (1976) noted that clear distinction between proteins and lipids may be difficult to make when membrane complexes and lipoproteins are involved. The intimate contact between these may obscure possible independent actions. Also, since phospholipid phase changes are immediately reversible, a phase change could transfer an activation signal to a membrane protein but be undetectable owing to its return to the gel phase at the termination of the activation treatment (Thevelein, 1981).

A further complication is that fungal spores may be activated by a variety of methods with different mechanisms, as discussed above. Also, given the diversity of fungi, there is no reason to expect that the mechanism of spore activation will be universal.

IV. SPORE GERMINATION

A. General Considerations

Sussman and Douthit (1973) define spore germination as "the first irreversible stage which is recognizably different from the dormant organism as judged by morphological, cytological, physiological, and biochemical criteria." However, as biochemical techniques become increasingly sensitive, one finds that many biochemical processes occur within minutes after spores are placed under germinating conditions. By the above definition, such spores would have to be considered germinated, even though germ tubes may not appear for several hours or even days. From a practical standpoint, it is easiest to consider germination morphologically; that is, a spore is germinated when it has formed a recognizable germ tube. Events associated with germination can then be considered as essential or nonessential and as occurring early or late in the overall process. It should be emphasized that spores of various fungi differ widely in the manner in which they reach the germ tube stage. For example, free-swimming zoospores encyst by forming a wall around the zoospore prior to germ tube formation; spores of many fungi undergo extensive spherical growth prior to the emergence of a germ tube; spores of yeast do not form a germ tube per se but, instead, form a daughter cell by budding; and slime mold spores germinate by swelling and releasing amoebas. Furthermore, some spores can germinate by one of two ways, depending on the environmental conditions; that is, they can form a germ tube (direct germination) or release free-swimming zoospores (indirect germination) which encyst and form a germ tube.

The germination of a fungal spore usually leads to the appearance of a germ tube, suggesting that most aspects of cell metabolism required for vegetative growth may be involved at some point in this process. That is, germ tube formation is not solely the rearrangement of existing material, but requires the de novo synthesis of many cellular components, including protein, RNA, polysaccharides, and lipids used in the synthesis and assembly of cell walls, membranes, and organelles. Since the synthesis of these components requires energy, the germinating spore must also have or rapidly develop an efficient means of generating energy. As many spores germinate in water alone, it is obvious that these spores contain sufficient reserves to supply the energy as well as the building blocks for these syntheses. Spores of other species will form a germ tube only if they are activated or provided with a carbon and/or nitrogen source or appropriate salts and/or vitamins. All spores require a means of generating energy and synthesizing protein for germ tube formation, however, since inhibitors of respiration (cyanide or azide) and protein synthesis (cycloheximide) usually prevent germ tube formation. The necessity for syntheses of lipids, RNA, DNA, and polysaccharides for germ tube formation has not been firmly established for many fungi, since specific inhibitors for some of these processes are not available, or the inhibitors may not enter the spore. Components which are usually synthesized late in the germination sequence (e.g., DNA) may not always be required for germ tube formation.

In most fungi, the germ tube continues to elongate, leading to the formation of vegetative hyphae under appropriate environmental conditions. A tacit assumption is that vegetative growth is simply an extension of the events associated with germ tube formation. However, several things suggest that this may be an oversimplification and that another "developmental program" may be required for growth. For example, in culture, spores of several fungi will form a germ tube and then stop growing. Also, in some cases, conditions which favor spore germination differ from those favoring growth (other than simple nutritional factors). Thus the transition from germ tube formation to vegetative growth may require the activation of new genes. Very few investigators have addressed this possibility experimentally.

Because of the diversity of spore types produced by fungi, it is not surprising that the ultrastructural and biochemical processes associated with germination differ widely among species and even between different spore types produced by a single species. Consequently, key events associated with the germination process for one spore may or may not apply to other spores. The following discussion will briefly describe some of the events that are associated with germination.

B. Ultrastructural Changes

There are numerous reports and review articles on ultrastructural changes associated with the germination of specific types of spores (e.g., Weber and Hess, 1976; Turian and Hohl, 1981) and it is difficult to generalize from this massive amount of data. In the majority of fungi, the spore remains filled with cytoplasm as the germ tube develops; in contrast, in a few fungi, most of the spore contents are transferred into the germ tube, leaving a residual spore resembling a large vacuole [e.g., aeciospores of *Cronartium fusiforme* (Walkinshaw et al., 1967); oospores of *Saprolegnia ferax* (Beakes, 1980a)].

In many fungi, mitochondria and nuclei increase in volume early in germination, going from irregular, shrunken structures to more rounded ones (Weber and Hess, 1975). In other fungi, a few large mitochondria fragment into numerous smaller mitochondria [e.g.,

P. blakesleeanus (Pambor, 1979)]. Likewise, the single mitochondrion in zoospores of *B. emersonii* undergoes subdivision during germination (Lovett, 1975). Finally, in some fungal spores no obvious changes in the appearance of mitochondria are detected during germination [e.g., *Botryodiplodia theobromae* (Brambl et al., 1978b)].

Ultrastructural changes in the cytomembrane have been reported during the germination of *P. blakesleeanus* (see the review by Furch, 1981) and *D. discoideum* spores (Hohl et al., 1978). Additional evidence that changes occur in the membrane during spore germination is provided by studies on *B. emersonii* zoospores. Van Brunt and Harold (1980) report that K^+, used to induce germination of these spores, rapidly depolarizes the membrane. Jen and Haug (1979) have shown that changes in the binding of concanavalin A to the zoospore surface also changes with germination.

One of the more striking ultrastructural events associated with spore germination is the development of the germ tube wall. Bartnicki-Garcia (1968) classified fungi into three categories based on their mechanism of germ tube wall formation. In type I fungi, the germ tube wall is merely an extension of the spore wall or one of its inner layers. Type II fungi include only those which form free-swimming zoospores. During encystment of these spores prior to germ tube formation, a wall is synthesized de novo around the naked protoplast. In type III germination, a new wall(s) is synthesized de novo under the spore wall, contiguous with the germ tube wall.

At least one other type of germination appears to exist for some fungi. The germ tube wall of *Geotrichum candidum* arthrospores forms from a newly synthesized inner wall as do type III fungi. However, the new inner wall does not completely encircle the cytoplasm but is confined to the region of germ tube emergence (Steele and Fraser, 1973).

The manner in which the germ tube penetrates the outer spore wall(s) may vary from organism to organism. Rust uredospores germinate through a germination pore by enzymatically digesting the germ pore wall (Hess et al., 1975). Whether enzymatic dissolution of the outer spore walls of other fungi aids in germination, as has been suggested for *N. crassa* conidia (Mahadevan and Rao, 1970), or whether the process is strictly physical is unknown. The outer spore walls often appear frayed rather than smooth where the germ tube emerges, suggesting that physical forces play a role, although the wall(s) may be weakened by prior endogenous enzyme activity.

C. Cell Walls

As noted above, fungal spores fall into three classes, depending on the origin of the germ tube wall. Obviously, new wall materials must be synthesized and assembled to encompass the newly developing germ tube. Surprisingly, there have been few comparative studies on the chemical compositions of spore walls versus germ tube walls. Although fungal walls consist primarily of various polysaccharides, they also contain proteins and/or glycoproteins, sporopollenin, and pigments (Gooday and Trinci, 1980). In some fungi (e.g., Mucorales) the germ tube wall is chemically different from the spore walls. Walls of sporangiospores of both *Mucor* (Bartnicki-Garcia and Reyes, 1964) and *Phycomyces* (Van Laere et al., 1977) are composed predominantly of glucose polymers, whereas the germ tube and vegetative cell walls contain high concentrations of glucosamine polymers (as chitin and chitosan). The synthesis of chitin-chitosan begins almost immediately after *Mucor* spores are placed under germinating conditions (Bartnicki-Garcia and Lippman, 1977). Storck and Price (1977) clearly demonstrated the importance of this chitosan synthesis to maintaining the integrity of the germinating spore. Ungerminated spores of the fungus are completely resistant to osmotic lysis following treatment with chitosanase. However, spores incubated under germination conditions for 3 h lysed after enzyme treatment, even though germ tubes are not visible until 4-5 h.

Extensive research has been conducted on the encystment of zoospores of *B. emersonii* (e.g., see the reviews by Lovett, 1975, Chapter 8, this volume). During encystment a chitinous wall is synthesized and assembled within minutes around the entire naked protoplast. As encystment does not require de novo protein synthesis, wall synthesis must occur via preexisting components prepackaged in the spore (see Selitrennikoff and Sonneborn, 1976, for a summary of their studies on the regulation of this process). Since the zoospore contains all of the machinery to synthesize a new wall, it is reasonable to ask what prevents encystment immediately after zoospores are formed. Although no answer is available, three possibilities have been suggested: (1) substrate deficiency, (2) enzyme latency, and (3) enzyme/substrate compartmentation (see Bartnicki-Garcia, 1981). Finally, Gottschalk and Sonneborn (1981) have evidence that a "zoospore maintenance factor" may exist in zoospores of *B. emersonii* which can prevent encystment. In contrast, to the zoospore system, other spores require the de novo synthesis of enzymes involved in wall formation or in the regulation of this process, since inhibitors of protein synthesis prevent new wall synthesis [e.g., *P. blakesleeanus* (Van Laere and Carlier, 1978)].

Germination of most fungi involves a polarization or localization of wall synthesis at the point of germ tube emergence. Autoradiographs showing polarization of wall synthesis have been published for *Mucor* (Bartnicki-Garcia and Lippman, 1977). While the mechanism underlying this process is not understood, it is reasonable to assume that polarization is brought about by a directional change in the pattern of migration of chitosomes and other vesicles involved in wall growth (Bartnicki-Garcia, 1981). Factors contributing to this process could be structural elements such as microfilaments and microtubules, chemical signals such as cAMP, an altered membrane potential, or electrical current.

D. Respiration

Once spores are placed under conditions favorable for germination, the rate of oxygen uptake increases dramatically so that the Q_{O_2} of germinated spores may be 10-30 times higher than in dormant spores. This increase in respiration is required for germ tube formation, since cyanide or azide usually inhibits both the increased oxygen uptake and spore germination. The specific activities of many enzymes involved in carbohydrate metabolism, such as the Embden-Meyerhof-Parnas pathway, the tricarboxylic acid cycle, and the mitochondrial cytochrome electron transport system, have been measured to determine whether fungal spores were deficient in one or more of these enzymes and if these enzymes showed increased activity during germination (see the review by Gottlieb, 1976). With a few exceptions, the major conclusion from these studies was that most of the enzymes are present in the dormant spore and that the specific activities of the majority of the enzymes increase during germination.

Following the discovery that mitochondria contain DNA (as well as the machinery to transcribe and translate this DNA) which codes for at least some mitochondrial proteins involved in the terminal electron transport system, there has been a renewed interest in the role that mitochondria play in the development of an active respiratory system during spore germination. Brambl and his colleagues (see the review by Brambl, 1981) have examined the development of a functional mitochondrial electron transport system during germination of *Botryodiplodia theobromae* spores. Conidia of this fungus, unlike spores of several other fungi, lack detectable cytochrome oxidase and oligomycin-sensitive ATPase activity. The activities of these two multicomponent enzymes appear only after spores are placed under germinating conditions for 60-120 min. Although both enzymes are comprised of polypeptides synthesized on both cytoplasmic and mitochondrial ribosomes, the appearance of the two enzymatic activities during germination is regulated differently.

Cytochrome oxidase consists of seven distinct polypeptides: The three largest subunits are products of mitochondrial ribosomes and the four remaining subunits are products of cytoplasmic ribosomes. In dormant spores of *B. theobromae*, the products of the cytoplasmic ribosomes are stored in the cytoplasm (probably in a precursor form); these enter the mitochondria and associate with the mitochondrial synthesized subunits to form a functional enzyme only after the spores are placed under germinating conditions. Thus the initial development of cytochrome oxidase activity with germination depends on the modification and rearrangement of preserved components of the enzyme stored in the dormant spore (Brambl, 1980). This process requires the synthesis of some unknown cytoplasmic ribosomal product, since assembly of a functional cytochrome oxidase does not occur in the presence of cycloheximide.

The oligomycin-sensitive ATPase of *B. theobromae* consists of 12 peptides, 11 of which are products of the cytoplasmic ribosomes and one being derived from mitochondrial ribosomes. Unlike the cytochrome oxidase components, most of these subunits are absent in the dormant spore and are synthesized de novo from preexisting spore mRNA and assembled into a functional enzyme complex in the first 45-60 min of germination.

E. Role of Small Molecules

As noted previously, small molecules play a key role in germination. For example, in some spores, self-inhibitors must be removed before germination begins. Also, many spores contain large quantities of polyols and trehalose, which are either secreted or metabolized during germination. In addition to these compounds, Brody and his colleagues (see Schmidt and Brody, 1976; Brody, 1981) have measured the concentrations of amino acids, pyridine nucleotides, and glutathione during the germination of *N. crassa* conidia. These investigators showed that the generation of reduced pyridine nucleotides and reduced glutathione are key early events in the germination of *N. crassa* conidia. Furthermore, they proposed that the generation of reducing power may be of general significance; that is, the generation of reducing power may play a critical role in the germination of bacterial spores and seeds of higher plants, as well as in fungal spores.

The hypothesis of a "trigger pathway" for *N. crassa* spore germination is based on three observations: (1) the rapid disappearance (within minutes) of a large glutamic acid pool in the spores; (2) a transient surge in the pyridine nucleotide redox ratio, that is, a temporary increase in the amounts of NADH and NADPH; and (3) the conversion of oxidized glutathione to its reduced form. The function of the increased reduced glutathione is unknown, although Brody (1981) suggests that it might reduce protein sulfhydryl groups, leading to the activation of certain critical enzymes. In *N. crassa* spores glutamic acid serves as the source of the reducing power; Brody suggests that polyols or other amino acids such as proline might serve the same function in other fungal spores. Important aspects of this hypothesis are that these events occur very early during germination and they are germination specific.

A few investigators have examined the changes in polyamines (putrescine, spermidine, and spermine) during fungal spore germination. Polyamine concentrations increase more or less in parallel with nucleic acid concentrations during the germination of *Aspergillus nidulans* conidia (Winther and Stevens, 1976), *B. emersonii* zoospores

(Mennucci et al., 1975), *R. stolonifer* sporangiospores (Nickerson et al., 1977), *Botryodiplodia theobromae* conidia (Nickerson et al., 1977), and *Saccharomyces cerevisiae* ascospores (Choih et al., 1978). However, spermidine and spermine syntheses are not required for yeast ascospore germination, since the spores germinate normally in the presence of methylglyoxyl-bis(guanylhydrazone), an inhibitor of spermidine and spermine syntheses.

The role of cyclic nucleotides in spore germination is presently not understood. During the germination of *Mucor* spores (under conditions leading to mycelial growth), cAMP and cGMP concentrations increased dramatically, followed by a decrease at about the time germ tubes emerged (Paznokas and Sypherd, 1975; Orlowski and Sypherd, 1976; Orlowski, 1980). Likewise, there was a transient increase in intracellular cAMP concentrations during *B. emersonii* zoospore germination (Gomes et al., 1978). In contrast, cAMP and cGMP levels remain constant during *N. crassa* conidia germination (Rosenberg and Pall, 1978).

F. Lipid Metabolism

Although many reports have appeared on the changes in the distribution of lipids with germination (see Weete, 1974, 1981; Reisener, 1976), the role of lipids as structural elements in the germination of fungal spores has largely been ignored. However, it is obvious that the synthesis and assembly of new membranes to surround organelles and developing germ tubes must occur during germination. That fatty acid synthesis is important in germination is implied from two studies on the effect of the fatty acid synthesis inhibitor cerulenin on germ tube development. The drug prevented germ tube development in both *Botryodiplodia theobromae* (Brambl, 1977; Brambl et al., 1978b) and *R. stolonifer* (Nickerson and Leastman, 1978). In both cases it was implied that at least part of the inhibitory effect was due to the inability of the spore to synthesize or assemble functional mitochondria.

Based on ultrastructural evidence, McKeen (1970) suggested an important function for lipids in the germination of *Erysiphe graminis hordei* conidia. These spores, like those of other powdery mildews, contain large lipid bodies and germinate at very low relative humidities. McKeen suggested that respiratory oxidation of the lipids during germination generates sufficient water for the spores to germinate.

G. Protein and Nucleic Acid Metabolism

There has been considerable interest in protein and nucleic acid syntheses during fungal development and this interest is reflected in the large number of reviews on the subject (Berry and Berry, 1975; Lovett, 1976; Lovett et al., 1977; Lovett and Sikkema, 1981; Van Etten et al., 1976, 1977, 1981; Brambl et al., 1978a; Staples and Huang, 1981).

A dogma among developmental biologists is that development results from differential gene expression. This view is also widely held among those studying fungal spore germination, though alternative opinions have been expressed (Wright, 1979; Bartnicki-Garcia, 1981). One of the earliest events associated with spore germination is a rapid increase in the rate of protein synthesis. Furthermore, cytoplasmic protein synthesis is essential for germ tube formation in all fungal spores (Lovett, 1976; Brambl et al., 1978a). This conclusion is supported by several lines of evidence: (1) Conidia from a mutant strain of *N. crassa* which is temperature sensitive for protein synthesis do not form germ tubes at the nonpermissive temperature (Loo, 1976); (2) conidia from some amino acid auxotrophs of *N. crassa* do not form germ tubes on unsupplemented media (Schmidt and Brody, 1976); and (3) germ tube formation in most spores is prevented by low concentrations of cycloheximide (Lovett, 1976). In contrast to the essentiality of cytoplasmic protein synthesis, mitochondrial protein synthesis is not required for germ tube formation in *Botryodiplodia theobromae*, *N. crassa*, and *S. cerevisiae* (see the review by Brambl, 1981).

The essentiality of nucleic acid (DNA and RNA) synthesis for fungal spore germination is difficult to assess. In most fungi, the synthesis of all classes of RNA [tRNA, rRNA, and polyA(+)RNA (mRNA)] begins shortly after spores are placed under germination conditions. The polyA(+)RNA synthesized early in germination is rapidly incorporated into polysomes (Mirkes and McCalley, 1976; Johnson et al., 1977; Freer et al., 1977; Van Etten and Rawn, 1979). However, translation of these newly synthesized mRNAs does not necessarily mean that the protein products are unique and/or essential for germ tube formation.

Zoospores of *B. emersonii* form germ tubes in the presence of actinomycin D (an inhibitor of RNA synthesis), and thus germ tube formation in this organism occurs in the absence of transcription (Lovett and Sikkema, 1981). Although many other fungi form germ tubes in the presence of actinomycin D or other RNA synthesis inhibitors, it is likely that many spores are impermeable to the drugs. In all likelihood, some fungal spores require RNA synthesis for germ tube formation, whereas others do not.

In the majority of fungi, DNA synthesis begins after the initiation of RNA and protein synthesis, but prior to or at about the time of germ tube emergence. In a few cases, nuclear DNA synthesis is not required for germ tube formation, for example, *Uromyces phaseolus* (Staples and Huang, 1981) and *N. crassa* (Loo, 1976), whereas in other fungi DNA synthesis is apparently required, for example, *Ceratocystis ulmi* (K. W. Nickerson, personal communication).

A number of investigators (see the review by Brambl et al., 1978a) have examined dormant spores to determine whether these were deficient in some component of the protein synthetic machinery, and whether this defect might regulate protein synthesis. As a group, these studies indicate that the lack of protein synthesis in the dormant spore does not result from the absence or inactivity of ribosomes, tRNAs, elongation factors, aminoacyl-tRNA synthetases, or mRNA, since all of these components can be isolated from dormant spores and are biologically active when tested in vitro. Other components of the protein synthetic apparatus, such as initiation and release factors, have not been examined in relation to spore germination. Suppression of initiation factor activity would provide a convenient way of regulating protein synthesis in the dormant spore.

Other mechanisms for regulating protein synthesis in fungal spores have been proposed, including compartmentalization of the protein synthetic components, the presence of inhibitor(s), involvement of a novel tRNA, and hydration-induced modifications of preexisting molecules. All of these models have been discussed in a recent review article (Brambl et al., 1978a). However, more experiments will have to be conducted before it is possible to decide whether any of these models are valid for even a single spore system.

The manner in which mRNA or its precursor is stored in dormant spores is unknown. In some spores, at least a portion of it is preserved in polysomes; however, in other spores polysomes have not been detected even though functional mRNA is present (see the review by Van Etten et al., 1981). Jaworski and Stumhofer (1981) have evidence that at least some mRNA is sequestered in ribonucleoprotein particles (distinct from ribosomes) in zoospores of *B. emersonii*. Furthermore, some of this mRNA is stored as nonadenylated RNA; the polyadenylate tail is then added during germination. In contrast, prepackaging of mRNA devoid of polyadenylate segments, which is subsequently adenylated during germination, does not occur in *R. stolonifer* (Roheim et al., 1974) and *Botryodiplodia theobromae* (Van Etten and Rawn, 1979).

Once it was established that many, if not all, fungal spores, contain prepackaged mRNA, it was logical to ask about the nature of the proteins encoded by the prepackaged mRNA; that is, Does the prepackaged mRNA code for the synthesis of specific proteins essential to initiate and/or complete germination or is the same complement of proteins synthesized throughout germ tube formation and vegetative growth? Experiments directed to answer these questions have been conducted for a few fungal spores [*Botryodiplodia theobromae* (Wenzler and Brambl, 1978; Brambl, 1981), *R. stolonifer* (Freer and Van Etten, 1978), *B. emersonii* (Lovett and Sikkema, 1981), and *D. discoideum* (Giri and Ennis, 1978)]. The results of these experiments indicate that prepackaged mRNA codes for the synthesis of numerous proteins; many of these proteins appear to be synthesized at all stages of germination. For example, at no time during the germination of *R. stolonifer* spores was a germination-specific protein synthesized which represented a major fraction of the total proteins synthesized (Freer and Van Etten, 1978). This inability to detect many spore germination-specific proteins may simply reflect the insensitivity of the methods being used; that is, at most only 200 or so proteins have been resolved and these undoubtedly represent the most abundant proteins. Proteins whose syntheses are required specifically for germination may be synthesized in smaller quantities. However, one must consider the alternative; that is, spore germination per se may not require the synthesis of many unique proteins. The lack of success at isolating temperature-sensitive mutants specific for spore germination in *N. crassa* supports this possibility (Schmidt and Brody, 1976). With the exception noted earlier, where some of the prepackaged spore mRNA of *Botryodiplodia theobromae* encodes for the synthesis of components of the FI-ATPase enzyme complex, little is known about the function of the proteins coded for by prepackaged spore mRNA.

A few investigators have compared the proteins synthesized in vivo with those synthesized in vitro in a cell-free system, using either total mRNA or polysomal mRNA as templates to determine if the proteins synthesized during spore germination are under transcriptional or translational control (Van Etten and Freer, 1978; Freer and Van Etten, 1978; Giri and Ennis, 1978; Dowbenko and Ennis, 1980). These studies suggest that the majority of proteins are under transcriptional control during spore germination. This observation presents a paradox, however, since mRNA is preserved in the dormant spore and thus protein synthesis in the spore must be controlled at the level of translation.

V. CONCLUSIONS

Even though much new information has accumulated in the past few years about fungal spore germination, very few definitive statements can be made about the process. To paraphrase Brambl (1981), the initiation and completion of spore germination undoubtedly involves an interaction of many physiological and biochemical processes and it is unlikely that germination is triggered by a single event (i.e., a master switch) from which all other steps follow in an ordered sequence. Instead, it is more likely that germination results from several essential and independent metabolic activities which must proceed in parallel. Finally, because of the wide diversity among fungi, one must be careful in assuming that results obtained for one spore will apply to other fungal spores.

ACKNOWLEDGMENTS

We thank Stuart Brody, Bob Brambl, Dave Cotter, B. Furch, Jim Lovett, Salomon Bartnicki-Garcia, Al Sussman, Dick Staples, and Vladimer Macko for providing preprints to their review articles which appeared in *The Fungal Spore: Morphogenetic Controls*, G. Turian and H. R. Hohl (Eds.), Academic, New York, and Johann Thevelein for providing valuable information. Research from our laboratory was supported by grants from the National Institutes of Health and the National Science Foundation. Published as Paper Number 6618. Journal Series, Nebraska Agricultural Experiment Station. This work was conducted under Nebraska Agricultural Experiment Station Project No. 21-17.

REFERENCES

Abe, H., Tanaka, T., and Saito, H. (1976). Structure of discadenine, a spore germination inhibitor from the cellular slime mold *Dictyostelium discoideum*. *Tetrahedron Lett.* *42*:3807-3810.

Adelman, T. G., and Lovett, J. S. (1974). Evidence for a ribosome-associated translation inhibitor during differentiation of *Blastocladiella emersonii*. *Biochim. Biophys. Acta.* *335*:236-245.

Allen, P. J. (1976). Control of spore germination and infection structure formation in the fungi. In *Encyclopedia of Plant Physiology*, Vol. 4. R. Heitefuss and P. H. Williams (Eds.). Springer-Verlag, New York, pp. 51-85.

Ballio, A., Di Vittorio, V., and Russi, S. (1964). The isolation of trehalose and polyols from the conidia of *Penicillium chrysogenum* Thom. *Arch. Biochem. Biophys.* *107*:177-183.

Bartnicki-Garcia, S. (1968). Cell wall chemistry, morphogenesis, and taxonomy of fungi. *Annu. Rev. Microbiol.* *22*:87-108.

Bartnicki-Garcia, S. (1981). Cell wall construction during spore germination in Phycomycetes. In *The Fungal Spore: Morphogenetic Controls*. G. Turian and H. R. Hohl (Eds.). Academic, New York, pp. 533-556.

Bartnicki-Garcia, S., and Hemmes, D. E. (1976). Some aspects of the form and function of Oomycete spores. In *The Fungal Spore: Form and Function*. D. J. Weber and W. M. Hess (Eds.). Wiley, New York, pp. 593-639.

Bartnicki-Garcia, S., and Lippman, E. (1977). Polarization of cell wall synthesis during spore germination of *Mucor rouxii*. *Exp. Mycol.* *1*:230-240.

Bartnicki-Garcia, S., and Reyes, E. (1964). Chemistry of spore wall differentiation in *Mucor rouxii*. *Arch. Biochem. Biophys.* *108*:125-133.

Beakes, G. W. (1980a). Electron microscopic study of oospore maturation and germination in an emaculate isolate of *Saprolegnia ferax*. 1. Gross changes. *Can. J. Bot. 58*:182-194.

Beakes, G. W. (1980b). Electron microscopic study of oospore maturation and germination in an emaculate isolate of *Saprolegnia ferax*. 2. Wall differentiation. *Can. J. Bot. 58*:195-208.

Beilby, J. P., and Kidby, D. K. (1980). Biochemistry of ungerminated and germinated spores of the vesicular-arbuscular mycorrhizal fungus, *Glomus caledonius*: Changes in neutral and polar lipids. *J. Lipid Res. 21*:739-750.

Berry, D. R., and Berry, E. A. (1975). Nucleic acid and protein synthesis in filamentous fungi. In *The Filamentous Fungi: Biosynthesis and Metabolism*, Vol. 2. J. E. Smith and D. R. Berry (Eds.). Edward Arnold, London, pp. 238-291.

Bianchi, D. E., and Turian, G. (1967). Lipid content of conidia of *Neurospora crassa*. *Nature 214*:1344-1345.

Brambl, R. (1977). Mitochondrial biogenesis during fungal spore germination. *Arch. Biochem. Biophys. 182*:273-281.

Brambl, R. M. (1980). Mitochondrial biogenesis during fungal spore germination: Biosynthesis and assembly of cytochrome c oxidase in *Botryodiplodia theobromae*. *J. Biol. Chem. 255*:7673-7680.

Brambl, R. (1981). Respiration and mitochondrial biogenesis during fungal spore germination. In *The Fungal Spore: Morphogenetic Controls*. G. Turian and H. R. Hohl (Eds.). Academic, New York pp. 585-604.

Brambl, R., Dunkle, L. D., and Van Etten, J. L. (1978a). Nucleic acid and protein synthesis during fungal spore germination. In *The Filamentous Fungi*, Vol. 3. J. E. Smith and D. R. Berry (Eds.). Edward Arnold, London, pp. 94-118.

Brambl, R., Wenzler, H., and Josephson, M. (1978b). Mitochondrial biogenesis during fungal spore germination: Effects of the antilipogenic antibiotic cerulenin upon *Botryodiplodia* spores. *J. Bacteriol. 135*:311-317.

Brody, S. (1981). Genetic and biochemical studies on *Neurospora* conidia germination and metabolism. In *The Fungal Spore: Morphogenetic Controls*. G. Turian and H. R. Hohl (Eds.). Academic, New York, pp. 605-626.

Charlang, G., Ng, B., Horowitz, N. H., and Horowitz, R. M. (1981). Cellular and extracellular siderophores of *Aspergillus nidulans* and *Penicillium chrysogenum*. *Mol. Cell. Biol. 1*:94-100.

Choih, S. J., Ferro, A. J., and Shapiro, S. K. (1978). Relationship between polyamines and macromolecules in germinating yeast ascospores. *J. Bacteriol. 133*:424-426.

Coco, A. R., and Kornfeld, J. M. (1978). Evidence of an endogenous germinant in the spores of a strain of *Dictyostelium purpureum*. *Eur. J. Appl. Microbiol. Biotechnol. 5*:129-132.

Cotter, D. A. (1973). Spore germination in *Dictyostelium discoideum*. I. The thermodynamics of reversible activation. *J. Theor. Biol. 41*:41-51.

Cotter, D. A. (1981). Spore activation. In *The Fungal Spore: Morphogenetic Controls*. G. Turian and H. R. Hohl (Eds.). Academic, New York, pp. 385-411.

Cotter, D. A., and Dahlberg, K. R. (1977). Isolation and characterization of *Dictyostelium discoideum* spore mutants with altered activation requirements. *Exp. Mycol. 1*:107-115.

Cotter, D. A., and George, R. P. (1975). Germination and mitochondrial damage in spores of *Dictyostelium discoideum* following supraoptimal heating. *Arch. Microbiol. 103*:163-168.

Cotter, D. A., and Raper, K. B. (1966). Spore germination in *Dictyostelium discoideum*. *Proc. Nat. Acad. Sci. USA 56*:880-887.

Cotter, D. A., Morin, J. W., and O'Connell, R. W. (1976). Spore germination in *Dictyostelium discoideum*. II. Effects of dimethylsulfoxide on postactivation lag as evidence for the multistate model of activation. *Arch. Microbiol. 108*:93-98.

Dahlberg, K. R., and Cotter, D. A. (1978a). Autoactivation of spore germination in mutant and wild type strains of *Dictyostelium discoideum*. *Microbios 23*:153-166.

Dahlberg, K. R., and Cotter, D. A. (1978b). Activators of *Dictyostelium discoideum* spore germination released by bacteria. *Microbios. Lett. 9*:138-146.

Dahlberg, K. R., and Franke, R. G. (1977). Spore germination in the Myxomycete *Fuligo septica*: Evidence for the existence of a soluble autocatalytic factor. *Mycologia 69*:96-108.

Dowbenko, D. J., and Ennis, H. L. (1980). Regulation of protein synthesis during spore germination in *Dictyostelium discoideum*. *Proc. Nat. Acad. Sci. USA 77*:1791-1795.

Eckert, J. W., Rahm, H. L., and Hall, G. (1968). Neutral carbohydrates of conidia of *Penicillium digitatum*. *Phytopathology 58*:1406-1411.

Emyanitoff, R. G., and Wright, B. E. (1979). Effect of intracellular carbohydrates on heat resistance of *Dictyostelium discoideum* spores. *J. Bacteriol. 140*:1008-1012.

Fahey, R. C., Mikolajczyk, S. D., and Brody, S. (1978). Correlation of enzymatic activity and thermal resistance with hydration state in ungerminated *Neurospora* conidia. *J. Bacteriol. 135*:868-875.

Fletcher, J. (1971). Fine-structural changes during germination of conidia of *Penicillium griseofulvum* Dierckx. *Ann. Bot. 35*:441-449.

Freer, S. N., and Van Etten, J. L. (1978). Changes in messenger RNAs and protein synthesis during germination of *Rhizopus stolonifer* sporangiospores. *Exp. Mycol. 2*:313-325.

Freer, S. N., Mayama, M., and Van Etten, J. L. (1977). Synthesis of polyadenylate-containing RNA during the germination of *Rhizopus stolonifer* sporangiospores. *Exp. Mycol.* *1*:116-127.

French, R. C., Graham, C. L., Gale, A. W., and Long, R. K. (1977). Structural and exposure time requirements for chemical stimulation of germination of uredospores of *Uromyces phaseoli*. *Agric. Food Chem.* *25*:84-88.

Fries, N. (1977). Germination of *Laccaria laccata* spores in vitro. *Mycologia* *69*:848-850.

Fries, N. (1978). Basidiospore germination in some mycorrhizae-forming hymenomycetes. *Trans. Br. Mycol. Soc.* *70*:319-324.

Furch, B. (1981). Spore germination:heat mediated events. In *The Fungal Spore: Morphogenetic Controls*. G. Turian and H. R. Hohl, (Eds.). Academic, New York, pp. 413-433.

Giri, J. G., and Ennis, H. L. (1978). Developmental changes in RNA and protein synthesis during germination of *Dictyostelium discoideum* spores. *Dev. Biol.* *67*:189-201.

Gomes, S. L., Mennucci, L., and Maia, J. C. C. (1978). Adenylate cyclase activity and cyclic AMP metabolism during cytodifferentiation of *Blastocladiella emersonii*. *Biochim. Biophys. Acta.* *541*:190-198.

Gooday, G. W., and Trinci, A. P. J. (1980). Wall structure and biosynthesis in fungi. In *The Eurkaryotic Microbial Cell*. G. W. Gooday, D. Lloyd, and A. P. J. Trinci (Eds.). Cambridge University, Cambridge, pp. 207-251.

Gottlieb, D. (1976). Carbohydrate metabolism and spore germination. In *The Fungal Spore: Form and Function*. D. J. Weber and W. M. Hess (Eds.). Wiley, New York, pp. 141-163.

Gottschalk, W. K., and Sonneborn, D. R. (1981). *Blastocladiella emersonii* zoospore maintenance factor: A quantitative bioassay used to characterize extracellular appearance and maintenance of factor and interactions between factor, zoospores, and salt. *Exp. Mycol.* 5:1-14.

Harper, D. B., Swinburne, T. R., Moore, S., Brown, A. E., and Graham, H. (1980). A role for iron in germination of conidia of *Colletotrichum musae*. *J. Gen. Microbiol.* *121*:169-174.

Hashimoto, Y., Tanaka, Y., and Yamada, T. (1976). Spore germination promotor of *Dictyostelium discoideum* excreted by *Aerobacter aerogenes*. *J. Cell Sci.* *21*:261-271.

Hawker, L. E., and Madelin, M. F. (1976). The dormant spore. In *The Fungal Spore: Form and Function*. D. J. Weber and W. M. Hess (Eds.). Wiley, New York, pp. 1-70.

Hecker, L. I., and Sussman, A. S. (1973). Localization of trehalose in the ascospores of *Neurospora*: Relation to ascospore dormancy and germination. *J. Bacteriol.* *115*:592-599.

Hess, S. L., Allen, P. J., Nelson, D., and Lester, H. H. (1975). Mode of action of methyl cis-ferulate, the self-inhibitor of stem rust unredospore germination. *Physiol. Plant Pathol.* 5:107-112.

Hess, S. L., Allen, P. J., and Lester, H. H. (1981). Release of a mannoprotein during pore dissolution in *Puccinia graminis* f. sp. *tritici* unredospores (submitted for publication).

Hohl, H. R., Buhlmann, M., and Wehrli, E. (1978). Plasma membrane alterations as a result of heat activation of *Dictyostelium discoideum* spores. *Arch. Microbiol.* *116*:239-244.

Hollomon, D. W. (1973). Protein synthesis during germination of *Peronospora tabacina* conidia: An examination of the events involved in the initiation of germination. *J. Gen. Microbiol.* *78*:1-13.

Hubbard, S. C., and Brody, S. (1975). Glycerophospholipid variation in choline and inositol auxotrophs of *Neurospora crassa*. Internal compensation among zwitterionic and anionic species. *J. Biol. Chem.* *250*:7173-7181.

Jahn, E. (1940). Myxomycetenstudieren. XVII. Die Erweckung und Keimung der Sporen von *Reticularia lycoperdon*. *Ber. Dtsch. Bot. Ges.* *58*:182-199.

Jaworski, A. J., and Stumhofer, P. (1981). Stage-specific synthesis of proteins complexed to ribonucleoprotein particles and ribosomes in zoospores of *Blastocladiella emersonii*. *Mol. Cell. Biol.* *1*:310-320.

Jen, C. J., and Haug, A. (1979). Changes in surface properties of developing zoospores of *Blastocladiella emersonii*: Binding of concanavalin A. *J. Gen. Microbiol.* *115*:145-151.

Johnson, S. A., Lovett, J. S., and Wilt, F. H. (1977). The polyadenylated RNA of zoospores and growth phase cells of the aquatic fungus *Blastocladiella*. *Dev. Biol.* *56*:329-342.

Kao, C. W., and Leu, L. S. (1980). Sporangium germination of *Peronophythora litchii*, the causal organism of Litchi downy blight. *Mycologia* *72*:737-748.

Keynan, A. (1978). Spore structure and its relations to resistance, dormancy and germination. In *Spores*, Vol. 7. G. Chambliss and J. C. Vary (Eds.). American Society for Microbiology, Washington, D.C., pp. 43-53.

Kobilinsky, L., and Beattie, D. J. (1977). Respiratory competence of *Dictyostelium discoideum* spores. *J. Bacteriol.* *132*:113-117.

Kornfeld, J. M. (1968). Developmental aspects of the cellular slime molds. *Bull. Inst. Cell Biol. Univ. Conn.* *9*:1-4.

Leef, J. L., and Mazur, P. (1978). Physiological response of *Neurospora* conidia to freezing in the dehydrated, hydrated, or germinated state. *Appl. Environ. Microbiol.* *35*:72-83.

Lingappa, B. T., and Sussman, A. S. (1959). Endogenous substrates of dormant, activated, and germinating ascospores of *Neurospora tetrasperma*. *Plant Physiol.* *34*:466-473.

Loo, M. (1976). Some required events in conidial germination of *Neurospora crassa*. *Dev. Biol.* *54*:201-213.

Lovett, J. S. (1975). Growth and differentiation of the water mold *Blastocladiella emersonii*: Cytodifferentiation and the role of ribonucleic acid and protein synthesis. *Bacteriol. Rev.* *39*:345-404.

Lovett, J. S. (1976). Regulation of protein metabolism during spore germination. In *The Fungal Spore: Form and Function*. D. J. Weber and W. M. Hess (Eds.). Wiley, New York, pp. 189-242.

Lovett, J. S., and Sikkema, W. D. (1981). Protein synthesis in aquatic fungi. In *The Fungal Spore: Morphogenetic Controls*. G. Turian and H. R. Hohl (Eds.). Academic, New York, pp. 315-333.

Lovett, J. S., Gong, C. S., and Johnson, S. A. (1977). Zoospore germination and early development of *Blastocladiella emersonii*. In *Eucaryotic Microbes as Model Developmental Systems*. D. M. O'Day and P. A. Horgen (Eds.). Marcel Dekker, New York, pp. 402-424.

McKeen, W. E. (1970). Lipid in *Erysiphe graminis hordei* and its possible role during germination. *Can. J. Microbiol.* *16*:1041-1044.

Macko, V. (1981). Inhibitors and stimulants of spore germination and infection structure formation in fungi. In *The Fungal Spore: Morphogenetic Controls*. G. Turian and H. R. Hohl (Eds.). Academic, New York, pp. 565-584.

Macko, V., Staples, R. C., Yaniv, Z., and Granados, R. R. (1976). Self-inhibitors of fungal spore germination. In *The Fungal Spore: Form and Function*. D. J. Weber and W. M. Hess (Eds.). Wiley, New York, pp. 73-100.

Mahadevan, P. R., and Rao, S. R. (1970). Enzyme degradation of conidial wall during germination of *Neurospora crassa*. *Ind. J. Exp. Biol.* *8*:293-297.

Mennucci, L., Rojas, S., and Camargo, E. P. (1975). Polyamines and ornithine decarboxylase activity during growth and differentiation in *Blastocladiella emersonii*. *Biochim. Biophys. Acta* *404*:249-256.

Mims, C. W. (1971). An ultrastructural study of spore germination in the Myxomycete *Arcyria cinera*. *Mycologia* *63*:586-601.

Mirkes, P. E. (1974). Polysomes, ribonucleic acid, and protein synthesis during germination of *Neurospora crassa* conidia. *J. Bacteriol.* *117*:196-202.

Mirkes, P. E., and McCalley, B. (1976). Synthesis of polyadenylic acid-containing ribonucleic acid during the germination of *Neurospora crassa* conidia. *J. Bacteriol.* *125*:174-180.

Morgan, W. M. (1978). Germination of *Bremia lactucae* oospores. *Trans. Br. Mycol. Soc.* *71*:337-340.

Nickerson, K. W., and Leastman, E. (1978). Cerulenin inhibition of spore germination in *Rhizopus stolonifer*. *Exp. Mycol.* *2*:26-31.

Nickerson, K. W., McCune, B. K., and Van Etten, J. L. (1977). Polyamine and macromolecule levels during spore germination in *Rhizopus stolonifer* and *Botryodiplodia theobromae*. *Exp. Mycol.* *1*:317-322.

Nickerson, K. W., Freer, S. N., and Van Etten, J. L. (1981). *Rhizopus stolonifer* sporangiospores: A wet-harvested spore is not a native spore. *Exp. Mycol.* *5*:189-192.

Onesirosan, P. T. (1976). Nutritional requirements for germination of the conidia of *Phyllosticta lycopersici*. *Mycopathologia* *59*:131-135.

Orlowski, M. (1980). Cyclic adenosine 3',5'-monophosphate and germination of sporangiospores from the fungus *Mucor*. *Arch. Microbiol.* *126*:133-140.

Orlowski, M., and Sypherd, P. S. (1976). Cyclic guanosine 3',5'-monophosphate in the dimophic fungus *Mucor racemosus*. *J. Bacteriol.* *125*:1226-1228.

Ouellette, G. B., and Ward, E. W. B. (1970). Low-temperature requirements for ascospore germination and growth of *Hypoxylon diathrauston*. *Can J. Bot.* *48*:2223-2225.

Pambor, L. (1979). Development of mitochondria in dormant and germinating sporangiospores of *Phycomyces blakesleeanus*. *Microbios. Lett.* *8*:143-153.

Paznokas, J. L., and Sypherd, P. S. (1975). Respiratory capacity, cyclic adenosine 3',5'-monophosphate and morphogenesis of *Mucor racemosus*. *J. Bacteriol.* *124*:134-139.

Pezet, R., and Pont, V. (1975). Elemental sulfur responsible for self-inhibition of spores of *Phomopsis viticola*. *Experientia* *31*:439.

Pugh, D., and Cawson, R. A. (1979). The induction of germ tubes in *Candida albicans* by an intrinsic factor. *Microbios.* *24*:73-79.

Reisener, H. J. (1976). Lipid metabolism of fungal spores during sporogenesis and germination. In *The Fungal Spore: Form and Function*. D. J. Weber and W. M. Hess (Eds.). Wiley, New York, pp. 165-185.

Roheim, J. R., Knight, R. H., and Van Etten, J. L. (1974). Synthesis of ribonucleic acids during the germination of *Rhizopus stolonifer* sporangiospores. *Dev. Biol.* *41*:137-145.

Rosenberg, G., and Pall, M. L. (1978). Cyclic AMP and cyclic GMP in germinating conidia of *Neurospora crassa*. *Arch. Microbiol.* *118*:87-90.

Rudolph, H., and Ochsen, B. (1969). Trehalose-Umsatz wärmeaktivierter Sporen von *Phycomyces blakesleeanus*. *Arch. Mikrobiol.* *65*:163-171.

Scarborough, G. A. (1975). Isolation and characterization of *Neurospora crassa* plasma membranes. *J. Biol. Chem.* *250*:1106-1111.

Schmidt, J. C., and Brody, S. (1975). *Neurospora crassa* conidial germination: Role of endogenous amino acid pools. *J. Bacteriol.* *124*:232-242.

Schmidt, J. C., and Brody, S. (1976). Biochemical genetics of *Neurospora crassa* conidial germination. *Bacteriol. Rev. 40*:1-41.

Selitrennikoff, C. P., and Sonneborn, D. R. (1976). Post-translational control of de novo cell wall formation during *Blastocladiella emersonii* zoospore germination. *Dev. Biol. 54*:37-51.

Smart, R. F. (1937). Influence of certain external factors on spore germination in the Myxomycetes. *Am. J. Bot. 24*:145-159.

Staples, R. C., and Huang, B. F. (1981). Gene activation during differentiation of rusts and anthracnose fungi. In *The Fungal Spore: Morphogenetic Controls*. G. Turian and H. R. Hohl (Eds.). Academic, New York, pp. 335-353.

Steele, S. D., and Fraser, T. W. (1973). Ultrastructural changes during germination of *Geotrichum candidum* arthrospores. *Can. J. Microbiol. 19*:1031-1034.

Storck, R., and Price, J. S. (1977). Sensitivity of germinating spores of *Mucor rouxii*. *Exp. Mycol. 1*:323-338.

Sussman, A. S. (1966). Dormancy and spore germination. In *The Fungi*, Vol. 2. G. C. Ainsworth and A. S. Sussman (Eds.). Academic, New York, pp. 733-764.

Sussman, A. S. (1969). The dormancy and germination of fungus spores. In *Dormancy and Survival*. Cambridge University, Cambridge, pp. 99-121.

Sussman, A. S. (1976). Activators of fungal spore germination. In *The Fungal Spore: Form and Function*. D. J. Weber, and W. M. Hess, (Eds.). Wiley, New York, pp. 101-137.

Sussman, A. S., and Douthit, H. A. (1973). Dormancy in microbial spores. *Annu. Rev. Plant Physiol. 24*:311-352.

Sussman, A. S., and Halvorson, H. O. (1966). *Spores: Their Dormancy and Germination*. Harper and Row, New York.

Thevelein, J. M. (1981). Investigation into the nature of the primary process of spore heat activation in *Phycomyces blakesleeanus* Burgeff. Ph.D. dissertation, Katholieke Univ. Leuven, Belgium.

Thevelein, J. M., Van Assche, J. A., and Heremans, K. (1979). Protein conformational change or membrane phase transition as the underlying mechanism of fungal spore heat activation. *Arch. Int. Physiol. Biochem. 87*:1013-1014.

Thevelein, J. M., Van Assche, J. A., and Heremans, K. (1980). Effect of high pressure on the heat activation in vivo of trehalase in the spores of *Phycomyces blakesleeanus*. *Eur. J. Biochem. 111*: 171-175.

Turian, G., and Hohl, H. R. (Eds.). (1981). *The Fungal Spore: Morphogenetic Controls*. Academic, New York.

Van Brunt, J., and Harold, F. M. (1980). Ionic control of germination of *Blastocladiella emersonii* zoospores. *J. Bacteriol. 141*:735-744.

Van Etten, J. L., and Freer, S. N. (1978). Polyadenylate-containing RNA in dormant and germinating sporangiospores of *Rhizopus stolonifer*. *Exp. Mycol.* *2*:301-312.
Van Etten, J. L., and Rawn, C. D. (1979). Polyadenylate-containing RNA in dormant and germinating *Botryodiplodia theobromae* pycnidiospores. *Can. J. Microbiol.* *25*:375-379.
Van Etten, J. L., Dunkle, L. D., and Knight, R. H. (1976). Nucleic acids and fungal spore germination. In *The Fungal Spore: Form and Function*. D. J. Weber and W. M. Hess (Eds.). Wiley, New York, pp. 243-300.
Van Etten, J. L., Dunkle, L. D., and Freer, S. N. (1977). Germination of *Rhizopus stolonifer* sporangiospores. In *Eukaryotic Microbes as Model Developmental Systems*. D. M. O'Day and P. A. Horgen (Eds.). Marcel Dekker, New York, pp. 372-401.
Van Etten, J. L., Freer, S. N., and McCune, B. K. (1979). Presence of a major (storage?) protein in dormant spores of the fungus *Botryodiplodia theobromae*. *J. Bacteriol.* *138*:650-652.
Van Etten, J. L., Dahlberg, K. R., and Russo, G. M. (1981). Nucleic acids. In *The Fungal Spore: Morphogenetic Controls*. G. Turian and H. R. Hohl (Eds.). Academic, New York, pp. 277-299.
Van Laere, A. J., and Carlier, A. R. (1978). Synthesis and proteolytic activation of chitin synthetase in *Phycomyces blakesleeanus* Burgeff. *Arch. Microbiol.* *116*:181-184.
Van Laere, A. J., Carlier, A. R., and Van Assche, J. A. (1977). Cell wall carbohydrates in *Phycomyces blakesleeanus* Burgeff. *Arch. Microbiol.* *112*:303-306.
Walkinshaw, C. H., Hyde, J. M., and Van Zandt, J. (1967). Fine structure of quiescent and germinating ascospores of *Cronartium fusiforme*. *J. Bacteriol.* *94*:245-254.
Weber, D. J., and Hess, W. M. (1975). Diverse spores of fungi. In *Spores*, Vol. 6. P. Gerhardt, R. N. Costilow, and H. L. Sadoff (Eds.). American Society for Microbiology, Washington, D.C., pp. 97-111.
Weber, D. J., and Hess, W. M. (1976). *The Fungal Spore: Form and Function*. Wiley, New York.
Weete, J. D. (1974). *Fungal Lipid Biochemistry*. Plenum, New York.
Weete, J. D. (1981). Lipids in fungal growth and reproduction. In *The Fungal Spore: Morphogenetic Controls*. G. Turian and H. R. Hohl (Eds.). Academic, New York, pp. 463-485.
Wenzler, H., and Brambl, R. (1978). In vitro translation of polyadenylate-containing RNAs from dormant and germinating spores of the fungus *Botryodiplodia theobromae*. *J. Bacteriol.* *135*:1-9.
Winther, M., and Stevens, L. (1976). Synthesis of polyamines in the germinating conidia of *Aspergillus nidulans*. *Biochem. Soc. Trans.* *4*:1126-1128.
Wright, B. (1979). Causality in biological systems. *Trends Biochem. Sci.* *4*:N110-N111.

10

Fungal Dimorphism

PETER R. STEWART *Australian National University, Canberra, Australia*

PETER J. ROGERS *School of Science, Griffith University, Brisbane, Queensland, Australia*

I. INTRODUCTION

The phenomenon of fungal dimorphism is expressed in cellular terms as the ability of cells of a species or strain to grow predominantly either as hyphae (exhibiting linear or apical growth) or as yeastlike spherical or ovoid cells (spherical growth). The older terminology used to describe the two growth phases, mycelium and yeast, reflects the macroscopic differences seen in liquid or plate cultures of an organism growing in the two alternative phases. The alternative terminology, hyphal and yeast forms, more accurately reflects differences at the cellular level and will be used in this review.

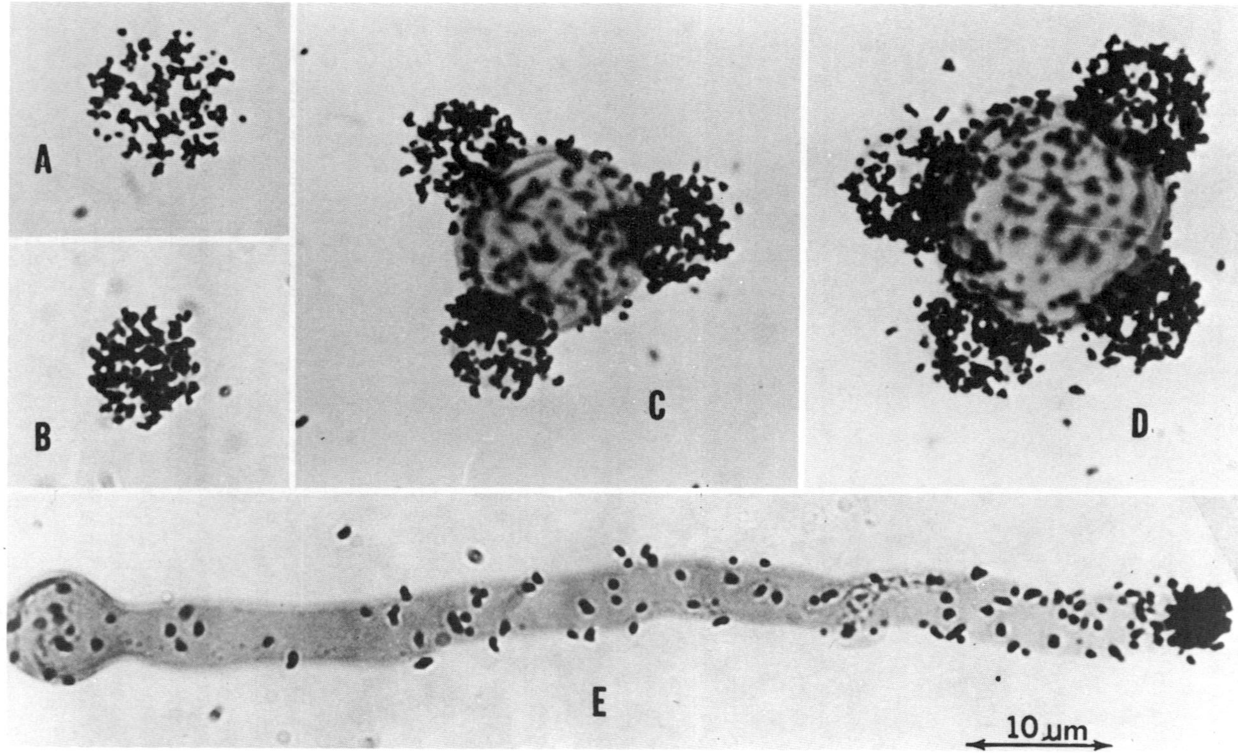

FIG. 1 Location of cell wall synthesis in yeast and hyphal forms of *Mucor rouxii*. Germinating spores (A,B), budding yeast cells (C,D), and hyphae (E) were exposed to [^{3}H]N-acetyl-D-glucosamine for 5 min, then radioautographed. Silver grain deposition indicates sites of chitin synthesis. (From Bartnicki-Garcia and Lippman, 1969.)

The transition between dimorphic states can often be controlled experimentally, usually by modifying environmental or cultural conditions. The best studied organisms in this respect are *Mucor racemosus* and *Mucor rouxii* (Sypherd et al., 1978).

In practical terms, an understanding of dimorphism is important for a range of reasons, primary among which are the prospects of a better understanding of the pathogenesis of the many clinical infections caused by dimorphic fungi (Rippon, 1980) and the insight it gives into eukaryotic differentiation. In the latter case, the special features offered are that growth and morphogenesis occur concurrently, and morphogenesis is reversible.

In molecular and ultrastructural terms, an understanding of fungal dimorphism requires an explanation of why growth is restricted more or less to the hyphal apex in linear growth, whereas it is generalized throughout the wall surface when spherical growth occurs. Radioautography of cells synthesizing cell wall in the two modes of growth dramatically illustrates this difference (Fig. 1).

In this chapter we consider fungal dimorphism in terms of the ultrastructural and biochemical correlates of morphological change. We have sought to identify in particular those observations and propositions which promote an understanding of the relationship between the ultrastructural and biochemical processes.

Much background and earlier information can be found in Bartnicki-Garcia (1973) and Stewart and Rogers (1978). Other recent reviews and surveys of fungal growth and differentiation which contain information relevant to dimorphism include Smith and Berry (1978), Farkâs (1979), Burnett and Trinci (1979), Sypherd et al. (1978), and Rippon (1980). The collections of essays and reviews in Smith and Berry (1978) and Burnett and Trinci (1979) are especially valuable sources.

II. ULTRASTRUCTURAL ASPECTS OF MORPHOGENESIS

A. General Considerations of Cell Wall Growth and Morphogenesis

An implied assumption in studies of fungal morphology is that cell shape is determined by the cell wall (and more specifically by a chitinous or cellulosic microfibrillar skeleton, depending on the taxonomic group), with the protoplast appressed (due to turgor pressure) closely to the wall. While a good deal of evidence can be adduced to support the importance of chitin microfibrils for cell shape and rigidity in the chitinous fungi, other components (β-glucans in particular) or factors (cross-linking by matrix substances) are likely to play a role in determining fungal cell shape.

A further assumption is that the "driving force" for cell expansion during fungal growth and differentiation derives from the turgor pressure of the protoplast. While there is little reason to doubt that under

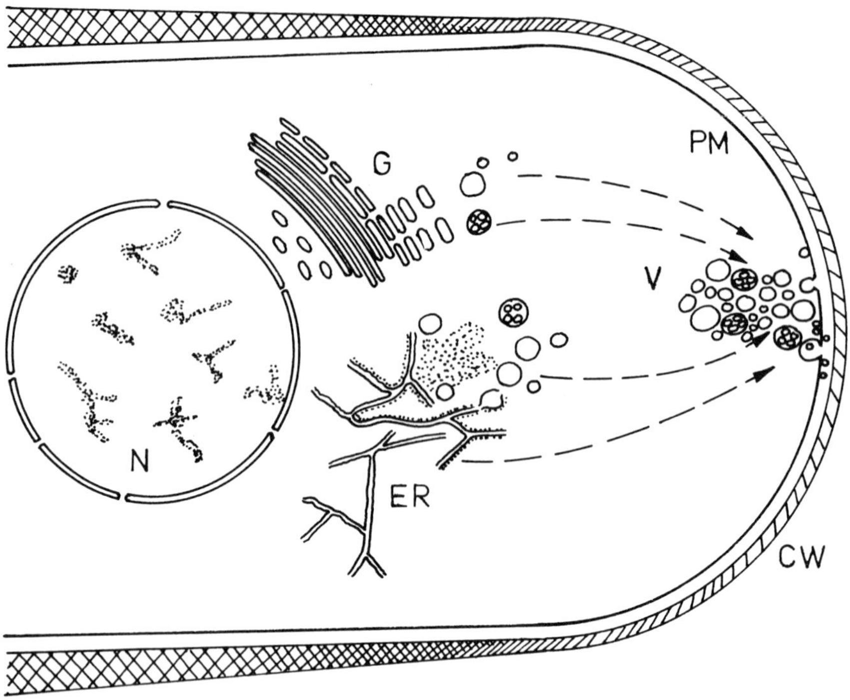

FIG. 2 Origin of wall-synthesizing elements. Vesicles containing wall polysaccharide synthetases, wall precursors, and lytic enzymes are assembled from component proteins and membranes synthesized by the endoplasmic reticulum and/or Golgi complex. The vesicles are then transported to the plasma membrane, delivering their contents to the actively growing wall at the hyphal apex. In the case of yeast phase growth, the vesicles would be distributed more generally to the expanding cell wall (N, nucleus; ER, endoplasmic reticulum; GC, Golgi complex; V, vesicles; PM, plasma membrane; CW, cell wall).

normal conditions turgor pressure keeps the plasma membrane closely appressed to the cell wall, whether this is sufficient to distort the wall, or to otherwise induce shape change, is not known with certainty. A prediction of this hypothesis would be that reduction of this pressure should affect cell growth and perhaps shape. The older literature (reviewed by Bartnicki-Garcia, 1973) indicates that hyphal tip elongation indeed ceases in hypertonic solutions. More recent data, however, present a less clear picture. Thus temperature-sensitive morphological mutants of *Aspergillus nidulans* and *Neurospora crassa* have been reported to grow, some with normal morphology, when cultured in hypertonic polyols or salts (Waldron and Roberts, 1974; Bainbridge et al., 1979). In wild-type *A. nidulans*, on the other hand, hypertonic medium disturbs normal hyphal growth, and increased branching and septation occur (Katz and Rosenberger, 1971). At this time it would seem prudent to conclude that the relationship of wall-membrane contact to normal growth and morphology is not clear.

Structural studies of fungal dimorphism appear currently to be directed at testing a model which has the following salient features (Fig. 2). Vesicles containing the necessary polysaccharide synthetases are assembled at or near either or both the Golgi complex (or an equivalent organelle) and the endoplasmic reticulum, and moved from internal tracts of the cytoplasm to sites of wall growth. Polysaccharide building blocks or partly polymerized precursors may also be included in these or other vesicles. Lysins, to weaken or cleave preexisting wall polymers, are also positioned at growth sites, perhaps in yet other vesicles.

At the wall growth site, vesicles fuse with or are transferred across the cell membrane and the synthetases begin or continue the synthesis of the major microfibrillar polysaccharide found in dimorphic fungi, chitin. Accretion or intussusception of new microfibrils is promoted by lysins, and cross-linking proteins or matrix polysaccharides consolidate the new microfibrils. A "growth unit" (Bartnicki-Garcia, 1973) thus functions (Fig. 3).

While this may be an oversimplified summary of an exceedingly complex process, it is intended only as a framework within which to discuss experimental evidence. The aspects of this model which are uncertain are the following:

1. How growth sites are "selected" within cells
2. How synthetases, lysins, and precursors are transported from sites of synthesis and positioned at sites of wall growth
3. How external or environmental signals are transmitted to the controlling elements in (1) and (2)

Growth site selection and the coupling of external signals to wall synthesis may be closely related phenomena. This coupling process is being studied, for example, in terms of membrane potential and cyclic nucleotides as localized regulators of biosynthetic activity. These are considered further in Sec. III below.

The transport of synthetases and ancillary components has not been directly addressed. However, a good deal is now known about chitin synthesis in fungi and its regulation, and these studies support certain models of spatial and temporal regulation of wall polysaccharide synthesis. In turn, this has bearing on the selection of growth sites. The chitin synthetase system and its control are now considered in some detail.

(A)

(B)

(C)

(D)

(E)

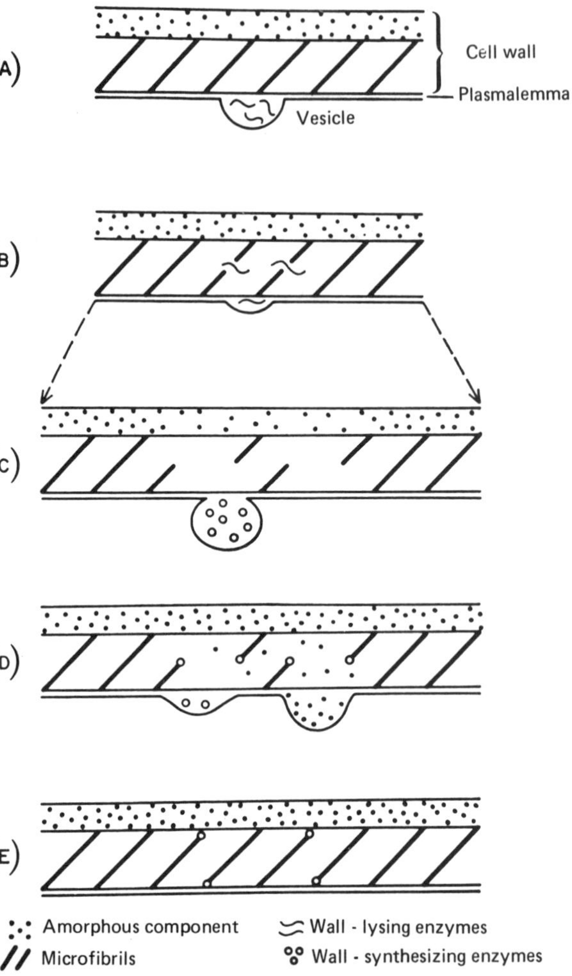

FIG. 3 Unitary model of wall growth. Hypothetical representation of the processes of wall lysis and weakening (A,B), wall expansion (C), and addition of new wall material (D) to form an extended "unit" of cell wall (E). (From Bartnicki-Garcia, 1973).

B. Chitin Synthesis

1. *Chitin Synthetase*

Chitin is a homopolymer of β-1,4 linked N-acetylglucosamine units, with the sugar in the chair form. Consequently it adopts a linear chain conformation similar to cellulose, with each sugar residue rotated by 180° with respect to its neighbors. Antiparallel linear association of chains occurs that is similar to though not as condensed as in paracrystalline α-cellulose. This form, known as α-chitin, is the form found in the chitinous fungi.

Paracrystalline aggregates of chitin chains appear in the electron microscope as microfibrils, 10-30 nm thick and with lengths of the order of micrometers. In the cell they are found in the inner cell wall, adjacent to the plasma membrane.

Massive deacetylation of chitin results in the formation of chitosan, which is a major component of cell walls in the Mucorales. In *M. rouxii* (yeast and hyphae) the amount of chitosan exceeds chitin more than threefold (Bartnicki-Garcia, 1968a).

Chitin is cross-linked to proteins and other polysaccharides (chitosan, glucans, and mannans primarily), and through these, the microfibrils associate laterally and are embedded in the amorphous matrix of the cell wall.

Chitin synthetase has been isolated from many fungi and yeasts. The enzyme catalyzes the reaction

$$(\beta\text{-}1,4\text{-GlcNAc})_n + \text{UDP-GlcNAc} \xrightarrow[\text{GlcNAc}]{\text{Mg}^{2+}} (\beta\text{-}1,4\text{-GlcNAc})_{n+1} + \text{UDP}$$

Chitin synthetase is allosteric, with Mg^{2+} and GlcNAc as cofactors. Primer requirement is uncertain. Preparations commonly carry tightly bound primer oligosaccharide through purification. GlcNAc probably acts as a primer. The enzyme is potently inhibited by polyoxin D, an analog of UDP-GlcNAc. This and other inhibitors are discussed below.

Chitin synthetase is usually membrane bound in the early stages of purification. In *Saccharomyces cerevisiae* (Cabib et al., 1979), *Phycomyces blakesleeanus* (Jan, 1974), and *Candida albicans* (Braun and Calderone, 1978), it is reported to be associated with plasma membrane. In a range of fungi (including *S. cerevisiae*) examined by Bartnicki-Garcia et al. (1979), it was found to be organized as chitosomes, as discussed in the next section.

In *S. cerevisiae*, most of the chitin synthetase exists as inactive zymogen, associated with the plasma membrane. Activation requires a limited proteolysis of the zymogen. The proteolytic enzyme responsible for this appears to be protease B, which is also involved in the regulation of tryptophan synthetase in this yeast. To prevent indiscriminate activation of chitin synthetase zymogen, an inhibitor is believed to bind to and inactivate protease released from specialized vesicles other than at growth sites (Cabib et al., 1979). Nevertheless, the protease may be responsible for slow inactivation, and thus turnover, of the synthetase (Hasilik, 1974). The inhibitor, like the protease, has been purified. It is polypeptide of molecular weight 8500 which forms a tight bimolecular complex with the protease (Cabib et al., 1979).

The zymogen of chitin synthetase bound to plasma membrane, like that from chitosomes of *M. rouxii* discussed below, can be solubilized, using digitonin, as a protein of molecular weight $\sim$500,000 (Duran and Cabib, 1978). The solubilized zymogen is activated by protease and an acidic phospholipid and remains capable of microfibrillar chitin synthesis. The chitin synthetase from *Coprinus cinereus* can be dissociated further, using high salt concentrations, to yield subunits with a molecular weight of about 150,000 (Gooday, 1977).

Proteolytic activation of chitin synthetase occurs in many other fungi. The extent of activation in vitro is dependent on the type of protease used (Bartnicki-Garcia et al., 1979).

2. *Chitin Synthesis and Chitosomes*

Isolation from a dimorphic fungus of a particulate fraction with chitin synthetase activity was first described by Bartnicki-Garcia and colleagues (Ruiz-Herrera and Bartnicki-Garcia, 1974; Ruiz-Herrera et al., 1975). Because the synthetase appeared to be associated with discrete membranous particles, these authors named the particles *chitosomes*. The purification and properties of chitosomes, and their occurrence among representative fungi have recently been reviewed (Bartnicki-Garcia et al., 1979).

Chitosomes are vesicular, probably with a surrounding membrane, and have a size range of 50-100 nm (Fig. 4a). Incubation of chitosomes in vitro with UDP-N-acetyl glucosamine, Mg^{2+}, and an appropriate activating protease results in the formation of microfibrils attached to the chitosomes (Fig. 4b,c). Coiled microfibrils, continuous with extruded microfibrils, are sometimes seen inside chitosomes, suggesting that the lumen of the chitosome is the site of synthesis of the polymers, and of their association into microfibrils.

Chitosomes from *M. rouxii* can be dissociated with digitonin into zymogenic subunits with a sedimentation coefficient of 16 S (compared to 105 S for intact chitosomes) and a diameter of 7-12 nm (Ruiz-Herrera et al., 1980). These dimensions correspond approximately to a spherical protein of 500,000 daltons. No change in the sedimentation coefficient occurs after proteolytic activation. Chitin microfibrils formed by chitosome subunits are generally much shorter than those formed by intact chitosomes (Ruiz-Herrera et al., 1980).

Removal of digitonin from dissociated chitosomes from *M. rouxii* results in reaggregation of the subunits, though not generally to form chitosomes. However, incubation of these aggregates with substrate and activators (Mg^{2+}, N-acetyl glucosamine, protease) results in the formation of microfibrils somewhat shorter than those formed by untreated chitosomes (Bartnicki-Garcia et al., 1979).

Chitosomes of the size and characteristics described above are isolated from *M. rouxii* either by the mechanical breakage of cells or by the more gentle lysis of protoplasts. There is no evidence for a continuum of chitosome particle sizes. These observations have been offered as evidence that chitosomes as isolated are not fragments of larger organelles or membrane structures (Bartnicki-Garcia et al., 1979).

So far as can be determined, chitin microfibrils are formed at the plasma membrane of the cell or just external to it. There is a large and diverse array of vesicles in the cytoplasm near regions of active apical wall growth (Grove, 1978). Bartnicki-Garcia et al. (1979) proposed that chitosomes represent one type in this mixture of cytoplasmic vesicles (see Fig. 2) and that chitin synthesis and microfibril formation in *M. rouxii* have the following features. Chitin synthetase, together with other proteins and the membrane of the chitosome, are assembled in the cytoplasm. The chitosomes migrate in a directed fashion to sites where wall synthesis is to occur. At the plasma membrane, they fuse with this membrane or pass through it and function as independent "organelles" outside or on the plasma membrane (but within the cell wall), generating chitin microfibrils which are attached to or embedded in the expanding face of the inner wall. Alternatively, they may remain just below the plasma membrane and extrude microfibrils through the plasma membrane to associate with the growing cell wall. The structural relationship of wall, plasma membrane, and chitosomes is not understood. A plant cell analog is discussed in Sec. II.F below.

Chitin synthetase has been identified in association with the plasma membrane in *S. cerevisiae* (Duran et al., 1975) and *C. albicans* (Braun and Calderone, 1978), and with cell wall from both yeast and hyphal phases of *M. rouxii* (Bartnicki-Garcia et al., 1979).

3. *Active and Zymogenic Chitin Synthetase in Dimorphic Fungi*

In *M. rouxii*, the regulation of chitin synthetase differs from that in *Saccharomyces*. No specific inhibitor of the protease which activates the synthetase zymogen has been identified. Instead, an endogenous inhibitor (17,500-dalton protein) competes with UDP-N-acetyl glucosamine ($K_1 \sim 0.6\ \mu M$) to inhibit the synthetase directly. This inhibitor has no effect on the proteolytic activation (by trypsin or rennilase) of the zymogen (Lopez-Romero et al., 1978).

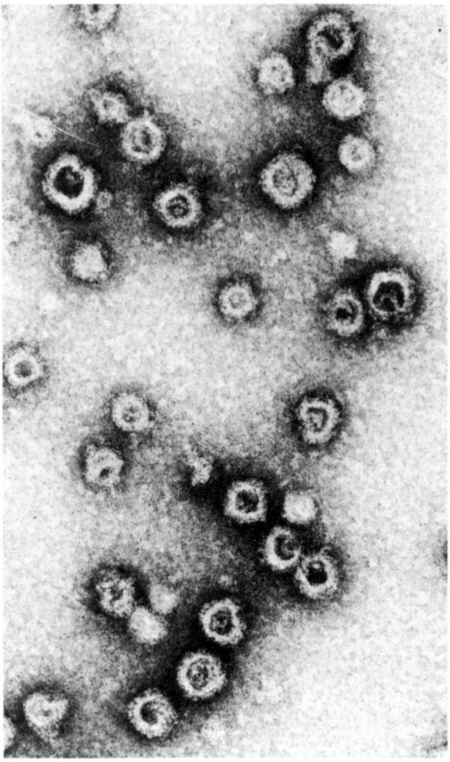

A

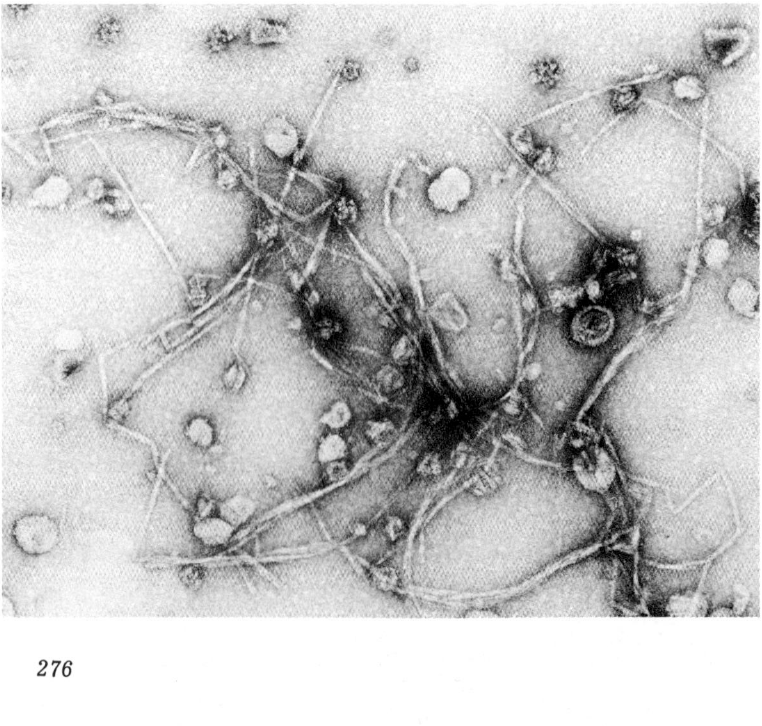

B

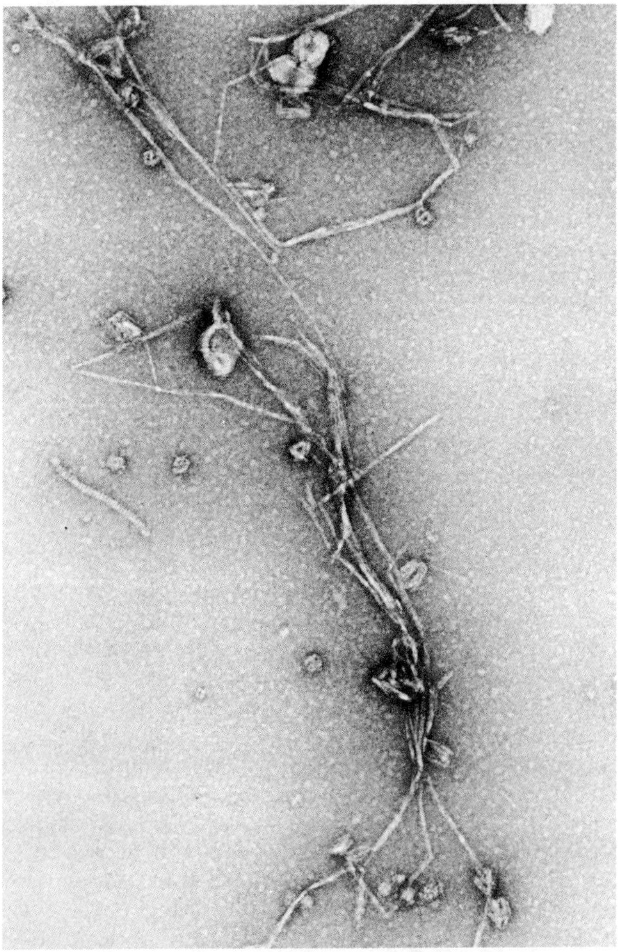

C

FIG. 4 Chitosomes from *Mucor rouxii*. (A) Chitosomes (negatively stained) prepared from *M. rouxii* hyphae were treated with a high concentration of trypsin to reveal their internal structure, a lumen surrounded by a shell consisting of a thin outer veil and a thick inner wall. Chitosomes are not in an active state when prepared in this way. (B,C) Chitosomes isolated in active form from *M. rouxii* yeast cells and hyphae, respectively. After isolation, these preparations were incubated with substrate and activators to form fibrils in vitro. The diameter of the largest vesicles is approximately 100 nm. (From Bartnicki-Garcia et al., 1979.)

In yeast phase cells of *M. rouxii*, the zymogen form of chitin synthetase predominates. In hyphae, the reverse is true. Both cell types have approximately the same total activity when the zymogen is totally activated (Ruiz-Herrera and Bartnicki-Garcia, 1976). These differences are reflected in vivo: Yeast phase cells incorporate labeled N-acetylglucosamine at about one-third the rate of hyphae (Domek and Borgia, 1981).

In *C. albicans*, there is also a greater proportion of active chitin synthetase in the hyphal form compared with the yeast form. The extent of zymogen activation (by trypsin) in both morphic forms is about sevenfold. However, a vacuolar fraction, which could be prepared from yeast but not from hyphae, is approximately equally effective at activating synthetase zymogen from either cell type (Braun and Calderone, 1978). Thus in the two morphic forms of *C. albicans*, the amounts and compartmentation of activating protease are different, and this may result in different patterns of zymogen activation.

Chiew et al. (1980) suggested that chitin synthesis in *C. albicans* is regulated by the first and last enzymes of the biosynthetic pathway (glutamine-fructose-6-phosphate aminotransferase and chitin synthetase); both enzymes are activated four- to fivefold during germ tube outgrowth.

In *Mucor* and *Candida* there is greater chitin synthetase activity in hyphal than in yeast forms. However, there are no consistent differences in the amounts of chitin present in the yeast and hyphal forms of dimorphic fungi. Thus, for example, hyphal walls of *Histoplasma capsulatum* and *Blastomyces dermatitidis* contain less chitin than do the yeast forms (Domer, 1971), whereas *C. albicans* hyphae contain three times more chitin than yeast cells (Chattaway et al., 1968), and yeast and hyphal forms of *M. rouxii* contain about equal amounts of the polysaccharide (Bartnicki-Garcia, 1968a). This may be a measure of the importance of lytic enzymes in determining the steady state concentrations of chitin in different cells and hyphae. A survey of wall composition in the yeast and hyphal forms of dimorphic fungi is given by Dow and Rubery (1977).

Seemingly, there are differences in the models of localized chitin synthesis put forward by the groups of Bartnicki-Garcia and Cabib, one emphasizing the chitosome as the vehicle of chitin synthetase or its zymogen, the other the plasma membrane as the site of active and inactive forms of the synthetase. Methodological differences may be important in determining the membranes with which chitin synthetase is found to be associated. Moreover, the considerable taxonomic differences between the organisms used by the two groups may be important. Thus, while chitin is a major and nonlocalized component of the cell wall of *M. rouxii* (Bartnicki-Garcia, 1973), it is found only in septa and bud scars in *S. cerevisiae* (Molano et al., 1980). Beyond this is the possibility that chitosomes are indeed a vehicle for chitin synthetase in yeasts and fungi generally but that they exist in some organisms (e.g., *Saccharomyces*) only transiently, while in others they are more stable entities, accumulating in the cytoplasm, from which they can be readily isolated.

A series of other matters have yet to be considered in regard to the regulation of chitin synthesis. For example, how does the protease gain access to the zymogen: by fusion of vesicles containing protease with the plasma membrane/chitosomes; by release of protease into the cytosol and diffusion to the zymogen? What is the critical determinant in the spatial distribution of active chitin synthetase in the cell: localization of zymogen, protease, protease inhibitor, or synthetase inhibitor?

Associated questions relate to the availability of substrates and the role of physiological effectors and inhibitors in determining chitin synthetase activity. Dow and Villa (1980) reported the release of oligomers of glucuronic acid in *M. rouxii* during cell growth. Do these affect the activity of chitin and other polysaccharide synthetases? Simonetti et al. (1974) found that exogenous N-acetylglucosamine is an inducer of the yeast to mycelial transition in *C. albicans*, and speculated that this might be a consequence of activation of chitin synthetase by this substance.

C. Inhibitors of Cell Wall Polymer Synthesis

Polyoxin D is a structural analog of UDP-N-acetylglucosamine, the sugar donor substrate of chitin synthetase (Endo et al., 1970). Polyoxin D is a competitive inhibitor of the enzyme, binding more tightly (K_i ∿1 μM) to the enzyme than does the substrate (K_m ∿500 μM). When added to intact cells or hyphae, concentrations of the inhibitor (∿10-100 μM) greater than anticipated from the K_i are required to inhibit chitin synthesis. This suggests a barrier to penetration of intact cells by the inhibitor. Nevertheless, inhibition of chitin synthetase by polyoxin D causes apical weakening and bursting or ballooning in cultures of *M. rouxii* hyphae (Bartnicki-Garcia and Lippman, 1973).

A number of other polyoxins, and a related analog, nikkomycin, have been identified and characterized to varying degrees (reviewed by Gooday, 1978). All either inhibit hyphal elongation or cause ballooning or distortion of hyphae in a range of fungi. An uncharacterized galactosamine glycan may act as an endogenous chitin synthetase inhibitor in *N. crassa* (Reissig, 1974).

Other inhibitors of amino-sugar incorporation into cell walls of fungi include kitizin (Maeda et al., 1970) and tunicamycin, which inhibit protein glycosylation (Kuo and Lampen, 1976). The synthesis of β-1,3-glucan, the other major fibrillar component in chitinous fungi, is inhibited by high concentrations of sorbose (Mishra and Tatum, 1972). Hyphal apices of *N. crassa* disintegrate in the presence of this sugar, and "colonial" growth occurs (Rizvi and Robertson, 1965;

Mahadevan and Tatum, 1965), apparently in response to inhibition of glucan synthetase.

In *S. cerevisiae*, 2-deoxy-2-fluoro-D-glucose inhibits cell wall glucan synthesis (Biely et al., 1973). There has been no systematic application of inhibitors of wall polymer synthesis in the study of morphogenesis in the dimorphic fungi.

Other potentially useful agents which may help to dissect the processes of synthesis, assembly, and deposition of microfibrils in fungal cell walls are calcofluor white and Congo red. Haigler et al. (1980) have used the optical brightener calcofluor white ST to distinguish the processes of sugar polymerization and polymer crystallization during the biogenesis of extracellular cellulose microfibrils by the bacterium *Acetobacter xylinum*. Herth (1980) has reported that calcofluor white and Congo red each inhibit chitin microfibril assembly in the alga *Ochromonas malhamensis*. Instead of long, laterally associated microfibrils, short rods or irregular networks of thinner fibrils are formed. Herth inferred from his observations that an interval occurs between synthesis of the chitin polymers and their association to form microfibrils. The dyes interact with the polymer during this interval. As a consequence, microfibril orientation in the wall is affected.

D. Determinants of Cell Wall Plasticity

Mention was made earlier of alternative models for cell wall elongation or expansion in response to internal pressure from the growing sacculus. These two proposals involve partial lysis of preexisting wall on the one hand, and microfibril orientation in the wall on the other.

That lytic enzymes are closely associated with the fungal cell wall and membrane is established (Gooday, 1978; Rosenberger, 1979). Moreover, inhibition of synthesis of chitin results in the ballooning and/or bursting of fungal apices, and mutants of *Aspergillus* with greatly decreased capacity to synthesize chitin show similar changes, unless osmotically stabilized (Katz and Rosenberger, 1971). Shimoda and Yanagishima (1971) found that inhibition of glucanases in *S. cerevisiae* by δ-gluconolactone resulted in cessation of auxin-induced wall growth.

However, to deduce from these observations that lysins act in a specific fashion at sites of new wall synthesis may be an undue extrapolation of the evidence. Cytochemical ultrastructural analysis of intact yeast and hyphal forms demonstrating a close and specific relationship between lytic vessels and growing wall is called for, together with biochemical characterization of carefully fractionated organelle preparations. Evidence of this sort has not yet been described even for fungi generally, though Nolan and Bal (1974) have demonstrated cellulase activity in vesicles and dictyosome cisternae in *Achlya bisexualis* hyphae.

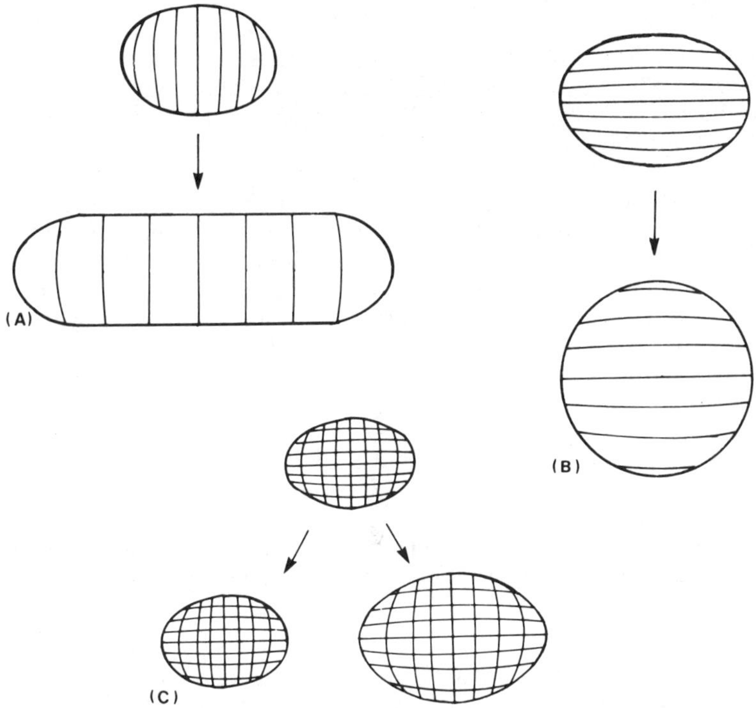

FIG. 5 Fibril net geometry in cell wall as a determinant of cell shape. Schematic and greatly simplified version of Roelofsen's (1965) fibril net model for determining cell shape. If fibrils are embedded in the walls of ovoid cells (A), (B), and (C) in the orientation shown, and if these fibrils are less deformable than the matrix material between the fibrils, then the cells will tend to expand as shown when internal pressure increases. Filamentous or spherical forms will thus result [cases (A) and (B)], and depending on the strength of the fibrils, either no expansion or uniform expansion results in case (C).

Roelofsen's (1965) proposal (Fig. 5) for a differentially deformable fibril net as the basis of determining cell wall plasticity has been partly tested in nondimorphic fungi. Roelofsen (1965) had found in *Phycomyces blakesleeanus* a predominantly transverse orientation of microfibrils in the inner (younger) cell wall layers of hyphae which was thus consistent with his model for filamentous growth (Fig. 5). On the outer surface, the older microfibrils were axial, which Roelofsen surmised was a consequence of stretching and reorientation as a result of growth (extension) of the hyphal filament.

Gooday (1979) found that the filamentous stipe cells of *Coprinus cinereus* exhibited predominantly a transverse orientation of chitin fibrils in preparations that were actively growing or which had just ceased growth. These cells elongate extensively without substantial change in diameter. The growth mode in this case is also consistent with Roelofsen's proposal.

On the other hand, Burnett and coworkers (Burnett, 1979) found that in *N. crassa* only in the apex of germ tubes (formed by spore outgrowth) is any orientation other than a random one seen. In this case the orientation was axially biased; this would not be consistent with the model for filamentous growth shown in Fig. 5.

These differences need to be examined more closely. They may indicate that there are differences in the mechanisms by which cells expand while maintaining a particular shape. It would be of particular interest to see this approach applied to dimorphic fungi in different morphogenetic states.

With the evidence available, it is not at present possible to distinguish between microfibril geometry and lytic weakening as the means by which turgor-driven extension or expansion could take place. However, there appears to be a decided preference among investigators in the field for the idea of lytic erosion and weakening.

E. Topology of Cell Wall Growth

How is the location of cell wall plasticization and microfibril synthesis determined? Ultimately, of course, the information encoded in the cell's genome, and its expression as proteins, determines where and when cell wall synthesis occurs. The synthesis of prospective "morphic" proteins is considered in some detail in Sec. III.C below.

However, in terms of distributing and directing synthetases, lysins, and wall precursor molecules to regions where cell wall growth is to proceed, orienting and moving mechanisms must operate. Three models have currency.

The first of these, exemplified by Farkâs (1979), is that biochemical "gradients" of inhibitors, activators, and substrates are established close to the cell membrane and wall. Where conditions are appropriate, wall lysis and synthesis take place. If this is localized, apical growth or budding occurs; if conditions are more generally favorable for the activity of these enzymes, then spherical growth occurs. No testable proposals have been made as to how the necessary "gradients" are established or maintained, though local synthesis of activators or inhibitors (and for that matter synthetases and lysins) could be one mechanism.

The second model, of which there are two variants, invokes the physical movement of synthetases, hydrolyases, and precursors (within or without corresponding vesicles) from their sites of synthesis and packaging, to wall growth sites. In the first of these, the mechanism of movement has primarily an electrochemical basis, involving the establishment of differential ionic gradients at and across the cell membrane in different parts of the cell in such a way that molecules or vesicles are electrophoretically moved to sites of cell wall synthesis (Jennings, 1979; Harold and Harold, 1980). Ionophores which destabilize or discharge ionic gradients have significant effects on morphogenesis and growth in fungi (Van Brunt and Harold, 1980). This topic is considered further in Sec. III.A.

The other variant of the physical transport model invokes cytoskeletal systems based on supramolecular structures which have been partially characterized in animal cells. In general terms, microtubules (tubulin) and microfilaments (actin, myosin) are seen as the basis for the construction of a chemomechanical system for the transport of vesicles and molecules to specific sites within cells. However, once again there is little direct experimental data available, even in relation to the presence of these structural elements in fungi, so that this model too remains simply speculative for the time being.

In this regard, it is worth considering the approach which has been taken and the recent progress made in regard to motile mechanisms acting in plant cell wall synthesis. These systems may provide useful models or analogs of ultrastructural interactions which occur at the plasma membrane-cell wall interface during wall formation in fungi.

F. Cellulose Synthesis in Plant Cells: An Analog of Chitin Synthesis and Morphogenesis?

Brown (1979) and colleagues have examined fibril formation during cell wall formation in a range of higher and lower plant cells. In the unicellular alga *Oocystis apiculata*, freeze fracture images of the outer layer of the plasma membrane (E fracture face) show impressions of newly synthesized microfibrils in the newly forming wall. These microfibrils are laid down more or less in parallel and are superimposed on a deeper or distal (presumably older) layer of microfibrils, also parallel in array but at an angle to the proximal layer (Fig. 6a). In some cases, microfibrils apparently torn through the membrane during fracture can be seen (Fig. 6b).

At the ends of the proximal microfibrils structured aggregates of particles are seen (Fig. 6c). These may be the sites of cellulose synthesis for the microfibril concerned and have been denoted as *terminal synthesizing complexes*. These complexes consist of three rows of subunits, with about 30 subunits in each row. Though seen in the outer half of the plasma membrane of *Oocystis*, the complexes may span the entire membrane, thus making them accessible to sugar nucleotide substrates and to underlying cytoplasmic structural or motile elements

The terminal complexes may thus be the cellulose-synthesizing analog of chitin synthetase aggregates or chitosomes in fungi.

A
B

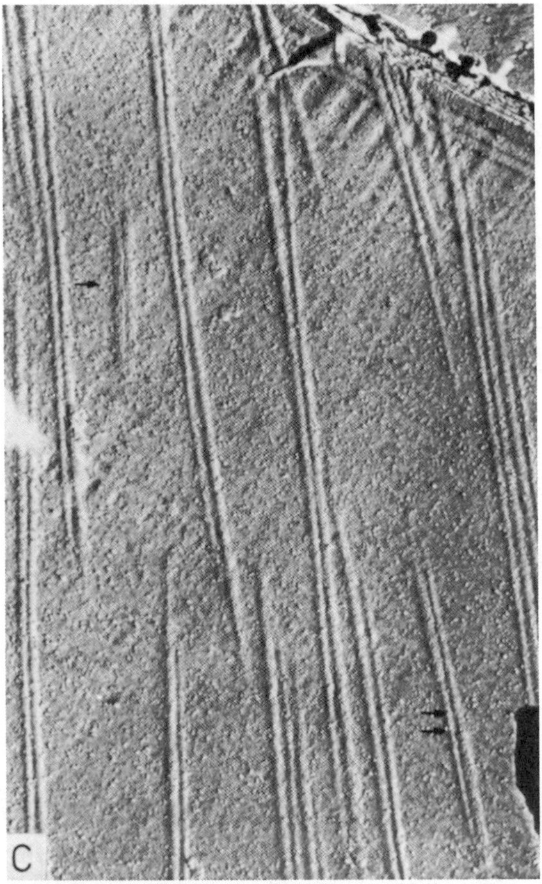

FIG. 6 Cellulose microfibril synthesis in *Oocystis*. (A) E fracture face through the plasma membrane of a vegetative *Oocystis* cell showing microfibrillar impressions and terminal synthesizing complexes (arrows). Previously deposited microfibrils lie beyond the membrane plane at an angle of about 60° to the newer microfibrils. (B) E fracture face through the plasma membrane of *Oocystis* showing microfibrils torn through the outer leaflet of the plasma membrane. The axes of deposition of distal and proximal layers of microfibrils in this case differ by about 90°. (C) Same as for (A), showing inactive paired terminal synthesizing complexes (single arrow) and a recently activated paired complex (double arrow). (From Brown, 1979.)

The directional relationships between microfibrils in the *Oocystis* wall, with alternate layers of microfibrils at an angle and newly synthesized microfibrils laid in parallel, indicate that there is a mechanism for orienting the direction of synthesis of the microfibrils and altering this direction as consecutive layers are formed in the cell wall. At the poles of *Oocystis* cells, microfibrils are arranged nearly concentrically, with abrupt termini in the tracks of individual microfibrils, suggesting that after circling the pole of the cell, the terminal synthesizing complex stops, reverses direction, and, at an acute angle to the original direction of movement, proceeds back toward the equatorial region of the cell. Moreover, terminal complexes appear to function in pairs, moving in *opposite* directions toward the two poles. This initial pairing of complexes is proposed to ensure symmetry in synthesis of the wall (Brown, 1979).

A symmetrical, limited excursion mechanism of this sort would generate spherical or ovoid cells, as in *Oocystis*. Elimination of one of each pair of terminal complexes, plus a more extended excursion of the complex, could result in unidirectional filamentous growth, provided that the wall was appropriately extensible. Thus two growth modes could result from switching between symmetrical and asymmetrical modes of wall growth.

How is the topology of cellulose microfibrils synthesis in the organism controlled? Brown (1979) suggested two possibilities. The first is that cytoplasmic microtubules, just below the plasma membrane, orient granules which occur in the inner leaflet of the plasma membrane. These granules, in turn, have been proposed to determine the direction in which the terminal complexes (in the outer leaflet) move as they lay down the microfibrils external to the plasma membrane. After a layer is laid in one orientation, a "switch" (of unknown character) operates to alter the orientation of the lower leaflet granules by about 90°. The direction of microfibril synthesis is then changed likewise.

The second alternative was suggested from studies using another unicellular alga, *Glaucocystis nostochinearum* (Brown, 1979). In this case, single rather than paired terminal synthesizing complexes are seen in the plasma membrane. Three "centers" of microfibril deposition occur at each pole of the cell, around each of which microfibril synthesis is proposed to proceed. Thus microfibril deposition occurs without the reversal which apparently occurs in *Oocystis*. The synthesizing complex returns to the equator and moves to the opposite pole. This unidirectional mechanism depends, it is thought, on protoplast rotation within the cell wall. Thus the terminal synthesizing complexes "spin" a web of microfibrils over the inside of the wall as the protoplast rotates within the cage of the forming wall. Movement of the location of the terminal complex in the plasma membrane would result in coverage of the entire cell wall. How changes in the relative rates of cellulose deposition are achieved and how this may relate to septation, branching, elongation, expansion, and other aspects of morphogenesis is not clear at this stage. Nevertheless, a comparison of the structural aspects of plant cell wall formation with the biochemical aspects of chitin synthesis revealed in yeasts and fungi, however tenuous, can obviously suggest new experimental approaches to fungal cell wall morphogenesis.

III. BIOCHEMISTRY OF DIMORPHISM

A. Cyclic AMP

The relationship between carbon metabolism and dimorphism has been subjected to considerable analysis. Respiratory metabolism was at one time thought to be an essential correlate of filamentous growth. This was later shown not to be the case (Rogers et al., 1974; Paznokas and Sypherd, 1975; for a review see Stewart and Rogers, 1978, and Sypherd et al., 1978). Subsequent studies have indicated that there is no essential relationship between carbon metabolism generally and dimorphic expression (Sypherd et al., 1978). However, a clear relationship exists between intracellular cAMP concentration and cell morphology, as detailed in Table 1 and in the following discussion.

Recent studies have tended to concentrate on the important question of how intracellular levels of this nucleotide are regulated by a plethora of environmental conditions and growth factors (Sypherd et al., 1978). In addition, the role of cAMP in intermediary metabolism (notably in polyamine synthesis and nitrogen assimilation), the characterization of cAMP-dependent protein kinases and their target substrates, and the way in which cAMP is involved with morphogenetic processes (in particular, apical and spherical growth) are continuing areas of interest.

The role of plasma membrane energization appears especially significant in view of the demonstrable effect it has on membrane-bound adenylate cyclase (Pall, 1977), and thereby the ease with which a wide range of unrelated physiological effectors of dimorphism can be rationalized on the basis of their effects on membrane functions (Fig. 7). The observed allosteric effects of cAMP on some of the biosynthetic and catabolic enzymes of *Mucor* – most notably the effect of cAMP on ornithine decarboxylase (Peters and Sypherd, 1979) – together with the identification of cAMP-dependent protein kinase (Moreno and Passeron, 1980) have clearly demonstrated the role of cAMP as a potential second messenger in the dimorphic process. Perhaps the least successful approach has been the search for phase-specific proteins and the characterization of specific messenger RNAs. However, it could reasonably be argued that total protein analyses complement investigations aimed at correlating enzyme induction with morphology, or at least may soon do so.

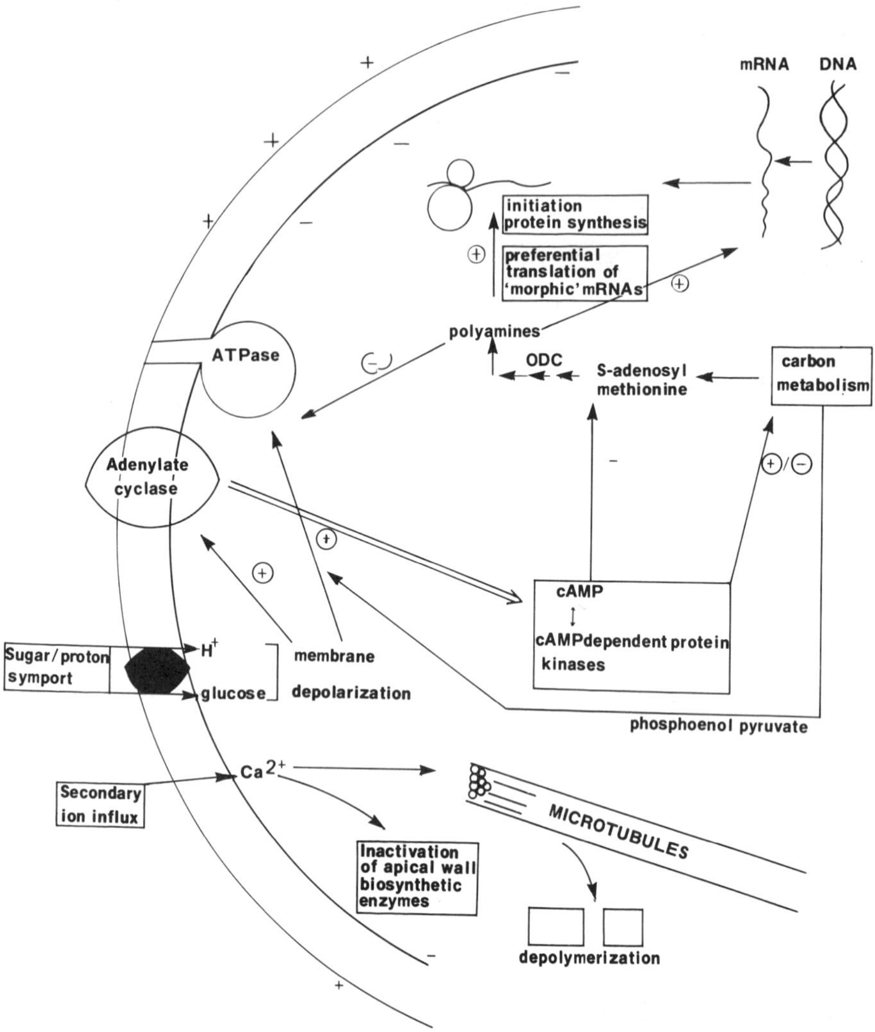

FIG. 7 Possible relationships between membrane energization, intermediary metabolism, gene expression, and dimorphism in the *Mucorales*.

Furthermore, since the cloning of structural genes in bacterial cells with appropriate vectors has become commonplace, those genetic elements that are responsible for the expression of key enzymes (e.g., ornithine decarboxylase) may well be identified before long. In fact,

TABLE 1 cAMP Concentrations in *Mucor* (Batch Cultures) as a Function of Time and Culture Conditions

Organism and culture conditions[a]	Cellular concentration of cAMP (pmol/mg protein): Initiation of culture	Lag phase (or spore swelling stage)	Early log phase	Mid log phase	Stationary phase	Reference
M. racemosus						
• → H, Air	50	550	200	160	90	Paznokas and Sypherd (1975)
• → o, CO_2	150	660	400	400	150	Paznokas and Sypherd (1975)
• → H, Low N_2 flow	25	100	40	15	15	Paznokas and Sypherd (1975)
• → o, High N_2 flow	40	160	100	25	25	Paznokas and Shepherd (1975)
M. rouxii						
• → H, Air	20	25	10	20	25	Cantore, Galvagno, and Passeron (1980)
M. genevensis						
• → H, Air, O_2	4	50	10	∿4	∿4	Orlowski (1980)
M. mucedo						
• → H, Air	4	28	8	∿4	∿4	Orlowski (1980)

[a] •, Spore; o, yeast cell; and H, hyphae.

it may ultimately be possible, judging from past experience in the field of molecular genetics, to identify and use *Mucor* cloning vectors in the dissection of morphogenetic processes in this organism (Cihlar and Sypherd, 1980).

1. cAMP in Spores and Yeast and Hyphal Forms

Endogenous cAMP levels have been correlated with morphology transitions in certain dimorphic fungi. The most detailed studies have been carried out with *M. racemosus* and *M. rouxii* (for a review see Sypherd et al., 1978; Moreno and Passeron, 1980; Cantore et al., 1980; Orlowski, 1980).

Both yeastlike and hyphal forms of *Mucor* can be grown in liquid culture under anaerobic conditions, depending on the gas used for sparging (Table 1). High rates of nitrogen sparging produce yeastlike cells, while low rates support hyphal growth. Aerobic growth occurs as hyphae and is supported by a wide range of sugars, amino acids, and organic acids.

Upon germination of sporangiospores under aerobic conditions the intracellular cAMP content increases to a maximum prior to germ tube appearance and thereafter declines (Table 1). Although it is often difficult to compare results from batch culture studies in which culture conditions and growth rates may differ, the substantial rise and fall in cellular cAMP concentration seen under aerobic conditions is not seen when nitrogen replaces air to generate anaerobiosis (Table 1). Carbon dioxide complicates a simple interpretation of the effect of anaerobiosis, since high cellular concentrations of cAMP are achieved when this gas replaces air. Furthermore, although the patterns of cellular cAMP change accompanying aerobic spore germination and hyphal growth in cultures of *M. genevensis* and *M. mucedo* are comparable to that described for *M. racemosus*, the concentrations of cAMP in these species are considerably lower. Comparatively small changes in cellular cAMP concentration may thus be related to morphogenesis.

The means by which morphogenesis in *Mucor racemosus* can be manipulated by varying gaseous environment, energy substrate, nitrogen source and endogenous cAMP content (via external dibutyryl-cAMP, dbcAMP) are summarized in Fig. 8.

Sporangiospores, when grown in the presence of cAMP analogs, exhibit one of two types of response which suggests, albeit thinly, a causal relationship between nitrogen metabolism and dimorphism (Wertman and Paznokas, 1981). cAMP, N^6-monobutyryl cAMP, 8-bromo-cAMP, or adenosine produced an increased synchrony and volume growth in spores already germinating. However, they did not initiate growth in the presence of a carbon source that would support growth, but was not itself capable of initiating germination. dbcAMP caused initiation of spore germination in the absence of substrate, and enabled germination of M-spores in substrates that otherwise would not. But inhibition of spore swelling, subsequent to the initiation of germination,

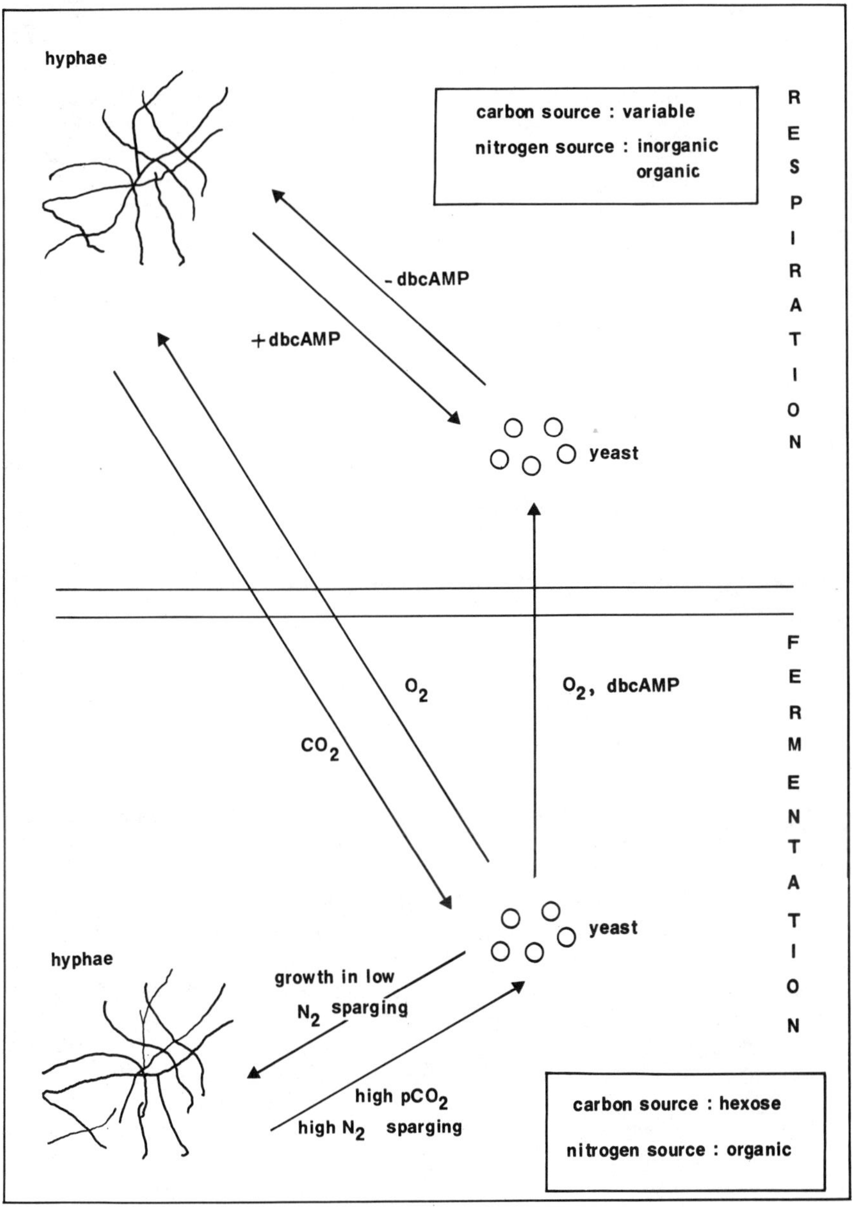

FIG. 8 Relationships between morphology, dbcAMP, and carbon metabolism in *M. racemosus*.

occurred with some substrates (cellobiose, maltose, galactose, leucine, lysine, 2 ketoglutarate). This could be eliminated by addition of alanine, aspartate, glutamate, or glutamine, but not by ammonium ions. These amino acids are present in high concentration in dormant spores of *Mucor*, as compared to the levels of leucine and lysine which are very low and which do not overcome the inhibition (Orlowski and Sypherd, 1978b). Furthermore, spores when germinated on glucose without an exogenous nitrogen source, mimic the dbcAMP-induced inhibition. Wertman and Paznokas (1981) propose that dibutyl cyclic AMP can interfere with ammonia assimilation but not with nitrogen metabolism from organic sources, and have suggested that this may be analogous to G_1 arrest in some mammalian cell lines caused by nutrient limitation; for example by ammonia, sulfate, phosphate, or biotin. Similar studies of the effects of nutrient limitation in *Mucor* may establish whether the observations above are specifically related to N-metabolism or whether they are part of a broader pleiomorphic effect. (See also Sec. III.D.6.)

In order to exclude the possibility that changes in cAMP are caused by nutritional adaptation, rather than being related to the yeast-to-hyphae (Y→H) transition, Orlowski and Ross (1981) have measured intracellular cAMP levels as a function of nutritional (supplemented versus minimal medium) and gaseous shifts ($N_2 \rightarrow$ air; $CO_2 \rightarrow$ air) both in wild type and in a nutritional mutant of *M. racemosus*. The mutant exhibits a morphogenetic response conditional on the exogenous methionine level (Y→H permissible only at high methionine concentrations). cAMP content changes with morphogenesis, but not in relation to the gaseous environment, growth rate, or rate of protein synthesis (Orlowski and Ross, 1981).

These studies are complemented by the observation that dbcAMP prevents the Y→H transition of *M. racemosus* under aerobic conditions (Paznokas and Sypherd, 1975) and can also cause H→Y reversal by budding of hyphae to yeast (Paveto et al., 1975). The effect is specific for dbcAMP and does not occur with other analogs of cAMP nor with cAMP plus butyrate (Wertman and Paznokas, 1981). Exogenous dbcAMP therefore seems to mimic intracellular cAMP in terms of effects on morphology. The inability of exogenous cAMP to prevent apical growth under aerobic conditions is undoubtedly a consequence of cell impermeability. Exogenous dbcAMP, used to induce yeast from hyphal forms, also lowers the rate of chitin plus chitosan synthesis in *M. racemosus* (Domek and Borgia, 1981; see also Sec. II.B.3 above).

Similar morphological findings have been reported for both *C. albicans* and *Histoplasma capsulatum*, although in these cases the relationship between morphology and intracellular cAMP concentration is not the same as that for *Mucor*. In *C. albicans*, a rise in cAMP accompanies germ tube emergence (Niimi et al., 1980), and cysteine (which suppresses germination) reverses this increase. Exogenous dbcAMP promotes germ tube emergence under conditions of nutrition and temperature which do not otherwise induce germination.

Likewise, in *H. capsulatum* cysteine induces development of yeast-like forms, with an attendant decrease in intracellular cAMP, whereas hyphal growth is characterized by higher cAMP levels (Maresca et al., 1977). The Y→H transition is also accomplished by increasing the temperature and is accompanied by an increase in both intra- and extracellular cAMP (Medoff et al., 1981; Sacco et al., 1981). These decline during the reverse H→Y transition.

2. *Adenylate Cyclase and cAMP Phosphodiesterase*

cAMP is formed by adenylate cyclase, while a specific phosphodiesterase causes hydrolysis of the phosphodiester linkage. The intracellular cAMP concentration may thus be dependent on the relative activities of these two enzymes. Both enzymes are detected in crude extracts of yeast and hyphal forms of the Mucorales (Paveto et al., 1975; Cantore et al., 1980; Orlowski, 1980). The adenylate cyclase is associated mainly with a membranous fraction in cell homogenates, while the phosphodiesterase is soluble. Attempts have been made to establish a link between the activity of the two enzymes and the cAMP concentrations observed during sporangiospore germination (Cantore et al., 1980; Orlowski, 1980). In both *M. genevensis* and *M. mucedo*, cyclase activity fluctuates during germination, with maximum activity coinciding with maximum intracellular cAMP concentrations (Orlowski, 1980). In comparison, phosphodiesterase activity remains fairly constant. For *M. rouxii*, variation in the activity of both enzymes occurs during germination (Cantore et al., 1980). The data do not conclusively correlate with the high cAMP levels observed prior to germination, and since extracellular cAMP levels continually increase throughout the course of germination, the *intracellular* concentration of the nucleotide is probably not determined solely by the relative activities of the two enzymes. A similar study in *Blastocladiella emersonii* supports this conclusion (Epstein and Silverman, 1978). Relatively high extracellular concentrations of cAMP are attained in cultures of *M. racemosus* (Paznokas et al., 1981).

In *H. capsulatum*, high concentrations of cAMP accumulate in the medium during the Y→H transition (Medoff et al., 1981) in a manner similar to the stepped pattern of intra- and extracellular cAMP accumulation in *Dictyostelium* (Gerisch et al., 1975). Wertman and Paznokas (1981) have discriminated between the effects of cAMP and dbcAMP during germination of *M. racemosus*. The former stimulates the rate of sporangiospore germination, while dbcAMP stimulates the initiation of spore germination. If, as is generally regarded, exogenous cAMP cannot freely enter the cells of spores, its effect may be mediated by surface receptors.

A cAMP-dependent protein kinase may control the activity of a soluble form of the phosphodiesterase in *M. rouxii* (Galvagno et al., 1979). The activity of the cyclase may be linked to membrane energization, as will now be discussed.

3. *Regulation of Adenylate Cyclase by Membrane Potential*

According to the chemiosmotic theory, membrane energization occurs when respiration or ATP hydrolysis is used to establish a transmembrane potential and/or transmembrane proton gradient. Recent findings in *Escherichia coli* and *Neurospora crassa* have shown that adenylate cyclase activity is regulated by the state of membrane energization.

The *E. coli* data show that both phosphorylation of sugar transport proteins of the membrane and depolarization of the membrane either by sugar-H^+ symport or by protonophores inhibits the cyclase (Peterkofsky and Gazdar, 1979). Conversely, in *Neurospora* (Pall, 1977; Trevillyan and Pall, 1979) depolarization of the membrane by glucose transport, amino acid uptake, ammonium ion uptake, addition of protonophores, or elevation of temperature leads to a rapid transient increase in the level of intracellular cAMP. The transport of glucose, several amino acids, and ammonium ions is known to occur in symport with protons (Slayman, 1977). In the presence of ATPase uncouplers and membrane antibiotics such as nystatin and phenethyl alcohol, a transient rapid increase in the cAMP levels occurs not only in *N. crassa*, but also in *S. cerevisiae* and *M. racemosus* (Trevillyan and Pall, 1979). The *crisp-1* mutant of *Neurospora*, which has very low cyclase activity, does not show the same increased cAMP levels as the wild type after nystatin treatment (Trevillyan and Pall, 1979). Germ tube induction in *C. albicans* (Shepherd et al., 1980) is also inhibited by nystatin and the antimembrane agent phenethyl alcohol. Dinitrophenol, an uncoupler of oxidative phosphorylation, is also effective, but cyanide and polyoxin D are not. Again, in this instance, the effects of dinitrophenol and nystatin may be rationalized on the basis of their ability to perturb membrane structure and thereby modify the activity of adenylate cyclase.

Orlowski and Ross (1981) have shown that phenethyl alcohol prevents the Y→H (N_2 → air) transition in *M. racemosus* (Terenzi and Storck, 1969), but that cAMP levels in yeast phase cells are typical of those obtained for aerobic hyphal cultures. In view of the pleiomorphic effects of this alcohol on membrane and cell wall organization in *Candida* (Pugh and Cawson, 1980), and the growth inhibitory effect it has on the Mucorales (Terenzi and Storck, 1969), it may be blocking morphogenesis at some point remote from the effects of cAMP.

Elevated temperatures increase the permeability of cell membranes to small ions. For pathogenic dimorphic fungi such as *Blastomyces dermatitides*, *Paracoccidioides brasiliensis*, and *Coccidioides immitis*, temperature is the most important external factor for H→Y conversion (for a review see Rippon, 1980). In the presence of all necessary nutrient factors, the dimorphic transition remains dependent on temperature. However, once the transition is initiated, temperature is no longer an essential factor. Pall's data (Pall, 1977; Trevillyan and Pall,

1979) suggest that the causative agent is cAMP and that membrane depolarization at high temperatures probably stimulates the cyclase activity.

The tendency of Mucorales to adopt yeastlike morphology at high hexose concentrations (Bartnicki-Garcia, 1968b; Rogers et al., 1974; Inderlied and Sypherd, 1978; Barrera and Corral, 1980) may be interpreted in the same manner. The development of yeast morphology by *M. racemosus* at high pCO_2 (Sypherd et al., 1978) may, if mediated by membrane potential, stem from the relatively high intracellular concentration of carbonic acid. At elevated concentrations, dissociation of this acid may neutralize the charge on the inner membrane, induce osmotic swelling because of the bicarbonate concentration, and cause greater penetration of the membrane by small ions. The effect of CO_2 and cations (in particular Ca^{2+}; see Kretsinger, 1977) on the osmotic properties of *Mucor* protoplasts may shed some light on this possibility.

4. Membrane Polarization and Cell Wall Biosynthesis

Trevillyan and Pall (1979) proposed that cAMP and/or membrane depolarization are responsible for changes in cell wall biosynthesis. The evidence for this is scant but based on the observation that exogenous cAMP increases cell wall biosynthesis in *Neurospora*, resulting in walls up to six times thicker than normal (Tu and Malhorta, 1977). dbcAMP also induces pinocytosis in spheroplasts of strains of *Neurospora* which are cell wall deficient, an effect caused also by inhibitors of cAMP phosphodiesterase (Trevithick and Galsworthy, 1977).

Harold and coworkers (Harold, 1977; Harold and Harold, 1980; Van Brunt and Harold, 1980), using *B. emersonii*, have examined the possibility that localized depolarization of the cell membrane is accompanied by a change in the orientation of cell wall synthesis. *Blastocladiella emersonii* has a life cycle which alternates between two distinct phases, the sessile vegetative cell and the motile but nongrowing zoospore. Upon germination to give the vegetative cell, the germ tube of the zoospore branches into a filamentous rhizoid attached to a large thallus containing up to 100 nuclei. The orientation of the thallus and the rhizoids is affected by the presence of ionophores (Van Brunt and Harold, 1980). When ionophore-coated fibers were placed in the growth medium, the orientation of the thallus was directed away from the fibers (Harold and Harold, 1980). These ionophores did not affect the general metabolism of the organism, and the authors therefore concluded that proton flux across the membrane was responsible for the directed growth. They also proposed that the proton motive force may be responsible for directing the intracellular flow of precursors and lytic and biosynthetic cell wall enzymes to sites of high membrane potential on the inner membrane surface. The intermediary relationship between cyclase activation, and proton and ion flux (e.g., Lazo et al., 1981) during dimorphism will no doubt receive greater attention in the future.

5. cAMP-Dependent Protein Kinases

If cAMP has some regulatory role during fungal dimorphism, then it must interact with other molecules, most probably proteins (see Fig. 7). cAMP-binding proteins have been reported present in a range of dimorphic fungi. Six proteins of 52,000-168,000 daltons have been described in *M. mucedo* (Orlowski, 1980), and two (51,000, 65,000) in *M. racemosus* (Orlowski, 1979). In *M. mucedo* two of the binding proteins appear during development. The most prominent binding protein (57,000 daltons) disappears during the Y→H transition (Orlowski, 1980).

In *M. genevensis*, the cAMP-binding proteins also decrease during the Y→H transition (Orlowski, 1979, 1980). Apparent molecular weights of 46,000-73,000 have been reported for these proteins. In *B. emersonii* the cAMP-binding proteins appear to be associated with protein kinase activity (Silverman, 1978), while Moreno and Passeron (1980) have reported the partial purification of a cAMP-dependent protein kinase from *M. rouxii*. Most of the characterized cAMP-dependent protein kinases consist of regulatory (R) and catalytic (C) subunits in a quaternary complex (R_2C_2) of low efficiency (Keely et al., 1975). Activation of the phosphotransferase activity by cAMP is thought to occur via dissociation of the C subunit from the holoenzyme after cAMP is bound to the R subunits. The *M. rouxii* protein kinase is fully dissociated only in the presence of high salt concentrations and cAMP (Moreno and Passeron, 1980). Removal of cAMP causes reassociation and formation of the holoenzyme. Hence during development of *Mucor* sporangiospores, the decline in intracellular cAMP immediately prior to germ tube emergence could be expected to inactivate the kinase by causing reassociation of the catalytic and regulatory subunits. The apparent molecular weight of the *M. rouxii* holoenzyme is 210,000; the R and C subunits have molecular weights of 80,000 and 40,000, respectively. These molecular weights are similar to the values of 85,000 and 35,000 for cAMP-binding proteins in *M. genevensis*, and the 73,000- and 39,000-dalton binding proteins in *M. mucedo* (Orlowski, 1980).

The target substrate of the *M. rouxii* protein kinase is not yet known, although the recently demonstrated cAMP dependence of a number of inducible catabolic and anabolic enzymes is suggestive. Galvagno et al. (1979), for example, proposed that a cAMP-dependent protein kinase may regulate a soluble form of cAMP-phosphodiesterase in *M. rouxii*. Crude preparations of cAMP-phosphodiesterase were stimulated up to twofold in the presence of Mg^{2+} ions, ATP and cAMP. The activation is apparently a catalytic process and is also observed after ion exchange and gel permeation chromatography of extracts containing phosphodiesterase. In the presence of ^{32}P-labeled ATP, label was incorporated into protein that co-chromatographed with phosphodiesterase activity. Only one form of the enzyme has been detected, with a K_m for cAMP of 1 μm, consistent with the intracellular concentrations of this nucleotide (Galvagno et al., 1979).

6. *cAMP and Nitrogen Metabolism*

In the original experiments which established a link between cAMP and dimorphism in the Mucorales, a complex medium, supplemented with yeast extract, bactopeptone, or casamino acids was used. Paznokas and coworkers (Wertman and Paznokas, 1981; Tripp and Paznokas, 1981a,b) have developed a minimal medium which supports the germination of spores of *M. racemosus* and also the vegetative growth of both hyphal and yeast forms. This has permitted the effects of cAMP on both carbon and nitrogen metabolism, and the interdependence between the two, to be evaluated, an approach which is impossible with supplemented medium. Spores prepared using minimal (M spores) or complex (C spores) medium sporulate when returned to either medium in liquid culture (Wertman and Paznokas, 1981). However, the M spores have a specific dependence on hexose and mannose for germination, whereas the C spores have a broader substrate tolerance.

In minimal medium C spores germinate with cellobiose as the carbon source (Fig. 9). Exogenous dbcAMP stimulates initial swelling, but prevents budding or germination of spores. The inhibition is overcome by glutamate. The observation that ammonia limitation in the presence of glucose also prevents germination has prompted Wertman and Paznokas (1981) to propose that dbcAMP can interfere with ammonia assimilation and that this may be an important factor in dimorphism. (See also Sec. III.A.1 above.)

Mucor racemosus possesses both NAD-dependent glutamate dehydrogenase and NADPH-dependent glutamate dehydrogenase (Peters and Sypherd, 1979). It also seems likely that ammonia fixation occurs via the glutamine synthetase-glutamate synthetase pathway (Sypherd et al., 1978). 2-Oxoglutarate dehydrogenase was not detected in *Mucor* extracts, but glutamate decarboxylase probably functions as summarized in Fig. 10 (O'Connell and Paznokas, 1980). A delicate balance between glutamate catabolism and the need for this amino acid for ammonia assimilation must exist. Peters and Sypherd (1979) have reported that NAD-dependent glutamate dehydrogenase, the catabolic enzyme, is under negative cAMP control, possibly mediated by a cAMP-dependent phosphorylation as reported for *Candida utilis* (Hemmings, 1978) and *S. cerevisiae* (Hemmings, 1980). In the presence of high dbcAMP, NAD-glutamate dehydrogenase activity would be decreased, possibly leading to a decrease in 2-oxoglutarate concentration. Glutamate levels would also fall if cAMP positively affected the aminobutyrate pathway of glutamate catabolism (Fig. 9). Together, these effects may lead to a restriction of ammonia assimilation, and hence explain the nutritional requirement of *M. racemosus* spores for glutamate in the presence of cAMP.

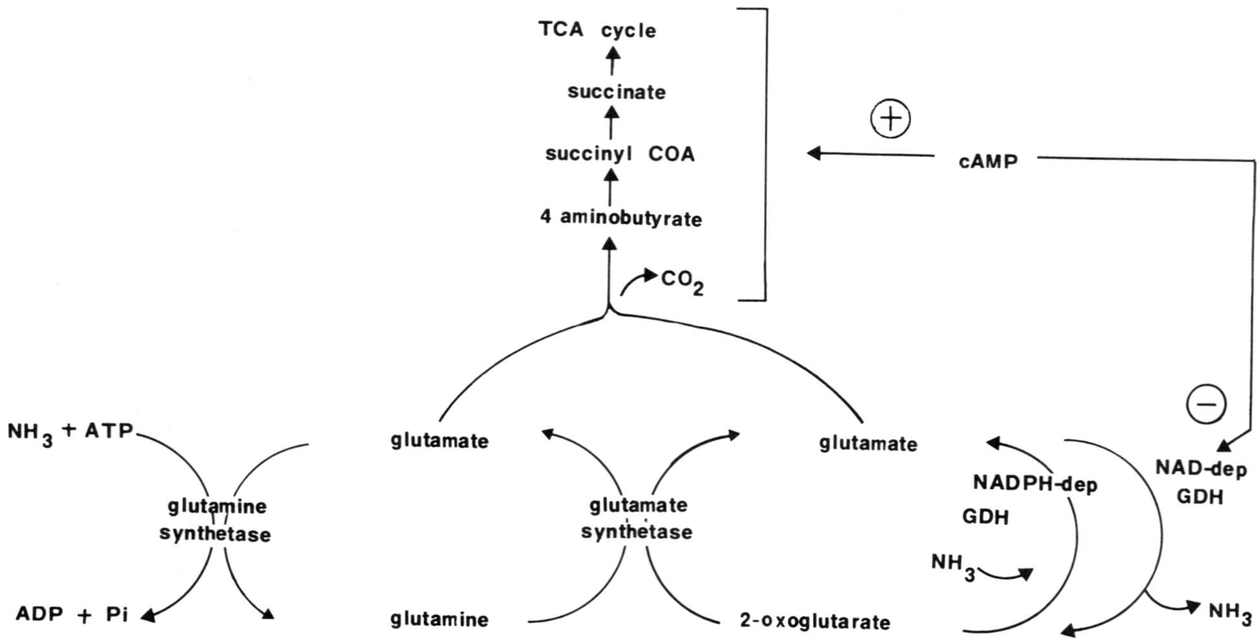

FIG. 9 The glutamate synthetase-glutamine synthetase scheme for ammonia assimilation and the aminobutyrate pathway (ornithine decarboxylase pathway) for glutamate metabolism.

7. *Polyamines*

Changes in the cellular concentration of polyamines have been related to a number of morphological and developmental changes in eukaryotic cells. During the Y→H (anaerobic → aerobic) transition in *M. racemosus* the intracellular cAMP level varies inversely with the endogenous level of putrescine (Peters and Sypherd, 1979), with the activity of ornithine decarboxylase (Inderlied et al., 1980), and with the intracellular S-adenosyl methionine concentration (Garcia et al., 1980) (Fig. 7). Ornithine decarboxylase is the initial enzyme in the biosynthetic pathway for polyamine synthesis, and S-adenosyl methionine acts as a methyl donor in the pathway. The increase in the activity of ornithine decarboxylase in *M. racemosus* is 50-fold or more. Exogenous putrescine prevents the enzyme increase but not the morphic change. The ornithine decarboxylase activity increase is dependent on protein synthesis, suggesting synthesis of new enzyme. However, in mammalian cell lines, inactivation of both ornithine decarboxylase and S-adenosyl methionine decarboxylase is mediated by a protein kinase (Insel and Fenno, 1978). Both mechanisms of regulation need to be considered in filamentous organisms.

M. Orlowski (1981a) has found that exogenous putrecine decreases the intracellular cAMP level in yeast forms of *M. racemosus* grown under carbon dioxide. The effect appears to be the result of a specific inhibition of adenylate cyclase, although the phosphodiesterase is also inhibited, albeit to a lesser extent.

A final comment on the polyamine-morphology relationship: Polyamines are known to be associated with changes in growth rate (Whitney and Morris, 1978) in a number of cell types. The increase in putrescine during the Y→H transition may be dependent on growth rate, rather than being directly related to morphogenesis. The analysis of the role of polyamines in transcriptional and translational events in fungi may require chemostatic control of the growth rates to clarify their part in dimorphism.

B. Catabolic Enzymes

Hyphal growth in members of the Mucorales is supported by a wide range of carbon sources and by ammonium salts as the sole source of nitrogen (Bartnicki-Garcia and Nickerson, 1962). Growth in the yeast mode requires the presence of a hexose and an organic nitrogen source (Bartnicki-Garcia, 1968b) and is characterized by a high rate of flux through the Embden-Meyerhof and pentose phosphate pathways, although fermentative metabolism is not essential (Inderlied and Sypherd, 1978).

In view of these differences, there must exist major physiological differences between the yeast and hyphal forms; these differences probably depend on differential expression of key enzymes in nitrogen and carbon metabolism. With this in mind, a number of workers have examined the activity of key enzymes of glucose metabolism in yeast and hyphal stages under different nutritional conditions.

1. *Pyruvate Kinase*

Three forms of pyruvate kinase have been detected in *M. rouxii* (Friendenthal et al., 1973). Type I isoenzyme predominates in the presence of glucose, while type II is present in the absence of glucose under aerobic conditions. Depending on the nature and the concentration of the sugar used for development of the aerobic hyphae, a hybrid form (type III) may also be detected.

Paznokas and Sypherd (1977) have provided evidence for the existence of two forms of pyruvate kinase in *M. racemosus*, but Hohn and Paznokas (1981) have identified five forms, two of which are immunologically distinct, while the other three appear to be hybrids. The fifth type of the enzyme is the sole form present in yeast cells grown anaerobically on glucose, but its presence does not correlate exclusively with yeast morphology. Another type is favored by growth under aerobic conditions.

While not providing special insight per se into fungal morphogenesis, the data provide an interesting comparison to recent studies in prokaryotic systems and a framework for future studies. Firstly, the pyruvate kinase activity in *Streptococcus lactis* is regulated by inorganic phosphate (Mason et al., 1981). At low glucose concentrations, inorganic phosphate increases in the cell, thereby inhibiting pyruvate kinase and causing the phosphoenol pyruvate concentration to rise dramatically. Secondly, the phosphoenol pyruvate-sugar phosphotransferase system is involved in the regulation of cell metabolism (Scholte and Postma, 1981). Both the rate of cAMP synthesis and metabolite transport (other than glucose) are affected by this system. The intracellular concentrations of metabolites in the filamentous fungi may also be able to signal the metabolic status of the cell to the transport carrier and adenylate cyclase in the plasma membrane. One of the pyruvate kinases may, for instance, function in a regulatory manner analogous to that seen in bacteria (Scholte and Postma, 1981) and interact either directly or indirectly with the adenylate cyclase. It is now possible with the advent of high-field NMR spectrometers to gather ^{13}C and ^{31}P spectra of intact cells in as little as 2 min (Schulman et al., 1979). Resonance lines attributable to inorganic phosphate pools, triose phosphates, and nucleotides can be detected. Application of this technique to the dynamics of dimorphism may enable correlation to be made between morphogenesis, inorganic phosphate levels, and the concentration of glycolytic intermediates such as phosphoenol pyruvate.

2. *Phosphofructokinase*

Phosphofructokinase is reportedly inhibited by ATP and citrate in *M. rouxii*; both types of inhibition are overcome by cAMP (Paveto and Passeron, 1977). Although the cellular ATP concentrations in yeast and hyphal forms are approximately the same, citrate concentration is considerably higher in hyphal forms than in yeast cells; citrate inhibition of the glycolytic pathway may therefore be greater in hyphal forms. The specific activity of the kinase is almost 50 times higher in yeast cells which have been grown under carbon dioxide on enriched medium. However, nutritional effects and the growth rate dependence of the enzyme activity have not been considered. No analysis of purified enzyme has yet been reported.

C. Growth Rate and Protein Synthesis

1. *Growth Rate and Rate of Protein Synthesis*

Studies with batch cultures of *M. rouxii* show that the Y→H transition is preceded and initially accompanied by a burst of protein synthesis (Orlowski and Sypherd, 1977, 1978). Rates of protein synthesis increased approximately fourfold upon exposure to inducing conditions and prior to hyphal germ tube appearance. This increase is due to a faster rate of amino acid addition to nascent peptide chains, and to an increase in the rate of ribosome movement along messenger RNA (Orlowski and Sypherd, 1977, 1978).

Theoretically, a change in the rate of initiation of translation can alter the relative proportions of gene products (Lodish, 1976). This arises because of differential mRNA ribosome binding constants. The temptation is to interpret the earlier data of Orlowski and Sypherd (1977, 1978) within this framework. However, since their experiments were conducted in batch cultures during which the nutritional status and the growth rates varied, it cannot be argued necessarily that morphogenesis is regulated by the rate of protein synthesis. Another alternative is that protein synthesis is a function simply of growth rate and not of development. This has been suggested in the case of baker's yeast (Boehlke and Friesen, 1975; Bonven and Gulløv, 1979) and has recently been proposed for *Mucor racemosus* (Orlowski and Ross, 1981). The step time of protein synthesis in *M. racemosus* correlates with the growth rate rather than with morphogenesis. In a Y→H (carbon dioxide → nitrogen) transition, the exponential growth rate increased from 0.18 to 0.43 h^{-1} and the rate of peptide chain elongation more than doubled. During a Y→H (nitrogen → air) transition, the growth rate remained almost unchanged (0.43 h^{-1} → 0.50 h^{-1}), as did the rate of peptide elongation. Orlowski (1981b) has found that in chemostat cultures of yeast phase *M. racemosus*, both the proportion of active ribosomes and the rate of polypeptide chain elongation increase with growth rate. Similarly, hyphal cultures exhibit the same phenomenon, but a much lower proportion of ribosomes are active in protein synthesis (Orlowski, 1981b). Taken together, the data indicate that translational control of gene expression in the manner proposed by Lodish (1976) cannot by itself account for dimorphism. However, it does not rule out translational control due to ribosomal modification.

Larsen and Sypherd (1979, 1980) have analyzed the degree of phosphorylation of the S_6 ribosomal protein in *M. racemosus*. The extent to which the three phosphorylation sites are occupied is generally correlated with higher rates of protein synthesis and intracellular levels of ATP, but not with morphogenesis (Larsen and Sypherd, 1980). It is not known whether any structural differences, including the degree of S_6 phosphorylation, exist between active and inactive ribosomes in hyphae. Higher concentrations of putrescine (see Sec. III.A.7 above) are present in hyphae as compared to yeast cells of *M. racemosus*. Polyamines may stimulate peptide chain elongation by stabilizing RNA-ribosome interactions (Abraham and Pihl, 1981) and by "ironing out" unfavorable conformations in messenger RNA (Jelenc and Kurland, 1979). A simple regulatory mechanism for the preferential synthesis of proteins from "morphic" messenger RNAs could therefore be based on the polyamine-stimulated translation of certain messenger RNA species within the cell (Fig. 7). Polyamines also stimulate RNA polymerase activity in other eukaryotic systems (Abraham and Pihl, 1981), and this may also provide a means for differential genetic expression during morphogenesis in fungi.

2. *"Morphic" Proteins*

Some general evidence for the existence of "morphic" proteins, that is, proteins that are present only in one or another of the morphological forms, has recently been presented.

Hiatt et al. (1980) analyzed the cellular proteins of *M. racemosus* during the Y→H transition, using two-dimensional gel electrophoresis. About 500 polypeptides were resolved, but relatively few of these – perhaps 10 – were specific to either morphology. The rate of synthesis of many peptides appeared to alter, but this could have been a consequence of the stimulation of growth rate (and protein synthesis) accompanying the Y→H change. None of the phase-specific peptides were functionally identified.

Similar studies have been undertaken with *C. albicans* (Manning and Mitchell, 1980), comparing one strain which produces only yeast cells at 24 and 37°C with another which grows as yeast at 24°C and as hyphae at higher temperatures. Those peptides which are characteristic of morphological changes and not of physiological adaptation to higher temperatures should thus be resolved. It appears that a number of yeast phase proteins (approximately 10) are present in hyphae in modified form (they are absorbed by anti-yeast phase immunoglobin) and that yeast phase cells contain many proteins not in the hyphae. However, further interpretation is difficult, because the strains examined had been isolated independently from clinical sources (and thus probably were not isogenic) and growth rate and nutrient differences between the cultures would have occurred. Moreover, these analyses apply only to cytosolic proteins. The most significant differences may occur in those proteins and enzymes which are membrane associated.

It is not clear whether any of the morphic proteins identified by Manning and Mitchell (1980) correspond to the morphic autoregulatory substance reported by Hazen and Cutler (1979) to be associated specifically with the yeast phase of *C. albicans*, or to the one protein shown by Dabrowa et al. (1970) to be specific to the yeast phase of *C. albicans*. An example of a phase-specific protein was described by Boguslawski et al. (1975): a 24,000-dalton protein present only in the mycelial cells of *Histoplasma capsulatum* that is capable of inhibiting RNA polymerase activity. However, no extension of this initial study appears to have been reported.

IV. GENETIC ANALYSIS OF DIMORPHISM

Genetic analysis of the processes and enzymes associated with dimorphism will greatly speed progress in understanding the fundamental molecular and structural features of dimorphism. Only some species of *Mucor* are heterothallic and thus susceptible to genetic analysis through sexual recombination. *Mucor racemosus* is in this class; *M. rouxii* is not.

Peters and Sypherd (1978) have described procedures for selecting nutritional and morphological mutants of *M. racemosus* by a freeze-thawing enrichment process which is in essence analogous to the penicillin enrichment method used with bacterial mutants.

Heterokaryon formation by fusion of protoplasts of auxotrophic strains of *M. racemosus* and analysis of prototrophic recombinants have been described by Genthner and Borgia (1978) and Lasker and Borgia (1980). In the latter case, the frequency of fusion was improved by using polyethylene glycol and calcium. About 5% of the input protoplasts then underwent fusion. At this level it becomes possible to select heterokaryons directly by morphological criteria, and thus to measure complementation between morphological mutants.

Genetic analysis of *C. albicans* has reached an interesting point. Olaiya and Sogin (1979) reported that when compared with a series of *S. cerevisiae* strains of varying ploidy (n to 4n), *C. albicans* strains contained DNA equivalent to the diploid of *S. cerevisiae*. Killing rates by ultraviolet irradiation and chemical mutagens supported this estimate. These workers suggested that *C. albicans* therefore exists normally as a diploid, a conclusion which would help explain the difficulty they encountered in isolating auxotrophic mutants of this organism. However, Saracheck et al. (1981) argued that comparative DNA contents as a measure of ploidy are questionable. These workers claimed, moreover, that auxotrophy can be induced in *C. albicans*, and the simple patterns of segregation of auxotrophic markers in progeny from fused spheroplast hybrids suggest that *C. albicans* normally exists as a haploid.

Olaiya et al. (1980) found that a series of strains of *C. albicans* unable to form germ tubes (and thus unable to develop the hyphal state) had half the DNA content of the parental strain, which is dimorphic. Increased killing rates when exposed to ultraviolet radiation or chemical mutagens also indicated a diminished amount of DNA in the derived strains. These data are consistent with recessive mutations controlling differentiation in *C. albicans*, but such a conclusion will need to await a decisive measure of ploidy.

Hybridization of *C. albicans* spheroplasts by fusion has been described recently by a number of laboratories (Poulter et al., 1981; Saracheck et al., 1981). This should permit parasexual genetic analysis and may also provide a means with which to study cytoplasm-nucleus interactions during differentiation.

REFERENCES

Abraham, A. K., and Pihl, A. (1981). Role of polyamines in macromolecular synthesis. *Trends Biochem. Sci. 6*:106-107.

Bainbridge, B. W., Valentine, B. P., and Markham, P. (1979). The use of temperature-sensitive mutants to study wall growth. In *Fungal Walls and Hyphal Growth.* J. H. Burnett and A. P. J. Trinci (Eds.). Cambridge University, Cambridge, pp. 71-91.

Barrera, C. R., and Corral, J. (1980). Effect of hexoses on the levels of pyruvate decarboxylase in *Mucor rouxii. J. Bacteriol. 142*:1029-1031.

Bartnicki-Garcia, S. (1968a). Cell wall chemistry, morphogenesis and taxonomy of fungi. *Annu. Rev. Microbiol. 22*:87-108.

Bartnicki-Garcia, S. (1968b). Control of dimorphism in *Mucor* by hexoses: Inhibition of hyphal morphogenesis. *J. Bacteriol. 96*:1586-1594.

Bartnicki-Garcia, S. (1973). Fundamental aspects of hyphal morphogenesis. *Symp. Soc. Gen. Microbiol. 23*:245-267.

Bartnicki-Garcia, S., and Lippman, E. (1969). Fungal morphogenesis: Cell wall construction in *Mucor rouxii. Science 165*:302-304.

Bartnicki-Garcia, S., and Lippman, E. (1973). The bursting tendency of hyphal tips of fungi: Presumptive evidence for a delicate balance between wall synthesis and wall lysis in apical growth. *J. Gen. Microbiol. 73*:487-500.

Bartnicki-Garcia, S., and Nickerson, W. J. (1962). Induction of yeast-like development in *Mucor* by carbon dioxide. *J. Bacteriol. 84*:829-840.

Bartnicki-Garcia, S., Ruiz-Herrera, J., and Bracker, C. E. (1979). Chitosomes and chitin synthesis. In *Fungal Walls and Hyphal Growth*. J. H. Burnett and A. P. J. Trinci (Eds.). Cambridge University, Cambridge, pp. 149-188.

Biely, P., Kovarik, J., and Bauer, S. (1973). Lysis of *Saccharomyces cerevisiae* with 2-deoxy-2-fluoro-D-glucose, an inhibitor of the cell wall glucan synthesis. *J. Bacteriol. 115*:1108-1120.

Boehlke, K. W., and Friesen, J. D. (1975). Cellular content of ribonucleic acid and protein in *Saccharomyces cerevisiae* as a function of exponential growth rate: Calculation of the apparent peptide chain elongation rate. *J. Bacteriol. 121*:429-433.

Boguslawski, G., Kobayashi, G. S., Schlessinger, D., and Medoff, G. (1975). Characterization of an inhibitor of ribonucleic acid polymerase from the mycelial phase of *Histoplasma capsulatum*. *J. Bacteriol. 122*:532-537.

Bonven, B., and Gulløv, K. (1979). Peptide chain elongation rate and ribosomal activity in *Saccharomyces cerevisiae* as a function of growth rate. *Mol. Gen. Genet. 170*:225-230.

Braun, P. C., and Calderone, R. A. (1978). Chitin synthesis in *Candida albicans*: Comparison of yeast and hyphal forms. *J. Bacteriol. 135*:1472-1477.

Brown, R. M. (1979). Biogenesis of natural polymer systems with special reference to cellulose assembly and deposition. In *Third Philip Morris Science Symposium*. E. M. Walk (Ed.). Philip Morris, New York, pp. 52-123.

Burnett, J. H. (1979). Aspects of the structure and growth of hyphal walls. In *Fungal Walls and Hyphal Growth*. J. H. Burnett and A. P. J. Trinci (Eds.). Cambridge University, Cambridge, pp. 1-25.

Burnett, J. H., and Trinci, A. P. J. (Eds.) (1979). *Fungal Walls and Hyphal Growth*. Cambridge University, Cambridge.

Cabib, E., Duran, A., and Bowers, B. (1979). Localized activation of chitin synthase in the initiation of yeast septum formation. In *Fungal Walls and Hyphal Growth*. J. H. Burnett and A. P. J. Trinci (Eds.). Cambridge University, Cambridge, pp. 189-201.

Cantore, M. L., Galvagno, M. A., and Passeron, S. (1980). Variations in the levels of cyclic nucleotide adenosine 3', 5'-monophosphate and in the activities of adenylate cyclase and cyclic adenosine 3', 5'-monophosphate phosphodiesterase during aerobic morphogenesis of *Mucor rouxii*. *Arch. Biochem. Biophys. 199*:312-320.

Chattaway, F. W., Holmes, M. R., and Barlow, A. J. E. (1968). Cell wall composition of the mycelial and blastospore forms of *Candida albiçans*. *J. Gen. Microbiol.* 5:367-376.

Chiew, Y. Y., Shepherd, M. G., and Sullivan, P. A. (1980). Regulation of chitin synthesis during germ-tube formation in *Candida albicans*. *Arch. Microbiol.* *125*:97-104.

Cihlar, R. L., and Sypherd, P. S. (1980). The organization of the ribosomal RNA genes in the fungus *Mucor racemosus*. *Nucl. Acids Res.* *8*:793-804.

Dabrowa, N., Howard, A., Landau, J. W., and Shechter, Y. (1970). Synthesis of nucleic acids and proteins in the dimorphic forms of *Candida albicans*. *Sabouraudia* *8*:163-169.

Domek, D. B., and Borgia, P. T. (1981). Changes in the rate of chitin-plus-chitosan synthesis accompany morphogenesis of *Mucor racemosus*. *J. Bacteriol.* *146*:945-951.

Domer, J. E. (1971). Monosaccharide and chitin content of cell walls of *Histoplasma capsulatum* and *Blastomyces dermatitidis*. *J. Bacteriol.* *107*:870-877.

Dow, J. M., and Rubery, P. H. (1977). Chemical fractionation of the cell walls of mycelial and yeast-like forms of *Mucor rouxii*: A comparative study of the polysaccharide and glycoprotein components. *J. Gen. Microbiol.* *99*:29-41.

Dow, J. M., and Villa, V. D. (1980). Oligoglucuronide production in *Mucor rouxii*: Evidence for a role for endohydrolases in hyphal extension. *J. Bacteriol.* *142*:939-944.

Duran, A., and Cabib, E. (1978). Solubilization and partial purification of yeast chitin synthetase. Confirmation of the zymogenic nature of the enzyme. *J. Biol. Chem.* *253*:4419-4425.

Duran, A., Bowers, B., and Cabib, E. (1975). Chitin synthase zymogen is attached to the yeast plasma membrane. *Proc. Nat. Acad. Sci. USA.* *72*:3952-3955.

Endo, A., Kakiki, K., and Misato, T. (1970). Mechanism of action of the antifungal agent polyoxin D. *J. Bacteriol.* *104*:189-196.

Epstein, P. M., and Silverman, P. M. (1978). Induction of cyclic AMP phosphodiesterase in *Blastocladiella emersonii* and its relation to cyclic AMP metabolism. *J. Bacteriol.* *135*:968-975.

Farkâs, V. (1979). Biosynthesis of cell walls of fungi. *Microbiol. Rev.* *43*:117-144.

Friendenthal, M., Roselino, E., and Passeron, S. (1973). Multiple molecular forms of pyruvate kinase from *Mucor rouxii*. Immunological relationship among the three isozymes and nutritional factors affecting the enzymatic pattern. *Eur. J. Biochem.* *35*:148-158.

Galvagno, M. A., Moreno, S., Cantore, M. L., and Passeron, S. (1979). Cyclic adenosine 3', 5'-monophosphate cyclase from *Mucor rouxii*: Regulation of enzymic activity by phosphorylation and dephosphorylation. *Biochem. Biophys. Res. Commun.* *89*:779-785.

Garcia, J. R., Hiatt, W. R., Peters, J., and Sypherd, P. S. (1980). S-adenosyl methionine levels and protein methylation during morphogenesis of *Mucor racemosus*. *J. Bacteriol.* *142*:196-201.

Genthner, F., and Borgia, P. T. (1978). Spheroplast fusion and heterokaryon formation in *Mucor racemosus*. *J. Bacteriol.* *134*:349-352.

Gerisch, G., Fromm, H., Huesgen, A., and Wick, U. (1975). Control of cell-contact sites by cyclic AMP pulses in differentiating *Dictyostelium* cells. *Nature* *255*:547-549.

Gooday, G. W. (1977). Biosynthesis of the fungal cell wall – Mechanisms and implications. *J. Gen. Microbiol.* *99*:1-11.

Gooday, G. W. (1978). The enzymology of hyphal growth. In *The Filamentous Fungi*, Vol. 3. J. E. Smith and D. R. Berry (Eds.). Edward Arnold, London, pp. 51-77.

Gooday, G. W. (1979). Chitin synthesis and differentiation in *Coprinus cinereus*. In *Fungal Walls and Hyphal Growth*. J. H. Burnett and A. P. J. Trinci (Eds.). Cambridge University, Cambridge, pp. 203-223.

Grove, S. N. (1978). The cytology of hyphal tip growth. In *The Filamentous Fungi*, Vol. 3. J. E. Smith and D. R. Berry (Eds.). Edward Arnold, London, pp. 28-50.

Haigler, D., Brown, R. M., Benziman, M. (1980). Calcofluor white ST alters the in vivo assembly of cellulose microfibrils. *Science* *210*:903-906.

Harold, F. M. (1977). Ion currents and physiological functions in microorganisms. *Annu. Rev. Microbiol.* *31*:181-203.

Harold, R. L., and Harold, F. M. (1980). Oriented growth of *Blastocladiella emersonii* in gradients of ionophores and inhibitors. *J. Bacteriol.* *144*:1159-1167.

Hasilik, A. (1974). Inactivation of chitin synthase in *Saccharomyces cerevisiae*. *Arch. Microbiol.* *101*:2905-2301.

Hazen, K. C., and Cutler, J. E. (1979). Autoregulation of germ tube formation by *Candida albicans*. *Infect. Immun.* *24*:661-666.

Hemmings, B. A. (1978). Phosphorylation of NAD-dependent glutamate dehydrogenase from yeast. *J. Biol. Chem.* *253*:5255-5258.

Hemmings, B. A. (1980). Phosphorylation and proteolysis regulate the NAD-dependent glutamate dehydrogenase from *Saccharomyces cerevisiae*. *FEBS Lett.* *122*:297-302.

Herth, W. (1980). Calcofluor white and Congo red inhibit chitin microfibril assembly of *Poterioochromonas stipitata*: Evidence for a gap between polymerization and microfibril formation. *J. Cell Biol.* *87*:442-450.

Hiatt, W. R., Inderlied, C. B., and Sypherd, P. S. (1980). Differential synthesis of polypeptides during morphogenesis of *Mucor*. *J. Bacteriol.* *141*:1350-1359.

Hohn, T. M., and Paznokas, J. L. (1981). Multiple forms of pyruvate kinase in *Mucor racemosus*. *Biochim. Biophys. Acta* (in press).

Inderlied, C. B., and Sypherd, P. S. (1978). Glucose metabolism and dimorphism in *Mucor*. *J. Bacteriol.* *133*:1282-1286.

Inderlied, C. B., Cihlar, R. L., and Sypherd, P. S. (1980). Regulation of ornithine decarboxylase during morphogenesis of *Mucor racemosus*. *J. Bacteriol.* *141*:699-706.

Insel, P. A., and Fenno, J. (1978). Cyclic AMP-dependent protein kinase mediates a cyclic AMP-stimulated decrease in ornithine and S-adenosyl methionine decarboxylase activities. *Proc. Nat. Acad. Sci. USA.* *75*:862-865.

Jan, Y. N. (1974). Properties and cellular localization of chitin synthetase in *Phycomyces blakesleeanus*. *J. Biol. Chem.* *249*:1973-1979.

Jelenc, C., and Kurland, C. G. (1979). Nucleoside triphosphate regeneration decreases with the frequency of translation errors. *Proc. Nat. Acad. Sci. USA.* *76*:3174-3178.

Jennings, D. M. (1979). Membrane transport and hyphal growth. In *Fungal Walls and Hyphal Growth*. J. H. Burnett and A. P. J. Trinci (Eds.). Cambridge University, Cambridge, pp. 279-294.

Katz, D., and Rosenberger, R. F. (1971). Lysis of an *Aspergillus nidulans* mutant blocked in chitin synthesis and its relation to wall assembly and wall metabolism. *Arch. Mikrobiol.* *80*:284-292.

Keely, S., Corbin, J., and Park, C. (1975). Regulation of adenosine 3', 5'-monophosphate-dependent protein kinase. Regulation of the heart enzyme by epinephrine, glucagon, insulin and 1-methyl-3-isobutylxanthine. *J. Biol. Chem.* *250*:4832-4840.

Kretsinger, R. H. (1977). Evolution of the informational role of calcium in eukaryotes. In *Calcium Binding Proteins and Calcium Function*. R. H. Wasserman (Ed.). Elsevier, New York, pp. 63-67.

Kuo, S. C., and Lampen, J. O. (1976). Tunicamycin inhibition of (^{3}H)-glucosamine incorporation into yeast glycoproteins: Binding of tunicamycin and interaction with phospholipids. *Arch. Biochem. Biophys.* *172*:574-581.

Larsen, A., and Sypherd, P. (1979). Ribosomal proteins in the dimorphic fungus, *Mucor racemosus*. *Molec. Gen. Genet.* *175*:99-109.

Larsen, A., and Sypherd, P. S. (1980). Physiological control of phosphorylation of ribosomal protein S6 in *Mucor racemosus*. *J. Bacteriol.* *141*:20-25.

Lasker, B. A., and Borgia, P. T. (1980). High-frequency heterokaryon formation by *Mucor racemosus*. *J. Bacteriol.* *141*:565-569.

Lazo, P. S., Barros, F., de la Peña, P., and Ramos, S. (1981). Ion gradients as candidates for transmembrane signalling. *Trends Biochem. Sci.* *6*:83-86.

Lodish, H. F. (1976). Translational control of protein synthesis. *Annu. Rev. Biochem.* *45*:39-72.

Lopez-Romero, E., Ruiz-Herrera, T., and Bartnicki-Garcia, S. (1978). Purification of inhibitory protein of chitin synthetase from *Mucor rouxii*. *Biochim. Biophys. Acta* *525*:338-345.

Maeda, T., Abe, H., Kakiki, K., and Misato, T. (1970). Studies on the mode of actions of the organophosphorus fungicide, kitizin. *Agric. Biol. Chem. Jap.* *34*:700-709.

Mahadevan, P. R., and Tatum, E. L. (1965). Relationship of the major constituents of the *Neurospora crassa* cell wall to wild type and colonial morphology. *J. Bacteriol.* *90*:1073-1081.

Manning, M., and Mitchell, T. G. (1980). Morphogenesis of *Candida albicans* and cytoplasmic proteins associated with differences in morphology, strain or temperature. *J. Bacteriol.* *144*:258-273.

Maresca, B., Medoff, F., Schlessinger, D., and Kobayashi, G. S. (1977). Regulation of dimorphism in the pathogenic fungus *Histoplasma capsulatum*. *Nature* *266*:447-448.

Mason, P. W., Carbone, D. P., Cushman, R. A., and Waggoner, A. S. (1981). The importance of inorganic phosphate in regulation to energy metabolism of *Streptoccus lactis*. *J. Biol. Chem.* *256*: 1861-1866.

Medoff, J., Jacobson, E., and Medoff, G. (1981). Regulation of dimorphism in *Histoplasma capsulatum* by cyclic adenosine 3', 5'-monophosphate. *J. Bacteriol.* *145*:1452-1455.

Mishra, N. C., and Tatum, E. L. (1972). Effect of L-sorbose on polysaccharide synthetases of *Neurospora crassa*. *Proc. Nat. Acad. Sci. USA.* *69*:313-317.

Molano, J., Bowers, F., and Cabib, E. (1980). Distribution of chitin in the yeast (*Saccharomyces cerevisiae*) cell wall: An ultrastructural and chemical study. *J. Cell Biol.* *85*:199-212.

Moreno, S., and Passeron, S. (1980). Further studies on cyclic adenosine 3':5'-monophosphate protein kinase from dimorphic fungus *Mucor rouxii*. *Arch. Biochem. Biophys.* *199*:321-330.

Niimi, M., Niimi, K., Tokanaga, J., and Nakayama, H. (1980). Changes in cyclic nucleotide levels and dimorphic transition in *Candida albicans*. *J. Bacteriol.* *142*:1010-1014.

Nolan, R. A., and Bal, A. K. (1974). Cellulase localization in hyphae of *Achlya bisexualis*. *J. Bacteriol.* *117*:840-843.

O'Connell, B. T., and Paznokas, J. L. (1980). Glyoxylate cycle in *Mucor racemosus*. *J. Bacteriol.* *143*:416-421.

Olaiya, A. F., and Sogin, S. J. (1979). Ploidy determination of *Candida albicans*. *J. Bacteriol.* *140*:1043-1049.

Olaiya, A. F., Steed, J. R., and Sogin, S. J. (1980). Deoxyribonucleic acid-deficient strains of *Candida albicans*. *J. Bacteriol.* *141*:1284-1290.

Orlowski, M. (1979). Changing pattern of cyclic AMP-binding proteins during germination of *Mucor racemosus* sporangiospores. *Biochem. J.* *182*:547-554.

Orlowski, M. (1980). Cyclic adenosine 3', 5'-monophosphate and germination of sporangiospores from the fungus *Mucor*. *Arch. Microbiol.* *126*:133-140.

Orlowski, M. (1981a). Regulation of internal cyclic AMP levels by polyamines in the fungus *Mucor*. *Abstracts Annu. Meet. Amer. Soc. Microbiol.* *I-154*:112.

Orlowski, M. (1981b). Growth-rate-dependent adjustment of ribosome function in the fungus *Mucor racemosus*. *Biochem. J.* *196*:403-410.

Orlowski, M., and Ross, J. F. (1981). Relationship of internal cyclic AMP levels, rates of protein synthesis and *Mucor* dimorphism. *Arch. Microbiol.* (in press).

Orlowski, M., and Sypherd, P. S. (1977). Protein synthesis during morphogenesis of *Mucor racemosus*. *J. Bacteriol.* *132*:209-218.

Orlowski, M., and Sypherd, P. S. (1978). Regulation of translation rate during morphogenesis in the fungus *Mucor*. *Biochemistry 17*: 569-575.

Pall, M. L. (1977). Cyclic AMP and plasma membrane potential in *Neurospora crassa*. *J. Biol. Chem.* *252*:7146-7150.

Paveto, C., and Passeron, S. (1977). Some kinetic properties of *Mucor rouxii* phosphofructokinase. Effect of cyclic adenosine 3', 5'-monophosphate. *Arch. Biochem. Biophys.* *178*:1-7.

Paveto, C., Epstein, A., and Passeron, S. (1975). Studies on cyclic adenosine 3', 5'-monophosphate levels, adenylate cyclase and phosphodiesterase activities in the dimorphic fungus *Mucor rouxii*. *Arch. Biochem. Biophys.* *169*:449-457.

Paznokas, J. L., and Sypherd, P. S. (1975). Respiratory capacity, cyclic adenosine 3', 5'-monophosphate and morphogenesis of *Mucor racemosus*. *J. Bacteriol.* *124*:134-139.

Paznokas, J. L., and Sypherd, P. S. (1977). Pyruvate kinase isozymes of *Mucor racemosus*: Control of synthesis by glucose. *J. Bacteriol.* *130*:661-666.

Paznokas, J. L., Tripp, M. L., and Wertman, K. F. (1981). Dibutyryl cyclic 3', 5'-AMP resistant mutants of *Mucor racemosus*. *Exp. Mycol.* (in press).

Peterkofsky, A., and Gazdar, C. (1979). *Escherichia coli* adenylate cyclase complex: Regulation by the proton electrochemical gradient. *Proc. Nat. Acad. Sci. USA.* *76*:1099-1103.

Peters, J., and Sypherd, P. S. (1978). Enrichment of mutants of *Mucor racemosus* by differential freeze-killing. *J. Gen. Microbiol.* *105*:77-81.

Peters, J., and Sypherd, P. S. (1979). Morphology associated expression of nicotinamide adenine dinucleotide dependent glutamate dehydrogenase in *Mucor racemosus*. *J. Bacteriol.* *137*:1134-1139.

Poulter, R., Jeffrey, K., Hubbard, M. J., Shepherd, M. G., and Sullivan, P. A. (1981). Parasexual genetic analysis of *Candida albicans* by spheroplast fusion. *J. Bacteriol.* *146*:833-840.

Pugh, D., and Cawson, R. A. (1980). Calcium, sequestering agents and nystatin – Interactions on cell wall morphology and fungistasis of *Candida albicans*. *Sabouraudia* *18*:157-159.

Reissig, J. L. (1974). Decoding of regulatory signals at the microbial surface. *Curr. Top. Microbiol. Immunol. 67*:43-96.
Rippon, J. W. (1980). Dimorphism in pathogenic fungi. *CRC Crit. Rev. Microbiol. 8*:49-97.
Rizvi, S. R. H., and Robertson, N. F. (1965). Apical disintegration of hyphae of *Neurospora crassa* as a response to L-sorbose. *Trans. Br. Mycol. Soc. 48*:468-477.
Roelofsen, P. A. (1965). Ultrastructure of the wall in growing cells and its relation to the direction of growth. *Adv. Bot. Res. 2*:67-149.
Rogers, P. J., Clark-Wlaker, G. D., and Stewart, P. R. (1974). Effects of oxygen and glucose on energy metabolism and dimorphism of *Mucor genevensis* grown in continuous culture: Reversibility of yeast-mycelium conversion. *J. Bacteriol. 119*:282-293.
Rosenberger, R. F. (1979). Endogenous lytic enzymes and wall metabolism. In *Fungal Walls and Hyphal Growth.* J. H. Burnett and A. P. J. Trinci (Eds.). Cambridge University, Cambridge, pp. 265-277.
Ruiz-Herrera, J., and Bartnicki-Garcia, S. (1974). Synthesis of cell wall microfibrils in vitro by a "soluble" chitin synthetase from *Mucor rouxii. Science 186*:357-359.
Ruiz-Herrera, J., and Bartnicki-Garcia, S. (1976). Proteolytic activation and inactivation of chitin synthetase from *Mucor rouxii. J. Gen. Microbiol. 97*:241-249.
Ruiz-Herrera, J., Sing, V. O., Van der Woude, W. J., and Bartnicki-Garcia, S. (1975). Microfibril assembly by granules of chitin synthetase. *Proc. Nat. Acad. Sci. USA. 72*:2706-2710.
Ruiz-Herrera, J., Bartnicki-Garcia, J. S., and Bracker, C. E. (1980). Dissociations of chitosomes by digitonin into 16S subunits with chitin synthetase (EC 2.4.1.16) activity. *Biochim. Biophys. Acta 629*:201-216.
Sacco, M., Maresca, B., Kumar, B. V., Kobayashi, G. S., and Medoff, G. (1981). Temperature and cyclic nucleotide-induced phase transitions of *Histoplasma capsulatum. J. Bacteriol. 146*:117-120.
Saracheck, A., Rhoads, D. D., and Schwarzhoff, R. H. (1981). Hybridization of *Candida albicans* through fusion protoplasts. *Arch. Microbiol. 129*:1-8.
Scholte, B. J., and Postma, P. W. (1981). Competition between two pathways for sugar uptake by the phosphoenolpyruvate-dependent sugar phosphotransference system in *Salmonella typhimurium. Eur. J. Biochem. 114*:51-58.
Schulman, R. G., Brown, T. R., Urgurbil, K., Ogawa, S., Cohen, S. M., and den Hollander, J. A. (1979). Cellular application of ^{31}P and ^{13}C nuclear magnetic resonance. *Science 205*:160-166.
Shepherd, M. G., Chiew, Y. Y., Ram, S. P., and Sullivan, P. A. (1980). Germ tube induction in *Candida albicans. Can. J. Microbiol. 26*:21-26.

Shimoda, C., and Yanagishima, N. (1971). Role of cell wall-degrading enzymes in auxin-induced cell expansion in yeast. *Physiol. Plant 24*:46-50.

Silverman, P. M. (1978). Cyclic AMP binding proteins and cyclic AMP-dependent protein kinase from *Blastocladiella emersonii*. *J. Bacteriol. 135*:976-980.

Simonetti, N., Strippoli, V., and Cassone, A. (1974). Yeast mycelial conversion induced by N-acetyl-D-glucosamine in *Candida albicans*. *Nature 250*:344-346.

Slayman, C. L. (1977). Energetics and control of transport in *Neurospora*. In *Water Relations in Membrane Transport in Plants and Animals*. T. K. Hodges, A. Kleinzeller, and S. G. Schultz (Eds.). Academic, New York, pp. 69-86.

Smith, J. E., and Berry, D. R. (Eds.) (1978). Developmental mycology. *The Filamentous Fungi*. Vol. 3. Edward Arnold, London.

Stewart, P. R., and Rogers, P. J. (1978). Fungal dimorphism: A particular expression of cell wall morphogenesis. In *The Filamentous Fungi*. Vol. 3. J. E. Smith and D. R. Berry (Eds.). Edward Arnold, London, pp. 164-196.

Sypherd, P. S., Borgia, P. T., and Paznokas, J. L. (1978). Biochemistry of dimorphism in the fungus *Mucor*. *Adv. Microbiol. Physiol. 18*:67-104.

Terenzi, H. F., and Storck, R. (1969). Stimulation of dimorphism in *Mucor rouxii*. *J. Bacteriol. 92*:1236-1244.

Trevillyan, J. M., and Pall, M. L. (1979). Control of cyclic adenosine 3', 5'-monophosphate levels by depolarizing agents in fungi. *J. Bacteriol. 138*:397-403.

Trevithick, J. R., and Glasworthy, P. R. (1977). Morphology of *slime* variants of *Neurospora crassa* growing on glass surfaces in liquid medium 1. under normal conditions and 2. in the presence of inhibitors. *Arch. Microbiol. 115*:109-118.

Tripp, M. L., and Paznokas, J. L. (1981a). Relationship between sporulation medium and germination ability of *Mucor racemosus* sporangiospores. *Exp. Mycol. 127*:35-43.

Tripp, M. L., and Paznokas, J. L. (1981b). Glucose initiated germination of *Mucor racemosus* sporangiospores. *J. Gen. Microbiol. 128*: 477-483.

Tu, J. C., and Malhorta, S. K. (1977). The significance of cAMP induced alterations in the cellular structure of *Phycomyces*. *Can. J. Microbiol. 23*:378-388.

Van Brunt, J., and Harold, F. M. (1980). Ionic control of germination of *Blastocladiella emersonii* zoospores. *J. Bacteriol. 141*:735-744.

Waldron, C., and Roberts, C. F. (1974). Cold-sensitive mutants in *Aspergillus nidulans*. I. Isolation and general characterization. *Mol. Gen. Genet. 134*:99-113.
Wertman, K. F., and Paznokas, J. L. (1981). Effects of cyclic nucleotides upon the germination of *Mucor racemosus* sporangiospores. *Exp. Mycol.* 5:314-322.
Whitney, P. A., and Morris, D. R. (1978). Polyamine auxotrophs of *Saccharomyces cerevisiae*. *J. Bacteriol. 134*:214-220.

11

The Hyphal Tip

GRAHAM W. GOODAY *Marischal College, University of Aberdeen, Aberdeen, Scotland*

I. INTRODUCTION

The key to hyphal growth lies at the tip. The fungal mycelium is visible to the naked eye as ramifications of individual hyphae, and each hypha has been defined, fixed in time and space, by the activities of its apex. Herein lies the fascination, but also the frustration, of investigating a tiny ordered part of a much larger whole. Thus, as discussed later, the bulk of the hyphal wall is laid down within the apical 1 μm dome; but this constitutes only a very small fraction of the biomass of the mycelium, much less than 1%, as Bull and Trinci (1977) list hyphal growth units (G, the length of hypha per tip for undifferentiated mycelium) for a range of fungi, and the average of 16 species is well over

100 μm. So the traditional techniques of biochemistry, of homogenization of uniform tissue or cells, liver, yeast, or bacteria, have had to be used very circumspectly in the investigation of hyphal growth. Likewise, the use of the wide range of mutants in the fungi has proved disappointing, as so many of the morphological mutants are pleiotropic in nature (Brody, 1973; Wright,1975); the use of inhibitors has also been disappointing, as their primary sites of action are often in doubt and are clouded by the widening ripples of their secondary and tertiary effects.

Rather, it is the techniques of microautoradiography and cytochemistry, pioneered at the light microscopic level by Zalokar (1959,1965) and which have now been developed for electron microscopic preparations, that have yielded clearer concepts of the mechanisms of apical growth, and it will be through precision measurements of vectorial activities of individual hyphae, such as those using the vibrating probe electrode by Stump et al. (1980), that our understanding will be advanced. Meanwhile, the technique of freeze substitution, particularly with the high-voltage electron microscope, will provide much clearer images of the cytology of apical growth.

II. PROPERTIES OF THE HYPHAL TIP

A. Behavior of the Hyphal Tip

1. *Chemotropism*

Chemotropism to nutrients, that is, the oriented growth of hyphae in gradients of nutrients, despite some belief to the contrary, has not been described for any member of the Zygomycotina, Ascomycotina, Basidiomycotina, or Deuteromycotina (Gooday, 1975). Thus Stadler (1952) summarized many experiments with *Rhizopus nigricans* and other fungi: "Outside the staling reaction, no true tropic response has been detected for the large number of compounds and mixtures tested." He concluded (Stadler, 1952,1953) that oriented growths that had been observed by previous workers were in fact the result of autotropisms (see later). It does occur, however, in members of the Mastigomycotina, for example, with hyphae of *Achlya bisexualis* growing toward sources of casein hydrolysate and amino acids (Musgrave et al. 1977), those of *Saprolegnia ferax*, *Saprolegnia mixta*, and *Achlya polyandra* growing toward sources of casein hydrolysate and malt extract (Fischer and Werner, 1955), and rhizoids of *Blastocladiella emersonii* growing toward sources of an amino acid mixture and of inorganic phosphate (Harold and Harold, 1980).

Under carefully controlled conditions, many fungal spores, when placed in pairs, germinate characteristically so that the germ tubes, and resultant vegetative hyphae, grow away from each other (Robinson, 1973a,1980; Robinson et al., 1968; Trinci et al., 1979). This is the phenomenon of autotropism.

Again under carefully controlled conditions, many fungal spores and hyphae when in a system such as one that is covered by a perforated plastic cover slip, will grow toward the air. Robinson (1973a,b,c) and Robinson and Griffith (1977) interpreted this as a positive chemotropism toward an optimum oxygen concentration, and interpreted the autotropisms described earlier as being chemotropisms toward oxygen; that is, each growing hyphal tip depletes the oxygen around itself and so causes neighboring tips to grow away from it toward higher oxygen concentrations.

True positive autotropisms, that is, very specific directed growths of hyphae toward others of the same species, are discussed in detail by Gooday (1975) and Carlile and Gooday (1978). These include those resulting in the commonly observed vegetative hyphal fusions in members of the Ascomycotina and Basidiomycotina, and those hormone-mediated attractions resulting in sexual fusions (Gooday, 1983).

2. *Spiral Growth*

When grown on agar media of very weak nutrient concentration, vegetative mycelia (particularly in young colonies) of many fungi exhibit spiral growth (Madelin et al., 1978; Trinci et al., 1979). Thus, of 157 isolates investigated by Madelin et al. (1978), 21 showed clockwise spiralling (looking down on the plate), and 39 anticlockwise spiralling. There was no taxonomic significance, so that of the members of Zygomycotina tested, 5 were clockwise, 13 anticlockwise, and in 19 cases spiralling was not detected.

Madelin et al. (1978) investigated the spiral growth of *Sordaria fimicola* in more detail and showed that it was the result of an interaction between the hyphal tips and the substratum (and not with extrinsic factors) by, among other experiments, allowing hyphae to grow from the upper surface of the agar plate onto the agar-coated underside of a cover slip supported at an angle to it. The direction of spiralling changed to the viewer, but remained clockwise with respect to the surface.

Beever (1980) described mutants of *Neurospora crassa* carrying the gene which he designated *coil-1* (linkage group IV R), with the phenotype of very pronounced clockwise spiralling. This spiralling was unaffected by light and temperature. The linear growth of the mutant strain J131 was indistinguishable from wild type, but the radius of spiral growth under identical conditions was reduced from about 1 mm to about 0.2 mm, so that complete circles were frequently formed. As discussed by Madelin et al. (1978), Trinci et al. (1979), and Beever

(1980), spiral growth must result from axial rotation of the hyphal tip, which in turn may be caused by an interaction between wall growth and spiral structures in the wall. Such spiral structures have yet to be described and may be difficult to find, as Madelin et al. (1978) calculated the required angle of lateral displacement of the apex over the extension zone of tips of *S. fimicala* to be only 0.7°, resulting in a rate of angular rotation of 500° mm^{-1}. Middlebrook and Preston (1952a,b) discussed the relationship between spirals and microfibrils and spiral growth, in particular with respect to the sporangiophore of *Phycomyces blakesleeanus*, and concluded that the microfibrils themselves are not necessarily the hypothetical spiral structure in the growth zone that gives rise to the spiral growth.

B. Dynamics of the Hyphal Tip

If you have never done so, look through a microscope at a leading hyphal tip of a fast-growing fungus such as *N. crassa*. As it forges its way across your field of view, you can contemplate the rate of deposition of material at the tip, the rate of supapical synthesis necessary to sustain this, and the difficulties of biochemically dissecting this dynamic system.

The growth of individual hyphae is analyzed in detail in Chapter 12, and so suffice it to say here that a hypha at the edge of a mature colony grows at a constant rate, at a constant diameter, and (except as discussed in the previous section) in a straight line. Deposition of wall material is strongly concentrated in the apical dome of the extension zone of the hypha, as shown by autoradiography (Gooday, 1971). Current models of hyphal tip growth (Prosser, 1979; Gooday and Trinci, 1980) view the apical dome as being composed of elastic unset material, which via the extension zone becomes transformed to the inelastic set material of the cylindrical wall of the hypha.

The hyphal tip is burst readily by hypotonic (Robertson, 1958,1965) and acidic solutions (Bartnicki-Garcia, 1973; Bartnicki-Garcia and Lippman, 1972; Dow and Rubery, 1975). Bartnicki-Garcia and Lippman (1972) interpreted this as the result of the low pH affecting the balance between the cell-regulated synthetic and lytic processes in the wall, whereas Dow and Rubery (1975), finding that calcium ions could antagonize the effect of acid on hyphae of *Mucor rouxii*, suggested that the bursting is a result of a physical response of matrix polymers in the wall, especially the polyglucuran mucoran.

C. Establishment and Maintenance of Polarity

From the isotropic fungal spore, polarity is established during germination to result in the outgrowth of the germ tube from a particular site (Bartnicki-Garcia and Lippman, 1977). This site may be affected by the presence of neighboring spores by negative autotropism, with, for example, spore pairs of *Geotrichum candidum* germinating at opposite ends away from each other (Robinson, 1973a,1980). Cytologically, this polarity is first manifest as an aggregation of vesicles at the site of germ tube emergence, and once established, the polarity of the hypha is maintained, being characterized by the transport and accumulation of these apical vesicles (discussed in detail later and in Chapter 11) and in the arrangement of organelles and components of the cytoskeleton (see later).

But what is the primary mechanism responsible for the polarity? Recent years have brought about an increasing awareness of the role of ionic currents generated within individual cells in the control of many of the cell's activities (Harold, 1977,1982; Jaffe, 1979; Jaffe and Nuccitelli, 1977). As Harold (1977) pointed out, these currents can simultaneously serve two functions; they can perform work while providing vectorial information such as the localization of the direction of growth. The aqueous medium of the cytosol of the hypha can then be seen as the ion conductor allowing rapid and directed power transmission from subapical regions to the apex, with electrophoretic movement of components, including the apical vesicles (see later). Such a suggestion has been mooted by Jaffe (1968), Bartnicki-Garcia (1973), and Jennings (1979). What experimental evidence can be brought to bear on it?

Two systems analogous to hyphal growth have been investigated by Jaffe and colleagues. Both are for cells considerably larger than those of fungi and are therefore more amenable to experimentation. The fertilized eggs of the brown algae, *Fucus* and *Pelvetia* species, about 100 μm in diameter, elongate at about 0.05 μm/min to give the rhizoid of the developing embryo. Nuccitelli and Jaffe (1974,1975,1976) have used their vibrating probe to investigate the current density around single developing eggs. The tip of the probe is a platinum ball, 10-30 μm in diameter, vibrating at several hundred times per second between two points 30 μm apart in the aqueous medium around the cell. Potential differences of 0.1 μV or less and equivalent current densities as low as 1 nA/cm^2 are then detectable. Numerous similar measurements showed that current entered the embryo's growing tip. This current was composed of two components: a steady one of about 1 $\mu A/cm^2$, which would maintain the polarization of growth, and a pulse current, reaching about 10 $\mu A/cm^2$, possibly regulating the turgor pressure. Weisenseel et al. (1975) have used the vibrating probe with germinating lily pollen grains. The pollen tubes elongate at about 8 μm/min, with cytoplasmic streaming of about 50 μm/min. As for the fucoid eggs, the authors detected a steady current of about 100 pA entering the growing pollen tube and leaving the pollen grain, and for some tubes that had elongated to more than 1 mm the authors also detected pulsating current spikes. By plotting currents around ungerminated grains, they could predict the site of tube emergence at the point of densest inward current. This steady current flow, once established, always continued for as long as growth continued, and growth was never seen without current. Experiments involving changes of the ionic medium indicated that most of the inward current was carried by potassium ions, entering uniformly along the tube, but Weisenseel and Jaffe (1976) suggested that the outward current from the grain was largely carried by protons. Jaffe (1979) and Harold (1980), however, placed much emphasis on the possible role of currents of calcium ions in regulating the directed growth, as, for example, autoradiography has shown concentrations of $^{45}Ca^{2+}$ at the growing tips for both fucoid embryos and pollen tubes.

The only equivalent published study to date using the vibrating probe with fungal cells is that of Stump et al. (1980), who have investigated vegetative and sporulating cells of the water mold *Blastocladiella emersonii*. The vegetative cells consist of thalli, 50-100 μm long, with basal tufts of equally long rhizoids. A current, estimated to be of the order of 1 $\mu A/cm^2$, was found that entered the rhizoids and left over the surface of the thalli. Experiments varying the ionic composition suggested that none of the common inorganic ions, that is, K^+,Na^+, Mg^{2+},Cl^-,PO_4^{3-},SO_4^{2-}, carried the bulk of the current. Cells in a medium lacking calcium quickly vacuolated and gave erractic electrical responses. The authors were left with the hypothesis that much of the current developed by the vegetative cells was carried by protons.

Vegetative cells of *B. emersonii* sporulate spontaneously in older cultures or can be induced to sporulate, for example, by replacement of the growth medium with buffered $CaCl_2$. The process, culminating in the discharge of zoospores (see Chapter 8), initially involves retraction of cytoplasm from the rhizoid and formation of a baseplate and an apical discharge papilla. Both spontaneous and induced sporulation produced a reversal of current flow from that in vegetative cells; that is, current entered at the discharge papilla and thallus body and left through the rhizoid. Stump et al. (1980) suggested that at least some of this reversed current is carried by calcium ions, as (1) Ca^{2+} is required for sporulation (Soll et al., 1969), (2) Ca^{2+} and protons are the only plausible ions carrying current in their system, and (3) $^{45}Ca^{2+}$ is taken up during sporulation at a rate comparable to that of the inward current. After further experimentation, however, F. M. Harold (personal communication) now seriously questions this role for calcium ions.

Further aspects of the ionic regulation of development in *B. emersonii* have been investigated by Van Brunt and Harold (1980). Thus the induction of germination by KCl was correlated with depolarization of the plasma membrane. The zoospores actively accumulated K^+ during germination and expelled Na^+. Ca^{2+} was also taken up but apparently mostly became associated with the surface of the cells. Harold and Harold (1980) have studied the orientations of growth of the resulting thalli and rhizoids in gradients of a range of ionophores and inhibitors. Their exemplary work shows how careful one needs to be in interpreting such experiments. The control of an untreated glass fiber had no effect on the orientation of thalli or rhizoids (Fig. 1a). A glass fiber coated with one of the calcium ionophores, A23187 or ionomycin, gave the initial appearance of orienting the growth of thalli (Fig. 1b), but closer inspection showed that there was open space next to the fiber where no cells had grown. So this orientation to the fiber was in fact negative autotropism to open space, with the thalli growing away from their close neighbors, either to oxygen or away from staling factors. This negative autotropism to open space is illustrated in Fig. 1c, where the thalli are growing toward a zone that had been cleared mechanically with a wooden stick. Further experiments, however, showed that the rhizoids, but not the thalli, became truly oriented toward glass fibers coated with gramicidin D (Fig. 2). This is a channel-forming ionophore allowing the passage of K^+, Na^+, and H^+. Of these three cations, Harold and Harold (1980) favored proton movements as being related to the observed orientations of rhizoids, as Na^+ is apparently not required for the growth of *B. emersonii*, and the powerful K^+ ionophore valinomycin had no effect on rhizoid orientation. Two proton-conducting uncouplers, tetrachlorosalicylanilide and 1799, when adsorbed onto beads of the ion exchange resin Dowex-1, also caused positive orientation of the growth of rhizoids. Harold and Harold (1980) suggested that the growing rhizoid tips are the sites of maximal proton entry of a proton current through the cell. In the presence of gradients of the ionophores, the rhizoid tips would grow so as to maximize the proton fluxes across themselves.

Ionic currents in the hyphae of *N. crassa* have been investigated by Slayman (1970), Slayman and Gradmann (1975), and Gradmann et al. (1978), who showed that the membrane potential is generated by an electrogenic proton-translocating ATPase integral in the plasma membrane, producing a proton current of the order of 20 $\mu A/cm^2$. The resulting membrane potential of up to about -200 mV was first directly measured by Slayman and Slayman (1962) with microelectrodes. In this study one finding, which apparently has never been reinvestigated in *N. crassa* or any other fungus, was that there was a gradient or membrane potential along the hypha, rising from under -100 mV beyond 3 mm behind the tip to about -30 mV at the tip. This finding could have been a result of greater injury at the tip (see discussions by Jaffe and Nuccitelli, 1977; Jennings, 1979), but would clearly be worth reinvestigation.

Using the marine fungus *Dendryphiella salina*, Galpin and Jennings (1975), Galpin et al. (1978), and Jennings (1979) have suggested a

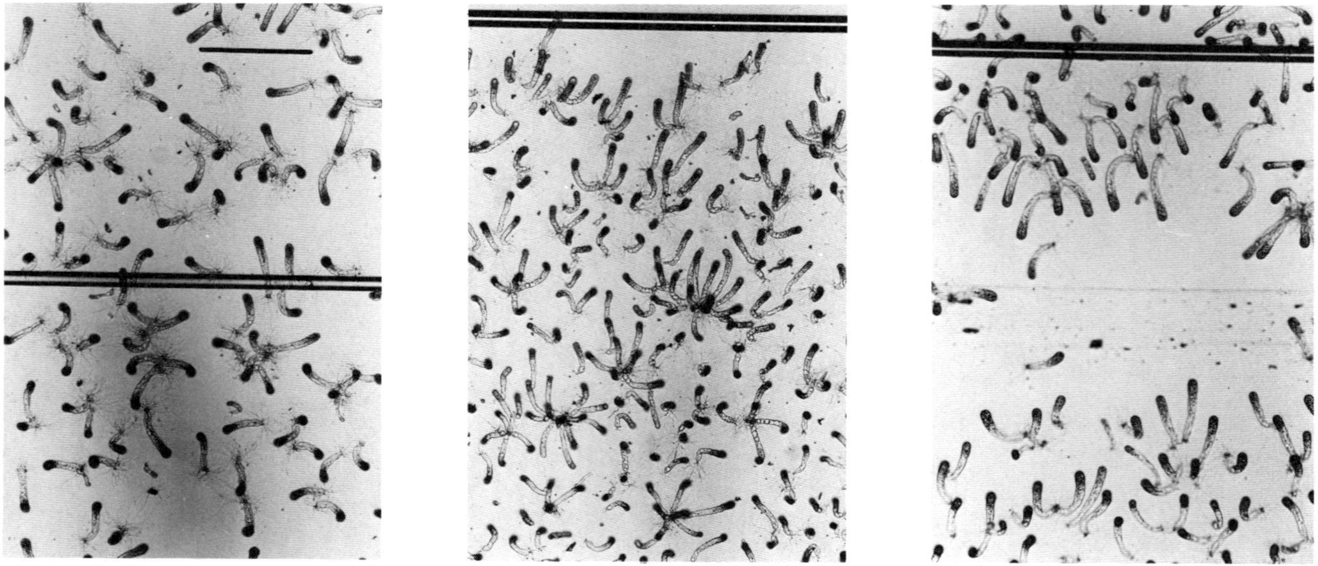

(a) (b) (c)

FIG. 1 Growth of *Blastocladiella emersonii* in the vicinity of glass fibers. (a) Control; random growth near a control fiber. (b) Fiber coated with ionomycin. There was oriented growth of thalli toward the fiber, but note the clear space adjacent to the fiber, where no growth has occurred. (c) Oriented growth to the space scraped in the culture dish; unaffected by a control fiber. All micrographs were made after inoculation (scale bar, 400 μm). (From Harold and Harold, 1980.)

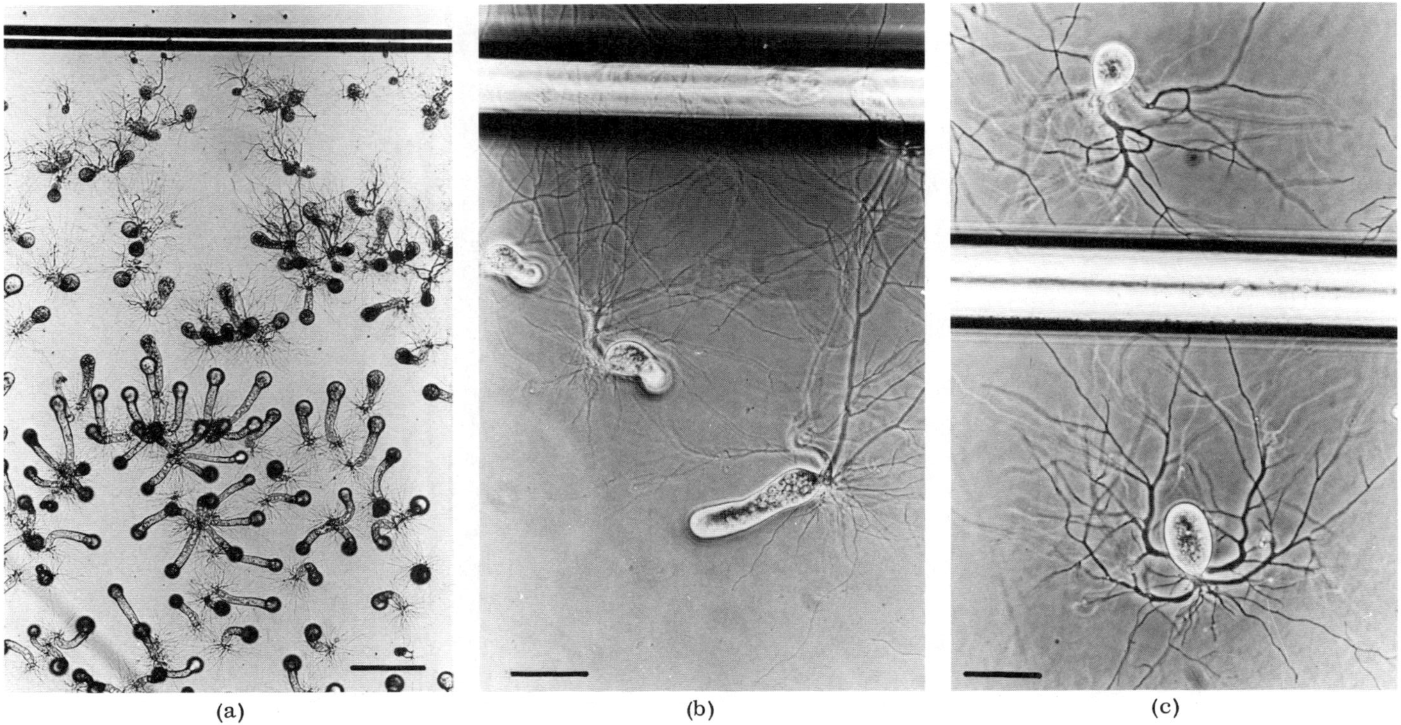

FIG. 2 Growth of *Blastocladiella emersonii* in gradients of gramicidin D. The procedure was as for Fig. 1, with glass fibers coated with the ionophore. Note the inhibition of growth near the fibers (a,b) but the growth of rhizoids toward them with (b) enlarged and thickened rhizoids growing onto the fiber. [scale bars: (a) 200 μm; (b and c) 40 μm]. (From Harold and Harold, 1980.)

TABLE 1 Sodium and Potassium Contents in Hyphae of *Dendryphiella salina*

Site	K	Na	K:Na
Distance behind apex (μm)[a]			
0	966	1208	0.8
1-50	1534	518	3.0
200	1128	753	1.5
Age (h)[b]			
0	17	14	1.2
6	30	7	4.3
24	40	13	3.1

[a]Hyphae grown in 10% seawater from spores on a carbon stub, freeze quenched, and analyzed by x-ray microanalysis. Values are x-ray counts per 200 s.

[b]Hyphal fragments grown in shake culture, digested in concentrated nitric acid, and analyzed by flame photometry. Values are micromoles per 100 mg dry weight.

Source: Compiled from Galpin et al. (1978).

role for a K^+,Na^+-ATPase in establishing and maintaining an ionic gradient in the hyphae. Galpin and Jennings (1975), using histochemical staining methods, could not detect the enzyme activity in young hyphal tips and germ tubes, whereas there was considerable staining in older regions of the hyphae. Galpin et al. (1978), using Edax energy dispersive x-ray analysis, detected a lower potassium content at hyphal tips than in the subapical regions (Table 1), mirrored by a higher sodium content. They suggested that the hyphal apex of *D. salina* is much less selective for potassium against sodium than the region 1-50 μm behind the apex, which in turn is more selective than older regions of the hypha. Gross potassium and sodium contents for mycelium showed a rise and fall, respectively, with increasing age (Table 1), and the authors suggested that these temporal changes might reflect the spatial changes observed by the x-ray analysis. Jennings (1979) proposed a model in which subapical membrane ATPase activities drive an internal potassium current, with a concomitant standing-flow osmotic gradient of water, toward the apex.

These measurements of sodium and potassium were made for a marine fungus, which Jennings (1979) suggested may have Na^+ currents instead of the H^+ currents of a terrestrial fungus such as *N. crassa*.

Galpin et al. (1978) investigated the contents of Na, P, S, Cl, K, and Ca along freeze-quenched hyphae of the terrestrial fungus *Chaetomium globosum*. From their measurements, only Cl and Ca showed a significant spatial gradient, both being about twice as abundant at the tip than in subapical regions, with the former presumably as Cl^- and the latter probably bound to negative charges of the cell wall.

III. CYTOLOGY OF THE HYPHAL TIP

A. General Survey

Our understanding of hyphal growth was revolutionized in the late 1960s with the discovery, by very careful fixation and oriented sectioning for electron microscopy, that all growing hyphae have distinct aggregations of vesicles at their apices (McLure et al., 1968; Girbardt, 1969; Grove et al., 1970; Grove and Bracker, 1970). This was amply confirmed during the following decade (see the excellent review by Grove, 1978) for vegetative hyphae, germ tubes, many different types of asexually and sexually differentiating hyphae, budding yeasts, and side-branch and clamp connection initials. This apical organization could be identified with the Spitzenkörper, a phase-dark region visible in the light microscope in many hyphal apices. The Spitzenkörper had been studied in *Polystictus* (=*Trametes*) *versicolor* by Girbardt (1957,1969), who observed that it was always present in growing hyphae, never in nongrowing hyphae, and became displaced in the tip toward the new direction of growth when hyphae changed direction.

It is timely that now, a decade later, there should be a major advance in technique for the visualization of hyphal tips. This is the use of freeze substitution for the preparation of material, especially then with the examination of thick sections in the high-voltage electron microscope at different angles. Using these techniques, Howard and Aist (1979,1980), Hoch and Howard (1980), and Howard (1981) have produced unparalleled images of hyphal tips containing an immense amount of new information (see Figs. 3-8).

B. The Wall

Electron microscopic studies of material prepared by chemical extraction and by the enzyme dissection technique of Hunsley and Burnett (1970) have shown that major lateral wall components, in particular chitin, cellulose, and other glucans, are present in the hyphal tip wall (Burnett, 1979). One exception is the glycoprotein reticulum of *N. crassa*, which is first visible in the subapical region (Hunsley and Burnett, 1970). Some components that are accessible at the tip are no longer so in subapical regions. Thus some tips stain preferentially with fluorescently labeled wheat-germ agglutinin and with fluorescent brightener calcofluor (Galun et al., 1976; Gull and Trinci, 1974), both of which

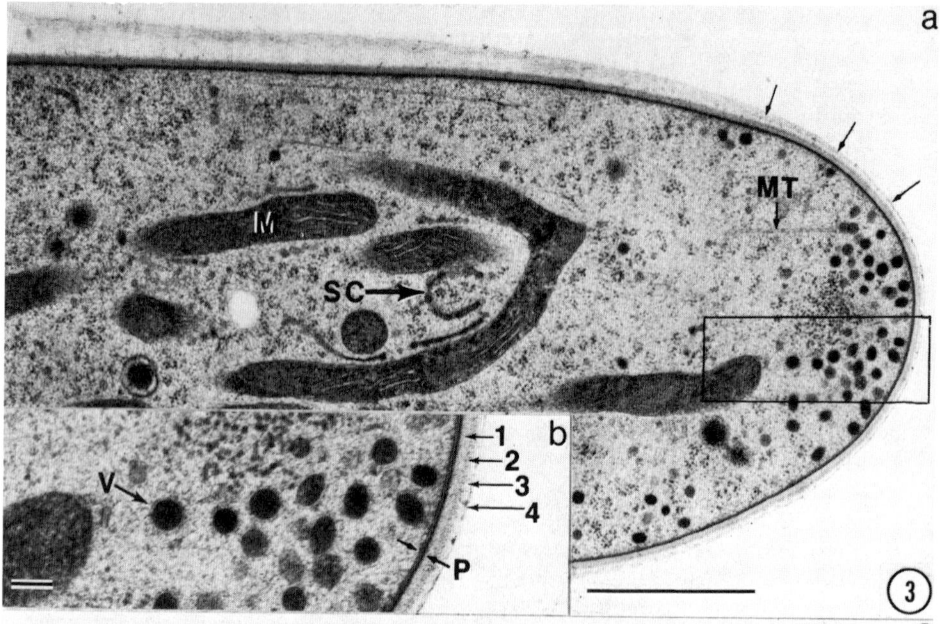

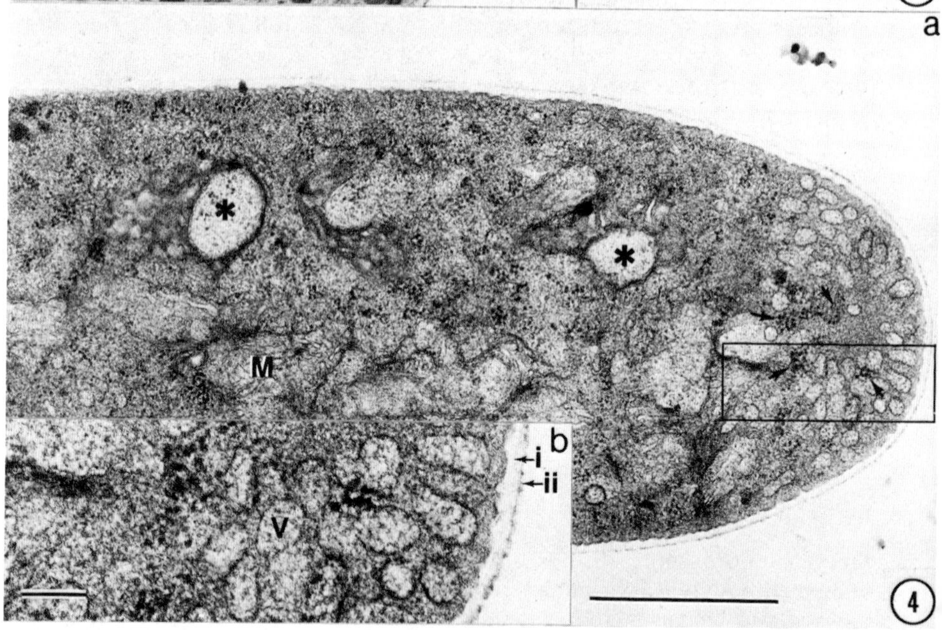

FIG. 3 Electron micrograph of a longitudinal section of a hyphal tip of *Fusarium acuminatum* fixed by freeze substitution. (a) Note the apical vesicles, microtubules (MT), the close association between mitochondria (M) and smooth cisternae (SC), and the apparent expansion of the outer cell wall layer (arrows at upper right). (b) Enlargement of the marked

are probably chiefly visualizing chitin. But Hunsley and Kay (1976) and Marchant and Smith (1968) described uncharacterized antigens specific for hyphal tips of *N. crassa* and *Fusarium culmorum*, respectively.

Trinci and Collinge (1975) have measured the thickness of the wall over the apical region in electron microscopic sections of *N. crassa* and *Geotrichum candidum* and reported a constant thickness of about 50 and 70 nm, respectively, over the apical dome and back for at least 50 μm. For *N. crassa*, it is much thicker much further back, about 275 nm at 3 mm behind the tip; so clearly much secondary wall growth eventually takes place. The innermost, chitinous layer remains constant, however, at about 20 nm, although the microfibrils themselves increase in diameter (Burnett, 1979). However, estimates of wall thickness by electron microscopic sectioning of conventionally fixed material will have to be reassessed in comparison with those of freeze-substituted material, as Howard and Aist (1979), examining tips of *Fusarium acuminatum*, showed a uniform two-layered wall of 25-40 nm thick over the apical 5 μm by the former technique (Fig. 4), but a four-layered wall of about 60 nm, with a considerable progressive increase in thickness of the outermost layer (arrows in Fig. 3) by the latter technique.

There have been no reliable reports of any special orientations of the microfibrillar elements of walls, chitin, and cellulose over the apical dome of vegetative hyphae; that is, preparations show an isotropic arrangement (Burnett, 1979). Burnett did record, however, a tendency for an axial arrangement of chitin microfibrils at the apices of germ hyphae of *N. crassa*.

C. Apical Vesicles and Microvesicles

It is unequivocal that the apical cluster of vesicles is involved in apical growth. Like the Spitzenkörper (as mentioned in Sec. III.A), it is always present in growing hyphae, and never in nongrowing hyphae. Vesicles are bounded by a single membrane. In the apical cluster they

---

FIG. 3 (Continued) area from (a) showing apical vesicles (V), the smooth profile of the plasma membrane (P), and the four-layered appearance of the cell wall (1,2,3,4) [scale bars: (a) 1.0 μm, (b) 100 nm]. (From Howard and Aist, 1979.)

FIG. 4 Electron micrograph of a longitudinal section of a hyphal tip of *Fusarium acuminatum* fixed conventionally. (a) Note the distortions of mitochondria (M) with enclaves of cytoplasm (*), and electronlucent vesicles and ribosomelike particles (arrows) at the cell apex. (b) Enlargement of the marked area from (a). The cell wall is seen as two layers (i,ii) with an irregular inner layer, and the apical vesicles (V) appear to have fibrous contents [scale bars: (a) 1.0 μm, (b) 100 nm]. (From Howard and Aist, 1979.)

occur in two size classes: In conventionally fixed material the apical vesicles range from below 100 to 300 nm, and the microvesicles from 30 - 100 nm (Gooday and Trinci, 1980). In freeze-substituted hyphae of *F. acuminatum* (Howard and Aist, 1979) the apical vesicles appeared spherical, 70-90 nm in diameter, and of two types, one staining more heavily than the other (Fig. 3a,b and Fig. 5), while microvesicles often appeared to be hexagonal and about 30 nm in diameter. In contrast, in conventionally fixed hyphae the apical vesicles appeared more oblong, 90-110 nm in diameter, of one type, and electron translucent with fibrous material (Fig. 4a,b), while microvesicles were not clearly recognizable.

By a serial thick-section high-voltage electron microscopic analysis of tip cells, Howard (1981) presented the ultrastructural equivalent of the Spitzenkörper as including a cluster of apical vesicles subtending a mass of ribosomes, with microvesicles lying in a meshwork of microfilaments within a lumen in the apical vesicle cluster (Fig. 6a-c). The microvesicles were at times hexagonal in cross section and typically coated with radiating spikes of electron-dense material, while others—termed *filasomes*-were surrounded by a mass filamentous material (see Fig. 11.5).

Freeze-substituted hyphae of the Basidiomycete *Laetisaria arvalis* showed the apical vesicles, 70-120 nm in diameter, predominantly localized in a cluster as the Spitzenkörper (Hoch and Howard, 1980). The contents appeared finely granular, and the outer surface often seemed to be decorated with a finely fibrillar matrix, as was the outer surface of presumed Golgi cisternae. Occasional smooth-profiled invaginations of the plasma membrane were observed that could represent fusion between vesicles and the membrane. Vesicles were not observed beyond about 30-40 μm behind the tip, except near sites of branch initiation; from the apex to 30-40 μm, there were few vesicles, these being concentrated near presumed Golgi activity. Microvesicles, mostly hexagonal and 34-40 nm in diameter, contained electron-dense material and were frequently associated with inflated cisternae of the presumed Golgi. Some were surrounded by filamentous material, about 5-6 nm wide, and can therefore be termed *filasomes*. Microvesicles and filasomes were closely associated with developing septa.

Reliable methods for the preparation of fractions of apical vesicles and microvesicles of sufficient purity for characterization of their compositions and possible enzyme activities have yet to be reported (the

---

FIG. 5 Median apical section of a hyphal tip of *Fusarium acuminatum*. Note the numerous microtubules (mt) and the network of microfilaments within the apical cluster of vesicles (cw, cell wall; F, filasome; GE, Golgilike endomembrane cisterna; and M, mitochondrion). The cell was pretreated with D_2O and DMSO for 45 min, with no apparent effect, as a control for other treatments; the cell was fixed by freeze substitution (scale bar, 1 μm). (From Howard and Aist, 1980.)

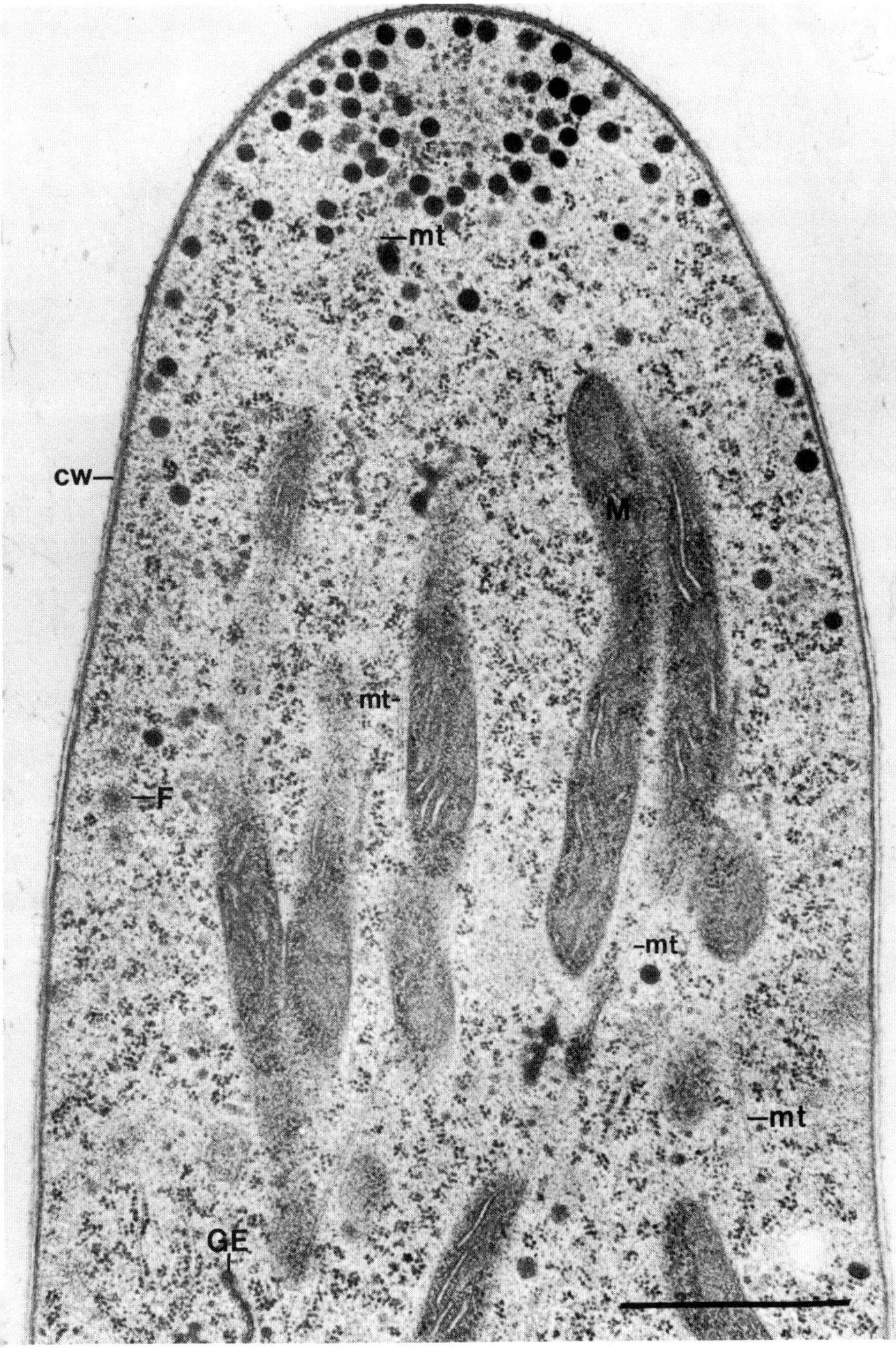
mt
cw
M
mt
F
mt
mt
GE

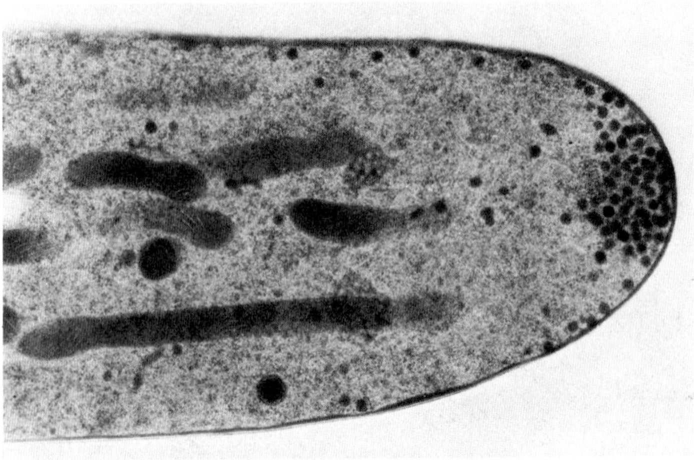
(a)

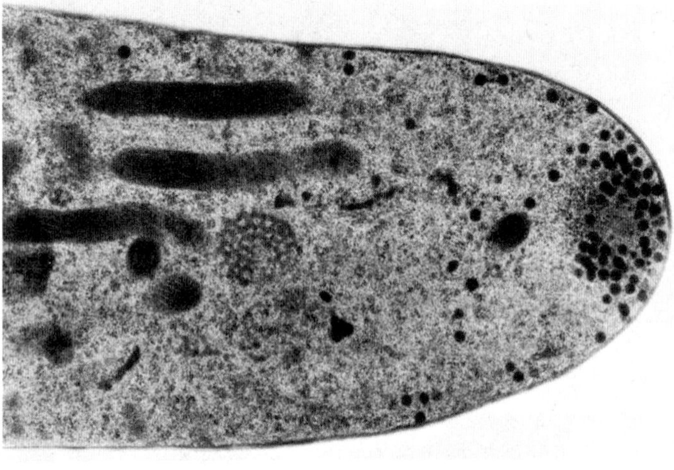
(b)

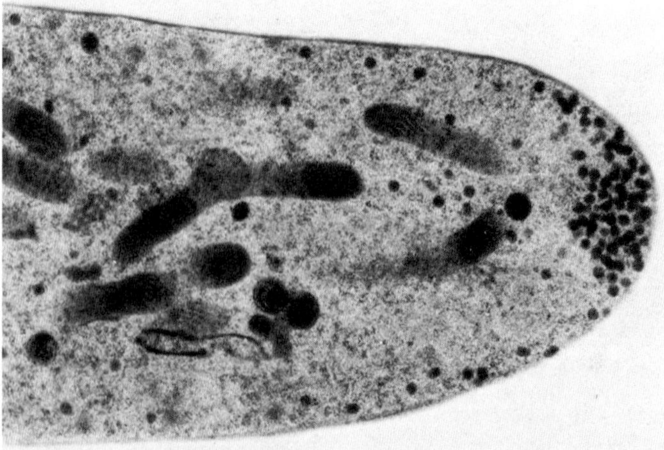
(c)

possible identity of chitosomes with some microvesicles is discussed later). Cytochemical staining has suggested the presence of polysaccharide in some apical vesicles and Golgi dictysomes of the Oomycetes *Achlya bisexualis*, *Achlya ambisexualis*, *Phytophthora palmivora*, and *Saprolegnia monoica* (Dargent, 1975; Hill and Mullins, 1980a; Meyer et al., 1976; Févre, 1979). Cytochemical staining of vesicles for enzyme activities is discussed later.

Girbardt (1969) has presented measurements of apparent membrane thicknesses in hyphae of *Polystictus versicolor* fixed with permanganate. The thicknesses clearly fell into two size classes, with the first (average, 2.81 nm) containing the mitochondrial membranes and the endoplasmic reticulum, and the second (average, 4.15 nm) containing the membranes of the Golgi cisternae, the apical vesicles, the microvesicles, and the plasma membrane (Table 2). Taken together with the above-mentioned observations of Hoch and Howard (1980), this suggests the ontogenetic sequence Golgi cisternae → apical vesicles and microvesicles → plasma membrane.

There have been several estimates of the distribution of vesicles along hyphae. Thus Girbardt (1969) and Collinge and Trinci (1974), respectively, gave values of 200-400 and up to 700 apical vesicles per cubic micrometer at the tips of *P. versicolor* and *N. crassa* (colonial mutant *cot 3*), these values falling to 10 and 25 at 15 μm and 2.6 and 20 at 40 μm behind the apex.

Collinge et al. (1978), expressing values as the percentage of protoplasmic volume occupied by vesicles, reported 42, 8, and 5% at the apex, at 2 μm and at 4 μm, respectively, for hyphae of *Penicillium chrysogenum*. Collinge and Trinci (1974) noticed an excellent correlation between observed vesicle concentration in hyphal tips of *N. crassa* and (1) predicted vesicle concentration ($V_p = R^2/r^2$, where V_p = predicted relative vesicle concentration in number of vesicles per unit volume of protoplasm, R = radius of the hypha in a region of constant diameter, and r = radius of the hypha at any given point), assuming that the number of vesicles per unit length of hyphal tip is constant,

---

FIG. 6 Three adjacent longitudinal serial sections (0.25 μm thick) of a hyphal tip of *Fusarium acuminata*. The Spitzenkörper is composed of a mass of apical vesicles, surrounding a lumen containing microvesicles, with a subtending cluster of ribosomes. Also visible are filasomes, located near the subapical cell wall; mitochondria, maintained at about 1 μm behind the tip; microfilaments; microtubules; and different views of smooth endomembrane cisternae. (For identification of the different organelles, consult the legends to Figs. 3-5 and the original paper). The specimen was prepared by freeze substitution and examined by high-voltage electron microscopy. The diameter of the hypha at the base of Fig. 6b (median section) is 3.5 μm. (From Howard, 1981.)

TABLE 2 Average Membrane Thickness in Hyphae of *Polystictus versicolor*

Membrane	Thickness (nm)[a]	Number of measurements
Endoplasmic reticulum	2.8 ± 0.6	100
Mitochondrial, inner	2.7 ± 0.6	178
Mitochondrial, outer	2.9 ± 0.5	210
Plasma membrane	4.2 ± 0.8	254
Apical vesicles	4.1 ± 0.7	97
Microvesicles	4.1 ± 0.8	201
Golgi cisternae	4.3 ± 0.8	76

[a]Unstained thin sections of *Polystictus* (= *Trametes*) *versicolor* fixed with potassium permanganate.
Source: Compiled from Girbardt (1969).

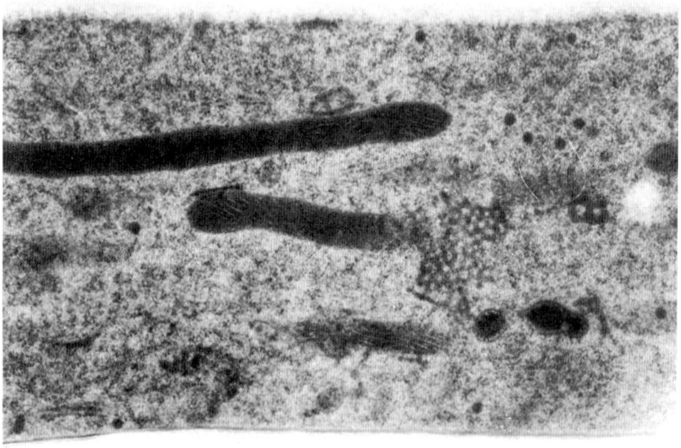

FIG. 7 Medium longitudinal thick section of the subapical region just behind that of Fig. 6b. Note especially the prominent fenestrated membrane cisternae of two apparent morphologies. The conditions and magnification are as for Fig. 6. (From Howard, 1981.)

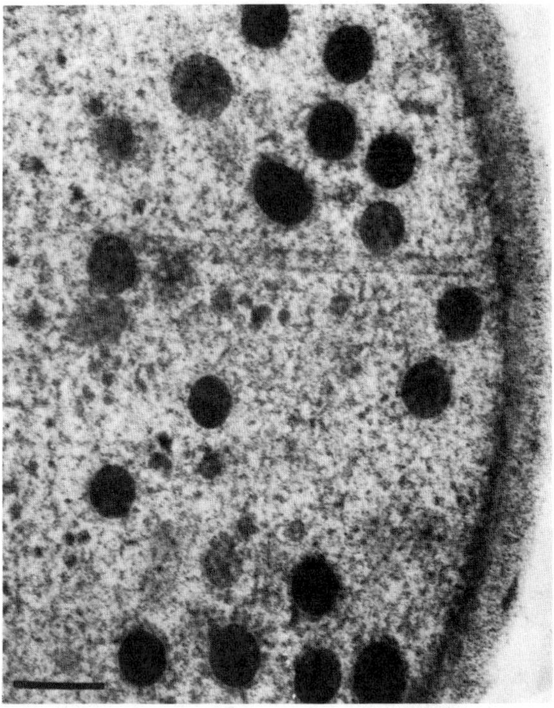

FIG. 8 Median longitudinal thin section of a hyphal apex of *Fusarium acuminatum*. Note the apical vesicles, the microvesicles, the network of microfilaments, and the prominent longitudinal microtubule, terminating with a dilated configuration at the plasma membrane (scale bar, 100 nm). (From Howard, 1981.)

and (2) the relative silver grain density over hyphal tip incubated for 1 min in [^{3}H]*N*-acetylglucosamine (Gooday, 1971). Collinge and Trinci (1974) and Collinge et al. (1978) estimated the required rates of vesicle fusion with the plasma membrane for *N. crassa* wild type, *cot 3*, and *Penicillium chrysogenum* as 38,000, 9,000, and 480 vesicles per minute, respectively (calculated from the hyphal extension rate and diameter and the vesicle diameter).

Howard and Aist (1980) also presented counts of vesicles, but in freeze-substituted preparations (Fig. 9). The very marked apical distribution was quickly attenuated after treatment with the antitubulin agent methyl benzimidazole-2-yl carbamate (MBC), clearly implicating microtubules in the transport of vesicles, either directly or indirectly (see Sec. III.D).

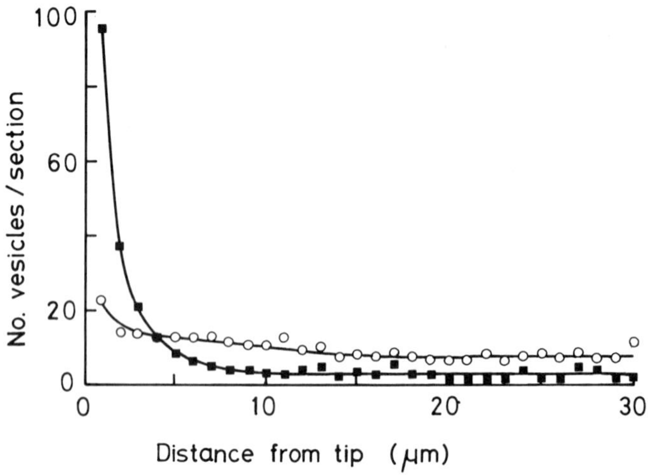

FIG. 9 Distribution of 70- to 90-nm-diameter vesicles in hyphal tip cells of *Fusarium acuminatum*: control (■), treated with 0.1% dimethyl sulfoxide (DMSO) for 45 min (with no observable effect on the growth or cytology of cells); (o) treated with methyl benzimidazole-2-yl carbamate (MBC) (1.0 μg/ml in 0.1% DMSO) for 45 min. The cells were fixed by freeze substitution and examined by high-voltage electron microscopy. The values are the average numbers of counts for control and treated cells, respectively; total average number of vesicles per 30 μm, 254 and 290. (From data of Howard and Aist, 1980.)

D. Cytoskeleton

The eukaryotic cell has at least three chemically distinct cytoskeletal systems: the microtubules, the microfilaments, and the intermediate filaments. These, in turn, are integrated with the microtrabecular lattice (Porter and Tucker, 1981). These cytoskeletal systems must have fundamental roles in the shape determination and movement of cytoplasmic components, and thus in apical hyphal growth. Only the first two components have been reasonably well characterized from fungal cells, however.

1. *Microtubules*

Cytoplasmic tubulins have been characterized from *Aspergillus nidulans* (Sheir-Neiss et al., 1978; Morris et al., 1979) and from *Saccharomyces cerevisiae* (Water and Kleinsmith, 1976; Clayton et al., 1979). Most interest in microtubules to date has centered on their roles during nuclear division (Morris, 1980). Evidence for their functioning in apical growth, however, is slowly amassing.

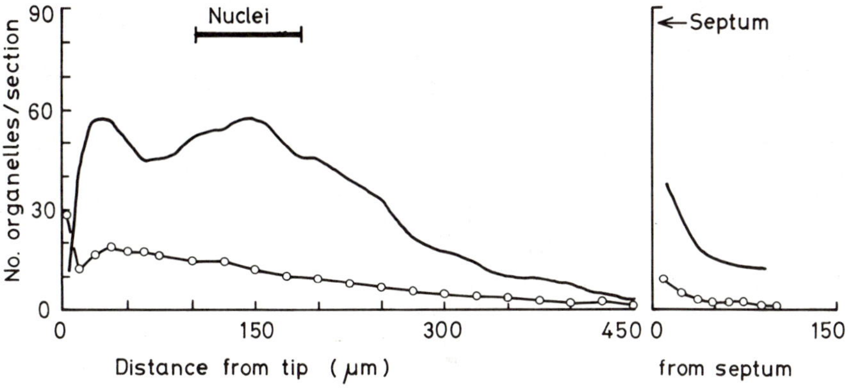

FIG. 10 Distribution of microtubules (———) and mitochondria (o-o-o-) along hyphae of *Trametes versicolor*, 25-40 min after nuclear division. The values are the averages of the numbers of organelles in a large number of transverse sections for seven apical cells and three first subapical cells. The nuclei were in the region indicated by the bar. (Redrawn from Girbardt, 1979a.)

Girbardt (1979a), using his method (Girbardt, 1965) of precisely locating and orienting hyphal tips for thin sectioning and microscopic examination, has mapped the occurrence of the microtubules in apical cells of *Trametes versicolor* (= *Polystictus versicolor*) (Fig. 10). The average microtubule density shows two maxima, both at nearly 60 microtubules per section, one about 30 μm behind the apex, the other at about 150 μm, in the region of the nucleus. The count declines to about 5 close to the septum, but then abruptly rises again to nearly 40 just behind the septum, in the subapical cell.

These microtubules run longitudinally through the hyphae of *T. versicolor*, as do those in other fungi, such as the rust *Uromyces phaseoli* var. *vignae* (Heath and Heath, 1978), and rhizoids of *B. emersonii* (Barstow and Lovett, 1974). The same orientation was observed in hyphae of *F. acuminatum* prepared for electron microscopy by freeze substitution (Howard and Aist, 1979) but, in contrast to conventionally fixed material, microtubules were more numerous, wider (about 24 nm in diameter compared to about 20 nm), and, most importantly, observed in the extreme apex (compare Figs. 3 and 4; see Fig. 6a-c). They were frequently observed in the Spitzenkörper region itself and, in the subapical region, were found associated with each other, with mitochondria, vesicles, and with spindle pole bodies (Howard, 1981).

In freeze-substituted preparations of both *F. acuminatum* (Fig. 8) and *Laetisaria arvalis* (Hoch and Howard, 1980), longitudinal microtubules have been observed at the apex, between the apical vesicles. In Fig. 8, one terminates with a dilated configuration at the plasma membrane.

Some antifungal agents interfere with the assembly of microtubules. Griseofulvin, an antibiotic used against dermatophytes, interacts not with tubulin itself, but with high-molecular-weight associated proteins (Roobol et al., 1977). The systematic fungicide methyl benzimidazol-2-yl carbamate (MBC), in contrast, binds directly to tubulin (Davidse and Flach, 1977; Davidse, 1979). Both agents have been used to investigate the role of microtubules, chiefly in nuclear behavior, but occasionally in hyphal growth. Griseofulvin was first isolated as "curling factor" (Brian, 1947), as affected hyphae grew into coiled and twisted patterns. It inhibited the colony radial growth rate of *Basidiobolus ranarum* in parallel to its effect of increasing the mitotic index (Gull and Trinci, 1973,1974). Unfortunately, it is unclear whether these effects on apical growth are the result of interactions between griseofulvin and cytoplasmic microtubules, nuclear microtubules, or both. Both griseofulvin and benzimidazole compounds related to MBC caused *B. emersonii* to grow as round little cells with stubby rhizoids all over ("ticks"), instead of ovoid thalli with single bushy rhizoids (Harold and Harold, 1980), clearly implicating microtubules in the regulation of shape in this fungus. Quinlan et al. (1980) reported a complete absence of cytoplasmic and spindle microtubules in cells of *S. cerevisiae* treated with 100 μM MBC.

Heath et al. (1981) have investigated the effect of a range of anti-tubulin agents on the formation of infection structures from uredospores of the rust fungus *Uromyces phaseoli* var. *vignae*. Colchicine (40 mM), colcemid (4 mM), and isopropylphenyl carbamate (0.7 mM) had little effect. Vincristine (50 μM), griseofulvin (100 μM) and nocodazole (a benzimidazole compound, 3.3 μM) treatments, however, markedly inhibited the migration of nuclei into the germ tubes; gave rise to a clear zone at the germ tube tips lacking mitochondria, and to the presence of vacuoles between the nuclei and the tip; totally inhibited protoplasmic streaming-saltations; and greatly increased Brownian motion. Heath et al. (1981) thus implicated microtubules in the movement of mitochondria, nuclei, and saltating particles (chiefly lipid droplets) in the hyphae.

Howard and Aist (1977,1980; Fig. 9) have directly investigated the effect of MBC on hyphal tips of *F. acuminatum*. The phase-contrast light microscopic examination showed that by 15 min after application of 5.2 μM MBC, the linear growth rate of tips was very much slowed, the Spitzenkörper had disappeared, and mitochondria were displaced from the hyphal tips. Later mitoses, as they occurred, were arrested at metaphase. Heavy water, D_2O, which stabilizes microtubules, partially counteracted the effects of MBC. Howard and Aist (1977) thus suggested that the effects on hyphal growth and tip cytology are caused by the effect of MBC on the cytoplasmic microtubules, and not, for example, as a secondary effect of MBC on nuclear microtubules. Extending this to an electron microscopic study of freeze-substituted material, Howard and Aist (1980) showed that after only a 10-min exposure to MBC, almost all cytoplasmic microtubules had disappeared, except near spindle pole bodies. They suggested that the gradual disappearance of the Spitzenkörper in MBC-treated material (Fig. 9) provides evidence for its role in the longitudinal array of cytoplasmic microtubules in the transport of vesicles to the apex.

2. *Microfilaments*

Microfilaments, 4 to 7 nm in diameter, have been found in hyphae of many fungi. In freeze-substituted hyphae of *L. arvalis* and *F. acuminatum*, they are seen within the lumen of the Spitzenkörper, as components of filasomes, associated with apical vesicles, as bundles free in the cytoplasm, and along forming septa and lateral hyphal walls (Hoch and Howard, 1980; Howard, 1981). Gull (1975), Patton and Marchant (1978), and Girbardt (1979b) described microfilaments in Basidiomycete hyphae, both as bundles in the cytoplasm and associated with developing septa. What is lacking for the fungi is chemical characterization of such microfilaments. Actin, myosin, and associated proteins have been characterized from a wide range of eukaryotic microbes and have even been reported for prokaryotes (Korn, 1980), and so we can expect that they are present in the Eumycota. The most definitive report to date is that by Allen and Sussman (1978), who described actinlike microfilaments formed in extracts of a morphological mutant of *N. crassa, snowflake*. This colonial mutant, growing at about 1/12 the rate of wild type, with cells about half the length and width, has bundles of anything up to several hundred microfilaments, about 7 nm in diameter and several micrometers long, in its cytoplasm (Allen et al., 1974). Supernatants from mycelia of this mutant formed such microfilaments after addition of ATP, $CaCl_2$, and KCl, which became decorated after addition of rabbit heavy meromyosin (Allen and Sussman, 1978).

The cytochalasins are a group of fungal metabolites that affect many different cell functions, possibly by interacting with actin, in many types of cells, including fungi (Thomas, 1978). Although their mode of action remains in doubt, they have characteristic effects on hyphal growth, for example, by inhibiting hyphal tip growth and resulting in irregular deposits of cell wall material in *N. crassa* (Allen et al,. 1980) and *Gilbertella persicaria* (Grove and Sweigard, 1980). When added to germlings of *B. emersonii*, the effect was similar to that of adding antitubulins–globose thalli with branched, thickened rhizoids (Harold and Harold, 1980).

E. Other Organelles

Mitochondria have a characteristic distribution in growing hyphae. They are absent in the extreme apex, but very abundant just subapically, being aligned longitudinally (Figs. 4-6, 9, and 10). In freeze-substituted preparations they often appear associated with microtubules (Howard and Aist, 1979). The synthetic processes of hyphal growth will require chemical energy and this concentration of mitochondria suggests a power plant producing ATP close to where it is required, but this also implies efficient transport of substrates for respiration to this region. One substrate required is oxygen, the suggested chemotropic attractant for hyphal apices (discussed earlier), and thus a model for bending toward the more oxygen-rich side would be a higher respiration rate in the subapical mitochondria on that side, with a consequent favoring of biosynthesis on that side, a consequent asymmetric arrangement of the Spitzenkörper, and a bending of the tip resulting from the new polarity. Girbardt (1970), using cinematography, described rapid changes in the shape of the elongated mitochondria moving toward the tips of hyphae of *P. versicolor*. Thus one mitochondrion, 15 μm long, showed the waxing and waning of five distinct dilations along its length during an 8-s period of observation. Girbardt suggested that these represent an intramitochondrial transport mediated by the outer mitochondrial membrane.

Most fungal cells do not exhibit the stacked Golgi cisternae characteristic of many other cells. Nevertheless, careful observation reveals that they do have membranous cisternae which are clearly homologous with Golgi bodies. Girbardt (1969) presented counts of the distribution of Golgi cisternae in hyphae of *P. versicolor*. They were absent in the Spitzenkörper, most abundant (at 2.0 per cubic micrometer) at 15 μm, falling to 0.6 and 0.1 per cubic micrometer at 40 and 500 μm behind the tip, respectively. Freeze-substituted material of *F. acuminatum* showed extensive subapical smooth endomembrane systems (Howard, 1981). These were often associated with mitochondria and appeared as tubules (Fig. 6a) or fenestrated sheets (Fig. 6b,c). Further observation showed that these were different components of the same organelle, termed "Golgi equivalent" by Howard, which fell into two classes: one of tubular and fenestrated cisternae about 35 nm wide, associated with microvesicles (20-50 nm in diameter), the other of tubular and fenestrated cisternae about 70 nm wide, associated with apical vesicles (70-90 nm in diameter). In plant and animal cells the Golgi cisternae are the sites of synthesis of many surface and extracellular materials. In plant cells, for example, the pectins of the wall are Golgi products. It is thus strongly tempting to implicate these Golgi equivalents in *F. acuminatum*, after subsequent transport via the microvesicles and apical vesicles, with the synthesis of some wall and plasma membrane components.

IV. ENZYMOLOGY

A. Synthetic Processes

The enzymology of hyphal growth has been reviewed by Gooday (1978), and that of aspects of wall formation by Farkâs (1979), Wessels and Sietsma (1979), Bartnicki-Garcia et al. (1979), Cabib et al. (1979), Gooday (1979), Fèvre (1979), and Gooday and Trinci (1980).

Most emphasis has been placed on the biosynthesis of the polysaccharides, chitin, and glucans. From light microscopic autoradiographic studies, it is clear that these are deposited in a very localized manner in growing hyphae, being mostly laid down in the apical 1 μm (e.g., Gooday, 1971; Galun et al., 1976; Bartnicki-Garcia and Lippman, 1977).

The weight of evidence is that the enzyme chitin synthetase, responsible for the biosynthesis of microfibrils, is *active* as an integral protein in the cell membrane (Vermeulen et al., 1979; Cabib et al., 1979). By, for example, electron microscopic autoradiographic studies (Valk and Wessels, 1977; Gooday, 1982), there is no evidence for intracellular polymerization of chitin. The enzyme, however, also exists as discrete intracellular particles, the chitosomes, now described from a wide range of fungi (Bartnicki-Garcia et al., 1978; Bartnicki-Garcia et al., 1979; Bartnicki-Garcia, 1980). The chitosomes contain *inactive* enzyme in a zymogenic form that can be activated by treatment with proteases. The activated chitosome, when incubated with substrate UDP-N-acetylglucosamine, will then manufacture chitin in the form of discrete microfibrils. Bartnicki-Garcia et al. (1979) concluded that the chitosome "is the microvesicular vehicle by which the cell delivers, to specific sites on the cell surface, individual, organized packets of chitin synthetase, each packet being responsible for the synthesis of one microfibril."

Most chitosomes are 40-70 nm in diameter. They are surrounded by a shell, giving the appearance of a unit membrane, but they contain very little phospholipid (sterols being the most abundant lipids) (Bartnicki-Garcia, 1980). Microvesicles looking very much like chitosomes can be seen in thin sections of conventionally fixed (Bracker et al., 1976) and freeze-substituted hyphae (Hoch and Howard, 1980; Howard, 1981). Howard (1981) pointed out that the microvesicular core of filasomes of *F. acuminatum* closely resembles the chitosome [but Hoch and Howard (1980) stressed that the filamentous component of filasomes is *not* chitin]. Thus it seems very likely that at least some of the microvesicles of the Spitzenkörper of chitin-forming fungi are chitosomes.

The following temporal and spatial organization of chitin synthetase in hyphal tips can thus be suggested: chitosome formation (by Golgi cisternae?); transport of chitosomes (with other microvesicles) to Spitzenkörper; insertion of chitosomes into apical membrane; activation of some chitin synthetase zymogen; activity of chitin synthetase with extrusion and subsequent crystallization of chitin microfibrils; and degradation of the now subapical chitin synthetase, but with some zymogen still in the membrane for possible later activation for microfibril thickening and branch and septum initiation. All these steps are irreversible.

Chitin synthetase preparations from hyphae of *Aspergillus nidulans* show considerable activation by proteolytic enzymes, particularly by a neutral protease isolated from the fungus (Campbell and Peberdy, 1979). There is a remarkable difference, however, in the proportion of active enzyme to zymogen in "early" and "late" protoplasts formed from these hyphae (Issac et al., 1978). The "early" protoplasts, formed preferentially from tips, had chitin synthetase activity with a much higher specific activity and only slight proteolytic activatability in comparison to "late" protoplasts, formed preferentially from older areas of hyphae. This finding is in accord with the model suggested above.

Much less is known of the biosynthesis of glucans in filamentous fungi. Cellulose microfibril synthesis in vitro has not been demonstrated unequivocally by any preparation from a cellulosic fungus, but Fèvre and Rougier (1981) described a fraction, separable by centrifugation from that producing the β-(1,3)-glucan, that produced a β-(1,4)-glucan from UDP-glucose. By analogy with plant wall synthesis, one would expect that cellulose microfibrils in the Oomycetes would be biosynthesized by an integral membrane system, that is, analogous to chitin.

Enzyme preparations have been made from *Phytophthora cinnamomi*, *Paracoccidiodes brasiliensis*, and *Saprolegnia monoica* that produce microfibrillar β-(1,3)-glucans from UDP-glucose (Wang and Bartnicki-Garcia, 1976; San Blas, 1979; Fèvre and Rougier, 1981). By cell fractionation techniques, Fèvre and Dumas (1977) and Fèvre and Rougier (1981) suggested that this activity is associated with Golgi dictyosomes and related membrane fractions in *Saprolegnia monoica*, in which Fèvre (1979) demonstrated polysaccharide contents by cytochemical staining. Hill and Mullins (1980a) have made similar observations for a UDP-glucose transferase activity in fractions from vegetative hyphae of *Achlya ambisexualis*. Thus it is probable that some of the Golgi-derived apical vesicles in all hyphal tips contain preformed wall matrix polysaccharides, such as β-(1,3)-glucans, being transported for insertion into the apical wall.

Marriott (1977), using purified membrane preparations from *Candida albicans*, suggested that the mannan/mannan-protein matrix components of the wall may be synthesized at two sites, with mannose units being polymerized with protein from GDP-mannose via lipid intermediates in intracellular membrane systems, but then with further mannose units being added to mannans from GDP-mannose at the plasma membrane.

Dow et al. (1981) have studied the biosynthesis of polyuronides and glycoproteins (both wall matrix components) by membrane fractions from hyphae of *Mucor rouxii*. The polyuronide-synthetic ability was richest in the lightest membrane fraction, while the glycoprotein-synthetic ability was richer in a heavier fraction; but in the absence of adequate enzyme markers, Dow et al. (1981) were unable to ascribe the different membrane fractions to particular cellular membrane systems.

B. Lytic Processes

The necessity for controlled apical wall lysis as a concomitant phenomenon of synthesis, eloquently suggested, at least as long ago as 1888, by Marshall-Ward (see Gooday, 1978) and championed more recently by Bartnicki-Garcia (1973), still remains unproven. "Indeed, teleologically speaking, the apex would seem to be a most dangerous location for a lytic entity!" (Burnett, 1979). Rosenberger (1979) reviewed the occurrence of endogenous lytic enzymes and their role in wall metabolism in fungi and voiced what is probably the current majority view, that "being surrounded by such a rigid 'exoskeleton', fungal hyphae can only grow by cutting and modifying the polymers making up their wall."

The most direct demonstration of the localization of a wall lytic enzyme has been the fluorescent antibody study of β-(1,3)-glucanase in *Sclerotium rolfsii* by Kritzman et al. (1978). Specific zones of immunofluorescence were clearly visible in the hyphal tips, and also in other sites of wall synthesis, that is, in clamp connections, new septa, and lateral branches. Light microscopic cytochemical staining for N-acetyl-β-D-glucosaminidase (which might be expected, together with chitinase, to participate in any chitin breakdown) in hyphae of *Mucor racemosus*, in contrast, showed a few localized sites of activity near the apices, but much greater activity further back in older hyphae (Hoch et al., 1979).

When male vegetative hyphae of the Oomycete *Achlya ambisexualis* are treated with the sex hormone antheridiol, they respond by forming many antheridial branches. This is associated with a marked increase in the cellulase activity of cells, which Thomas and Mullins (1969) suggested as being responsible for the observed thinning of the wall at the sites of antheridial branches, resulting in restructuring of the wall. Mullins and Ellis (1974) and Mullins (1979) studied this process with freeze-etched electron microscopic preparations and found dense localizations of vesicles at the sites of hormone-induced branching (Fig. 11), with many raised sites on the plasma membrane giving the appearance of sites of exocytosis. Nolan and Bal (1974) used carboxymethyl cellulose followed by treatment with Benedict's solution to localize cellulase activity in material subsequently fixed for electron microscopy. Staining was visible in Golgi dictyosomes and in vesicles which appear to be about 150 nm in diameter. Hill and Mullins (1979) described a membrane-bound latent cellulase activity, together with the previously characterized soluble activity, in growing vegetative hyphae of *Achlya ambisexualis*. Hill and Mullins (1980b) and Fèvre (1977,1979) presented results of studies of the subcellular distributions of cellulase activities

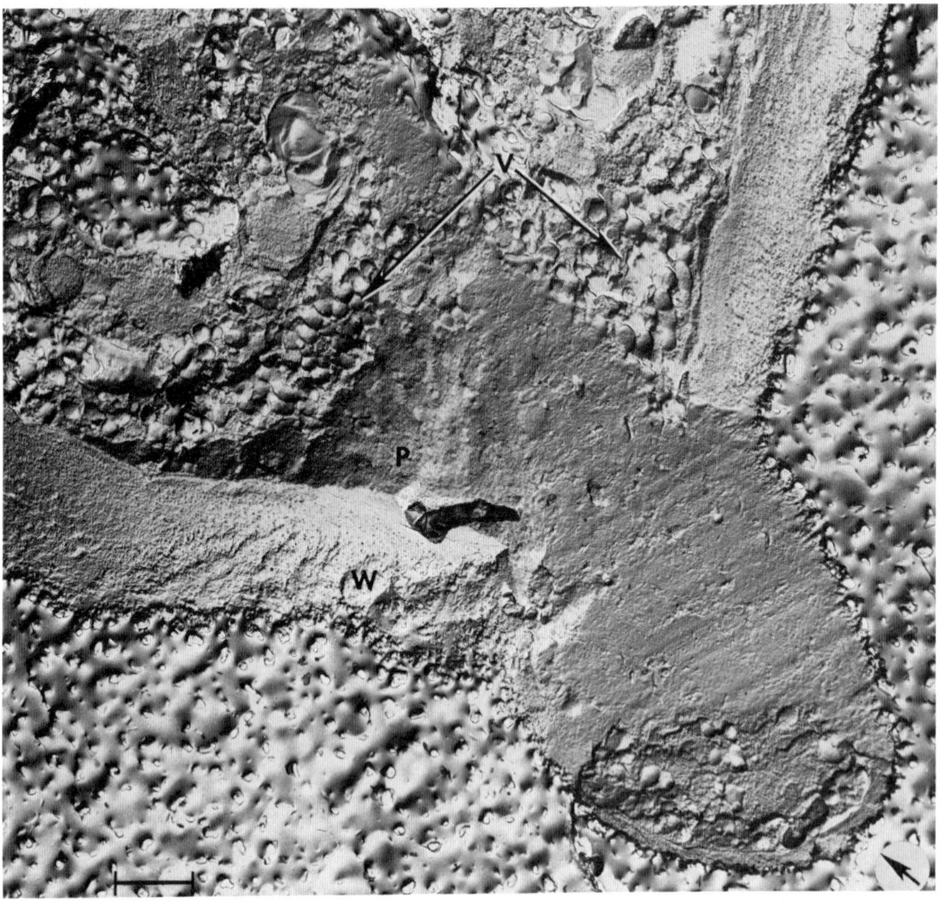

FIG. 11 A freeze-etched preparation of a vegetative hypha of *Achlya ambisexualis* (upper left) responding to treatment with antheridiol by production of an antheridial hypha (lower right). Note the numerous vesicles (V), the plasma membrane (P), and the wall (W). The arrow signifies the direction of fracture (scale bar, 0.5 μm). (Courtesy of Mullins and Ellis, 1974.)

from *Achyla ambisexualis* and *Saprolegnia monoica*, respectively. In the former case the maximum activity was associated with a membrane fraction with a density of 1.19 g/cm^3, whereas in the latter case it was associated, together with the maximum activity of β-(1,3)-glucanase, with membranes of density 1.16 g/cm^3; but in both cases the fractions contained Golgi dictyosomes and Golgi-derived vesicles.

There is less evidence for the involvement for chitinase activities in the growth of the chitinous fungi than that just discussed for cellulose activities in the cellulosic fungi. Certainly chitinase activities can be detected in these fungi (Berkeley, 1979) and implicated in extreme cases of autolysis accompanying spore release in puffballs and inky caps and in the intercalary elongation of stipe cells of *Coprinus* speccies (Gooday, 1979; Kamada et al., 1980). Gooday suggested a regulatory role of a "N-acetylglucosamine cycle," with chitinase and N-acetylglucosaminidase locally degrading the preexising chitin microfibrils to N,N'-diacetylchitobiose and N-acetylglucosamine, both of which would then locally activate chitin synthetase, resulting in the insertion of new chitin microfibrils and wall expansion, before then being recycled to provide further substrate UDP-N-acetylglucosamine for further chitin synthesis. W. H. Leith and G. W. Gooday (unpublished data) have characterized a membrane-bound chitinase activity from *Mucor mucedo* that is associated with the chitin synthetase activity. The result is that incubations of this mixed-membrane fraction with UDP-N-acetylglucosamine give N,N'-diacetylchitobiose as a product, by degradation of the chitin produced by the chitin synthetase. ["Nascent" chitin is very susceptible to chitinase attack (Molano et al., 1979)]. Such a membrane-bound activity is a strong contender for a morphogenetic lytic enzyme involved in the balance of synthesis/lysis necessary for orderly hyphal growth. Prick and Diekmann (1979) described a lectin from the wall of *N. crassa* that binds to chitin and inhibits chitinase. It is easy to see that such molecules might have regulatory roles during hyphal growth, in locally masking wall components from attack by morphogenetic enzymes.

By prelabeling hyphae of *A. nidulans* with radioactive substrates, Polachek and Rosenberger (1975,1978) found that newly formed apical wall components were better substrates for autolytic enzymes than were lateral walls. They suggested that wall components acquire resistance to autolysis soon after their deposition. Using a similar technique, Dow and Villa (1980) found that oligomers of glucuronic acid were rapidly released to the medium by growing hyphae of *M. rouxii*. They suggested that these were produced by lysis of the wall matrix component poly-D-glucuronic acid at the apex during hyphal elongation.

Hill and Mullins (1980a) presented results of electron microscopic cytochemical staining of hyphal tips of *Achlya ambisexualis*. Of the larger apical vesicles, about 150 nm in diameter, some, but not all, and no Golgi cisternae, stained for inosine diphosphatase (IDPase, equivalent to nucleoside diphosphatase) (Figs. 12 and 13). Nucleoside diphosphatase activity is often associated with chitin synthetase preparations, and Gooday (1979) suggested that it may play a role in recycling the inhibitory uridine diphosphate during polysaccharide synthesis. The observation that this enzyme activity is associated with vesicles, some of which are also implicated in polysaccharide synthesis, adds weight to this suggested role. A few of the larger apical vesicles stained for

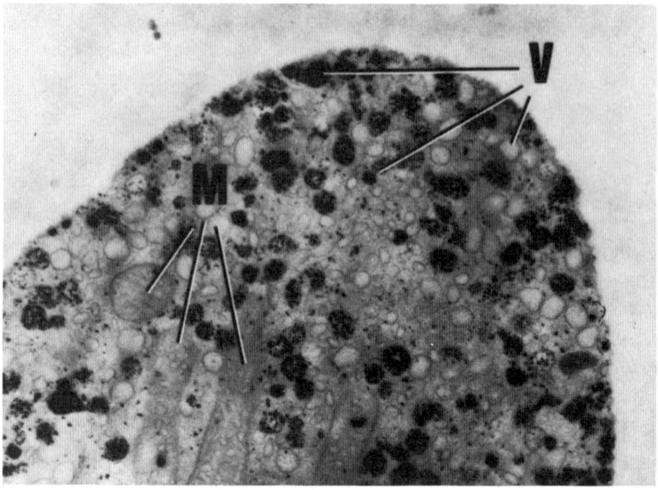

FIG. 12 Electron microscopic localization of inosine diphosphatase activity in the hyphal tip of *Achlya ambisexualis*. Note the staining of some, but not all, of the cytoplasmic vesicles (V) (X 19,000). (From Hill and Mullins, 1980a.)

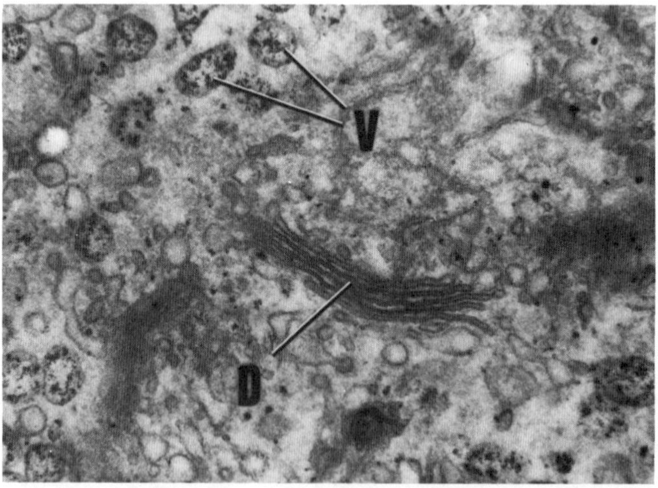

FIG. 13 Electron microscopic localization of inosine diphosphatase activity in the subapical region of *Achlya ambisexualis*. Note the staining of some, but not all, of the cytoplasmic vesicles (V) and the lack of staining in the dictyosome (D). Mitochondria are labeled M (X 38,000). (From Hill and Mullins, 1980a.)

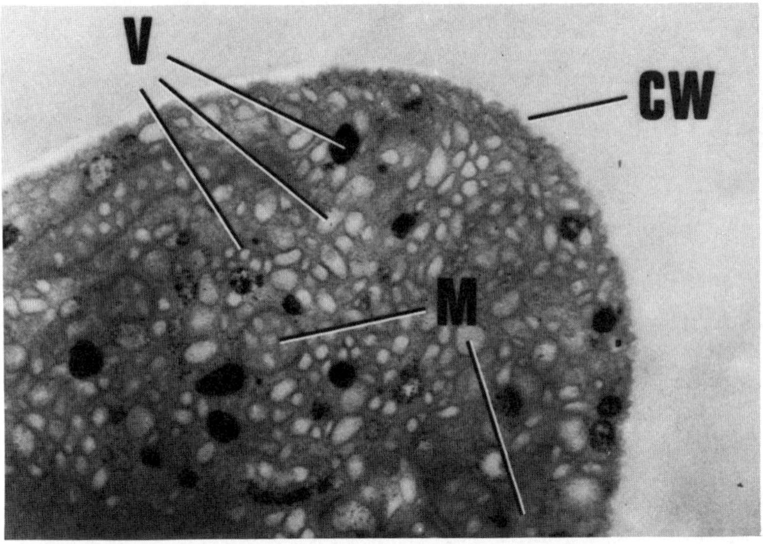

FIG. 14 Electron microscopic localization of acid phosphatase activity in the hyphal tip of *Achlya ambisexualis*. Note the staining of some, but not all, of the cytoplasmic vesicles (V). Mitochondria are labeled M; the cell wall (CW) does not show with this preparation technique (X 21,000). (From Hill and Mullins, 1980a.)

acid phosphatase, as did occasional Golgi cisternae (Figs. 14 and 15), whereas no vesicles, but a few Golgi cisternae, stained for alkaline phosphatase activity. In *Phytophthora palmivora*, Meyer et al. (1976) detected staining in some Golgi cisternae for both acid and alkaline phosphatase activities.

C. Redox Processes

From its reaction to redox dyes, Turian (1976,1978) suggested that the Spitzenkörper region of *Allomyces arbuscula*, *N. crassa*, and *Monilia* (*Sclerotinia*) *fructigena* is strongly reduced with respect to the subapical cytoplasm. Thus it reduced neutral red, methylene blue, phenazine methosulfate, ferricyanide, and Janus green B. Turian suggested that the apex, lacking mitochondria, is a localized site of glycolysis, giving lactate (*Allomyces arbuscula*) or ethanol as a product and maintaining its reducing power as high concentrations of NADH or–SH groups on proteins, so that the resulting redox gradient in the hyphae could then be a means of maintaining an electrochemical polarity within the cell, and perhaps a driving force for the tipwards transport of materials.

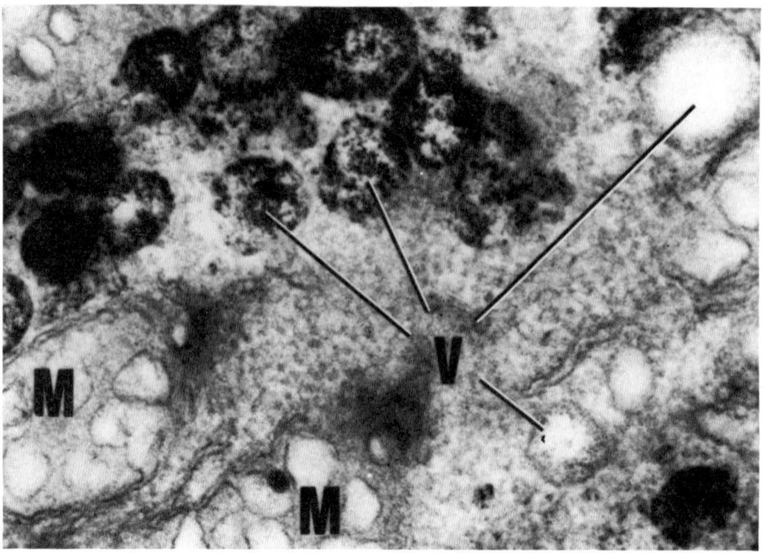

FIG. 15 Electron microscopic localization of acid phosphatase activity in the subapical region of *Achlya ambisexualis*. Note the staining of some, but not all, of the cytoplasmic vesicles (V). Mitochondria are labeled M (X 61,000). (From Hill and Mullins, 1980a.)

V. CONCLUSIONS AND DIRECTIONS

Tip growth in a fungus can be summarized as follows:

1. The major contributors to apical growth are the apical vesicles and microvesicles, formed from Golgi equivalents, which deposit wall material, membrane materials, and enzymes at the tip.
2. The apical vesicles are a mixed population, some containing polysaccharide and synthetic enzymes, some wall lytic enzymes, and some regulatory enzymes.
3. The microvesicles are a mixed population, some being chitosomes.
4. The extension zone is rich in cytoskeletal elements, microtubules, and microfilaments.
5. There is probably always an ionic current, entering the tip and leaving in subapical regions.

What is the driving force? A bioelectrical current is certainly very appealing as the work force, acting either electrophoretically on the vesicles or via an electro-osmotic mass flow. We eagerly await precise measurements of current densities in fungal hyphae so as to calculate the feasibility of such mechanisms.

In view of the large distances involved in the tipwards transport of materials in growing hyphae, the relatively short microtubules and microfilaments of the cytoskeleton probably participate little in the mass flow of vesicles and other materials to the apex; rather, their role may be in locally guiding and aligning these components during their transport and deposition. Precise counting and mapping of the cytoskeletal elements in freeze-substituted preparations of hyphal tips should help to give a clearer picture of their role.

ACKNOWLEDGMENTS

I thank the many authors who have sent me reprints and preprints, particularly Drs. Fèvre, Girbardt, Harold, Howard, Mullins, and Rougier.

REFERENCES

Allen, E. D., Lowry, R. J., and Sussman, A. S. (1974). Accumulation of microfilaments in a colonial mutant of *Neurospora crassa. J. Ultrastruct. Res. 48*:455-464.

Allen, E. D., and Sussman, A. S. (1978). Presence of actin-like protein in mycelium of *Neurospora crassa. J. Bacteriol. 135*:713-716.

Allen, E. D., Aiuto, R., and Sussman, A. S. (1980). Effects of cytochalasins on *Neurospora crassa*. I. Growth and ultrastructure. *Protoplasma 102*:63-75.

Barstow, W. E., and Lovett, J. S. (1974). Apical vesicles and microtubules in rhizoids of *Blastocladiella emersonii*: Effect of actinomycin D and cycloheximide on development during germination. *Protoplasma 82*:103-117.

Bartnicki-Garcia, S. (1973). Fundamental aspects of hyphal morphogenesis. In *Microbial Differentiation, Symposium of the Society for General Microbiology*, Vol. 23, J. O. Ashworth and J. E. Smith (Eds.). Cambridge University, Cambridge, pp. 245-267.

Bartnicki-Garcia, S. (1980). Chitosomes and the origin of chitin microfibrils. In *Advances in Protoplast Research*. L. Ferenczy and G. L. Farkâs (Eds.). Pergamon, London, pp. 475-484.

Bartnicki-Garcia, S., and Lippman, E. (1972). The bursting tendency of hyphal tips of fungi: Presumptive evidence for a delicate balance between wall synthesis and wall lysis in apical growth. *J. Gen. Microbiol. 73*:487-500.

Bartnicki-Garcia, S., and Lippman, E. (1977). Polarization of cell wall synthesis during spore germination of *Mucor rouxii*. *Exp. Mycol. 1*:230-240.

Bartnicki-Garcia, S., Bracker, C. E., Reyes, E., and Ruiz-Herrera, J. (1978). Isolation of chitosomes from taxonomically diverse fungi and synthesis of chitin microfibrils in vitro. *Exp. Mycol. 2*:173-192.

Bartnicki-Garcia. S., Ruiz-Herrera, J., and Bracker, C. E. (1979). Chitosomes and chitin synthesis. In *Fungal Walls and Hyphal Growth*. J. H. Burnett and A. P. J. Trinci (Eds.). Cambridge University, Cambridge, pp. 149-168.

Beever, R. E. (1980). A gene influencing spiral growth of *Neurospora crassa* hyphae. *Exp. Mycol. 4*:338-342.

Berkeley, R. C. W. (1979). Chitin, chitosan and their degradation. In *Microbial Polysaccharides and Polysaccharases*. R. C. W, Berkeley, G. W. Gooday, and D. C. Ellwood (Eds.). Academic, London, pp. 205-236.

Bracker, C. E., Ruiz-Herrera, J., and Bartnicki-Garcia, S. (1976). Structure and transformation of chitin synthetase particles (chitosomes) during microfibril synthesis *in vitro*. *Proc. Nat. Acad. Sci. USA 73*;4570-4574.

Brian, P. W. (1947). Studies on the biological activity of griseofulvin. *Ann. Bot. N. S. 13*:59-77.

Brody, S. (1973). Metabolism, cell walls, and morphogenesis. In *Developmental Regulation: Aspects of Cell Differentiation*. S. J. Coward (Ed.). Academic, New York, pp. 107-154.

Bull, A. T., and Trinci, A. P. J. (1977). The physiology and metabolic control of fungal growth. *Adv. Microbiol. Physiol. 15*: 1-84.

Burnett, J. H. (1979). Aspects of the structure and growth of hyphal walls. In *Fungal Walls and Hyphal Growth*. J. H. Burnett and A. P. J. Trinci (Eds.). Cambridge University, Cambridge, pp.1-25.

Cabib, E., Duran, A., and Bowers, B. (1979). Localized activation of chitin synthetase in the initiation of yeast septum formation. In *Fungal Walls and Hyphal Growth*. J. H. Burnett and A. P. J. Trinci (Eds.). Cambridge University, Cambridge, pp. 189-201.

Campbell, J. M., and Peberdy, J. F. (1979). Proteases of *Aspergillus nidulans* and the possible role of a neutral components in the activation of chitin synthase zymogen. *FEMS Microbiol. Let. 6*:65-69.

Carlile, M. J., and Gooday, G. W. (1978). Cell fusion in Myxomycetes and fungi. In *Cell Surface Reviews, Membrane Fusion*, Vol. 5, G. Poste and G. L. Nicolson (Eds.). Elsevier, Amsterdam, pp. 219-265.

Clayton, L., Pogson, C. I., and Gull, K. (1979). Microtubule proteins in the yeast *Saccharomyces cerevisiae*. *FEBS Lett. 106*:67-70.

Collinge, A. J., and Trinci, A. P. J. (1974). Hyphal tips of wild-type and spreading colonial mutants of *Neurospora crassa*. *Arch. Microbiol.* *99*:353-368.

Collinge, A. J., Miles, E. A., and Trinci, A. P. J. (1978). Ultrastructure of *Penicillium chrysogenum* hyphae from colonies and chemostat cultures. *Trans. Br. Mycol. Soc.* *70*:401-408.

Dargent, R. (1975). Sur l'ultrastructure des hyphes en croissance de l'*Achlya bisexualis* Coker. Mise en evidence d'une activité phosphatasique alcaline dans l'appareil de Golgi et au niveau des vésicules cytoplasmiques apicales. *C. R. Acad. Sci. Ser. D.* *282*:1601-1604.

Davidse, L. C. (1979). Mode of action and selectivity of benzimidazole compounds. In *Systemfungizide, Abhandlungen der Akademie der Wissenschaften der DDR*, H. Lyr and C. Polter (Eds.). Akademie-Verlag, Berlin, pp. 277-286.

Davidse, L. C., and Flach, W. (1977). Differential binding of methyl benzimidazol-2-yl carbamate to fungal tubulin as a mechanism of resistance to this antimitotic agent in mutant strains of *Aspergillus nidulans*. *J. Cell. Biol.* *72*:174-193.

Dow, J. M., and Rubery, P. M. (1975). Hyphal tips bursting in *Mucor rouxii*: Antagonistic effects of calcium ions and acid. *J. Gen. Microbiol.* *91*:425-428.

Dow. J. M., and Villa, V. D. (1980). Oligoglucuronide production in *Mucor rouxii*: Evidence for a role for endohydrolases in hyphal extension. *J. Bacteriol.* *142*:939-944.

Dow, J. M., Carreon, R. R., and Villa, V. D. (1981). Role of membranes of mycelial *Mucor rouxii* in synthesis and secretion of cell wall matrix polymers. *J. Bacteriol.* *145*:272-279.

Farkâs, V. (1979). Biosynthesis of cell walls of fungi. *Micbiol. Rev.* *43*:117-144.

Fèvre, M. (1977). Subcellular localization of glucanase and cellulase in *Saprolegnia monoica* Pringsheim. *J. Gen. Microbiol.* *103*:287-295.

Fèvre, M. (1979). Glucanases, glucan synthases and wall growth in *Saprolegnia monoica*. In *Fungal Walls and Hyphal Growth*. J. H. Burnett and A. P. J. Trinci (Eds.). Cambridge University, Cambridge, pp. 225-263.

Fèvre, M., and Dumas, C. (1977). β-Glucan synthetases from *Saprolegnia monoica*. *J. Gen. Microbiol.* *103*:297-306.

Fèvre, M., and Rougier, M. (1981). β-1-3 and β-1-4- synthesis by membrane fractions from the fungus *Saprolegnia*. *Planta* *151*:232-241.

Fischer, F. G., and Werner, G. (1955). Eine Analyse des Chemotropismus einiger Pilze, insbesondere der Saprolegniaceen. *Hoppe-Seyler's Z. Physiol. Che,.* *300*:211-236.

Galpin, M. F. J., and Jennings, D. H. (1975). Histochemical study of the hyphae and the distribution of adenosine triphosphatase in *Dendryphiella salina*. *Trans. Br. Mycol. Soc. 65*:151-153.
Galpin, M. F. J., Jennings, D. H., Oates, K., and Hobot, J. (1978). Localization by Xray microanalysis of soluble ions, particularly potassium and soidum, in fungal hyphae. *Exp. Mycol. 2*:258-269.
Galun, M., Braun, A., Frensdorff, A., and Galun, E. (1976). Hyphal walls of isolated lichen fungi. Autoradiographic localization of precursor incorporation and binding of fluoresein-labelled lectins. *Arch. Microbiol. 108*:9-16.
Girbardt, M. (1957). Der Spitzenkörper von *Polystictus versicolor (L). Planta 50*:47-59.
Girbardt, M. (1965). Eine Zielschnittmethode für Pilzzellen. *Microskopie 20*:254-264.
Girbardt, M. (1969). Die Ultrastruktur der Apikalregion von Pilzhyphen. *Protoplasma 67*:413-441.
Girbardt, M. (1970). Morphologische Korrelate von Transport Vorgängen in Zellen und Organellen. In *Stoff Transport und Stoffverteilung in Zellen Höherer Pflanzen*. K. Mothes, E. Müller, A. Nelles, and D. Neumann (Eds.). Akademie-Verlag, Berlin, pp. 25-37.
Girbardt, M. (1979a). Verteilung and experimentelle Beeinflussung cytoplasmatischer Microtubuli in der Spitzenzelle von *Trametes (= Polystictus) versicolor*. In *Systemfungizide, Abhandlungen der Akademie der Wissenschaften der DDR*, No. 2. H. Lyr and C. Polter (Eds.). Akademie-Verlag, Berlin, pp. 265-276.
Girbardt, M. (1979b). A microfilamentous septal belt (FSB) during induction of cytokinesis in *Trametes versicolor* (L. ex Fr.). *Exp. Mycol. 3*:215-228.
Gooday, G. W. (1971). An autoradiographic study of hyphal growth. *J. Gen. Microbiol. 67*:125-133.
Gooday, G. W. (1975). Chemotaxis and chemotropism in fungi and algae. In *Primitive Sensory and Communication Systems*, M. J. Carlile (Ed.). Academic, London, pp. 155-204.
Gooday, G. W. (1978). The enzymology of hyphal growth. In *The Filamentous Fungi, Developmental Mycology*. Vol. 3. J. E. Smith and D. R. Berry (Eds.). Edward Arnold, London, pp. 51-77.
Gooday, G. W. (1979). Chitin synthesis and differentiation in *Coprinus cinereus*. In *Fungal Walls and Hyphal Growth*. J. H. Burnett and A. P. J. Trinci (Eds.). Cambridge University, Cambridge, pp. 203-223.
Gooday, G. W. (1982). Metabolic control of fruitbody morphogenesis in *Coprinus cinereus*. In *Basidium and Basidiocarp: Evolution, Cytology, Function, and Development*. K. Wells and E. K. Wells (Eds.). Springer-Verlag, New York.

Gooday, G. W. (1983). Hormones and sexuality. In *Secondary Metabolism and Differentiation in Fungi.* J. W. Bennett and A. Ciegler (Eds.). Marcel Dekker, New York.

Gooday, G. W., and Trinci, A. P. J. (1980). Wall structure and biosynthesis in fungi. In *The Eukaryotic Microbial Cell.* G. W. Gooday, D. Lloyd, and A. P. J. Trinci (Eds.). Cambridge University, Cambridge, pp. 207-251.

Gradmann, P., Hansen, U. P., Long, W. S., Slayman, C. L., and Warncke, J. (1978). Current-voltage relationships for the plasma membrane and its principle electrogenic pump in *Neurospora crassa*: 1. Steady-state conditions. *J. Membr. Biol. 39*:333-367.

Grove, S. N. (1978). The cytology of hyphal tip growth. In *The Filamentous Fungi, Developmental Mycology*, Vol. 3. J. E. Smith and D. R. Berry (Eds.). Edward Arnold, London, pp. 28-50.

Grove, S. N., and Bracker, C. E. (1970). Protoplasmic organization of hyphal tips among fungi: Vesicles and Spitzenkörper. *J. Bacteriol. 104*:989-1009.

Grove, S. N., and Sweigard, J. A. (1980). Cytochalasin A inhibits spore germination and hyphal tip growth in *Gilbertella persicaria. Exp. Mycol. 4*:239-250.

Grove, S. N., Bracker, C. E., and Morré, D. J. (1970). An ultrastructural basis for hyphal tip growth in *Pythium ultimum. Am. J. Bot. 57*:245-266.

Gull, K. (1975). Cytoplasmic microfilament organization in two basidiomycete fungi. *J. Ultrastruct. Res. 51*:226-232.

Gull, K., and Trinci, A. P. J. (1973). Griseofulvin inhibits fungal mitosis. *Nature 244*:292-294.

Gull, K., and Trinci, A. P. J. (1974). Effects of griseofulvin on the mitotic cycle of the fungus *Basidiobolus ranarum. Arch. Mikrobiol. 95*:57-65.

Harold, F. M. (1977). Ion currents and physiological function in microorganisns. *Annu. Rev. Microbiol. 31*181-203.

Harold, F. M. (1982). Pumps and currents: A biological perspective. *Curr. Top. Membr. Transp. 26*:485-516.

Harold, R. L., and Harold, F. M. (1980). Oriented growth of *Blastocladiella emersonii* in gradients of ionophores and inhibitors. *J. Bacteriol. 144*:1159-1167.

Heath, I. B., and Heath, M. C. (1978). Microtubules and organelle movements in the rust fungus *Uromyces phaseoli* var. *vignae. Cytobiologie 16*:393-411.

Heath, I. B., Heath, M. C., and Herr, F. (1981). Motile systems in fungi. In *Eukaryotic and Prokaryotic Flagella.* W. B. Amos and J. G. Duckett (Eds.). Cambridge University, Cambridge.

Hill, T. W., and Mullins, J. T. (1979). Association of latent cellulase activity with a plasma membrane fraction from vegetative hyphae of *Achlya ambisexualis. Mycologia 61*:1227-1239.

Hill, T. W., and Mullins, J. T. (1980a). Hyphal tip growth in *Achlya*. I. Cytoplasmic organization. *Can. J. Microbiol.* *26*:1132-1140.

Hill, T. W., and Mullins, J. T. (1980b). Hyphal tip growth in *Achlya*. II. Subcellular localization of cellulase and associated enzymes. *Can. J. Microbiol.* *26*:1141-1146.

Hoch, H. C., and Howard, R. J. (1980). Ultrastructure of freeze-substituted hyphae of the basidiomycete *Laetisaria arvalis*. *Protoplasma* *103*:281-297.

Hoch, H. C., Hanssler, G., and Reisener, H. -J. (1979). Cytochemical localization of N-acetyl-β-D-glucosaminidase in hyphae of *Mucor racemosus*. *Exp. Mycol.* *3*:164-173.

Howard, R. J. (1981). Ultrastructural analysis of hyphal tip cell growth in fungi: Spitzenkörper, cytoskeleton and endomembranes after freeze-substitution. *J. Cell. Sci.* *48*:89-103.

Howard, R. J., and Aist, J. R. (1977). Effects of MBC on hyphal tip organization, growth, and mitosis of *Fusarium acuminatum*, and their antagonism by D_2O. *Protoplasma* *92*:195-210.

Howard, R. J., and Aist, J. R. (1979). Hyphal tip ultrastructure of the fungus *Fusarium*: Improved preservation by freeze-substitution. *J. Ultrastruct. Res.* *66*:224-234.

Howard, R. J., and Aist, J. R. (1980). Cytoplasmic microtubules and fungal morphogenesis: Ultrastructural effects of methyl benzimidazole-2-yl carbamate determined by freeze-substitution of hyphal tip cells. *J. Cell Biol.* *87*:55-64.

Hunsley, D., and Burnett, J. H. (1970). The ultrastructural architecture of the walls of some hyphal fungi. *J. Gen. Microbiol.* *62*:203-208.

Hunsley, D., and Kay, D. (1976). Wall structure of the Neurospora hyphal apex: Immunofluorescent localization of wall surface antigens. *J. Gen. Microbiol.* *95*:233-248.

Issac, S., Ryder, N. S., and Peberdy, J. F. (1978). Distribution and activation of chitin synthase in protoplast fractions released during the lytic digestion of *Aspergillus nidulans* hyphae. *J. Gen. Microbiol.* *105*:45-50.

Jaffe, L. (1968). Localization in the developing *Fucus* egg and the general role of localizing current. *Adv. Morphog.* *7*:295-328.

Jaffe, L. F. (1979). Control of development by ionic currents. In *Membrane Transduction Mechanisms*. R. A. Cone and J. E. Dowling (Eds.). Raven Press, New York. pp. 199-231.

Jaffe, L. F., and Nuccitelli, R. (1977). Electrical controls of development. *Annu. Rev. Biophys. Bioeng.* *6*:445-476.

Jennings, D. H. (1979). Membrane transport and hyphal growth. In *Fungal Walls and Hyphal Growth*. J. H. Burnett and A. P. J. Trinci (Eds.). Cambridge University, Cambridge, pp. 279-294.

Kamada, T., Fujii, T., and Takemaru, T. (1980). Stipe elongation during basidiocarp maturation in *Coprinus macrorhizus:* Changes in activity of cell wall lytic enzymes. *Trans. Mycol. Soc. Jap. 21*:359-367.

Korn, E. D., (1980). Biochemistry of actomyosin-dependent cell motility in *Acanthamoeba castellanii.* In *The Eukaryotic Microbial Cell.* G. W. Gooday, D. Lloyd, and A. P. J. Trinci (Eds.). Cambridge University, Cambridge, pp. 253-272.

Kritzman, G., Chet, I., and Henis, Y. (1978). Localization of β-(1,3)-glucanase in the mycelium of *Sclerotium rolfsii. J. Bacteriol. 134*:470-475.

McLure, W. K., Park, D., and Robinson, P. M. (1968). Apical organization in the somatic hyphae of fungi. *J. Gen. Microbiol. 50*:177-182.

Madelin, M. F., Toomer, D. K., and Ryan, J. (1978). Spiral growth of fungus colonies. *J. Gen. Microbiol. 106*:73-80.

Marchant, R., and Smith, D. G. (1968). A serological investigation of hyphal growth in *Fusarium culmorum. Arch. Microbiol. 63*:85-94.

Marriott, M. S. (1977). Mannan-protein location and biosynthesis in plasma membranes from the yeast form of *Candida albicans. J. Gen. Microbiol. 103*:51-59.

Meyer, R., Parish, R. W., and Hohl, H. R. (1976). Hyphal tip growth in *Phytophthora.* Gradient distribution and ultra histochemistry of enzymes. *Arch. Microbiol. 110*:215-224.

Middlebrook, M. J., and Preston, R. D. (1952a). Spiral growth and spiral structure. III. Wall structure in the growth zone of *Phycomyces. Biochim. Biophys. Acta 9*:32-48.

Middlebrook. M. J., and Preston, R. D. (1952b). Spiral growth and spiral structure. IV. Growth studies and mechanical constants in the cell wall. *Biochim, Biophys. Acta 9*:115-126.

Molano, J., Polachek, I., Duran, A., and Cabib, E. (1979). An endochitinase from wheat germ. Activity on nascent and preformed chitin. *J. Biol. Chem. 254*:4901-4907.

Morris, N. R. (1980). Chromosome structure and the molecular biology of mitosis in eukaryotic micro-organisms. In *The Eukaryotic Microbial Cell.* G. W. Gooday, D. Lloyd and A. P. J. Trinci (Eds.). Cambridge University, Cambridge, pp. 41-75.

Morris, N. R., Lai, M. H., and Oakley, C. E. (1979). Identification of a gene for α-tubulin in *Aspergillus nidulans. Cell 16*:437-442.

Mullins, J. T. (1979). A freeze-fracture study of hormone-induced branching in the fungus *Achlya. Tissue Cell 11*:585-595.

Mullins, J. T., and Ellis, E. A. (1974). Sexual morphogenesis in *Achlya*: Ultrastructural basis for the hormonal induction of antheridial hyphae. *Proc. Nat. Acad. Sci. USA 71*:1347-1350.

Musgrave, A., Ero, L., Scheffer, R., and Oehlers, E. (1977). Chemotropism of *Achlya bisexualis* germ hyphae to casein hydrolysate and amino acids. *J. Gen. Microbiol. 101*:65-70.

Nolan, R. A., and Bal, A. K. (1974). Cellulase localization in hyphae of *Achlya ambisexualis*. *J. Bacteriol. 117*:840-843.

Nuccitelli, R., and Jaffe, L. F. (1974). Spontaneous current pulses through developing fucoid eggs. *Proc. Nat. Acad. Sci USA 71*: 4855-4859.

Nuccitelli, R., and Jaffe, L. F. (1975). The pulse current pattern generated by developing fucoid eggs. *J. Cell Biol. 64*:636-643.

Nuccitelli, R., and Jaffe, L. F. (1976). The ionic components of the current pulses generated by developing fucoid eggs. *Dev. Biol. 49*:518-531.

Patton, A. M., and Marchant, R. (1978). An ultrastructural study of septal development in hyphae of *Polyporus biennis*. *Arch. Microbiol. 118*:271-277.

Polachek, I., and Rosenberger, R. F. (1975). Autolytic enzymes in hyphae of *Aspergillus nidulans*: Their action on old and newly formed walls. *J. Bacteriol. 121*:332-337.

Polachek, I., and Rosenberger, R. F. (1978). Distribution of autolysins in hyphae of *Aspergillus nidulans*: Evidence for a lipid-mediated attachment to hyphal walls. *J. Bacteriol. 135*:741-747.

Porter, K. R., and Tucker, J. B. (1981). The ground substance of the living cell. *Sci. Am. 244*:40-51. (No. 3)

Prick, G., and Diekmann, H. (1979). A chitin-binding lectin in *Neurospora crassa*. *FEMS Microbiol. Lett. 6*:427-429.

Prosser, J. I. (1979). Mathematical modelling of mycelial growth. In *Fungal Walls and Hyphal Growth*. J. H. Burnett and A. P. J. Trinci (Eds.). Cambridge University, Cambridge, pp. 359-384.

Quinlan, R. A., Pogson, C. I., and Gull, K. (1980). The influence of the microtubule inhibitor, methyl benzimidazol-2-yl-carbamate (MBC) on nuclear division and the cell cycle in *Saccharomyces cerevisiae*. *J. Cell Sci. 46*:341-352.

Robertson, N. F. (1958). Observations on the effect of water on the hyphal apices of *Fusarium oxysporum*. *Ann. Bot. N.S. 22*:159-173.

Robertson, N. F. (1965). The fungal hypha. *Trans. Br. Mycol. Soc. 48*:1-8.

Robinson, P. M. (1973a). Autotropism in fungal spores and hyphae. *Bot. Rev. 39*:367-384.

Robinson, P. M. (1973b). Oxygen-positive chemotropic factor for fungi? *New Phytol. 72*:1349-1356.

Robinson, P. M. (1973c). Chemotropism in fungi. *Trans. Br. Mycol. Soc. 61*:303-313.

Robinson, P. M. (1980). Autotropism in germinating arthrospores of *Geotrichum candidum*. *Trans. Br. Mycol. Soc. 75*:151-153.

Robinson, P. M., and Griffith, P. J. (1977). Effect of restricted aeration on chemotropism, morphogenesis and polarity of lateral branch induction in *Geotrichum candidum* Link ex Pers. *Trans. Br. Mycol. Soc. 68*:311-314.

Robinson, P. M., Park, D., and Graham, T. A. (1968). Autotropism in fungal spores. *J. Exp, Bot. 19*:125-134.

Roobol, A., Gull, K., and Pogson, C. I. (1977). Evidence that griseofulvin binds to a microtubule associated protein. *FEBS Lett. 75*:149-153.

Rosenberger, R. F. (1979). Endogenous lytic enzymes and wall metabolism. In *Fungal Walls and Hyphal Growth*. J. H. Burnett and A. P. J. Trinci (Eds.). Cambridge University, Cambridge, pp. 265-277.

San-Blas, G. (1979). Biosynthesis of glucans by cellular fractions of *Paracoccidiodes brasiliensis*. *Exp. Mycol. 3*:249-258.

Sheir-Neiss, G., Lai, M. H., and Morris, N. R. (1978). Identification of a gene for β-tubulin in *Aspergillus nidulans*. *Cell 15*: 638-647.

Slayman, C. L. (1970). Movement of ions and electrogenesis in micro-organisms. *Am. Zool. 10*:377-392.

Slayman, C. L., and Gradmann, D. (1975). Electrogenic proton transport in the plasma membrane of *Neurospora*. *Biophys. J. 15*:968-971.

Slayman, C. L., and Slayman, C. W. (1962). Measurement of membrane potentials in *Neurospora*. *Science 136*:876-877.

Soll, D. R., Bromberg, R., and Sonneborn, D. R. (1969). Zoospore germination in the water mold, *Blastocladiella emersonii*. I. Measurement of gernmination and sequence of subcellular morphological changes. *Dev. Biol. 20*:218-235.

Stadler, D. R. (1952). Chemotropism in *Rhizopus nigricans*. *J. Comp. Physiol. 39*:449-474.

Stadler, D. R. (1953). Chemotropism in *Rhizopus nigricans*: II. The action of plant juices. *Biol. Bull. 104*:100-108.

Stump, R. F., Robinson, K. R., Harold, R. L., and Harold, F. M. (1980). Endogenous electrical currents in the water mold *Blastocladiella emersonii* during growth and sporulation. *Proc. Nat. Acad. Sci. USA 77*:6673-6677.

Thomas, D. des S. (1978). Cytochalasin effects in plants and eukaryotic microbial systems. In *Cytochalasins—Biomedical and Cell Biological Aspects*. S. W. Tanenbaum (Ed.). Elseview, Amsterdam, pp. 257-276.

Thomas, D. des S., and Mullins, J. T. (1969). Cellulase induction and wall extension in the water mold *Achlya ambisexualis*. *Physiol. Plant. 22*:347-353.

Trinci, A. P. J., and Collinge, A. J. (1975). Hyphal wall growth in *Neurospora crassa* and *Geotrichum candidum*. *J. Gen. Microbiol. 91*:355-361.

Trinci, A. P. J., Saunders, P. T., Gosrani, R., and Campbell, K. A. S. (1979). Spiral growth of mycelial and reproductive hyphae. *Trans. Br.Mycol. Soc. 73*:283-293.

Turian, G. (1976). Reducing power of hyphal tips and vegetative tips and vegetative apical dominance in fungi. *Experientia 32*: 989-991.

Turian, G. (1978). Spitzenkörper, centre of reducing power in growing hyphae of two septomycetous fungi. *Experientia 34*:1277-1279

Valk, P. van der, and Wessels, J. G. H. (1977). Light and electron microscopic autoradiograph of cell-wall regeneration by *Schizophyllum commune* protoplasts. *Acta Bot. Neerl. 26*:43-52.

Van Brunt, J., and Harold, F. M. (1980). Ionic control of germination of *Blastocladiella emersonii* zoospores. *J. Bacteriol. 141*:735-744.

Vermeulen, C. A. Raeven, M. B. J. M., and Wessels, J. G. H. (1979). Localization of chitin synthase activity in subcellular fractions of *Schizophyllum commune* protoplasts. *J. Gen. Microbiol. 114*87-97.

Wang, M. L., and Bartnicki-Garcia, S. (1976). Synthesis of β-1-3-glucan microfibrils by a cell-free extract of *Phytophthora cinnamomi*. *Arch. Biochem. Biophys. 175*:351-354.

Water, R. D., and Kleinsmith, L. J. (1976). Identification of α- and β-tubulin in yeast. *Biochem. Biophys. Res. Comm. 70*:704-708.

Weisenseel, M. H., and Jaffe, L. F. (1976). The major growth current through lily pollen tubes enters as K^+ and leaves as H^+. *Planta 133*:1-7.

Weisenseel, M. H., Nuccitelli, R., and Jaffe, L. F. (1975). Large electrical currents traverse growing pollen tubes. *J. Cell Biol. 66*:556-567.

Wessels, J. G. H., and Sietsma, J. H. (1979). Wall structure and growth in *Schizophyllum commune.* In *Fungal Walls and Hyphal Growth.* J. H. Burnett and A. P. J. Trinci (Eds.). Cambridge University, Cambridge, pp. 27-48.

Wright, B. E. (1975). Usefulness of developmental mutants in the analysis of biochemical differentiation. In *Microbiology–1975.* D. Schlessinger (Ed.). American Society for Microbiology, Washington, D. C., pp. 500-507.

Zalokar, M. (1959). Growth and differentiation of *Neurospora* hyphae. *Am. J. Bot. 46*:602-610.

Zalokar, M. (1965). Integration of cellular metabolism. In *The Fungi, The Fungal Cell,* Vol. 1, G. C. Ainsworth and A. S. Sussmann (Eds.). Academic, New York, pp. 326-377.

12
Hyphal Growth Patterns

J. I. PROSSER *Marischal College, University of Aberdeen, Aberdeen, Scotland*

INTRODUCTION

The hyphal growth form is characteristic of two groups of microorganisms, the filamentous prokaryotic Actinomycetes and filamentous fungi, where it is most highly developed. It confers on these organisms several advantages over the unicellular growth of yeasts and the majority of bacteria. Firstly, material may be transported along hyphae in both directions, unless restricted by septa, and growth may proceed without changing the surface area/volume ratio. Growth of spherical cells requires continual adjustment to a reduction in this ratio until cell division occurs. Polarization of growth at the hyphal apex allows organized growth away from the point of inoculum at a rate which is usually greater than the rate of diffusion of substrates, such that hyphae at the edge of a colony are continually encountering fresh medium. It also enables growth into solid media which may proceed at the same rate as surface growth, allowing utilization of substrates which would otherwise only become available by diffusion.

These advantages are of particular significance for growth over solid surfaces, which may be themselves substrates or inert surfaces between two sources of nutrients. In liquid culture the translocation of nutrients is not of such importance and taxic response may be more efficient than the tropic responses of hyphae.

For maximum efficiency, hyphal growth must proceed in two or three dimensions in solid and liquid media, respectively. This is achieved by branching, which allows the coverage of large surface areas and communication and translocation between different regions. In many ways branch formation is equivalent to cell division in unicells, but it provides a much greater regulation of growth. Thus branching frequency will increase in areas of high nutrient concentration, while low concentrations will lead to sparsely branched colonies, but with resources channeled such that extension is still near the maximum rate. Unicells cannot control their growth in this manner and colonies are formed by cells piling up and pushing each other apart. Supply of substrates soon becomes dependent on diffusion and secondary metabolites accumulate, restricting growth away from the inoculum source, and such colonies are typically small and discrete.

The morphology of fungal colonies, particularly at the colony periphery, to a large extent reflects the nature and extent of hyphal branching. Discussion of hyphal growth patterns therefore requires consideration of the factors controlling the formation of branches and their subsequent growth. The growth kinetics of vegetative hyphae of "typical" colonies will be discussed, followed by a description of colony morphology. "Atypical" colonies resulting from mutations, environmental changes, and paramorphogens will then be discussed. This treatment will not attempt to be exhaustive, but will concentrate on studies which give information on the mechanisms of branch formation. Finally, the various proposed mechanisms will be discussed. Most of the relevant work involves the study of colony growth on agar where growth may be restricted to two dimensions. Growth in liquid culture is complicated by its three-dimensional nature and provides information on morphology rather than kinetics. The relationship between growth in the two media has been studied and shown to be equivalent, and the same mechanisms for growth and branching are thought to operate.

II. COLONY GROWTH

Vegetative fungal hyphae show some degree of variability between species in features such as hyphal diameter, septation, intercalary and apical compartment length, and tip shape. Despite this, the hyphal and colony growth kinetics of all fungi which have been studied in detail are essentially the same. The following section describes such kinetics, and differences between species will generally be of a quantitative rather than qualitative nature.

A. Germ Tube Growth

Smith (1924) found that germ tubes of *Botrytis cinerea* increased in length exponentially immediately following spore germination and Plomley (1959) found similar but slightly confusing results with *Chaetomium globosom*. Trinci (1969,1971a) has now shown that the growth of germ tube hyphae of a range of fungi occurs in four phases. Following a lag phase, the spore germinates and the extension rate is proportional to the germ tube length; that is, growth is exponential. In *Rhizopus stolonifer* this exponential phase continues up to a length of 40 μm and is followed by a deceleration phase and a final linear phase, which may potentially continue indefinitely. The specific growth rate during the exponential phase may be up to nine times that for mycelial growth in the equivalent liquid medium because of utilization of endogenous spore reserves along with nutrients in the medium.

Though some aerial hyphae exhibit intercalary growth, extension during surface growth is always by incorporation at the hyphal tip of materials transported from distal regions. The deceleration phase begins when translocation becomes impossible along the whole length of the hypha. Subsequent linear growth is supported by a constant length of hypha behind the tip.

B. Branching

1. *Significance for Colony Growth Kinetics*

A linear rate of extension imposes a restriction to overall growth, which in unicells proceeds at an exponential rate, and this is solved by branch formation. Smith (1924) observed an exponential increase in the combined lengths of parent and branch hyphae, though individual hyphae were extending linearly, and Plomley (1959) observed an exponential increase in total mycelial length in *C. globosum*. He also noted that hyphae grew into the agar such that young colonies may be considered hemispherical. Trinci (1969) found that hyphal density decreased exponentially with agar depth in *Aspergillus nidulans*.

Trinci (1974) studied early colony growth kinetics of *A. nidulans*, *Neurospora crassa*, *Mucor hiemalis*, and *Geotrichum candidum*, preventing submerged growth by inoculating onto a cellophane membrane

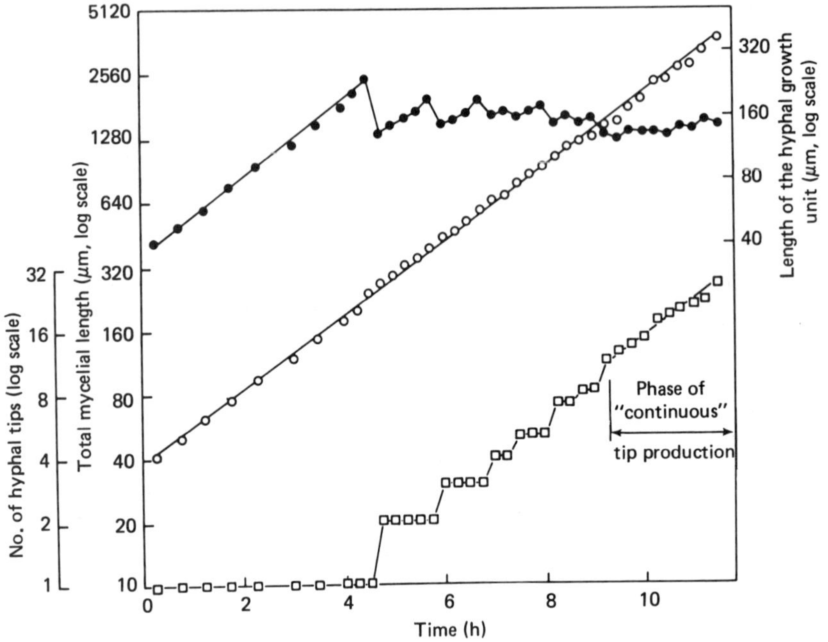

FIG. 1 Growth of a mycelium of *Geotrichum candidum* on solid medium at 25°C (□, number of tips; ○, total mycelial length; ●, hyphal growth unit length). (From Trinci, 1974.)

placed on the agar surface. All four species showed similar kinetics, with the results for *G. candidum* illustrated in Fig. 1. Four features are exhibited: (1) Total mycelial length increases exponentially for at least 11 h; (2) the first branch is formed before the germ tube hypha is growing linearly, though possibly not before the deceleration phase; (3) branch production is initially discontinuous, but after 10 branches have been formed the number increases exponentially with a specific rate equal to that for the total mycelial length; and (4) this specific rate is found to be equal to the specific growth rate during exponential growth in liquid medium, measured in terms of dry weight increase. Also plotted is the hyphal growth unit, defined as the total mycelial length divided by the number of tips. This will be discussed further below. Exponential growth is therefore achieved by the exponential production of branches, which themselves ultimately grow at a linear rate.

In fact, though branches in young colonies typically attain the linear extension rate of their parents, there is some variation in initial

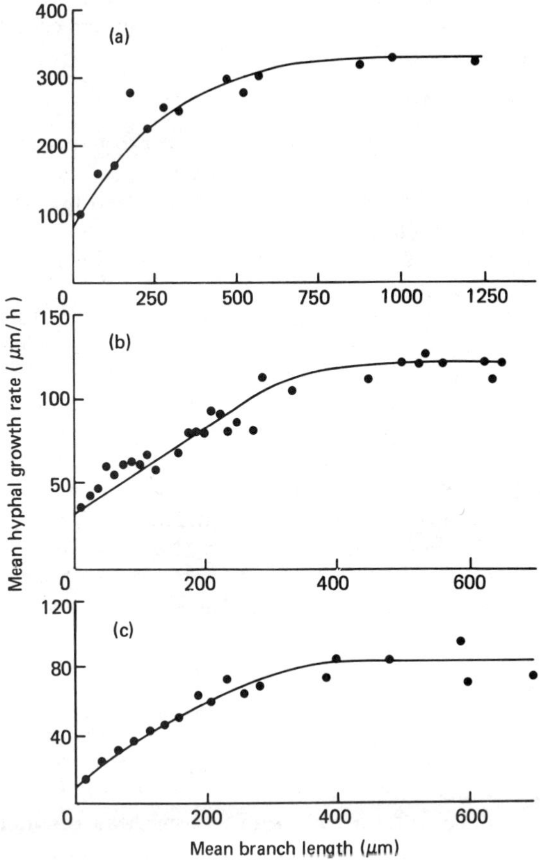

FIG. 2 Relationship between mean branch length and mean extension rate of (a) *Mucor hiemalis*, (b) *Geotrichum candidum*, and (c) *Aspergillus nidulans* at 25°C. (From Trinci, 1974.)

rates. Smith (1924) found that some branches grew faster than their parents, while the growth of others ceased. Trinci (1974) attributed the cessation of growth of some branches of *M. hiemalis* to competition between adjacent hyphae. There is, however, a general correlation between branch length and extension rate, as shown in Fig. 2. This implies exponential growth, but the fact that the y intercepts are all significantly greater than zero indicates some initial contribution to growth by the parent hypha. This has been confirmed for *A. nidulans* (Fiddy and Trinci, 1976a) and for *G. candidum* (Fiddy and Trinci, 1976b), where the combined length of branches and intercalary compartments increased exponentially up to a length of 325 μm.

2. Effect on Parental Extension Rate

Lateral Branching: Branches are typicaly formed laterally in acropetal succession some distance from the tip, and in most cases their formation does not reduce the rate of extension of the parent hypha. One exception is *Basidiobolus ranarum* (Robinow, 1963), where branch formation is preceded by a 30-50% decrease in extension rate. This, however, is associated with mitosis rather than branch initiation itself, which occurs when extension rate is increasing. In general, therefore, the supply of material by a parental hypha to a lateral branch does not affect the extension rate of the former.

Apical Branching: Emerson (1955) has described apical branching in *Allomyces* and Trinci (1970) has found branches initiated within 4 μm of hyphal tips during normal growth of *Geotrichum lactis* and *A. nidulans*. The latter species showed both subapical branching and true dichotomous branching, with an 18% decrease in extension rate prior to branch formation. Each branch then quickly reached the parental extension rate. In *G. lactis* branching was always dichotomous but was not associated with a reduction in extension rate. In both species three branches were occasionally initiated.

C. The Peripheral Growth Zone

Colony radial growth kinetics are essentially those of germ tube hyphae. In *A. nidulans* Trinci (1969) found a lag phase, an exponential phase until the hyphal front had extended 100 μm from the inoculum circumference, and a deceleration phase until a radius of 1.1 mm was reached. Subsequent expansion took place at a constant linear rate. However, exponential mycelial growth throughout a colony cannot be supported indefinitely. Eventually growth at the center decelerates as a result of nutrient limitation or inhibition by secondary metabolites. This occurs more quickly where branching density is high and is followed by colony differentiation as described by Yanagita and Kogane (1962) and Park and Robinson (1966).

Increase in total mycelial length is then restricted to the colony margin, but well before this time colony extension will have been limited to growth in a smaller marginal region, the peripheral growth zone (Trinci, 1971b). This is defined as the maximum length of hypha required to support tip growth and represents a marginal ring of constant width within which growth is exponential, with a specific rate equal to that in liquid culture. Radial extension results entirely from growth within this zone. Its length may be estimated by cutting diagonally through a hyphal front such that hyphae are severed at different distances from their tips. The peripheral growth zone is then equivalent to the shortest length of hypha that permits unaltered subsequent growth; alternatively, it may be inferred as the hyphal length at which the maximum extension rate is reached.

In fungi which form complete septa, for example, *G. candidum*, the peripheral growth zone length is that of the apical compartment. However, the constant rate of extension in this species, despite variation in apical compartment length during growth, suggests that the rear of the apical compartment contributes little to tip extension. Where septa contain pores, for example, in *N. crassa*, the peripheral growth zone may extend for several intercalary compartments. Aseptate species generally have large peripheral growth zones, for example, 8.6 mm in *R. stolonifer*, compared to 0.423 mm in *G. lactis*. Its physical representation in such species awaits elucidation and, in particular, quantification of the mechanisms involved in translocation of material along hyphae.

As a physiological unit, the peripheral growth zone has been well characterized in many fungi and clarifies the relationship between the colony radial growth rate on solid medium (K_r) and the specific growth rate in liquid culture (μ). This may be represented in the equation

$$K_r = w\mu \qquad (1)$$

where w is the peripheral growth zone length. This relationship explains the effect of branching on colony radial growth (Trinci, 1971b, 1973a). Thus temperature, which affects growth rate but not branching frequency, does not affect w. The ratio K_r/μ therefore remains constant at different temperatures. Supplementation of media with sorbose does not affect specific growth rate, but increases branching frequency. Material is redistributed into new branches rather than into existing leading hyphae, reducing w and consequently K_r/μ. The value of w also varies with glucose concentration (Trinci, 1969; Fiddy and Trinci, 1975), but here the effect is not on branching frequency but on the subsequent growth of branches. In an attempt to quantify the effect of branching on K_r, Morrison and Righelato (1974) modified Eq. (1) as follows:

$$K_r = \mu G k \qquad (2)$$

where G is the hyphal growth unit and k is a constant. They found reasonable constancy for k in several strains of *Penicillium chrysogenum* with different branching frequencies.

TABLE 1 Diameters and Extension Rates of Hyphae in Colonies of *Neurospora crassa*[a] and *Coprinus disseminatus*[b]

	Hyphal diameter and standard deviation where stated (μm)	Linear extension rate and standard deviation where stated (μm/min)
N. crassa		
Leading hyphae	13.1	38.0 (7.2)
Primary branches	8.9	27.4 (6.3)
Secondary branches	6.3	15.2 (4.9)
Germ tube hyphae	3.7	2.6 (1.3)
C. disseminatus		
Leading hyphae	3.88 (0.574)	4.31 (0.53)
Primary branches	4.69 (0.448)	2.81 (0.41)
Leading hyphae		3.82
Primary branches		2.55
Secondary branches		0.67

[a]Data from Steele and Trinci (1975b).

[b]Data for two colonies compiled from Butler (1961). Hyphal diameters were given for only one colony.

III. COLONY MORPHOLOGY

A. Qualitative Analysis

1. *Hyphal Characteristics*

Although the morphology and growth characteristics of vegetative hyphae vary continuously throughout colony development, it is convenient to consider separately hyphae in young colonies and hyphae at the margin of mature colonies. The former case encompasses initial growth following spore germination, when total mycelial length is increasing exponentially and growth is not affected by nutrient limitation or inhibitory effects of secondary metabolites. In such colonies all hyphae typically have similar diameters, extension zones, and maximum extension rates. Branches are subtended on the average at an angle of 90° and apical and subapical branching occurs rarely.

At the margin of mature colonies a hierarchy of vegetative hyphae exists. Smith (1924), Plomley (1959), and Butler (1961) noted that in

Botrytis cinerea, Chaetomium globosum, and *Coprinus disseminatus*, respectively, branches had smaller diameters than parent hyphae and that they, in turn, subtended thinner branches. Butler also found that the extension rates of leading hyphae, primary branches, and secondary branches were in the proportion 100:66:18. Trinci (1970) observed that primary branches in *G. lactis* and *A. nidulans* had extension rates 31 and 21% less than their respective parent hyphae. Steele and Trinci (1975a,b) obtained similar results for *N. crassa* and their data and those of Butler (1961) are summarized in Table 1.

The maximum extension rate of leading hyphae of *N. crassa* is 250% greater than that of germ tube hyphae, and even secondary branches extend at a faster rate. This effect is not so marked in other species. In *A. nidulans* leading hyphae extend 59% faster than germ tube hyphae, while in *M. hiemalis* and *G. candidum* the increases are only 16 and 10% greater, respectively. There is a general correlation between hyphal diameter and maximum extension rate, but Butler observed that primary branches grew slower than leading hyphae of the same diameter. Possible reasons for this apical dominance will be discussed below. In addition, faster-growing, wider leading hyphae subtended primary branches which were faster growing and wider than other primary branches. The factors which determine the diameter of newly formed branches are not understood.

Apical and subapical branching occur more frequently at the periphery of mature colonies and hyphae are oriented radically outwards with branch angles of less than 90°. Figure 3 shows two stages in the development of a colony of *Coprinus sterquilinus* (Buller, 1931). Branches in the younger colony (Fig. 3a) are subtended at approximately 90° from the parent hypha, but may then grow away from the colony center as neighboring hyphae are approached. In the older colony, marginal hyphae are all radially oriented with branch angles less than 90°. Although this is a common feature of hyphal growth, the situation is more complex in *Coprinus*. Most of the hyphae in the younger colony are haploid, while those in the older colony are dikaryotic. Buller (1933) found that branch angles in *Coprinus fimetarius* ranged from 70 to 90° in homokaryons and from 20 to 40° in dikaryons. Branches in *Coprinus* are always formed immediately behind septa and these differences in branch angle may be involved with the formation of clamp connections, which occurs in dikaryons but not monokaryons. Figure 3b also demonstrates the occurrence of anastomoses in older regions, such that the mycelium is linked in a continuous network, and later-formed hyphae tend to be thinner and lack directional orientation.

2. *Mycelial Development*

The development of these differences between hyphae in mature and young colonies has been followed in *N. crassa* by C. J. Morris and J. I. Prosser (unpublished results). The first parameter to change is the branch angle, which remains at 90° for 21 h of growth in defined

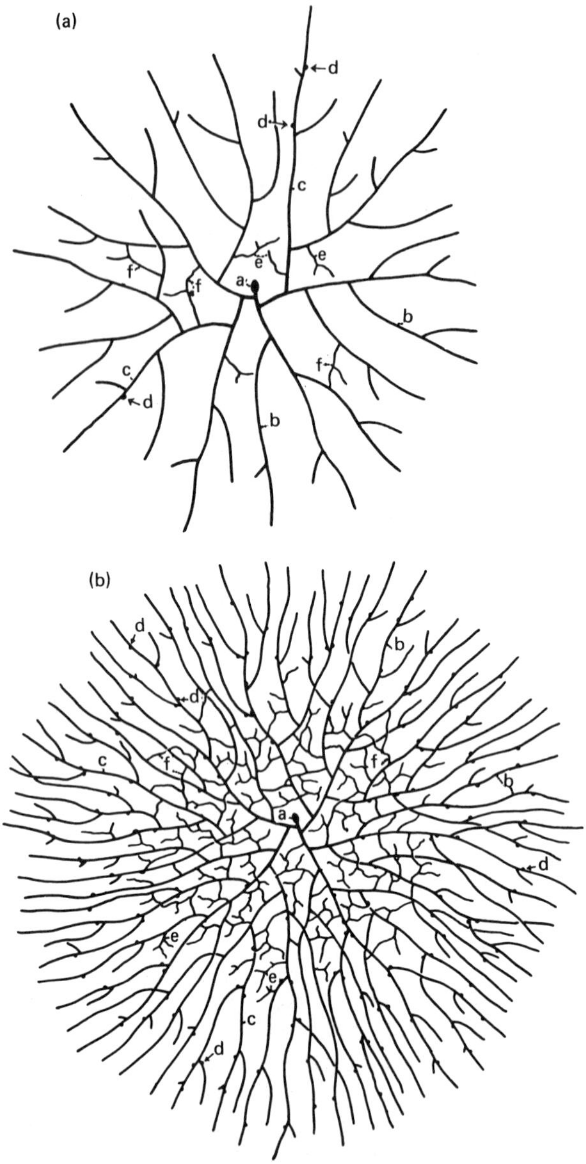

FIG. 3(a),(b) Two stages of colony development in *Coprinus sterquilinus*; a, basidiospore, b, leading hyphae, c,d, clamp connections, e, meandering lateral hyphae, and f, anastomoses. (From Buller, 1931.)

medium and then decreases continuously to reach 75° at 36 h. At 28 h the ratio between the extension rates of leading, primary, and secondary branches begins to increase and reaches 100:45:17 by 36 h. After 31 h similar increases are seen in the ratios between the diameters and apical compartment lengths of different branches.

The reason for these changes is unclear. Establishment of a nutrient concentration gradient at the colony margin would encourage radial growth into areas of higher nutrient concentration. Leading hyphae would then have an advantage over primary branches in terms of nutrient supply, which could enforce and maintain apical dominance. This does not explain the slower growth of hyphae in young colonies where nutrients are in excess and when their smaller diameter, and consequently greater surface area/volume ratio, should permit relatively greater rates of nutrient uptake. In addition, the spreading nature of fungal colonies, as opposed to the discrete form of bacterial colonies, is assumed to result from extension at a rate faster than that of substrate diffusion such that nutrients in the peripheral growth zone are not at limiting levels. Finally, chemotropism toward nutrients is a rare feature of hyphal growth (Chapter 11) and cannot be proposed as a general mechanism.

Alternatively, a reduction in branch angle could result from a negative chemotropic response to secondary metabolites accumulating within the colony. This would not occur in young colonies but does not, in itself, explain increased hyphal diameters and extension rates in mature hyphae. However, if such factors also reduced branching frequency, but not growth rate, biomass could be redirected to the increasing growth of individual hyphae. This would, of course, increase peripheral growth zone length, but comparisons between lengths in mature and young colonies are not available because of difficulties of measurement in the latter.

Apical dominance may also result from competition between hyphae or from other internal control mechanisms, but these must explain the absence of dominance in young colonies. Steele and Trinci (1975a) used the terms *undifferentiated* and *differentiated* to describe, respectively, mycelia in young colonies and those at the margin of mature colonies. Others (Park and Robinson, 1966) described changes in hyphal diameter as part of the hyphal aging process, including other features such as vacuolation. Differentiation, if it occurs as such, must take place at the hyphal tip and continual development of translocation systems may be envisaged during hyphal maturation. Apical dominance could then be established, and the faster extension rates of mature hyphae explained. However, much more work is required before this important feature of hyphal growth may be understood.

3. *Circular Morphology*

Ryan et al. (1943) noted that during tube growth of *N. crassa*, hyphae at the colony periphery were oriented at right angles to the marginal front. This behavior is reflected in plate cultures by radial orientation of hyphae in circular colonies. This is beautifully illustrated by the colonies shown in Fig. 3, which also demonstrate achievement of circular morphology early in growth. This occurs despite any initial asymmetry following spore germination. In addition, hyphal density at the colony margin remains constant during growth. Ryan et al. (1943) suggested that this density is maintained by some hyphae extending beyond the margin, enabling them to branch. Branching reduces apical dominance and consequently the rate of extension of the leading hyphae and allows branches to catch up and form a new margin. There is, however, no evidence for this process and it is thought that the margin is maintained by primary, and occasionally secondary, branches increasing in diameter and accelerating to become leading hyphae. Again, however, this process has never actually been reported and requires primary branches to grow faster than leading hyphae for a short time. Our understanding of this process therefore requires knowledge of the process by which hyphae develop in mature colonies.

Proposed mechanisms for the regulation of branching to produce circular colonies are generally based on the growth of hyphae into uncolonized and nutrient-rich areas or on autotropism. The arguments against nutrient gradient chemotropism have been discussed above and apply equally here. Possible autotropic mechanisms involve negative chemotropic responses to staling products or to reduced oxygen concentrations surrounding growing hyphae (Robinson, 1973). Hutchinson et al. (1980) found autotropic responses between growing hyphal tips of *M. hiemalis*, but only over distances of 10-20 μm. This range is much smaller than interhyphal distances at the colony margin, and autotropism is therefore unlikely to be significant in spatial regulation. Using a stochastic model, described in more detail below, Hutchinson et al. (1980) demonstrated that a circular colony is the most likely shape, merely on the basis of observed statistical variation in the hyphal extension rate, internode length, and branch angle. Such statistical variation may also explain the maintenance of hyphal density at the margin.

B. Quantitative Morphometry

1. *Hyphal Growth Unit*

A number of methods are used to quantify different aspects of branching patterns. The simplest involve measuring internode lengths, branch angles, distance from the tip to the first branch, the number of branches, and mean branch lengths (Metz et al., 1981; Ho, 1978). The most important parameter, however, is the hyphal growth unit (Caldwell and Trinci, 1973), which represents the ratio between the total hyphal length and total number of tips. It therefore represents the mean length of the hypha required to support tip growth, while the peripheral growth zone represents the maximum length. Whereas the peripheral growth zone and the specific growth rate are related by the maximum rate of extension, the hyphal growth unit (G) is related to specific growth rate by the mean extension rate (Steele and Trinci, 1975a). This is calculated as

$$E = \frac{2(H_t - H_0)}{B_0 + B_t} \tag{3}$$

where E is the mean extension rate, H_0 and H_t represent the total hyphal lengths 0 and 1 h later, and B_0 and B_t represent the respective numbers of hyphal tips. This is related to the specific growth rate by the equation

$$E = G\mu \tag{4}$$

Following spore germination (Fig. 1), G increases exponentially, then oscillates as branches are produced, and finally stabilizes during continuous branch production. The final value is a measure of branch density and varies with such factors as temperature, sorbose, and inhibitors in the same manner as the peripheral growth zone (Bull and Trinci, 1977). In addition to defining morphology, the hyphal growth unit has significance in regulating branch initiation, which will be discussed below.

2. *Branch and Length Ratios*

Horton (1945) developed a system for quantitatively analyzing branching systems which has been applied in modified form (Strahler, 1957) to a wide variety of biological and nonbiological situations. It involves the classification of branches into different orders, as illustrated in Fig. 4. Primary branches are defined as those which subtend no branches. Secondary branches subtend only primary branches, and tertiary branches subtend both primaries and secondaries. The ordering may be continued further, but mycelial systems above the fourth order are usually too complex to analyze. A secondary branch extends to the tip of its longest component primary branch.

Horton (1945) analyzed river drainage systems with this method of classification and found a logarithmic relationship between the number of branches of a particular order and the order number. Similarly, a logarithmic plot of average branch length against order number gave a straight line, but with a slope of opposite sign. The antilogarithms of these slopes are termed branch ratio and length ratio, respectively.

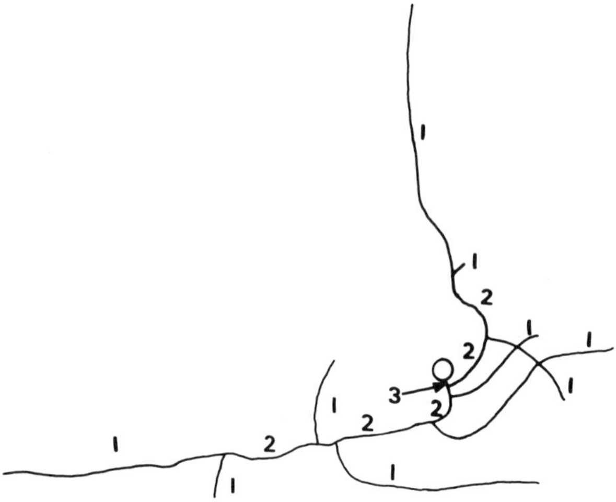

FIG. 4 Tracing of a mycelium of *Candida albicans* after 9.5 h of growth, showing primary, secondary, and tertiary branches, using the method of Strahler (1957). The branch and length ratios of this mycelium are 3.0 and 2.0, respectively. (Courtesy of N. A. R. Gow and G. W. Gooday, unpublished results.)

(A further parameter, diameter ratio, may also be calculated from a plot of average branch diameter against order number, but this has not been attempted for fungal mycelia.) A branch ratio of 2 indicates that there are twice as many primary branches as secondary branches, and in fact this is the minimum possible value, arising from a pure dichotomous system. Larger values indicate increasing asymmetry. A length ratio of 2 indicates that secondary branches are, on the average, twice as long as primary ones. Different values for length ratio result from differences in extension rate and branching frequency between branches of different order.

The logarithmic relationship resulting from this classification eliminates scaling factors and enables a comparison of widely diverse branching systems. Examples of values obtained are given in Table 2. Gull (1975) was the first to analyze fungal colonies with this technique, and his results for *Thamnidium elegans* are shown in Fig. 5. Values of branch and length ratios fall within the range found for other systems. Ho (1978) calculated branch, length, and diameter ratios in a number of Phycomycetes with a view to their taxonomic significance. However, these parameters were defined differently and appear to only have been calculated between leading and primary branches, so that the logarithmic relationship was not tested.

TABLE 2 Branch and, Where Available, Length Ratios for a Number of Branching Systems

Branching system	Branch ratio	Length ratio
Ash tree (*Fraxinus* sp)[a]	6.5	3.4
Fir tree (*Abies concolor*)[a]	4.8	2.7
Pine tree (*Pinus taeda*)[a]	4.8	2.5
Tulip (*Liviodendron tulipifera*)[a]	4.7	3.6
River drainage basin	4.51	
Portal vein	4.08	
Candida albicans[b]	3.86	2.34
Thamnidium elegans (third-order system)[c]	3.8	4.0
Bile ducts	3.59	
Pulmonary arteries	3.42	
Hepatic vein	3.39	
River drainage basin	3.38	
Glacier	3.21	
Purkinje cell	3.18	
Streptomyces viridochromogenes[d]	3.1	2.0
Bronchial tree	2.81	
Thamnidium elegans (fourth-order system)[c]	2.6	2.7
Bronchioles	2.32	

[a]Data from Leopold (1971).
[b]Unpublished data of N. A. R. Gow and G. W. Gooday.
[c]Data from Gull (1975).
[d]Unpublished data of E. J. Allan and J. I. Prosser.
Source: Data from Barker et al., 1973, except where stated.

N. A. R. Gow and G. W. Gooday (unpublished data) have found logarithmic relationships between the numbers and average lengths of branches and branch order in mycelia of *Candida albicans*, and E. J. Allan and J. I. Prosser (unpublished data) have found similar behavior in the filamentous Actinomycete *Streptomyces viridochromogenes*. The relationship therefore seems to hold in microbial branching systems and

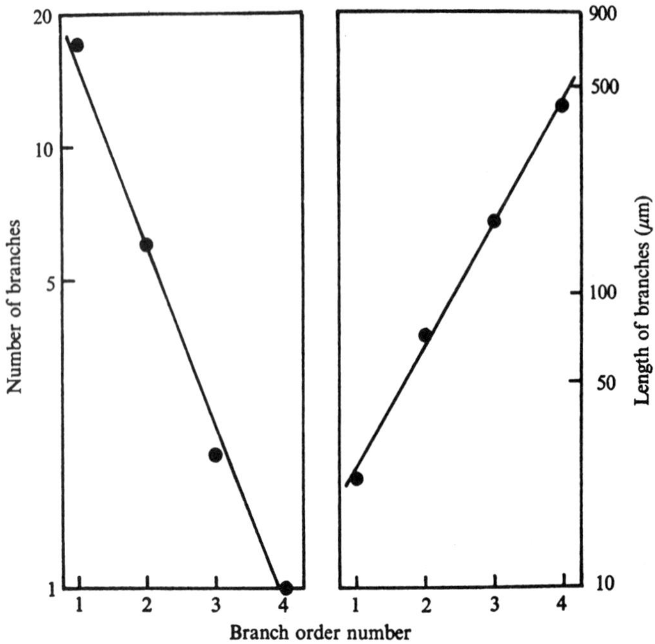

FIG. 5 Numbers and average lengths of branches as a function of order number for a fourth-order mycelial system of *Thamnidium elegans*. (From Gull, 1975.)

values for ratios are given in Table 2. These latter results, and close inspection of the results of Gull (1975) (Fig. 5) indicate some deviation from a straight-line relationship in plots of branch length versus order number. The majority of these plots curve at the highest order number. This appears to be due to the small number of branches that will be of the highest-order number in a colony at any particular time. As the colony develops and higher-order branches are formed, those of lower order increase in number to fall on the original straight line.

Branching in all the systems listed in Table 2 functions to maximize the surface area accessed while minimizing the total length of filament required. The area may be a source of material, for example, supplying trees with sunlight and tributaries with water or a sink in the case of pulmonary and arterial systems. Leopold (1971) has demonstrated that logarithmic relationships are the most probable and represent the most efficient way to cover surfaces. Their existence in fungal colonies indicates optimum efficiency in accessing nutrients from a surface while minimizing the amount of mycelium synthesized.

The significance of actual values for branch and length ratios is less clear. The only system to have been analyzed systematically is the bronchial tree. Horsfield (1980) varied length and diameter ratios with a fixed branch ratio and calculated values which resulted in minimal resistance to flow within the system. The ratios obtained corresponded closely to observed values for dog and human lungs. Until such an analysis is attempted for hyphal growth patterns, this technique will offer no mechanistic information, but it is still a useful descriptive and comparative branching parameter.

IV. VARIATION IN COLONY MORPHOLOGY

The colony morphology described in the previous section may be modified in morphological mutants by paramorphogenic compounds and altering growth conditions. These changes in colony morphology usually result from changes at the hyphal level and such techniques offer great potential in solving the problems of regulation of branch formation. Unfortunately, the complexity of the situation and the difficulties in ascribing effects to precise causes have meant that this potential is not yet fully realized.

A. Morphological Mutants

1. Branching Mutants

The genetic control of vegetative growth and branching has been studied most extensively in *N. crassa* and the work has recently been reviewed by Scott (1976a) and Mishra (1977). Mutants of *N. crassa* with altered colony morphology have been classified into six groups by Garnjobst and Tatum (1967), and all reflect changes in hyphal growth or branching. Colonial mutants form discrete, densely branched colonies, rather than the diffuse, spreading colonies of the wild type. Spreading colonial and semicolonial mutants also exhibit increased branching frequency, but colony extension occurs indefinitely, though at a reduced rate. Trinci (1973a) found that the maximum specific growth rates of a number of spreading colonial mutants and those of the wild-type strains were approximately equal. Variations in radial growth rate resulted from differences in peripheral growth zone length, which ranged from 493 to 6800 μm. These were associated with changes in hyphal diameter, reduced apical and intercalary compartment length, changes in septal plugging, and changes in the hyphal growth unit (Trinci, 1973b). In addition, increased branching density imparted some properties of bacterial colonies, in that the radial growth rate of one strain was proportional to the agar depth, a characteristic of bacterial but not fungal growth.

A fourth group of mutants contains temperature-sensitive colonial strains and those affected by other environmental factors. Finally, two miscellaneous groups contain a variety of strains with variable and sometimes inconsistent hyphal morphologies. These include fanlike branching in *shallow* mutants, multiple branching in *frost* mutants, curled hyphae in *ropy* mutants, and formation of hyphal bundles by interhyphal fusions in *sponge* mutants.

Analysis of these mutants in attempts to determine the precise branching mechanisms has implied roles for the cell wall , the cell membrane, various enzymes, and cyclic AMP. A number of morphological changes are associated with differences in cell wall composition. These include changes in glucose:glucosamine ratios (de Terra and Tatum, 1961), changes in glucan levels (Mahadevan and Tatum, 1965), and alterations in peptide composition (Wrathall and Tatum, 1974). Strangely, however, chitin levels were not reduced, although Katz and Rosenberger (1970) have isolated a temperature-sensitive mutant of *A. nidulans* deficient in chitin owing to its inability to synthesize N-acetylglucosamine.

The cell membrane has been implicated in morphological control by isolation of inositol-requiring colonial mutants showing differences in cell membrane structure (Fuller and Tatum, 1956; Shatkin and Tatum, 1961). In addition, Hubbard and Brody (1975) observed morphological effects on a double mutant requiring both inositol and choline associated with changes in membrane structure controlled experimentally by varying the supply of inositol and choline. The rôle of cAMP and membrane-bound adenylate cyclase has been studied by Scott (1976b) and Scott and Solomon (1975). They found that colonial morphology could be induced by inhibitors of adenylate cyclase and of phosphodiesterase. *Frost* mutants have membranes deficient in linolenic acid and addition of this restores wild-type morphology, as well as levels of cAMP and adenylate cyclase.

A number of mutants have deficiencies in various enzymes and work has concentrated on glucose-6-phosphate dehydrogenase, 6-phosphogluconate dehydrogenase, and phosphoglucomutase, all of which have important roles in cell wall and membrane synthesis. Unfortunately, the genetic control of these enzymes appears to be very complex and they are all involved in central carbon metabolism. It is therefore difficult to discover specific cause-effect relationships associated with these morphological mutants.

Morphological mutants are found in other species, notably *A. nidulans*. Bainbridge and Trinci (1969) described the growth of a colonial mutant of *A. nidulans*. As in *N. crassa*, colonial morphology resulted from an increase in branching frequency, rather than from a change in the specific growth rate, and the effects of glucose concentration on hyphal density in liquid and solid media are described. Clutterbuck (1978) discussed the occurrence of mutants with reduced branching frequency and the mitotic mutants of Morris (1976) which fail to form septa will be discussed below.

2. *Rhythmic Growth*

Morphological rhythms in fungi involve fruit body formation, spore release, biochemical changes, and periodic variations in vegetative growth, the latter resulting in alternate bands of dense and sparse mycelia. Rhythmic growth of vegetative hyphae may be induced by environmental conditions, but most studies have concentrated on mutants of three species: the periodic colonial "clock" mutant of *N. crassa* (Sussman et al., 1964), the "wavy" or "vague" mutant of *Ascobolus immersus* (Chevaugeon, 1959; Chevaugeon and van Nguyen, 1969), and three "wavy" mutants of *Podospora anserina* (van Nguyen, 1967; Chevaugeon and van Nguyen, 1969; Lysek, 1972), *circulosa, undulata,* and *zonata*. A more comprehensive list is given by Lysek (1978). The periodicity of the rhythms may vary with nutrients and environmental factors such as temperature and light, but circadian rhythms of vegetative growth are rare.

The cycle of growth in all these mutants may be divided into four stages: (1) Hyphal initials grow out from the dense margin of the previous cycle and show monopodial growth. (2) Dichotomous branching then increases, reducing the colony extension rate and decreasing the average internode length. Primary branches accelerate to the colony margin and become leading hyphae, to be overtaken by branches growing from behind; that is, growth is sympodial. (3) The increased branching density, the appearance of fanlike patterns, and the reduction in extension rate produce a band of dense hyphae which become vaculoate and exhibit signs of aging and death. (4) From this band appear a few younger healthy hyphae, some of which become new initials as the cycle is repeated. Figure 6 illustrates the difference between wild-type and rhythmic mutant hyphae.

In *Podospora anserina* the alternate cycles of monopodial and sympodial growth result from differences in oxygen sensitivity between substrate and surface hyphae (Lysek, 1972,1974). Surface hyphae grow as in colonial mutants, branching densely at the margin and thereby reducing radial growth rate. Substrate hyphae grow at a constant rate, eventually overtake the dense surface hyphal front, and themselves become surface hyphae. They now extend at a declining rate, leaving behind a band of more sparsely branched hyphae, and are subsequently overtaken by substrate hyphae. The periodicity of this cyclic process depends on the relative extension rates of substrate and surface hyphae.

Rhythmic growth can be induced in some wild-type strains of *N. crassa* by supplementation of media with sorbose, which also reduces band width in the periodic *clock* mutant (Sussman et al., 1964). No such induction has been reported in *Ascobulus* and *Podospora*, but in these species wild-type morphology may be restored by changes in medium constituents. Primary effects of mutations giving rise to rhythmic growth are not known, but Chevaugeon and van Nguyen (1969)

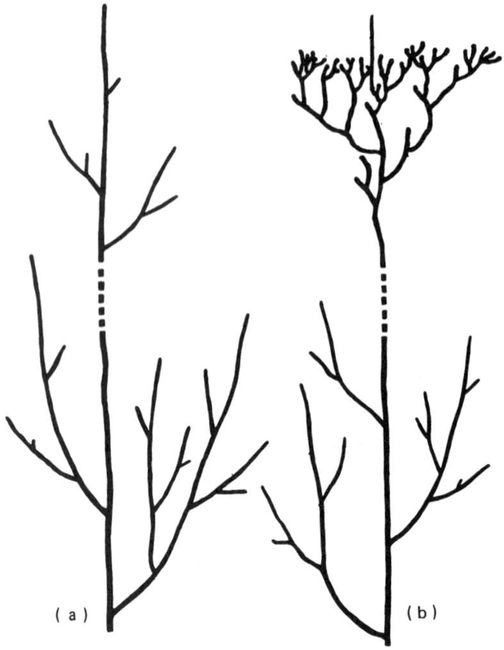

FIG. 6 Illustration of branching patterns in (a) wild-type and (b) *wavy* mutant strains of *Ascobolus immersus*. (From Chevaugeon and van Nguyen, 1969.)

presented arguments for the internal hormonal control of branching in *Podospora*. These are based on studies showing a requirement for interhyphal contact for maintenance of synchrony of rhythms between different regions of the colony.

Rhythmic growth of vegetative hyphae may lead to zonation of reproductive structures and in *Sclerotium rolfsii* is associated with sclerotium formation. Henis et al. (1973) demonstrated stimulation and inhibition of both lateral branch formation and sclerotium formation by the same factors. Kritzman et al. (1977) used L-threonine to "initiate" extensive branching and induce rhythmic growth. They found corresponding rhythmic production of sclerotia and also peaks in malate synthetase, which is thought to be a key enzyme in sclerotium formation.

3. *Sectoring*

The typical circular shape of a fungal colony is frequently modified by spontaneous occurrence of regions of faster growing hyphae which give rise to bulges or sectors. Sectoring can be increased by irradiation of

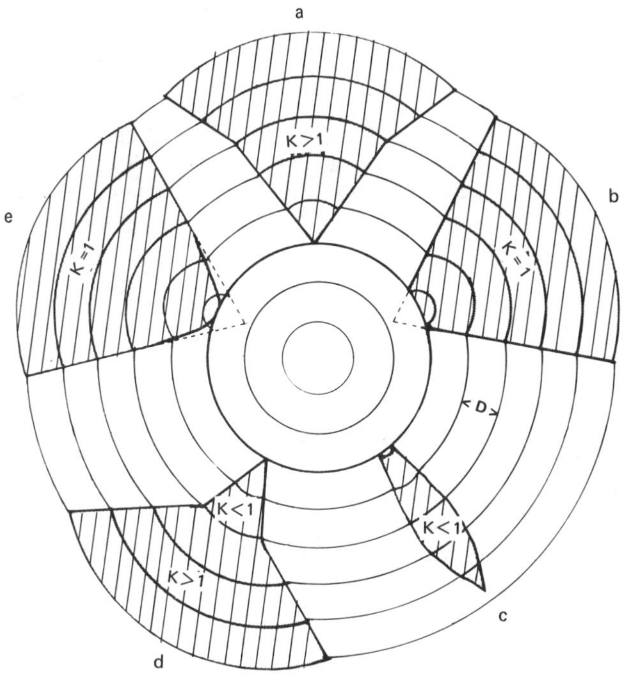

FIG. 7 Different sector types found in fungal plate colonies. The growth rate of different regions is represented by K and the parent mycelium has a value of K = 1. D represents the radial growth increment per unit time. (From Pontecorvo and Gemmell, 1944.)

colonies and has been studied by Pontecorvo and Gemmell (1944). The different morphologies are illustrated schematically in Fig. 7 and may be explained in terms of relative hyphal growth rates.

The simplest sector, type (a), results from emergence of hyphae with a higher growth rate and has boundaries which are equiangular spirals. The depth of these spirals depends on the difference in growth rates and on the colony diameter. Types (b) and (e) are similar to type (a), but have straight edges and occur less frequently. They arise when mutant hyphae emerge with unchanged growth rate but with an advantage over parent hyphae, either through differences in metabolism not associated with growth rate or through reduction in parental competition resulting from hyphal death after irradiation. Mutant hyphae with reduced growth rates will normally be overgrown, but if such a mutant has an initial advantage over the parent, then sectors of type (c) will form. Finally, sectors of type (d) are formed when the relative growth rate changes during colony development. This could be due to different effects on parent and mutant hyphae as staling products accumulate or to more specific interactions between the two strains.

B. Paramorphogens

The range of compounds affecting fungal growth and branching is apparently limitless and their precise effects vary between species. The discussion here will be limited to two compounds, sorbose and cyclic adenosine monophosphate (cAMP), which have proposed mechanistic roles in branch regulation, and to autotoxic staling compounds formed during growth.

1. *L-Sorbose*

A range of sugar analogs have been used as paramorphogens (Moore, 1981), but the most frequently used is sorbose. Tatum et al. (1949) found that sorbose induces colonial morphology in *Neurospora*, with reversion to wild-type morphology by glucose and other sugars. Sorbose has been shown to inhibit enzymes involved in cell wall synthesis and to cause changes in cell wall composition similar to those in colonial mutants.

However, Trinci and Collinge (1973) showed that effects were different in solid and liquid media. In the latter, the hyphal growth unit length, hyphal diameter, and wall thickness were unaffected by 2% sorbose. In surface cultures wall thickness and, to a lesser extent, hyphal diameter were increased and the hyphal growth unit greatly reduced. The growth rate was not significantly affected by sorbose. Its effect, therefore, is on the peripheral growth zone length and branching frequency, but the absence of this effect in liquid culture suggests the primary mechanism to be environmental, rather than endogenous; but, as pointed out by Moore (1981), the primary metabolic effect remains difficult to understand.

2. *Cyclic AMP*

Mishra (1976) found induction of colonial and semicolonial growth in *N. crassa* by cAMP and its dibutyryl derivative, and this along with its regulatory role in other organisms, has led to its proposed role in branch formation. As with sorbose induction, its effect is reduced by increasing glucose concentration and low levels of cAMP are associated with morphological effects in colonial mutants. Colonial and semicolonial growth may be induced by effectors of cAMP and by treatment with theophylline, which inhibits phosphodiesterase and can therefore lead to an accumulation of cAMP. One interesting effect involves the *N.*

crassa patch mutant, which is sorbose resistant but which gives colonial growth on treatment with cAMP, suggesting a different mechanism for the two compounds.

The role of cAMP in morphogenesis is not clear, but it is believed to affect cell wall synthesis by regulation of carbohydrate metabolism. Tu and Malhotra (1977), studying *Phycomyces blakesleeanus*, found that cAMP increased branching frequency and hyphal diameter and that this was associated with an increase in wall thickness and a parallel decrease in glycogen content. They suggested that cAMP regulates conversion of glycogen into wall material and also chitin synthetase, which may explain changes in wall thickness but not necessarily increased branch production. Jones and Bu'Lock (1977) found changes in wall structure, reduced extension rates, and increased branching in *Mucor mucedo* and *M. hiemalis* induced by dibutyryl cAMP. In addition, septation in apical regions increased. They suggested that cAMP acts by destroying apical dominance and encouraging "nongrowth" activities, but no specific mechanism has been proposed. Galpin et al. (1977) suggested that increased branching in *Dendryphiella salina*, induced by sorbose, results from its effect on intracellular levels of cAMP. Although cAMP had no direct effect, possibly through lack of uptake, theophylline, as in *N. crassa*, increased branching frequency. Another inhibitor of phosphodiesterase, imidazole, showed no such effect and the results are therefore inconclusive in terms of the precise roles of sorbose and cAMP.

3. *Staling Compounds*

The production of paramorphogenic secondary metabolites during colony growth is of obvious importance when considering the development of branching patterns. This process of staling was first studied by Brown (1923), who observed that in certain "staling fungi," for example, *Sphaeropsis malorum*, and a *Fusarium* sp., radial extension decreased as the colony aged, while growth of other "nonstaling" fungi, for example, *Alternaria grossulaviae*, was unaffected. Brown attributed some of the staling effects to ammonia and found that they could be alleviated by carbon dioxide. Pratt (1924a,b) also implicated carbon dioxide, suggesting bicarbonates as staling factors. Park (1961) described the changes in hyphal structure and orientation during aging of a *Fusarium oxysporum* colony, including a reduction in branching frequency. This effect could be relieved by dilution of the staling compounds with water and by reducing their rate of production with heat or cold shock.

Plunkett (1966) found that perfusion of colony margins of *M. hiemalis* with sterilized filtrate of a 14-h malt extract culture induced greater lateral branch formation than perfusion with fresh malt extract. This was found to be due to an inhibitor of branching in the fresh malt extract, which is removed during growth, and was thought to be related to amino acid content. Analysis of a range of amino acids showed varying effects, both increasing and decreasing internode length, and their possible effects on cell wall synthesis were discussed. Robinson (1972) found that an ether extract of culture filtrate of *Fusarium oxysporum*, characterized as succinic acid, induced subapical and lateral branching in *F. oxysporum, Aspergillus niger*, and *G. candidum*. In addition, fractions from a steam-distilled culture, possibly fatty acids, also induced branch formation. The situation is obviously very complex, with a range of compounds giving variable and opposing effects on branching, but their significance in colony morphogenesis must not be ignored.

C. Effects of Growth Rate and Environmental Factors

Fungal growth can occur at much lower initial nutrient concentrations than those of bacteria (Bull and Trinci, 1977), partly because of their ability to regulate biomass density and thus avoid depletion of nutrients from an area before their "escape." Nutrient concentration, however, affects colony morphology in a number of ways. Madelin et al. (1978) found increased induction of spiral growth in *Sordaria fimicola* as the concentration of malt extract agar was reduced from 2%, where growth was normal, to 0.02%. Spiral growth is discussed by Gooday (Chapter 11) and may also be induced by incubation temperature, pH, and by altering medium rigidity through varying agar content.

Katz et al. (1972) investigated the effect of growth rate on branching frequency by altering medium composition to give growth rates in the range 0.11-0.36 h^{-1}. They found that branching frequency increased in proportion to the specific growth rate and that the hyphal growth unit length decreased. Morrison and Righelato (1974) found a correlation between specific growth rate and branching frequency in several strains of *Penicillium chrysogenum* and a reduction in hyphal growth unit from 60 to 43 μm in media giving growth rates of 0.14 and 0.24 h^{-1}, respectively. In continuous culture an increase in growth rate from 0.05 to 0.12 h^{-1} decreased hyphal growth unit length from 52 to 36 μm.

As mentioned previously, glucose concentration has little effect on internode length during growth on solid media, although hyphal density increases. This is due to its effect on the subsequent growth of branches and not their formation. The work of Robinson and Smith (1976,1979,1980) and Smith and Robinson (1980) shows the situation in *G. candidum* to be more complex. In glucose-limited chemostat cultures lateral branching increases with specific growth rate. At a dilution rate of 0.025 h^{-1} branching is rare, while at 0.3 h^{-1} each intercalary compartment produces a branch. Apical branching occurs only above the critical dilution rate, during washout. The apical polarity exhibited in continuous culture is similar to that during growth on solid media.

In addition, the hyphal growth unit length decreases as the growth rate increases, but hyphal diameter also increases. The hyphal growth unit volume, however, does not vary with specific growth rate.

On solid media, initial glucose concentration does not greatly affect hyphal density up to 50 mg/liter. Above that concentration apical branching occurs and hyphal density increases sharply. The apical branches produced usually differed in diameter and there was a general increase in hyphal diameter with initial glucose concentration. However, at all glucose concentrations the average diameters of narrow and wide branches and the parent hypha were in the proportions 3:4:5; that is, the parental cross-sectional area was the sum of the cross-sectional areas of the two branches. Finally, Robinson and Smith proposed a model for apical branching involving an increase in the diameter and volume of the apical cell. A branch is formed upon attainment of a critical volume, which varies little with glucose concentration.

V. MECHANISMS AND MODELS

Bull and Trinci (1977) have suggested three regulatory mechanisms involved in colony formation: regulation of hyphal polarity, branch initiation, and spatial distribution of hyphae. The first of these has been discussed by Gooday (Chapter 11) and the last has been discussed above when considering the circular morphology of colonies. Regulation of branch initiation requires mechanisms for determining when and where a branch will be formed and its study has involved a number of approaches.

Unfortunately, work on morphological mutants and paramorphogens has provided little information on precise mechanisms. Indeed, the wide range of paramorphogenic compounds suggests that branch initiation may not be controlled by a precise biochemical mechanism. Proposed mechanisms involving the regulation of cellular metabolism are at such a level as to cause a wide range of effects, while changes in cell wall and cell membrane structures cannot in themselves provide mechanisms for spatial and temporal control.

A. Enzyme Studies

Gooday (Chapter 11) has reviewed current ideas regarding the balance between lytic and synthetic processes occurring during tip growth. Branch initiation must involve similar processes, suggesting a role for lytic and synthetic enzymes as morphogens. Fèvre (1968,1972) demonstrated suppression of branching in *Saprolegnia monoica* following treatment with inhibitors of protein and DNA synthesis. Treatment with glucono-δ-lactose and coumarin, inhibitors of glucanase and cellulose biosynthesis, respectively, produced a similar effect and also increased cell wall thickness Fèvre and Rougier, 1980). A correlation was found between cellulase and glucanase levels and the level of branching, which has also been reported in *Achlya ambisexualis* and *Dictychus monosporus* (Mullins, 1973). Kritzman et al. (1978) also reported a correlation between synthetic and lytic enzymes and lateral branch formation in *Sclerotium rolfsii*.

The localization of these enzymes within hyphae was discussed by Fèvre (1979) and Rosenberger (1979) and a model for control of branching has been proposed. Enzymes are transported to the hyphal tip in membrane-bound vesicles. Some release their contents at the tip, while others are trapped intact and remain associated with the wall as the hypha extends. The wall-bound enzymes may then be activated at a later stage to release their contents, partially lyse the cell wall, and allow the synthesis and incorporation of new material and development of a new growing tip. Rosenberger (1979) postulated that the activators may themselves be transported to their site of action within vesicles. This idea of wall-bound enzymes is consistent with suggestion of Collinge et al. (1978), that any part of the hypha has the potential for branch formation.

B. Kinetic Studies

Robertson (1958) found that arrest of hyphal extension by osmotic shock leads to regrowth by apical branching, and generally a reduction in extension rate, which is not associated with a reduction in specific growth rate, will result in increased branching frequency. On the basis of studies of the effect of different growth rates on branching, Katz et al. (1972) proposed a model for branch initiation which states that a new branch is formed when "the capacity of the hypha to elongate exceeds that of the existing tips." Thus when a hypha reaches its maximum rate of extension, a branch will be initiated to allow continued exponential growth. In fact, branches are usually formed in germ tubes at the beginning of the deceleration phase and before attainment of linear growth. In mature hyphae control must be precise to avoid deceleration in the extension rate of the parent hypha. At the hyphal level, then, branch initiation is controlled by the length of the peripheral growth zone.

Caldwell and Trinci (1973) have applied a similar approach at the mycelial level in their concept of the hyphal growth unit. The hyphal growth unit remains constant during mycelial growth, suggesting initiation of branching each time the ratio between the total hyphal length and the number of branches exceeds the hyphal growth unit. The hyphal growth unit may then be considered as a unit cell with branch formation equivalent to cell division in unicells. It has no morphological or structural identity, but as a physiological unit it has been used to explain colonial growth caused by mutation or paramorphogens and the effects of temperature, inhibitors, and medium composition.

Robinson and Smith (1979) pointed out the need to consider the hyphal growth unit volume rather than length, where hyphal diameter varies and used this concept to explain apical branch formation in *G. candidum*. The hyphal growth unit volume remains constant as specific growth rate is increased, this being achieved by an increase in the hyphal extension rate and an increase in the number of intercalary cells producing branches. Eventually a growth rate is reached at which all cells produce lateral branches. In order to maintain the hyphal growth unit volume at higher growth and extension rates, branches must be formed apically, and this is observed experimentally at high specific growth rates.

C. The Duplication Cycle

The concept of the hyphal growth unit allows the prediction of when a branch will be formed, but gives no information on spatial regulation. A more comprehensive mechanism for branch initiation has been hypothesized from investigations into mechanisms of tip growth (Bartnicki-Garcia, 1973) and from descriptions of a duplication cycle in filamentous fungi. This was first proposed for *A. nidulans* (Fiddy and Trinci, 1976a) and its existence in other fungi is discussed by Trinci (1978,1979).

In *A. nidulans* the cycle may be divided into three stages. First, multiple septation occurs in an apical compartment, reducing its length by half. The new apical compartment then extends at a linear rate while nuclei migrate at a slightly lower rate to the hyphal tip. When the ratio of cytoplasmic volume to the nuclear number reaches a critical value, nuclear division occurs. The cycle is then restarted by a further round of septation. The period of the cycle is equal to the organism's doubling time and the duplication cycle is equivalent to the cell cycle of unicells, but without cell separation.

Fiddy and Trinci (1976a) also found an association between septation and branch initiation with a lag of 50 ± 40 min. This association has been observed in many species, including *N. crassa* (Steele and Trinci, 1977), *Coprinus sterquilinus* (Buller, 1931), *Schizophyllum commune* (Niederpruem et al., 1971), *Basidiobolus ranarum* (Bracker, 1967), and *G. candidum* (Fiddy and Trinci, 1976b). In species which form complete septa, such as *Geotrichum*, the lag between septation and branching is smaller and shows less variation (28 ± 9 min) than in species where septa have a central pore, such as *Aspergillus*.

These observations may be explained by considering the mechanism of tip extension. This involves the transport of vesicles from distal hyphal regions toward the tip, where they fuse to give extension. Such vesicles are believed to contain wall precursors and/or enzymes involved in wall synthesis. Trinci (1979) suggested that a branch will be initiated in regions where such vesicles accumulate, leading to wall softening and the formation of a "new" tip. The model is therefore consistent with that of Rosenberger (1979) discussed above. Such an accumulation of vesicles has been observed behind septa in *N. crassa* (Trinci and Collinge, 1974) and *Achyla ambisexualis* (Nolan and Bal, 1974) and may be a general mechanism in septate fungi. The looser coupling between septation and branch initiation in fungi forming septa with a central pore may result from continued passage of material through the central pore following septation. This would increase both the time taken for accumulation and the variability in the length of the lag period.

This hypothesis may be extended to explain apical branching if such an accumulation takes place at the hyphal tip. Evidence for this is provided by a temperature-sensitive mutant of *A. nidulans* (Trinci and Morris, 1979) which produces septa at 25°C but not at 37°C. Growth at 37°C is by apical and lateral branching, but at 25°C only lateral branches are formed. This may be interpreted as being due to accumulation of vesicles at the hyphal tip at 37°C giving apical branching. In the presence of septa these vesicles would have been "trapped" before reaching the tip and would have initiated lateral branches.

Extension of this hypothesis to aseptate fungi is more difficult. Such fungi branch in the same manner as septate fungi, and even in the latter branch initiation is not always associated with septation. It is possible to envisage other mechanisms for vesicle accumulation, but we first need a greater understanding of the mechanism of transport of vesicles along hyphae and the control of apical polarity. At some distance from the tip, this polarity will break down and the vesicle flow rate will be slower than the tip extension rate. Accumulation could then occur, followed by branch production, but the spatial and temporal regulation would be less precise than in septate fungi. Even there, however, the problem of regulation will not be entirely solved until we fully elucidate the mechanism controlling spatial regulation of septum formation. Possibly the bioelectrical currents referred to in Chapter 11 hold the key to these problems.

The feasibility of this hypothesis for branch regulation in quantitative terms has been demonstrated by a mathematical modeling approach (Prosser and Trinci, 1979). Essentially this represents the processes of vesicle production, flow, absorption, and accumulation in the form of differential equations. Septation and the duplication cycle are incorporated into the model and branching occurs in hyphal regions where vesicles accumulate to a certain critical level. By providing values for rates of vesicle flow, production, and absorption and for other constants required by the model, it is possible to make predictions on hyphal growth kinetics which may be compared with experimental data.

Such a comparison was carried out with the experimental data on *G. candidum* of Trinci (1974) and Fiddy and Trinci (1976b). Figure 8 shows model predictions for increases in the total mycelial length, the number of branches, and the hyphal growth unit length. These fit

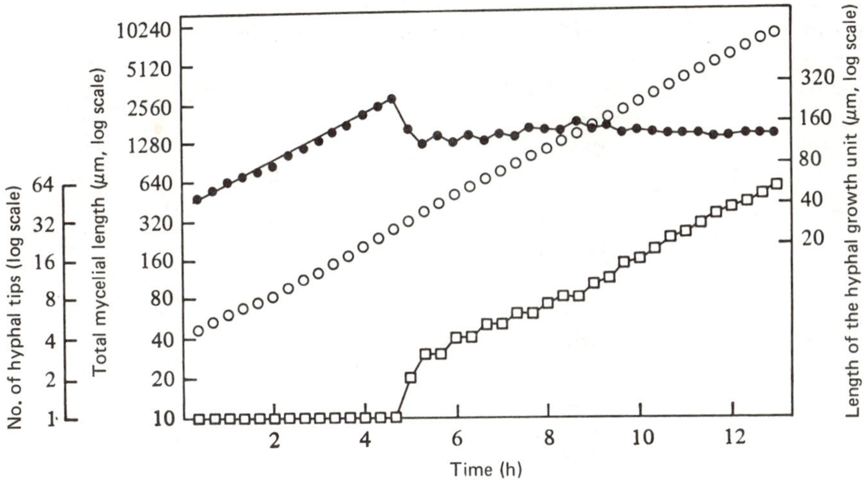

FIG. 8 Simulated growth of a mycelium of *Geotrichum candidum* (□, number of tips; o, total mycelial length; •, hyphal growth unit length). (From Prosser and Trinci, 1979.)

closely with experimental data (Fig. 1). Experimental and predicted values of various growth parameters are given in Table 3 and again show close agreement. The model therefore confirms that spatial and temporal regulation of growth and branching by vesicle production is feasible, both qualitatively and quantitatively, and this approach may also be used to determine which factors within the model exert the greatest control over the system.

D. Empirical and Stochastic Models

The model described above is the only available model of hyphal growth and branching, but a number of mathematical models have been proposed based on empirical growth laws and statistical probability. Koch (1975) modeled mycelial growth using a modified form of the logistic equation frequently used to describe population growth. This assumes that overall growth rate in a particular region will decrease as hyphal density increases up to a maximum level termed the carrying capacity. Koch found that the rate of growth was controlled by the ratio between the carrying capacity and the initial density of hyphae inoculating a particular region. This latter factor will depend on the branching characteristics of the mycelium.

Model predictions were compared with experimental data on growth in liquid and solid cultures and explained the different growth kinetics observed. In particular, this model clarified differences between

TABLE 3 Experimental Growth Parameters for *Geotrichum candidum* and Those Predicted by Simulation of a Mathematical Model

Parameter	Experimental	Predicted
Specific rate of increase in total mycelial length (h^{-1})	0.43	0.41
Specific rate of increase in number of branches (h^{-1})	0.40	0.43
Linear extension rate (mm/h)	132	141
Final hyphal growth unit length (μm)	150	140
Mean hyphal extension rate (μm/h)	48	56

Source: Prosser and Trinci, 1979.

two-dimensional surface growth and pellet formation which it showed to result from differences in the initial hyphal density inoculating regions surrounding the colony. It would be useful to apply this model to the effects of varying branching frequency on colony formation in plate cultures.

Cohen (1967) proposed a model applicable to a range of branching systems. This is based on a number of growth and branching rules. The former define the extent and direction of growth of a hypha, while the latter define the probability of the formation, position, and growth angle of a branch. Branching patterns obtained by iteration of these rules are realistic and modification of the rules is possible to determine controlling factors. It is possible to incorporate negative and positive chemotropic effects, and an example of this is illustrated in Fig. 9. The model also has facility for considering differences in extension rate between parent and branch hyphae and its application to fungal mycelial growth would be valuable.

Hutchinson et al. (1980) adopted an empirical approach to modeling the growth of *M. hiemalis*. They analyzed colonies to determine the variation in three parameters–the extension rates of individual hyphae, internode lengths, and branch angles–and determined for each the mean and the probability distribution around the mean. They then used this information to "reconstruct" a colony. This simulation is initiated with a small length of hypha, equivalent to a germ tube, whose extension over a time interval is estimated on the basis of the observed mean and probability distribution. The mean of the internode length and its associated variation are then applied to determine whether a branch will be formed in the current time interval. If this is so, the branch angle is determined by a similar process. By continuous iteration of these procedures a colony is developed and simulated, and observed colonies of the same age are illustrated in Fig. 10.

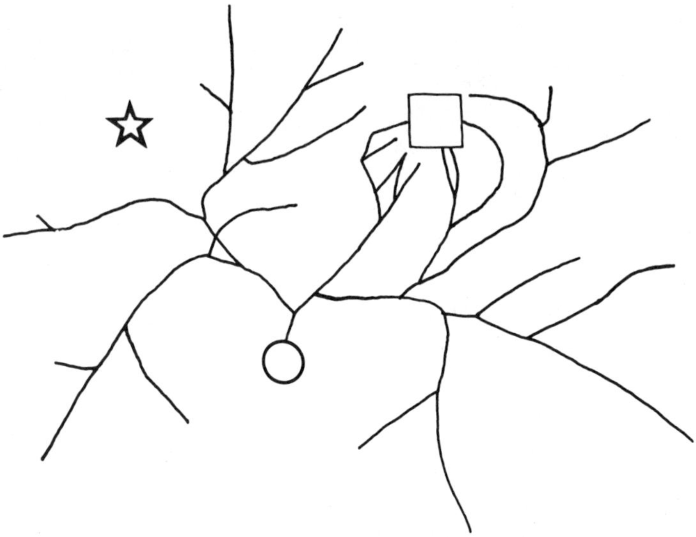

FIG. 9 Simulated branching pattern with growth directed away from a repulsing center (☆) and toward an attractant (□). (From Cohen, 1967.)

The author's main aim was to demonstrate that the circular shape of a colony can be generated by—and, in fact, is the most likely result of statistical variations observed in these three parameters. This is confirmed by the colony in Fig. 10b and mechanisms such as autotropism are not essential. Apart from relative overall size, the main difference in morphology between the simulated and observed colonies results from the perfectly straight simulated filaments, as opposed to the slightly wavy hyphae seen experimentally.

A similar approach was adopted by Adupa (1981) in modeling the mycelial growth of *Penicillium claviforme*. At each time interval four rules are applied. The first governs hyphal extension rate, with an underlying linear or exponential rate subjected to random variation based on experimentally observed probability distributions. The second rule determines the direction of growth, which again changes on the basis of observed variation. The final two rules govern branch initiation and branch angle in a similar manner.

Colonies simulated in this way showed good agreement with observed mycelia in terms of growth kinetics, but, mainly through incorporation of the second rule, mycelial morphology was much more realistic than that of previous models. The major discrepancy reported was in the orientation of growth of a hypha close to a neighbor, suggesting a need for consideration of hyphal interference. The model demonstrates

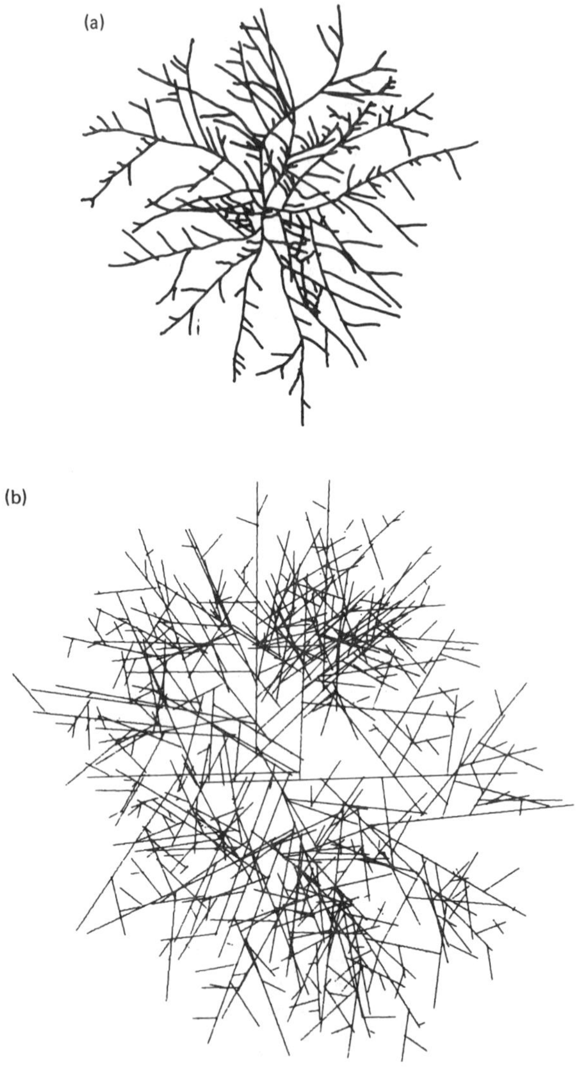

FIG. 10 (a) Tracing of a mycelium of *Mucor hiemalis* after approximately 10 h of growth. (b) Simulated growth after 10 h. (From Hutchinson et al., 1980.)

successful simulation of mycelial growth from observed variations in a few morphological parameters and suggests that investigation of the effects of variations in the parameters would be useful. For example,

Adupa (1981) found that reduction in the relative extension rate of lateral branch hyphae changed the number of growing tips and reduced the total hyphal length supporting their growth. Prosser (1979), however, discussed the disadvantages of using empirical models to predict and explain effects outside the range of data used for their construction.

VI. CONCLUSIONS

The variety of hyphal branching patterns and subsequent variation in colony morphology result from differences in hyphal extension rate, branching frequency, branch angle, and branch diameter. Of these four factors, the control of extension rate is best understood. This is due to formulation of the concept of the peripheral growth zone and to our knowledge of the mechanisms of apical growth.

We are beginning to understand the mechanisms involved in branch initiation and frequency. The hyphal growth unit concept allows some prediction of the timing of branch formation in a mycelium and explains many observed variations in branching frequency. Initiation of branching by vesicle accumulation in conjunction with the events of the duplication cycle provides a mechanism for both spatial and temporal regulation in septate fungi, but not in aseptate species. The requirement here, as in other aspects of fungal growth, is for a greater knowledge of the mechanisms of vesicle transport along hyphae. Such knowledge may explain not only normal branching patterns, but also those exhibited by morphological mutants and those caused by paramorphogens.

The factors determining branch angle and diameter are least understood, and our ignorance is most obvious when considering the development of a germ tube of, for example, *N. crassa* to form a mature colony. The morphology and kinetics of hyphae in the margin of the mature colony differ significantly from those in young colonies and also show hierarchical variation among themselves. This variation involves all four of the factors listed above and it is still not clear whether this development is due to environmental changes or is under endogenous control, nor whether it occurs to the same extent in liquid culture. A knowledge of the mechanisms controlling hyphal diameter and directional growth and of the effects of paramorphogens on cytological events is necessary for a greater understanding of the generation of hyphal growth patterns.

ACKNOWLEDGMENTS

I would like to thank N. A. R. Gow and Dr. G. W. Gooday for use of unpublished data.

REFERENCES

Adupa, R. L. (1981). Studies on the growth of *Penicillium claviforme* (Bainier). Ph. D. thesis, University College of North Wales, Bangor.

Bainbridge, B. W., and Trinci, A. P. J. (1969). Colony and specific growth rates of normal and mutant strains of *Aspergillus nidulans*. *Trans. Br. Mycol. Soc.* *53*:473-475.

Barker, S. B., Cumming, G., and Horsfield, K. (1973). Quantitative morphometry of the branching structure of trees. *J. Theor. Biol.* *40*:33-43.

Bartnicki-Garcia, S. (1973). Fundamental aspects of hyphal morphogenesis. In *Microbial Differentiation,* 23rd Symposium of the Society for General Microbiology. J. M. Ashworth and J. E. Smith (Eds.). Cambridge University, Cambridge, pp. 245-267.

Bracker, C. E. (1967). Ultrastructure of fungi. *Annu. Rev. Phytopathol.* *5*:343-374.

Brown, W. (1923). Experiments on the growth of fungi on culture media. *Ann. Bot.* *37*:105-129.

Bull, A. T., and Trinci, A. P. J. (1977). The physiology and metabolic control of fungal growth. *Adv. Microbiol. Physiol.* *15*:1-84.

Buller, A. H. R. (1931). *Researches on Fungi,* Vol. 4. Longmans Green, London.

Buller, A. H. R. (1933). *Researches on Fungi*, Vol. 5. Longmans Green, London.

Butler, G. M. (1961). Growth of hyphal branching systems in *Coprinus disseminatus*. *Ann. Bot. N. S.* *25*:341-352.

Caldwell, I. Y., and Trinci, A. P. J. (1973). The growth unit of the mould *Geotrichum candidum*. *Arch. Mikrobiol.* *88*:1-10.

Chevaugeon, J. (1959). La zonation du thalle, phénomène periodique autonome chez l'*Ascobolus immersus*. *C. R. Acad. Sci.* *248*:1381-1384.

Chevaugeon, J., and Nguyen, A. van (1969). Internal determination of hyphal growth rhythms. *Trans. Br. Mycol. Soc.* *53*:1-14.

Clutterbuck, A. J. (1978). Genetics of vegetative growth and asexual reproduction. In *The Filamentous Fungi,* Vol. 3. J. E. Smith and D. R. Berry (Eds.). Edward Arnold, London, pp. 240-256.

Cohen, D. (1967). Computer simulation and biological pattern generation processes. *Nature* *216*:246-248.

Collinge, A. J., Fletcher, M. H., and Trinci, A. P. J. (1978). Physiology and cytology septation and branching in a temperature sensitive colonial mutant (*cot 1*) of *Neurospora crassa*. *Trans. Br. Mycol. Soc.* *71*:107-120.

Emerson, R. (1955). The biology of water molds. In *Aspects of Synthesis and Order in Growth.* D. Rudnick (Ed.). Academic, New York, pp. 171-208.

Fèvre, M. (1968). Action de la D atinomycine et de la puromycine sur la croissance et la ramification du mycélium jeune de *Saprolegnia monoica* Pringsheim. *C. R. Acad. Sci.* *267*:293-295.

Fèvre, M. (1972). Contribution to the study of the determination of mycelium branching of *Saprolegnia monoica* Pringsheim. *Z. Pflanzenphysiol.* *68*:1-10.

Fèvre, M. (1979). Glucanases, glucan synthases and wall growth in *Saprolegnia monoica.* In *Fungal Walls and Hyphal Growth.* J. H. Burnett and A. P. J. Trinci (Eds.). Cambridge University, Cambridge, pp. 225-263.

Fèvre, M., and Rougier, M. (1980). Hyphal morphogenesis of *Saprolegnia*: Cytological and biochemical effects of coumarin and glucono-δ-lactone. *Exp. Mycol.* *4*:343-361.

Fiddy, C., and Trinci, A. P. J. (1975). Kinetics and morphology of glucose limited cultures of moulds grown in a chemostat and on solid media. *Arch. Microbiol.* *103*:191-197.

Fiddy, C., and Trinci, A. P. J. (1976a). Mitosis, septation, branching and duplication cycle in *Aspergillus nidulans.* *J. Gen. Microbiol.* *97*:169-184.

Fiddy, C., and Trinci, A. P. J. (1976b). Nuclei, septation, branching and growth of *Geotrichum candidum.* *J. Gen. Microbiol.* *97*:185-192.

Fuller, R. C., and Tatum, E. L. (1956). Inositol-phospholipid in *Neurospora* and its relation to morphology. *Am. J. Bot.* *43*:361-365.

Galpin, M. F. J., Jennings, D. H., and Thornton, J. D. (1977). Hyphal branching in *Dendryphiella salina*: Effect of various compounds and the further elucidation of the effect of sorbose and the role of cyclic AMP. *Trans. Br. Mycol. Soc.* *69*:175-182.

Garnjobst, L., and Tatum, E. L. (1967). A survey of new morphological mutants in *Neurospora crassa.* *Genetics* *57*:579-604.

Gull, K. (1975). Mycelium branching patterns of *Thamnidium elegans.* *Trans. Br. Mycol. Soc.* *64*:321-324.

Henis, Y., Okon, Y., and Chet, I. (1973). The relationship between early hyphal branching and formation of sclerotia in *Sclerotium rolfsii.* *J. Gen. Microbiol.* *79*:147-150.

Ho, H. H. (1978). Hyphal branching systems in *Phytophthora* and other Phycomycetes. *Mycopathologia* *64*:83-86.

Horsfield, K. (1980). Are diameter, length and branching ratios meaningful in the lung? *J. Theor Biol.* *87*:773-784.

Horton, R. E. (1945). Erosional development of streams and their drainage basins: Hydrophysical approach to quantitative morphometry. *Bull. Geogr. Soc. Am.* *56*:275-370.

Hubbard, S. C., and Brody, S. (1975). Glycerophospholipid variation in choline and inositol auxotrophs of *Neurospora crassa.* *J. Biol. Chem.* *250*:7173-7181.

Hutchinson, S. A., Sharma, P., Clarke, K. R., and MacDonald, I. (1980). Control of hyphal orientation in colonies of *Mucor hiemalis*. *Trans. Br. Mycol. Soc.* *75*:177-191.

Jones, B. E., and Bu'Lock, J. D. (1977). The effect of N^6,O^2-dibutyryl adenosine-3',5'-cyclic monophosphate on morphogenesis in Mucorales. *J. Gen. Microbiol.* *103*:29-36.

Katz, D., and Rosenberger, R. F. (1970). A mutation in *Aspergillus nidulans* producing hyphal walls which lack chitin. *Biochim. Biophys. Acta* *208*:452-460.

Katz, D., Goldstein, D., and Rosenberger, R. F. (1972). Model for branch initiation in *Aspergillus nidulans* based on measurements of growth parameters. *J. Bacteriol.* *109*:1097-1100.

Koch, A. L. (1975). The kinetics of mycelial growth. *J. Gen. Microbiol.* *89*:209-216.

Kritzman, G., Chet, I., and Henis, G. (1977). The relationship between rhythmic hyphal growth and circadian formation of sclerotia in *Sclerotium rolfsii* Sacc. *Can. J. Microbiol.* *23*:959-963.

Kritzman, G., Chet, I., and Henis, G. (1978). Localization of β-(1,3)-glucanase in the mycelium of *Sclerotium rolfsii*. *J. Bacteriol.* *134*:470-475.

Leopold, L. B. (1971). Trees and streams: The efficiency of branching patterns. *J. theor. Biol.* *31*:339-354.

Lysek, G. (1972). Rhythmic mycelial growth in *Podospora anserina*. VI. An attempt to elucidate the mechanism of a clock mutant. *Arch. Mikrobiol.* *87*:129-137.

Lysek, G. (1974). Zonierungen und Hexenringe, Morphologische Differenzierungen bei Pilzen. *Naturwiss. Rundsch.* *27*:449-455.

Lysek, G. (1978). Circadian rhythms. In *The Filamentous Fungi*, Vol. 3. J. E. Smith and D. R. Berry (Eds.). Edward Arnold, London, pp. 376-388.

Madelin, M. F., Toomer, D. K., and Ryan, J. (1978). Spiral growth of fungus colonies, *J. Gen. Microbiol.* *106*:73-80.

Mahadevan, P. R., and Tatum, E. L. (1965). Localization of structural polymers in the cell wall of *Neurospora crassa*. *J. Cell Biol.* *35*:295-302.

Metz, B., Bruijn, E. W. de, and Suijdam, J. C. van (1981). Method for quantitative representation of the morphology of molds. *Biotechnol. Bioeng.* *23*:149-162.

Mishra, N. C. (1976). The effect of cyclic adenosine monophosphate on the growth of *Neurospora crassa*. *Naturwissenschaften* *63*:485-486.

Mishra, N. C. (1977). Genetics and biochemistry of morphogenesis in *Neurospora*. *Adv. Genet.* *19*:341-405.

Moore, D. (1981). Effects of hexose analogues on fungi: Mechanisms of inhibition and of resistance. *New Phytol.* *87*:487-515.

Morris, N. R. (1976). Mitotic mutants of *Aspergillus nidulans*. *Genet. Res.* *26*:237-254.

Morrison, K. B., and Righelato, R. C. (1974). The relationship between hyphal branching, specific growth rate and colony radial growth rate in *Penicillium chrysogenum*. *J. Gen. Microbiol. 81*: 517-520.

Mullins, J. T. (1973). Lateral branch formation and cellulase production in the water molds. *Mycologia 65*:1007-1014.

Nguyen van, H. (1967). *Etude de rythmes internes de croissance chez le Podospora anserina. Theses.* Masson, Paris.

Niederpruem, D. J., Jersild, R. A., and Lane, P. L. (1971). Direct microscopic studies of clamp connection formation in growing hyphae of *Schizophyllum commune*. I. The dikaryon. *Arch. Mikrobiol. 78*:268-280.

Nolan, R. A., and Bal, A. K. (1974). Cellulase localization in hyphae of *Achlya ambisexualis*. *J. Bacteriol. 117*:840-843.

Park, D. (1961). Morphogenesis, fungistasis and cultural staling in *Fusarium oxysporum* Synder & Hansen. *Trans. Br. Mycol. Soc. 44*:377-390.

Park, D., and Robinson, P. M. (1966). Aspects of hyphal morphogenesis in fungi. In *Trends in Plant Morphogenesis*. E. G. Cutter (Ed.) Longmans, London pp. 27-44.

Plomley, N, J. B. (1959). Formation of the colony in the fungus *Chaetomium*. *Aust. J. Biol. Sci. 12*:53-64.

Plunkett, B. E. (1966). Morphogenesis in the mycelium: Control of lateral hypha frequency in *Mucor hiemalis* by amino-acids. *Ann. Bot. N. S. 30*:133-152.

Pontecorvo, G., and Gemmell, A. R. (1944). Colonies of *Penicillium notatum* and other moulds as models for the study of population genetics. *Nature 154*:532-534.

Pratt, C. A. (1924a). The staling of fungal cultures. I. General and chemical investigations of staling by *Fusarium*. *Ann. Bot. 38*:563-595.

Pratt, C. A. (1924b). The staling of fungal cultures. II. The alkaline metabolic products and their effect on the growth of fungal spores. *Ann. Bot. 38*:599-615.

Prosser, J. I. (1979). Mathematical modelling of mycelial growth. In *Fungal Walls and Hyphal Growth*. J. H. Burnett and A. P. J. Trinci (Eds.). Cambridge University, Cambridge, pp. 359-384.

Prosser, J. I., and Trinci, A. P. J. (1979). A model for hyphal growth and branching. *J. Gen. Microbiol. 111*:153-164.

Robertson, N. F. (1958). Observations on the effect of water on the hyphal apices of *Fusarium oxysporum*. *Ann. Bot. N. S. 22*:159-173.

Robinow, C. F. (1963). Observations on cell growth, mitosis and division in the fungus *Basidiobolus ranarum*. *J. Cell. Biol. 17*: 123-152.

Robinson, P. M. (1972). Isolation and characterization of a branching factor produced by *Fusarium oxysporum*. *Trans. Br. Mycol. Soc. 59*:320–322.

Robinson, P. M. (1973). Oxygen-positive chemotropic factor for fungi? *New Phytol. 72*:1349–1356.

Robinson, P. M., and Smith, J. M. (1976). Morphogenesis and growth kinetics of *Geotrichum candidum* in continuous culture. *Trans. Br. Mycol. Soc. 66*:413–420.

Robinson, P. M., and Smith, J. M. (1979). Development of cells and hyphae of *Geotrichum candidum* in chemostat and batch culture. *Trans. Br. Mycol. Soc. 72*:39–47.

Robinson, P. M., and Smith, J. M. (1980). Apical branch formation and cyclic development in *Geotrichum candidum*. *Trans. Br. Mycol. Soc. 75*:233–238.

Rosenberger, R. F. (1979). Endogenous lytic enzymes and wall metabolism. In *Fungal Walls and Hyphal Growth*. J. H. Burnett and A. P. J. Trinci (Eds.). Cambridge University, Cambridge, pp. 265–277.

Ryan, F. J., Beadle, G. W., and Tatum, E. L. (1943). The tube method of measuring the growth rate of *Neurospora*. *Am. J. Bot. 30*:784–799.

Scott, W. A. (1976a). Biochemical genetics of morphogenesis in *Neurospora*. *Annu. Rev. Microbiol. 30*:85–104.

Scott, W. A. (1976b). Adenosine 3':5'-cyclic monophosphate deficiency in *Neurospora crassa*. *Proc. Nat. Acad. Sci. USA 73*:2995–2999.

Scott. W. A., and Solomon, B. (1975). Adenosine 3',5'-cyclic monophosphate and morphology in *Neurospora crassa*: Drug induced alterations. *J. Bacteriol. 122*:454–463.

Shatkin, A. J., and Tatum, E. L. (1961). The relationship of m-inositol to morphology in *Neurospora crassa*. *Am. J. Bot. 48*:760–771.

Smith, J. H. (1924). On the early growth of the individual fungus hypha. *New Phytol. 23*:65–78.

Smith, J. M., and Robinson, P. M. (1980). Development of somatic hyphae of *Geotrichum candidum*. *Trans. Br. Mycol. Soc. 74*:159–165.

Steele, G. C., and Trinci, A. P. J. (1975a). Morphology and growth kinetics of hyphae of differentiated and undifferentiated mycelia of *Neurospora crassa*. *J. Gen. Microbiol. 91*:362–368.

Steele, G. C., and Trinci, A. P. J. (1975b). The extension zone of mycelial hyphae. *New Phytol. 75*:583–587.

Steele, G. C., and Trinci, A. P. J. (1977). Relationship between intercalary compartment length and hyphal growth unit length. *Trans. Br. Mycol. Soc. 69*:156–158.

Strahler, A. N. (1957). Quantitative analysis of watershed geomorphology. *Trans. Am. Geophys. Union 38*:913–920.

Sussman, A. S., Lowry, R. J. and Durkee, T. (1964). Morphology and genetics of a periodic colonial mutant of *Neurospora crassa*. *Am. J. Bot.* *51*:243-252.

Tatum, E. L., Barratt, R. W., and Cutter, V. M. (1949). Chemical induction of colonial paramorphs in *Neurospora* and *Syncephalastrum*. *Science* *109*:509-511.

Terra, N. de, and Tatum, E. L. (1961). Colonial growth of *Neurospora*. *Science* *134*:1066-1068.

Trinci, A. P. J. (1969). A kinetic study of the growth of *Aspergillus nidulans* and other fungi. *J. Gen. Microbiol.* *57*:11-24.

Trinci, A. P. J. (1970). Kinetics of apical and lateral branching in *Aspergillus nidulans* and *Geotrichum lactis*. *Trans. Br. Mycol. Soc.* *55*:17-28.

Trinci, A. P. J. (1971a). Exponential growth of the germ tubes of fungal spores. *J. Gen. Microbiol.* *67*:345-348.

Trinci, A. P. J. (1971b). Influence of the peripheral growth zone on the radial growth rate of fungal colonies. *J. Gen. Microbiol.* *67*:325-344.

Trinci, A. P. J. (1973a). Growth of wild type and spreading colonial mutants of *Neurospora crassa* in batch culture and on agar medium. *Arch. Mikrobiol.* *91*:113-126.

Trinci, A. P. J. (1973b). The hyphal growth unit of wild-type and spreading colonial mutants of *Neurospora crassa*. *Arch. Mikrobiol.* *91*:127-136.

Trinci, A. P. J. (1974). A study of the kinetics of hyphal extension and branch initiation of fungal mycelia. *J. Gen. Microbiol.* *81*:225-236.

Trinci, A. P. J. (1978). The duplication cycle and development in moulds. The *Filamentous Fungi*, Vol. 3. J. E. Smith and D. R. Berry (Eds.). Edward Arnold, London, pp. 132-163.

Trinci, A. P. J. (1979). The duplication cycle and branching in fungi. In *Fungal Walls and Hyphal Growth*. J. H. Burnett and A. P. J. Trinci (Eds.). Cambridge University, Cambridge, pp. 319-358.

Trinci, A. P. J. and Collinge, A. (1973). Influence of L-sorbose on the growth and morphology of *Neurospora crassa*. *J. Gen. Microbiol.* *78*:179-192.

Trinci, A. P. J. and Collinge, A. J. (1974). Occlusion of the septal pores of damaged hyphae of *Neurospora crassa* by hexagonal crystals. *Protoplasma* *80*:56-67.

Trinci, A. P. J., and Morris, N. R. (1979). Morphology and growth of a temperature-sensitive mutant of *Aspergillus nidulans* which forms aseptate mycelia at non-permissive temperatures. *J. Gen. Microbiol.* *114*53-59.

Tu, J. C., and Malhotra, S. K. (1977). The significance of cAMP induced alterations in the cellular structure of *Phycomyces*. *Can. J. Microbiol.* *23*:378-388.

Wrathall, C. R., and Tatum, E. L. (1974). Hyphal wall peptides and colonial morphology in *Neurospora crassa*. *Biochem. Genet.* *12*:59-68.

Yanagita, R., and Kogane, F. (1962). Growth and cytochemical differentiation of mold colonies. *J. Gen. Appl. Microbiol.* *8*:201-213.

13
Morphogenesis of Sclerotia

RODERIC COOKE *University of Sheffield, Sheffield, England*

I. INTRODUCTION

There are two major reasons for the current interest in sclerotium-forming fungi. First, many such species are important plant pathogens, and an understanding of the processes of sclerotium production may lead to the development of new methods for disease control. Second, they are considered by some to be suitable organisms through which to study a morphogenetic event of fundamental importances in mycology, that is, the transformation of normal vegetative hyphae to a different physical and physiological form. Recent reviews on the morphogenesis of sclerotia indicate that progress in this field has been slow and that, in many instances, experimentation has been so empirical as to present formidable problems of interpretation (Chet and Henis, 1975; Willetts, 1978). This situation compares unfavorably with the rapid advances being made in other areas of fungal morphogenesis. For this reason, no attempt is made here at an exhaustive literature review, the intention being rather to consider some of the more recent approaches that have been made to the problem of sclerotial morphogenesis.

II. STRUCTURE, FUNCTION, AND ORIGINS

Sclerotium structure and function have been discussed in detail a number of times so that only a brief summary is necessary here (Coley-Smith and Cooke, 1971; Willetts, 1971,1972). "Sclerotium" is an imprecise term referring to a diverse array of hyphal aggregates that serve as persistent resting stages in conditions inimical to mycelial growth. When favorable conditions return, they germinate, giving rise to either vegetative hyphae or sexual or asexual spore-bearing organs. Their evolutionary origins are probably as varied as their structure and behavior. As well as arising from strictly vegetative beginnings, many are either degenerate or much-modified sexual or asexual fruit-bodies, or at least have close structural affinities with them (Willetts, 1972; Macfarlane et al., 1978; Christen, 1979; Luttrell, 1980; Tu, 1980).

Sclerotia of those fungi so far studied in detail develop initially in one of four distinct ways (Table 1). In most cases the sclerotium begins as the aggregation of small groups of hyphae to form a primordium. Primordia are of three kinds; loose, terminal, and strand. Loose primordia arise from the irregular branching and increased septation of neighboring hyphae. The constituent cells of these branches then swell, so that the resulting sclerotium is composed of a poorly organized mass of cells which may be pseudoparenchymatous. Terminal primordia are formed by the dense branching of a single hyphal tip, or the tips of closely associated hyphae, to produce a mycelial knot. The constituent hyphae of this become appressed, and may also anastomose, so that the final sclerotium has a firm and compact structure. Similar compact sclerotia are formed from strand primordia, which are initiated by the growth, septation, interweaving, and fusion of parallel side branches from one or more main hyphae. Here, sclerotia are formed in positions lateral to main hyphae or hyphal strands. While most fungi possess one of these three primordial types, the sclerotia of some do not arise from such localized initials but, rather, are the result of the differentiation of an extensive, preexisting vegetative mycelium or asexual fruiting structure.

Sclerotia may be roughly divided into those which have a rind of flattened, dead, and usually pigmented cells, and those that do not. Rindless sclerotia usually have little differentiation of tissues into distinct types. By contrast, in some sclerotia with a rind, several tissue types may occur, each occupying a distinct zone within the sclerotium. The limited information available suggests that sclerotia with a rind are produced most frequently from terminal and strand primordia, although rindless sclerotia also develop in this way (Table 1). Sclerotium-forming species are found in all three subdivisions of septate fungi, but there is little evidence to suggest that any particular pattern of early sclerotium development, or final anatomy, is restricted to a single group of fungi.

TABLE 1 Types of Development and Differentiation of Sclerotia Found in Species from the Three Subdivisions of Septate Fungi

	Mode of Origin			
Subdivisions	Preexisting mycelium	Loose primordia	Terminal primordia	Strand primordia
Ascomycotina	*Claviceps* spp. R.	*Mycosphaerella ligulicola*	*Sclerotinia fructicola*	*Sclerotinia gladioli* R
	Cordyceps spp.		*S. sclerotiorum* R	*S. minor* R
			S. trifoliorum R	
			Pyronema domesticum R	
Deuteromycotina			*Sclerotium cepivorum* R	*Macrophomina phaseoli*
			Botrytis cinerea R,D	*Verticillium dahliae*
			B. allii R,D	*Aspergillus alliaceus* R
				Phymatotrichum omnivorum R
				Colletotrichum lindemuthianum R
				C. coccodes R,D
Basidiomycotina		*Rhizoctonia solani*		*Sclerotium rolfsii* R,D
				Typhula spp. R,D

[a]R, sclerotia with a rind; D, marked differentiation into tissue types.

III. BASIC CONCEPTS

Sclerotium development from primordia may be divided into three arbitrary, overlapping phases, each being controlled, or strongly influenced, by a complex of endogenous and exogenous factors. First, there is induction of primordia through the initiation of primary metabolic events not found in normal vegetative hyphae. Under any given set of conditions, induction establishes the number of primordia to be formed on a mycelium. This number, and also the position of each with respect to its neighbors, depends largely on the intensity of endogenous and exogenous inductive stimuli, but may also be regulated by nutrient availability. Normally, primordia either develop into sclerotia or remain as they are and, if they do develop, sclerotium number, size, and form are established during the growth phase. Throughout this, mature sclerotium biomass, and perhaps also number, is greatly influenced by environmental factors. Characteristically, there is rapid influx of nutrients into sclerotia, and nutrient availability is probably the most important factor governing growth. Finally, as growth proceeds, it is at first accompanied by, and then entirely replaced by, a process of maturation. Nutrient influx ceases and a complex of endogenous physical and physiological events bring the sclerotium to a state in which it is fitted for its survival role. These include dehydration, cell wall thickening, and deposition of storage compounds. Maturation may be influenced by environmental effects that alter metabolic rate, but is mostly independent of nutritional factors.

Studies on sclerotium morphogenesis have been made almost exclusively on those types that develop from primordia and, as with other problems in fungal morphogenesis, the key to understanding sclerotium induction is to identify the primary metabolic events that transform vegetative hyphae to primordia. Unfortunately, few investigations have come close to doing this and most have been concerned with the expression of induction, that is, factors controlling the numbers, position, and size of mature sclerotia, rather than with the process of induction itself. In addition, work has been concentrated for the most part on only two species. *Sclerotium rolfsii* (Basidiomycotina) and *Sclerotinia sclerotiorum* (Ascomycotina), so that generalizations from such a narrow base must obviously be made with caution. Furthermore, the necessity to carry out experiments on these plant pathogens in axenic culture removes the possibly important, but at present unquantifiable, influence of the host. In this regard, it might be noted that many pathogens, for instance, *Claviceps* species and *Sclerotinia curreyana*, fail to form sclerotia in axenic culture.

IV. INDUCTION OF PRIMORDIA

Profuse branching of main or lateral hyphae is the *sine qua non* of primordium formation, and is accompanied by retardation of linear hyphal growth at and around primordium sites. Treatment of mycelia with ethanol or acetate, which inhibit formation of lateral hyphae, reduces sclerotium formation. By contrast, antifungal antibiotics such as benomyl and griseofulvin, which inhibit linear extension of main hyphae, promote sclerotium formation (Chet and Henis, 1975; Tu and Jarvis, 1979a,b). It is tempting to assume that inhibition of linear growth precedes induction of primordia, and that the inhibitory factors involved are also responsible for, or closely linked to, stimulation of branching. This notion has provided the basis for the most plausible hypothesis for the mechanism of primordium induction but, as will be seen later, the physiological connection between branching and slowing of linear growth remains to be established.

Three broad approaches to the problem of induction have been used: studies on the possibility of endogenous control of this process, examination of the effects of exogenous stimuli, and investigations on metabolic changes during primordium formation.

A. Endogenous Factors

Different species have differing capabilities for sclerotium production and even different isolates of the same species can vary very widely in this respect. However, despite this great variety of behavior, most have some capacity to form sclerotia in the dark at constant temperature. This indicates that, although numbers of primordia may be markedly influenced by the application of exogenous environmental stimuli, the ability to form them has some degree of autonomy. There is clear evidence as to the important of endogenous factors in primordium induction. Here the term *endogenous* is used, rather broadly, to refer not only to factors generated in and acting within the cells of the mycelium, but also to changes that occur naturally in the external milieu as a result of mycelial growth.

In culture, a basic pattern of colony growth and primordium formation is common to all species. There is first a period of extension growth followed, after a certain interval, by profuse hyphal branching in discrete areas, or in a distinct zone, at some distance behind the colony margin. Primordia then appear within these areas of branching. In *S. rolfsii* and *Sclerotinia sclerotiorum*, onset of branching is marked by a drastic decline in the rate of advance of the colony margin, but this then gradually rises to its former level after primordia have been formed (Humpherson-Jones and Cooke, 1977b; Kritzman et al., 1977). When these two species are grown in the dark at constant temperature, primordium induction has a marked endogenous rhythm, resulting in the formation of successive zones of sclerotia with a corresponding regular fluctuation in the rate of colony extension (Figs. 1 and 2). As well as a generalized reduction in hyphal extension during primordium induction, in *Sclerotinia sclerotiorum* there are more localized effects, in that the greatest decrease in growth rate occurs in that area of the colony

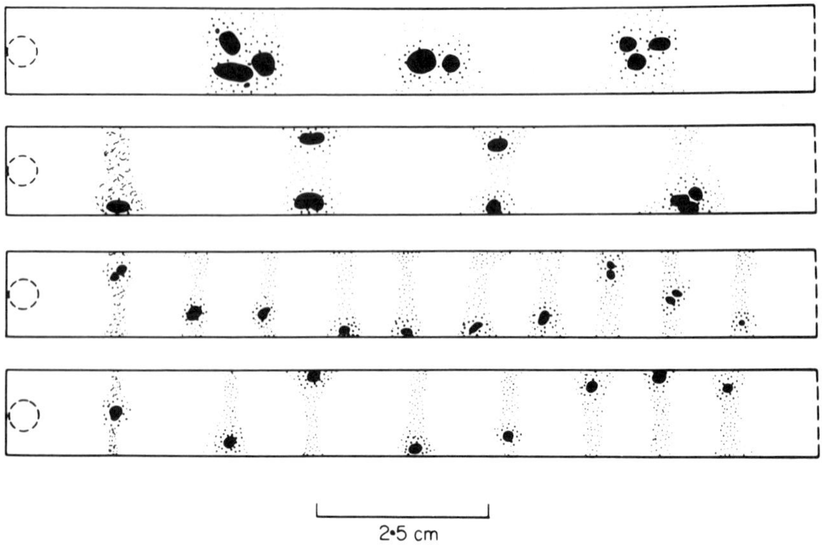

FIG. 1 Surface view diagrams of rhythmic sclerotium production by four isolates of *Sclerotinia sclerotiorum* growing in tubes in the dark at 25°C. The inoculum disk is to the left; growth proceeds from left to right; the sclerotia are black; and the piled, branched mycelium is stippled. (From Humpherson-Jones and Cooke, 1977b.)

margin which is in a position immediately in front of any developing sclerotium (Fig. 3). Primordium induction occurs on hyphae that have reached a particular age, and is located in a medium whose composition has been much modified by the passage of the colony margin through it and by subsequent hyphal development within it. Induction also occurs when colony growth is impeded by a mechanical barrier. Here, not only does hyphal extension cease, but conditions are also created in which metabolites rapidly accumulate within hyphae and the surrounding medium.

These behavior patterns suggest a number of physiological bases for endogenous primordium induction, and a number of hypotheses have been erected. In considering these, it should be noted that, for the most part, they await confirmation by experiment and that probably no single explanation of induction is presently wholly acceptable. Recognition of the relationship between cessation or retardation of colony growth and sclerotium formation has led to the suggestion that, in *S. rolfsii*, induction of primordia depends on internal metabolites or morphogens similar to those which also maintain normal linear growth (Wheeler and Waller, 1965). The two processes are therefore in competition for these metabolites, with the balance being strongly in favor of hyphal

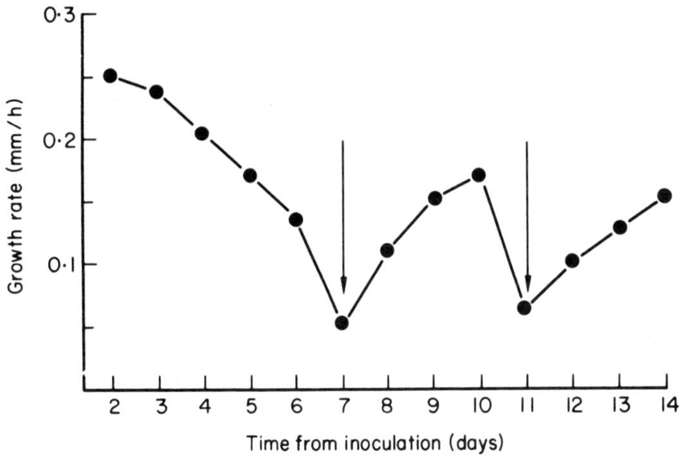

FIG. 2 Changes in the rate of hyphal extension at the colony margin during sclerotium formation by a rhythmic isolate of *Sclerotinia sclerotiorum*. The arrows indicate the time of appearance of primordia behind the margin. (From Humpherson-Jones and Cooke, 1977b).

extension, so that induction will occur only when, for some yet undetermined reason, they accumulate in sufficient amounts within hyphae to permit it to take place. This could explain induction of primordia when marginal hyphae meet a mechanical barrier to growth, but it is difficult to see how competition between the processes of linear growth and primordium formation could spontaneously alter in favor of the latter. It has similarly been postulated for *S. rolfsii* that an endogenous protein, synthesized within mature hyphae, is translocated to the colony margin, where it inhibits growth and consequently triggers a chain of unidentified processes which result in primordium induction (Geiger and Goujon, 1970). Thermolabile primordium-inducing morphogens have been detected in cell-free extracts of *S. rolfsii*, but have not been identified.

While little is known of the nature and role of internal inductive factors, greater progress has been made with respect to the morphogenetic effects of metabolites released from hyphae during growth. It has been demonstrated that in *S. rolfsii* and *Sclerotinia sclerotiorum*, primordia can be induced by the application of staled media (Bedi, 1958; Liu and Wu, 1971). In *Sclerotinia sclerotiorum*, induction occurs in response to two unidentified organic acids which accumulate in the medium during growth. Similar compounds are probably produced by *S. rolfsii*, but not by the closely related species *Sclerotium delphinii* (Humpherson-Jones and Cooke, 1977c). The latter fungus is, however, sensitive to the staling products of *S. rolfsii*. These behavioral differences indicate

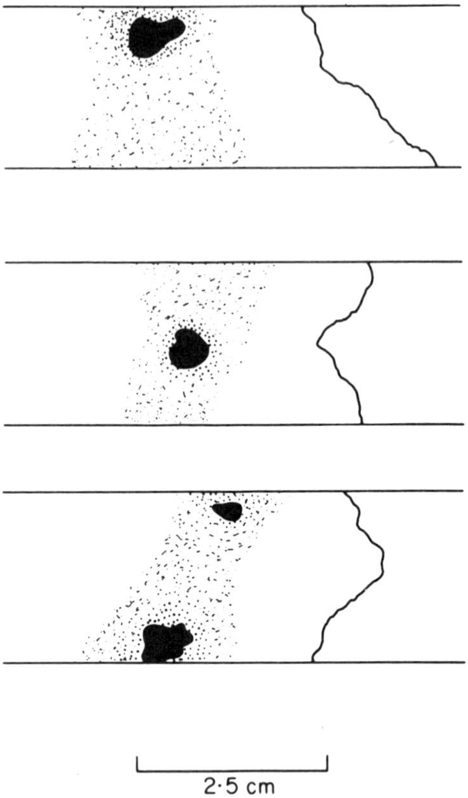

FIG. 3 Sclerotium formation in tubes by *Sclerotinia sclerotiorum* showing the shape of the colony edge in relation to the position of the most recently formed sclerotium. Growth proceeds from left to right; the sclerotia are black; and the piled, branched mycelium is stippled. (From Humpherson-Jones and Cooke, 1977b.)

the probably widely varying role of staling compounds among sclerotium-forming fungi. It is often overlooked that accumulation of staling products within media and removal of nutrients from them are coincidental processes that are difficult, if not impossible, to separate. By analogy with the induction of sporulation in many fungi, nutrient depletion alone may provide a sufficient stimulus for primordium formation. This may be the case, particularly among sclerotium-forming pathogens developing on their hosts, where sclerotia normally appear when exploitation of host tissues is nearing completion.

When considering the capacity for endogenous control of induction, it is important to realize that genetic constitution has a fundamental influence. For example, monoascospore isolates of *Sclerotinia sclerotiorum* vary widely in this ability, some forming abundant sclerotia in a regular manner in the dark and at constant temperature, others requiring strong exogenous stimuli before primordia appear (Humpherson-Jones and Cooke, 1977b,c). Comparisons between the physiologies of such contrasting isolates might throw more light on endogenous control than has hitherto been possible. In the absence of sexual spores from which to obtain genetically distinct isolates, use may be made of cytoplasmic factors that control sclerotium formation to the same effect (Humpherson-Jones and Cooke, 1977d).

B. Exogenous Factors

Numerous kinds of exogenous stimuli, both physical and biochemical, have been applied to cultures of sclerotium-forming fungi, often with a view to obtaining information on changes in metabolism during induction, and the results of such investigations have been adequately summarized previously (Chet and Henis, 1975; Willetts, 1978). It is sufficient to point out that, for the most part, results have been confusing, useful information having emerged in only a minority of cases. The latter will be discussed later in connection with sclerotium metabolism, but here consideration of exogenous factors is restricted to the effects of radiation.

Few studies have been made on the photobiology of sclerotium-forming fungi and what little information is available is in some respects conflicting. This may be due partly to the already mentioned physiological variation to be found between different isolates of the same species, and partly to differences in techniques used by different investigators. For instance, in some isolates of *Botrytis cinerea*, *Botrytis squamosa*, *Verticillium dahliae*, and some *Aspergillus* species, induction of primordia is suppressed by light, particularly if the source is intense or has high blue or ultraviolet emissivity (Paul, 1927; Tarurenko, 1954; Becker, 1956; Page, 1956; Kaiser, 1962; Bennett et al., 1978). However, in other isolates of these species and of *Fusarium oxysporum*, *Rosellinia necatrix*, *Sclerotinia sclerotiorum*, *S. delphinii*, and *S. rolfsii*, light enhances primordium formation (Carlile, 1956; Heath and Eggins, 1965; Trevethick and Cooke, 1973; Sztejnberg et al., 1980). Finally, some isolates of *Aspergillus* species and *V. dahliae* are indifferent to light (Rudolph, 1962; Brandt, 1964).

Light-enhanced primordium induction has been studied in detail in several isolates of *Sclerotinia sclerotiorum*, *S. rolfsii*, and *S. delphinii* (Humpherson-Jones and Cooke, 1977a). Numbers of sclerotia produced in response to white light increase with irradiance (Fig. 4). Primordia are induced on dark-grown colonies when these are subsequently exposed to white light, the degree of response decreasing with the duration of previous dark incubation, sensitivity being greatest during a period well before primordia form naturally either in continuous light or

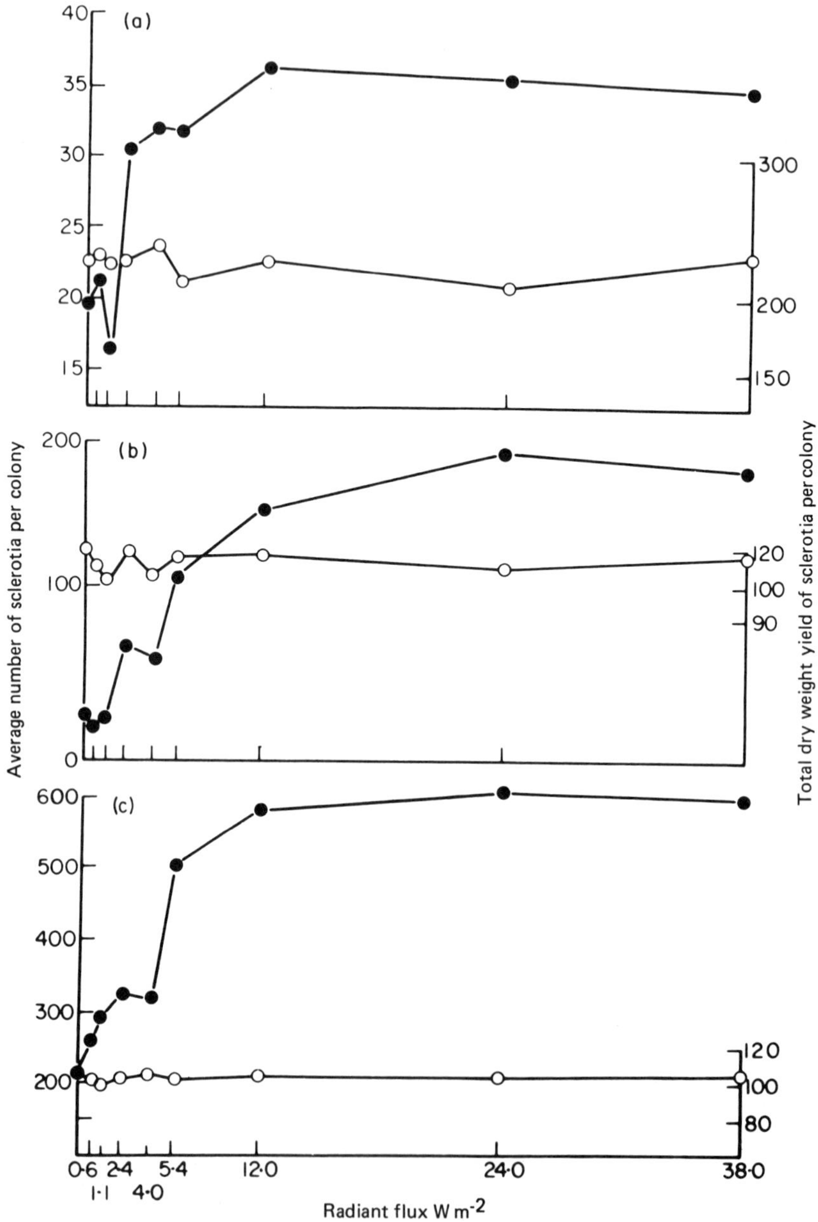

FIG. 4 Effect of continuous light of varying irradiance on sclerotium number (•) and total dry weight yield of sclerotia in mg (o) on agar-grown colonies of (a) *Sclerotinia sclerotiorum*, (b) *Sclerotium delphinii*, and (c) *Sclerotium rolfsii*. (From Humpherson-Jones and Cooke, 1977a.)

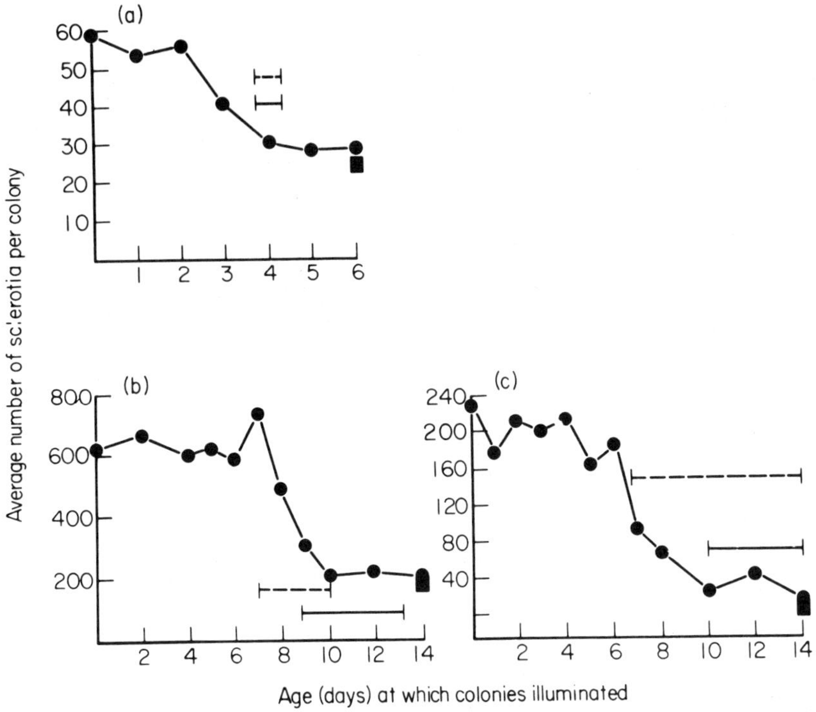

FIG. 5 Sensitivity of dark-grown colonies of different ages to subsequent exposure to continuous light of irradiance 38.0 W/m^2: (a) *Sclerotinia sclerotiorum*, (b) *Sclerotium rolfsii*, and (c) *Sclerotium delphinii*. Unbroken and broken horizontal bars indicate the duration of sclerotium formation in controls grown in continuous dark and continuous light, respectively. The solid square indicates the final number of sclerotia formed in continuous dark. (From Humpherson-Jones and Cooke, 1977a.)

continuous dark (Fig. 5). As the time for natural—that is endogenously controlled—induction approaches, colonies become less sensitive until, upon induction, they cease to respond to the light stimulus. The most effective wavelengths are those in the blue portion of the spectrum (420-460 nm; peak, 435 nm) and those in the near-ultraviolet portion (peak, 365 nm), which is in accord with action spectra for sporulation in many fungi. It might, however, be noted that yellow, red, and infrared light have been found to be stimulatory in some isolates of *B. cinerea* and *B. squamosa* (Bjornsson, 1959; Tan and Epton, 1973).

Despite the variability of response of different species and isolates to light, the photobiology of primordium induction may provide a means of exploring the first phase of sclerotium formation through an investigation of the link between photoreception and early changes in metabolism. However, an examination of the current state of knowledge concerning primary metabolic events during induction indicates the severity of the problems to be overcome.

C. Metabolism During Induction

Attempts to elucidate metabolic patterns during primordium formation have, for the most part, been based on observations on the inhibitory or stimulatory effects of biochemicals and metabolites on sclerotium production. Having assessed these effects, deductions have then been made with regard to those metabolic processes most likely to be influenced by the applied compounds, and the possible relationships of these processes to primordium induction. The dangers and shortcomings inherent in such an approach are obvious. More rarely, the metabolism of primordia or primordium-bearing mycelia has been studied directly to more advantage.

A great variety of compounds have been shown to either induce or inhibit primordium formation (Table 2). Some of these may be acting directly via metabolic pathways involved in induction, but others are probably acting indirectly by affecting physiological processes remote from induction itself. Direct effects have been attributed to relatively few compounds and, even for these, their proposed mode of action is highly speculative. Principal among them is the stimulatory role of sulfhydryl group (–SH) antagonists, for example, iodoacetic acid, potassium iodate, and metal chelates, in primordium formation in *S. rolfsii*. Cysteine and other sulfur-containing amino acids, inhibits primordium formation. Cysteine is a precursor of coenzyme A, which, in turn, affects the entry of carbohydrates into the Krebs cycle. It has been suggested that sulfhydryl group antagonists block this system through their effect on cysteine metabolism and that the activity of the Krebs cycle is thus reduced (Willetts, 1978). This reduction may then be followed by an increase in activity of the pentose phosphate pathway, the resultant shift in reducing potential triggering primordium induction. This appears to be a reasonable proposal and is supported, indirectly, by evidence from metabolic studies on *Sclerotinia sclerotiorum* (Wong and Willetts, 1974). Activity of the polyphenoloxidase tyrosinase is very low in mycelia of this species, but high in primordia. Similarly, while laccase is active in both mycelia and primordia, it is very much higher in the latter and in young sclerotia. Polyphenoloxidases may inactivate glyceraldehyde-3-phosphate dehydrogenase, so suppressing glycolysis and increasing activity of the pentose phosphate pathway. It may be of significance here that in *Sclerotinia sclerotiorum* and *Sclerotinia libertiana* the phenolic metabolite

TABLE 2 Effects of Some Chemicals, Antibiotics, and Metabolites on the Production of Sclerotial Primordia

Species	Stimulation	Inhibition
Sclerotium rolfsii	Iodoacetic acid Potassium iodate Metal chelates (Chet et al., 1966)	Cysteine Methionine Glutathione
	Griseofulvin Trichomycin Bacitracin (Chet and Henis, 1975)	Mercaptoethanol (Christias, 1975) Dimethyl sulfoxide (Melhuish and Bean, 1971)
		Ethanol Acetate Threonine Lactose (Henis et al., 1973; Kritzman et al., 1976)
Sclerotinia libertiana	Indole acetic acid (Marukawa et al., 1975a)	p-Amino benzoic acid Excess ammonium salts
	Sclerin + cyclic AMP Sclerin + DOPA (Marukawa and Satomura, 1977)	
Sclerotinia sclerotiorum	Sclerin (Marukawa et al., 1975b)	

sclerin—which stimulates polyphenoloxidase activity—induces primordia, while p-amino benzoic acid—which inactivates polyphenoloxidases—inhibits their formation (Marukawa et al., 1975a,b).

A major problem involved in examining the primary metabolic events of induction is the rapidity with which this process takes place. For example, it has been shown that in sclerotium formation by *Rhizoctonia solani* it is during the first 2 h that the critical reorientation of cellular metabolism probably takes place (Hashiba and Staples, 1976). Here, rapid RNA synthesis is accompanied by a marked decrease in the permeability of primordial cells. The key to induction may well lie in studies on early changes in nucleic acid synthesis rather than in investigations of grosser metabolic processes.

V. PHYSIOLOGY OF GROWTH AND MATURATION

The processes of growth and maturation are not entirely distinct from one another, but overlap to a considerable degree. Growth is characterized by nutrient influx, which depends on movement through the hyphae that connect the primordium to the mycelium. While maturation begins during the growth phase, it is completed long after all connection has been lost between the sclerotium and its underlying mycelium. The growth phase is remarkably short, about 32 h and 72 h in *R. solani* and *Sclerotinia sclerotiorum*, respectively, but maturation occurs over a period of at least 10-15 days in these two species (Cooke, 1970; Hashiba and Staples, 1976). Essentially, maturation involves deposition or redistribution of polysaccharide reserves, melanization of the outer sclerotium tissues, and dehydration to a level which favors survival under adverse conditions (Cooke, 1971; Saito, 1974; Waters et al., 1975). These events occur against a background of declining general metabolic activity and are peripheral to the major, and highly dynamic, processes that govern rapid development of primordia into sclerotia. It is therefore growth that will be mainly considered here.

Given that sufficient nutrients are available to them, primordia growing in culture or on host plants rapidly increase in size and, at the same time, invariably exude copious amounts of fluid. Mechanisms for maintenance of translocation into sclerotia and the nature and function of sclerotium exudates have come to occupy a central position in considerations of the postinduction stages of morphogenesis. Initially, source-sink movement of nutrients, particularly carbohydrates, is probably largely controlled by the high respiration rate of primordia, but, as the latter increase in size and translocation rates correspondingly rise, other systems assume more importance. One obvious mechanism is the conversion of incoming carbohydrates to storage or other compounds within sclerotia. For instance, one-way flow of carbohydrates into developing sclerotia of *Claviceps purpurea, Sclerotinia sclerotiorum,* and *Sclerotinia trifoliorum* appears to be maintained largely through mannitol and trehalose synthesis within their tissues. Similarly, in sclerotia of *Coprinus cinereus*, a fall in levels of reducing sugars in the surrounding medium is associated with accumulation of glycogen in the tissues (Cooke, 1969; Dickerson et al., 1976; Jirjis and Moore, 1976).

Exudates produced during growth are of two kinds: sugar-rich honeydews, found typically in *Claviceps* species, and the sugar-poor droplets found on the sclerotium surface in most other sclerotium-forming species. During the early stages of ergot formation in *Claviceps*, honeydew serves to drastically decrease the water potential of the developing sclerotium and surrounding host tissues. This causes host sap to flow into infected tissues, so maintaining the translocation of host sugars toward the fungus (Mower and Hancock, 1975a,b). The physiological role of the fluids produced by other fungi is more problematical.

In *Sclerotinia sclerotiorum* and *S. rolfsii*, low-sugar exudation begins with the appearance of minute liquid droplets on primordia. As the latter increase in size, so do the droplets, which then, at some point, coalesce to form relatively few large drops on the sclerotium surface. During maturation, exudate drops decrease in size and finally disappear. Small individual droplets and larger coalesced drops are covered with a continuous, membranous sac of possibly colloidial material. Upon the eventual loss of liquid from these sacs, they collapse and their remnants persist on the sclerotium surface (Colotelo et al., 1971). The interior of the young sclerotium of *Sclerotinia sclerotiorum* contains a liquid matrix, and exudate may originate from this as it is squeezed out to the surface owing to the decrease in free space within the developing tissues (Colotelo, 1973,1974).

There are two obvious, but as yet unconfirmed, physiological roles for low-sugar exudation: First, it could be a mechanism for rapid dehydration of maturing tissues; and second, its evaporation could create water flow into sclerotia from the supporting mycelium and so at the same time enhance nutrient translocation. However, additional functions are suggested by observations that, although its carbohydrate content is low, the exudate contains a wide range of organic and inorganic compounds together with enzymes (Table 3). Furthermore, the nature and amounts of these substances change during growth and maturation. For example, in *Sclerotinia sclerotiorum* there is a decrease in proteins, lipids, potassium, and magnesium, but an increase in sodium (Colotelo, 1973). In addition to these changes, the composition of exudate does not always reflect the constitution of the sclerotium (Cooke, 1969; Christias, 1980). For instance, sclerotia of *Sclerotinia trifoliorum* are rich in mannitol and trehalose, but it is only the latter which occurs in exudate. Similarly, alanine is abundant in sclerotia of *S. rolfsii*, but, while exudates contain a wide range of other amino acids that are also present in the sclerotium, alanine is absent from among them. Changes in the composition of exudates have been taken to indicate their reabsorption

TABLE 3 Range of Substances Found in Low-Sugar Exudates from Sclerotia

Soluble carbohydrates (Cooke, 1969)	Organic acids (Colotelo, 1973)
Glucose	α-Ketoglutaric acid
Inositol	Glyoxylic acid
Mannitol	Malic acid
Trehalose	Oxalic acid
	Succinic acid
Amino acids (Christias, 1980)	Enzymes (Colotelo, 1973, 1978)
Glutamic acid in concentrations of 0.1–1.5 μg/ml, of exudate	Acid phosphatase
Glycine	β-Glucosidase
Aspartic acid	Catalase
Serine	Cellulase
Lysine	Peroxidase
Leucine	Polygalacturonase
Arginine	Polyphenoloxidase
Threonine	Acid and alkaline protease
Alanine	RNase
	Metal salts (Colotelo, 1973)
Histidine in trace amounts only	K
Valine	Mg
Isoleucine	Na
Tyrosine	
Phenylalanine	

and further metabolism, while the apparently selective nature of exudation suggests that it is involved in the maintenance of some kind of internal physiological balance. The latter could also, of course, be effected by reabsorption, but the nature of the internal processes so affected is not known.

Honeydews have an additional important function in that they are attractive in insects which subsequently disperse the conidia that are normally produced within these exudates. The sclerotia of some fungi that produce low-sugar exudates may have originated from conidial stromata, so that their exudates may be the relic of an ancient insect dispersal mechanism.

VI. SCLEROTIUM DEVELOPMENT AND GENERAL ORGANOGENESIS

It might be expected that strong affinities exist between the physiology of sclerotium induction and growth and these same processes in the development of other vegetative organs, for instance, mycelial strands and rhizomorphs. This might also be true of reproductive structures—for instance, coremia and sporodochia—in which there is a great deal of vegetative development before spores are formed upon them. Even some sexual fruit-bodies may have some early developmental characteristics in common with sclerotia. For example, sclerotium primordia of *C. cinereus* can develop into either sclerotia or carpophores, depending on environmental conditions. Scerlotia have also been observed to form from the already differentiated carpophore tissues of *Typhula* (Christen, 1979; Moore et al., 1979). There are, unfortunately, too few data available for detailed comparisons to be made between sclerotium morphogenesis and that of other organs, and there is certainly too little evidence for the establishment of metabolic similarities between them. A recent valiant attempt to present a unified view of organogenesis served mainly to emphasize differences rather than similarities in fundamental events (Watkinson, 1979). It should not be overlooked that the morphogenesis of vegetative structures other than sclerotia is based on the intertwining of hyphae that are rapidly growing apically, and sometimes indefinitely, while, by contrast, sclerotia develop through the rapid swelling of a finite group of cells established during a transient primordial stage.

The difficulties of comparative morphology are well illustrated by a parallel consideration of factors inducing sclerotia and rhizomorphs. This is particularly illuminating, since it has been suggested that a sclerotium resembles a static rhizomorph, and the latter a continually elongating sclerotium. Rhizomorphs of *Armillaria mellea* are induced by ethanol, indole acetic acid, and p-amino benzoic acid, but of these only indole acetic acid induces sclerotia, the remaining two compounds being inhibitory (Weinhold and Garraway, 1966; Garraway, 1969,1970). These behavioral differences may reflect the contrasting role of phenolic compounds in sclerotium and rhizomorph morphogenesis. In culture, phenolic compounds accumulate during the growth of both sclerotium-forming species and of *A. mellea*, and these inhibit the apical growth of main hyphae. This leads to sclerotium induction through stimulated hyphal branching, but rhizomorph development must necessarily be inhibited, since it depends on the apical extension of main hyphae. Application of ethanol or p-amino benzoic acid to *A. mellea* possibly redirects phenol metabolism so that inhibitory compounds either accumulate more slowly or are not formed at all. By contrast, in, for instance, *Sclerotinia sclerotiorum*, p-amino benzoic acid inhibits the action of the phenolic metabolite sclerin, which is a sclerotium inducer. Ethanol inhibits branch formation, so that in *A. mellea* such inhibition may favor rhizomorph initiation through the diversion of resources from branch formation toward enhancement of apical growth. In sclerotium-forming fungi its effect is to prevent the degree of branching required to establish primordia.

While details of organ induction may differ widely in detail, the genesis of vegetative and some reproductive structures takes place against a common background–that of general nutrient depletion during mycelial growth. Within this impoverished milieu, there is then localized accumulation of materials at a number of sites. This accumulation requires a change in translocatory patterns within the mycelium so that, general ly, movement of materials is directed away from the advancing edge of the colony. Primordia then develop at these sites, growth being to some extent autonomous but resources also being drawn from the remaining resources of the medium. This pattern of behavior is reasonably clear, but an understanding of the key physiological events at each stage has not yet been reached.

REFERENCES

Becker, J. G. (1956). Physico-chemical aspects of microsclerotia formation in *Verticillium albo-atrum*. Ph.D. thesis, Purdue University.

Bedi, K. S. (1958). The role of stale products in the formation of sclerotia of *Sclerotinia sclerotiorum* (Lib.) de Bary. *Indian Phytopathol. 2*:29-36.

Bennett, J. W., Fernholz, F. A., and Lee, L. S. (1978). Effect of light on aflatoxins, anthraquinones, and sclerotia in *Aspergillus flavus* and *A. parasiticus*. *Mycologia 70*:104-116.

Bjornsson, I. P. (1959). Response of certain fungi particularly *Trichoderma* sp. to light. *J. Wash, Acad. Sci. 43*:317-325.

Brandt, W. H. (1964). Morphogenesis in *Verticillium*: Effects of light and ultraviolet radiation on microsclerotia and melanin. *Can. J. Bot. 42*:1017-1023.

Carlile, M. J. (1956). A study of the factors influencing non-genetic variation in a strain of *Fusarium oxysporum*. *J. Gen. Microbiol. 14*:643-654.

Chet, I., and Henis, Y. (1975). Sclerotial morphogenesis in fungi. *Annu. Rev. Phytopathol. 13*:169-192.

Chet, I., Henis, Y., and Mitchell, R. (1966). The morphogenetic effect of sulphur-containing amino acids, glutathione and iodoaacetic acid on *Sclerotium rolfsii* Sacc. *J. Gen. Microbiol. 45*:341-546.

Christen, A. A. (1979). Formation of secondary sclerotia in sporophores of species of *Typhula*. *Mycologia 71*:1267-1269.
Christias, C. (1975). Specific inhibition of sclerotium formation by 2-mercaptoethanol and related sulfhydril compounds in *Sclerotium rolfsii*. *Can. J. Microbiol. 21*:1541-1547.
Christias, C. (1980). Nature of the sclerotial exudate of *Sclerotium rolfsii* Sacc. *Soil Biol. Biochem. 12*:199-210.
Coley-Smith, J. R., and Cooke, R. C. (1971). Survival and germination of fungal sclerotia. *Annu. Rev. Phytopathol. 9*: 65-92.
Colotelo, N. (1973). Physiological and biochemical properties of the exudate associated with developing sclerotia of *Sclerotinia sclerotiorum* (Lib.) de Bary. *Can. J. Microbiol. 19*:73-79.
Colotelo, N. (1974). A scanning electron microscope study of developing sclerotia of *Sclerotinia sclerotiorum*. *Can. J. Bot. 52*: 1127-1130.
Colotelo, N. (1978). Fungal exudation. *Can. J. Microbiol. 24*:1173-1181.
Colotelo, N., Sumner, J. L., and Voegelin, W. S. (1971). Presence of sacs enveloping the liquid droplets on developing sclerotia of *Sclerotinia sclerotiorum* (Lib.) de Bary. *Can. J. Microbiol. 17*:300-301.
Cooke, R. C. (1969). Changes in soluble carbohydrates during sclerotium formation by *Sclerotinia sclerotiorum* and *S. trifoliorum*. *Trans. Br. Mycol, Soc. 53*:77-86.
Cooke, R. C. (1970). Physiological aspects of sclerotium growth in *Sclerotinia sclerotiorum*. *Trans. Br. Mycol. Soc. 54*:361-365.
Cooke, R. C. (1971). Physiology of sclerotia of *Sclerotinia sclerotiorum* during growth and maturation. *Trans. Br. Mycol. Soc. 56*:51-59.
Dickerson, A. G., Mantle, P. G., and Nisbet, L. J. (1976). Carbon assimilation by *Claviceps purpurea* growing as a parasite. *J. Gen- Microbiol. 97*:267-276.
Garraway, M. O. (1969). The influence of compounds related to the shikimic acid pathway on rhizomorph initiation and growth in *Armillaria mellea*. *Phytopathology 59*:1027.
Garraway, M. O. (1970). Rhizomorph initiation and growth in *Armillaria mellea* promoted by *o*-amino benzoic and *p*-amino benzoic acids. *Phytopathology 60*:861-865.
Geiger, J. P., and Goujon, M. (1970). Mise en évidence dans les extraits de thalle d'un facteur morphogène responsable de l'apparition des sclérotes du *Corticium rolfsii* (Sacc.) Curzi. *C. R. Acad. Sci. Ser. D 271*:41-44.
Hashiba, T., and Staples, R. C. (1976). Nucleic acid metabolism in developing sclerotia of the rice sheath blight fungus (*Rhizoctonia solani* Kühn). *J. Gen. Microbiol. 96*:239-245.

Heath, L. A. F., and Eggins, H. O. W. (1965). Effect of light, temperature and nutrients on the production of conidia and sclerotia by forms of *Aspergillus japonicus*. *Experientia 21*:385-387.

Henis, Y., Okon, Y., and Chet, I. (1973). The relationship between early hyphal branching and sclerotial formation in *Sclerotium rolfsii*. *J. Gen. Microbiol. 79*:147-150.

Humpherson-Jones, F. M., and Cooke, R. C. (1977a). Morphogenesis in sclerotium-forming fungi. I. Effects of light on *Sclerotinia sclerotiorum, Sclerotium delphinii* and *S. rolfsii*. *New Phytol. 78*: 171-180.

Humpherson-Jones, F. M., and Cooke, R. C. (1977b). Morphogenesis in sclerotium-forming fungi. II. Rhythmic production of sclerotia by *Sclerotinia sclerotiorum* (Lib.) de Bary. *New Phytol. 78*:181-187.

Humpherson-Jones, F. M., and Cooke, R. C. (1977c). Induction of sclerotium formation by acid staling compounds in *Sclerotinia sclerotiorum* and *Sclerotium rolfsii*. *Trans. Br. Mycol. Soc. 68*:413-420.

Humpherson-Jones, F. M., and Cooke, R. C. (1977d). Changes in sclerotium formation of *Sclerotinia sclerotiorum* in culture. *Trans. Br. Mycol. Soc. 68*:459-461.

Jirjis, R. I., and Moore, D. (1976). Involvement of glycogen in morphogenesis of *Coprinus cinereus*. *J. Gen. Microbiol. 95*:348-352.

Kaiser, W. J. (1962). Influence of light on the production of sclerotia by *Verticillium albo-atrum*. *Phytopathology 45*:465.

Kritzman, G., Okon, Y., Chet, I., and Henis, Y. (1976). Metabolism of L-threonine and its relationship to sclerotium formation in *Sclerotium rolfsii*. *J. Gen. Microbiol. 95*78-86.

Kritzman, G., Chet, I., and Henis, Y. (1977). The relationship between rhythmic hyphal growth and circadian formation of sclerotia in *Sclerotium rolfsii* Sacc. *Can. J. Microbiol. 23*:959-963.

Liu, T. M., and Wu, L. C. (1971). The effect of amino acids on the growth and morphogenesis of *Sclerotium rolfsii* Sacc. *Plant Prot. Bull, 13*:87-96.

Luttrell, E. S. (1980). Host-parasite relationships and development of the ergot sclerotium in *Claviceps purpurea*. *Can. J. Bot. 58*: 942-958.

Macfarlane, T. D., Kuo, J., and Hilton, R. N. (1978). Structure of the giant sclerotium of *Polyporus mylittae*. *Trans. Br. Mycol. Soc. 71*:359-365.

Marukawa, S., and Satomura, Y. (1977). Conditions for reversal of sclerotial formation in mutant of *Sclerotinia libertiana* Fuckel. *Agric. Biol. Chem. 41*:655-660.

Marukawa, S., Funakawa, S., and Satomura, Y. (1975a). Some physical and chemical factors on formation of sclerotia of *Sclerotinia libertiana* Fuckel. *Agric. Biol. Chem. 39*:463-468.

Marukawa, S., Funakawa, S., and Satomura, Y. (1975b). Role of sclerin on morphogenesis in *Sclerotinia sclerotiorum* de Bary (including *S. libertiana* Fuckel). *Agric. Biol. Chem. 39*:645-650.

Melhuish, J. H., and Bean, G. A. (1971). Effect of dimethyl sulphoxide on the sclerotia of *Sclerotium rolfsii*. *Can. J. Microbiol. 17*:429-431.

Moore, D., Elhiti, M. M. Y., and Butler, R. D. (1979). Morphogenesis of the carpophore of *Coprinus cinereus*. *New Phytol. 83*: 695-722.

Mower, R. L., and Hancock, J. G. (1975a). Sugar composition of honeydews. *Can. J. Bot 53*:2813-2825.

Mower, R. L., and Hancock, J. G. (1975b). Mechanism of honeydew formation by *Claviceps* species. *Can. J. Bot. 53*:2826-2834.

Page, O. T. (1965). The influence of light and other environmental factors on the mycelial growth and sclerotial production by *Botrytis squamosa*. *Can J. Bot. 34*:881-890.

Paul. W. R. C. (1927). A comparative morphological and physiological study of a number of strains of *Botrytis cinerea* Pers. with special reference to their virulence. *Trans. Br. Mycol. Soc. 14*: 118-135.

Rudolph, E. D. (1962). The effect of some physiological and environmental factors on sclerotial Aspergilli. *Am. J. Bot. 49*:71-78.

Saito, I. (1974). Ultrastructural aspects of the maturation of sclerotia of *Sclerotinia sclerotiorum* (Lib.) de Bary. *Trans. Mycol. Soc. Jap. 15*:384-400.

Sztejnberg, A., Madar, Z., and Chet. I. (1980). Induction and quantification of microsclerotia of *Rosellinia necatrix*. *Phytopathology 70*:525-527.

Tan, K. K., and Epton, H. A. S. (1973). Effect of light on the growth and sporulation of *Botrytis cinerea*. *Trans. Br. Mycol. Soc. 61*:145-157.

Tarurenko, E. (1954). The influence of light on the development of mould fungi. *Mikrobiologya 23*:29-33.

Trevethick, J., and Cooke, R. C. (1973). Non-nutritional factors influencing sclerotium formation in some *Sclerotinia* and *Sclerotium* species. *Trans. Br. Mycol. Soc. 60*:559-566.

Tu, J. C. (1980). The ontogeny of the sclerotia of *Colletotrichum coccodes*, *Can. J. Bot. 58*:631-636.

Tu, J. C., and Jarvis, W. R. (1979a). Response of *Colletotrichum lindemuthianum* to benomyl. *Can. J. Plant Pathol. 1*:12-16.

Tu, J. C., and Jarvis, W. R. (1979b). Ontogeny, organization and longevity of sclerotium-like structures produced by *Colletotrichum lindemuthianum* in the presence of benomyl. *Can. J. Plant. Pathol. 1*:17-22

Waters, H., Moore, D., and Butler, R. D. (1975). Morphogenesis of aerial sclerotia of *Coprinus lagopus*. *New Phytol. 74*:207-213.

Watkinson, S. C. (1979). Growth of rhizomorphs, mycelial strands, coremia and sclerotia. In *Fungal Walls and Hyphal Growth*. J. H. Burnett and A. P. J. Trinci (Eds.). Cambridge University, Cambridge, pp. 93-113.

Weinhold, A. R., and Garraway, M. O. (1966). Nitrogen and carbon nutrition of *Armillaria mellea* in relation to the growth-promoting effects of ethanol. *Phytopathology 56*:108-112.

Wheeler, B. E. J., and Waller, J. M. (1965). The production of sclerotia by *Sclerotium rolfsii*. II. The relationship between mycelial growth and initiation of sclerotia. *Trans. Br. Mycol. Soc. 48*:303-314.

Willetts, H. J. (1971). The survival of fungal sclerotia under adverse environmental conditions. *Biol. Rev. 46*:387-407.

Willetts, H. J. (1972). The morphogenesis and possible evolutionary origins of fungal sclerotia. *Biol. Rev. 47*:515-536.

Willetts, H. J. (1978). Sclerotium formation. In *The Filamentous Fungi, Developmental Mycology*, Vol. 3. J. E. Smith and D. R. Berry (Eds.). Edward Arnold, London, pp. 197-213.

Wong, A. L., and Willetts, H. J. (1974). Polyacrylamide-gel electrophoresis of enzymes during morphogenesis of sclerotia of *Sclerotinia sclerotiorum*. *J. Gen. Microbiol. 81*:101-109.

14

Asexual Differentiation in the Fungi

KNUD ALLERMANN and JØRGEN OLSEN *Institute of Plant Physiology, University of Copenhagen, Copenhagen, Denmark*

JOHN E. SMITH *Strathclyde University, Glasgow, Scotland*

I. INTRODUCTION

In the natural classification of fungi much reliance has been placed on the perfect or sexual phase of development, resulting in the clearly defined groupings of Zygomycotina, Mastigomycotina, Ascomycotina, and Basidiomycotina. In contrast, the Deuteromycotina, or Fungi Imperfecti is a subdivision of the fungi which has been created based en-

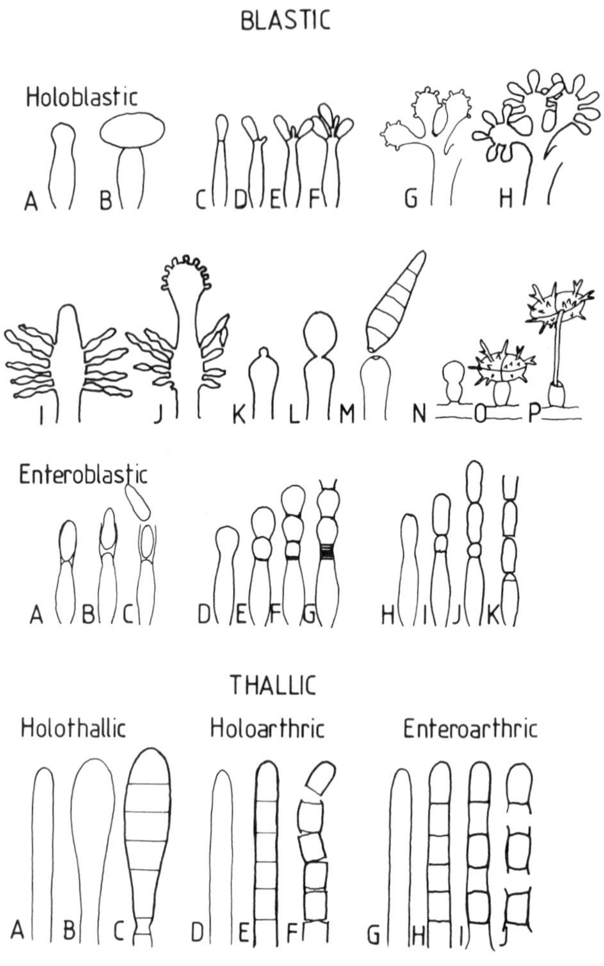

FIG. 1 Summary of modes of conidial and conidiogeneous cell development examined in the text. The following symbols denote the three major criteria which distinguish the developmental categories: CW, conidial wall differentiation; OA, order of production and arrangement of conidia on conidiogenous cell; and CC, process of conidiogenous cell proliferation. Classification of ontogenic processes based on modes of conidial wall differentiation (CW) under blastic and thallic development is indicated in the diagram. A list of representative species illustrated here is presented below. (From Cole and Samson, 1979.)

CW = holoblastic. (A and B) *Nigrospora sphaerica* (Sacc.) Mason; OA = terminal, solitary; CC = determinate. (C-F) *Tritirachium oryzae* tirely on the use of morphological characteristics of asexual mechanisms of reproduction. The Deuteromycotina have been divided into three form classes: Blastomycetes (imperfect yeasts and yeastlike fungi), Hyphomycetes (mycelial forms in which conidiogeneous structures are borne on hyphae or on aggregations of hyphae but not enclosed in fructifications), and Coelomycetes (conidia and conidiogenous cells produced in or on specialized fructifications).

During the Kananaskis Conference on Taxonomy of Fungi Imperfecti (Kendrick, 1971a), it was proposed that two basic modes of conidiogenesis exist in the Deuteromycotina; these were termed blastic and thallic development. In the blastic mode of conidium development there is a marked enlargement of a recognizable conidium initial before the initial is delimited by a septum and the conidium differentiates from part of the cell. In the thallic mode of conidium development any enlargement of the recognizable conidium initial occurs only after the initial has been delimited by a septum or septa and the conidium differentiates from a whole cell.

This chapter will be concerned with the ontogeny of blastic and thallic conidia of the Deuteromycotina. The asexual spores of the Zygomycotina and the Mastigomycotina will not be considered, owing in part to the authors' inexperience with these groups, but also to the lack of space with which to adequately deal with them (for further information on these groups see Cole and Samson, 1979).

(Vincens) de Hoog; OA = terminal, solitary; CC = proliferous (sympodial). (G and H) *Botrytis cinerea*; OA = synchronous, botryose; CC = determinate. (I and J) *Gonatobotryum apiculatum*; OA = 1° conidia synchronous, botryose; 2° conidia asynchronous, catenulate; CC = proliferous (percurrent). (K-M) *Deightoniella torulosa* (Syd.) M. B. Ellis: OA = terminal, solitary; CC = porogenous. (N-P) *Spegazzinia tessarthra* (Berk and Curt) Sacc.; OA = terminal, solitary; CC = basauxic.

CW = enteroblastic. (A-C) *Phialophora lagerbergii* (Melin & Nannf.) Conant; OA = basipetal, solitary; CC = phialidic. (D-G) *Scopulariopsis brevicaulis*: A = basipetal, solitary; CC = proliferous (annellidic). (H-K) *Cladobotryum varium*; OA = basipetal, solitary; CC = retrogressive.

CW = holothallic. (A-C) *Microsporum gypseum*; OA = terminal, solitary; CC = determinate.

CW = holoarthric. (D-F) *Geotrichum candidum* Link ex Pers; OA = random, chains; CC = determinate.

CW = enteroarthric. (G-J) *Sporendonema purpurascens*; OA = random, chains; CC = determinate.

There have been many authoritative reviews on asexual sporulation in fungi and these should be consulted for a deeper understanding of the ultrastructural and biochemical complexities of this type of differentiation (Smith and Galbraith, 1971; Smith and Anderson, 1973; Turian, 1974, 1975, 1976; Lovett, 1975; Smith, 1978).

II. BLASTIC CONIDIOGENESIS

A. Morphology

Blastic conidial development occurs by the "blowing out" and de novo growth of part of the hyphal element, the conidiogenous cell (for examples and references see Cole and Samson, 1979) (Fig. 1). This conidiogenous cell is basically a differentiated part of a hypha arising either terminally or laterally from the vegetative hypha. Conidiogenous cells can be morphologically indistinguishable from vegetative hyphae or can manifest varying degrees of morphological differentiation. Polarized growth and cytodifferentiation of the fertile locus of the conidiogenous cell is similar in function to that of the hyphal tip.

Wall growth of conidia developing from the conidiogenous cell may be isotropic, producing a globose cell, may remain polarized, producing a cylindrical cell (with or without septa), or may vary between the isotropic and polarized conditions, producing a wealth of conidial shapes. In all such cases conidial growth will involve differentiation of a new wall layer (secondary wall) adjacent to the plasmalemma and inside the outer wall (primary wall), which is continuous with the outer conidiogenous cell wall. The latter may stay intact and surround the mature conidium (holoblastic conidial formation), or it may rupture (enteroblastic conidial formation).

A conidiogenous cell can produce a single conidium or numerous conidia. During conidiation the conidiogenous cell may terminate extension growth after the first conidium is initiated (determinate), continue to elongate (proliferous), or it may shorten (retrogressive).

All conidia are delimited by a septum at the base, but can continue growing even after secession. Secession of conidia will occur either by schizolytic splitting of the basal septum or rhexolytic rupture of a supporting cell (Cole and Samson, 1979).

Within the Deuteromycotina there are three main arrangements of blastic conidia arising from conidiogenous cells (which may or may not be part of a larger structure, the conidiophore, containing several to many conidiogenous cells):

1. *Solitary.* Conidia are formed singly, terminally, and/or laterally from conidiogenous cells or in basipetal succession from the apex of specially differentiated conidiogenesis cells, that is, phialides, annellides, or retrogressive conidiogenous cells.

2. *Botryose.* Conidia arise in clusters normally from the swollen apex of a conidiogenous cell, either synchronously or asynchronously.
3. *Catenulate.* Conidia develop in acropetal chains, either simple or branched, from the apex and/or sides of conidiogenous cells.

One of the most studied examples of blastic conidial formation is the genus *Aspergillus*, particularly *Aspergillus niger* and *A. nidulans*.

The genus *Aspergillus* is characterized by phialidic development. The phialide is a conidiogenous cell producing a basipetal succession of conidia. The fertile apex of the conidiogenous cell ruptures during or after differentiation of the primary, holoblastic conidium forming the collarette (Hanlin, 1976); further conidia are enteroblastic and are produced in basipetal succession with little apparent change in the length of the conidiogenous cell. In the aspergilli, formation of primary phialoconidia on conidiophore heads occurs more or less synchronously, followed by chains of secondary conidia. Conidia remain in chains owing to the intact septum between adjacent cells and the continuous wall layer.

It is intriguing to speculate on the factors that determine the change from an entirely vegetative mycelium to a mycelium which will be a mixture of vegetative and simple or complex conidiating structures. It has generally been accepted that conidiation is stimulated by factors which restrict vegetative growth (Smith, 1978). The phases of growth and conidiation, while antagonistic, are not necessarily incompatible and can be considered as cellular processes competing for limiting metabolic intermediates, rather than as mutually exclusive phenomena. Indeed, most cultures will maintain some vegetative growth, even during conidiation. In particular, in surface cultures the peripheral hyphae will move forward vegetatively, while the mycelium behind rapidly forms conidiating structures.

An organism must be genetically competent for conidiation, though successful conidiation will require suitable environmental conditions. In the Deuteromycotina many erstwhile asexual fungi have later been found capable of expressing sexuality in the presence of unique combinations of environmental conditions.

B. Genetics and Competence of Phialoconidiation

The genetics of phialoconidiation have recently been reviewed (Clutterbuck, 1978). Mutants defective in conidiogenesis in *A. nidulans* have been extensively documented. Only two mutants blocked after conidiophore formation have been found; *bristle* and *abacus*. *Bristle* mutants are blocked in the formation of conidiophore vesicles and subsequent structures while the conidiophore continues to elongate many times the normal length. *Abacus* mutants have normal vesicles, but the sterigmata develop abnormally into rods with periodic swellings instead of the usual chain of conidia. Each mutant is believed to be at a separate single locus.

There are three additional variants which, although developmentally aberrant, do produce some conidia: *medusa, stunted*, and *wet* mutants. The number of genes involved in conidiation has been estimated to be 45-150, though most genes involved in conidiogenesis function before conidiophore formation.

It is now recognized that a growing mycelium must become competent before conidiation is inducible (Champe et al., 1981), competence being defined as the acquired responsiveness to a stimulus that induces development. The possibility now arises for identifying genes which may function as master switches for developmental processes.

These studies have arisen from the earlier observations that spore-originated colonies of *A. nidulans* must grow vegetatively for a specific period of time before the colony is inducible to conidiate. In surface culture studies wild-type *A. nidulans* growing at 37°C will develop conidiophore vesicles at 24 h after conidial inoculation (Axelrod, 1972). Synchrony is quite precise from colony to colony and the medium has little effect on the time scale of conidiophore development (Pastushok and Axelrod, 1976). Conidiation time is temperature dependent and varies inversely with colony density, being about 6 h earlier for a confluent culture mat than for a single colony at 37°C.

Under normal medium conditions when *A. niger* is inoculated into liquid-submerged conditions, the ensuing vegetative growth will remain vegetative. However, if a spore-originated colony grown in liquid medium is transferred to a solid surface culture, conidiation will occur, the transfer from liquid to solid medium acting as the inducing stimulus for conidiation. Transfer from the liquid culture can be made any time after 20 h of growth, and conidiation will commence 4 h later. Thus, as with surface cultures, responsiveness to induction, or competence, is acquired at the critical time of 20 h, conidiation then following induction after a fixed maturation period of 4 h (Axelrod et al., 1973). Pastushok and Axelrod (1976) have shown that the phenomenon of conidiation at 24 h (and to become competent at 20 h) is endogenously controlled, since this time is not affected by continuous replacement of the medium.

In vegetative mycelia certain metabolic changes may be related to the acquisition of competence: inducible enzymes become less inducible, DNA extractability increases and there is a progressive decrease in the transport of carbohydrates (Champe et al., 1981). The loss of transport capability with increasing mycelial age and the reversibility achieved through medium replacement could infer that this parameter may be endogenously programmed and could then act as a timing mechanism for conidiogenesis (Hall and Axelrod, 1978; Kurtz and Champe, 1979).

Finally, using thermosensitive aconidial mutants, temperature shift experiments have suggested that most of these mutants are blocked at various times during the maturation, that is, after induction. Only one mutant has been found which appears to be blocked before induction and is therefore unable to become competent (Champe et al., 1981).

C. Phialoconidiation in Submerged Culture

Although most filamentous fungi will remain entirely vegetative in submerged culture, it is now possible with many species of *Aspergillus* and *Penicillium* to routinely achieve conidiation in submerged agitated cultures by manipulation of medium components, aeration, and/or pH (Righelato et al., 1968; Galbraith and Smith, 1969; Ng et al., 1973; Smith and Anderson, 1973; Martinelli, 1976; Pitt and Poole, 1981; Broderick and Greenshields, 1981). Many studies have been made with simple flask culture techniques, while others have been carried out in sophisticated batch or continuous fermenter cultures which can produce large volumes of controlled conidiating cultures that may then be subjected to biochemical analysis (for references see Smith, 1978).

1. *Batch Cultures*

Induction of conidiation in *Penicillium* sp. by high concentrations of calcium ions in submerged cultures not undergoing nutrient depletion has long been recognized (for references see Pitt and Poole, 1981). Pitt and Poole (1981) have reexamined this phenomenon and verified with *Penicillium notatum* that induction of conidiation by calcium ions in submerged culture occurred without limitation of the major nutrients. A high degree of synchronous conidiation could be achieved in 36 h when Ca^{2+} was added simultaneously after inoculation, or 12 h after Ca^{2+} was added to a 36-h culture. Conidiation was much more sensitive to reduced oxygen tension than vegetative growth and spore production declined drastically at dissolved oxygen tensions below 90 mmHg. Oxygen consumption was consistently greater for conidiating cultures than for vegetative cultures, thus confirming the need for a highly oxidative metabolism to support spore formation (Smith, 1978). The physiological and biochemical reasons for the calcium effect are not yet known.

A technique of medium replacement in the batch fermenter cultivation of *A. niger* has made it possible to show that conidiation can be divided into at least four distinct phases (Anderson and Smith, 1971a), and these phases could then be studied for physiological and biochemical changes associated with the various stages of the conidiation process (for references see Smith, 1978). By careful medium control it was possible to achieve a high degree of synchrony of conidiphore elongation, metulae and phialide development, and subsequent conidia formation.

However, interpretation of developmental responses to specific cultural conditions is complicated in any batch system because of the transient nature of the environmental conditions during growth. Thus it is not possible to deduce from a batch culture system whether conidiation results from nutrient limitation or from the limitation of growth rate imposed by these conditions.

2. *Continuous Cultivation*

In a stirred tank reactor under continuous operating conditions with *A. niger*, maximum conidiation occurred at growth rates just above zero and the critical growth rates above which steady state vegetative growth occurred (Ng et al., 1973). When citrate was supplied as the carbon source, limitation of this nutrient gave good conidiation, albeit with a conidiphore of reduced complexity. When glucose was the carbon source, conidiation did not occur at any limitation. In these studies conidiation in *A. niger* appeared to be determined by an interaction between the growth rate and the nature of the carbon and nitrogen sources in the culture medium.

Studies using a continuous tower fermenter system have shed more light on the factors regulating conidiation in *A. niger* and *Aspergillus ochraceus* (Broderick, 1981; Broderick and Greenshields, 1981). Conidiation of both organisms was induced by restricting growth by nutrient limitation while under continuous cultivation. Shock carbon limitation did not produce any conidiation, but the gradual decrease of sucrose or starch supply to *A. niger* produced limited conidiation.

Gradual nitrate limitation did not produce conidiation, while a shock decrease in nitrate concentration resulted in heavy conidiation of both organisms. Conidiation occurred over a wide range of organism-specific growth rates and it was not possible to define a specific growth rate below which conidiation was induced in every system. Thus there was no overall threshold effect of growth rate upon conidiation. Instead, for each specific system there appeared to be a critical growth rate above which vegetative growth occurred and below which conidiation could be induced.

Although growth rate may be a central factor in determining conidiation, conidiation is by no means due to a decreased growth rate alone. Movement away from balanced growth is a prerequisite for conidiation, but unbalanced growth does not always cause conidiation. The nature of the unbalanced growth is the important factor, as shown by the different effects that nitrogen and carbon nutrient limitation have upon growth and conidiation (Ng et al., 1973; Broderick and Greenshields, 1981; Broderick, 1981). Conidiation therefore depends upon a delicate equilibrium between the growth rate and the physiochemical and nutritional nature of the environment.

D. Conidiophore Morphology Under Submerged Conditions

In batch submerged cultures, most studies have shown that when conidiation does occur, the conidiophore apparatus is very similar to that of subaerial conidiation, though frequently smaller and lacking conidial chains (owing to medium agitation). However, an interesting feature of stirred-tank chemostat culture studies with *Penicillium chrysogenum* (Righelato et al., 1968) and *A. niger* (Ng et al., 1973) was the considerable reduction in the complexity of the conidiophore apparatus. Conidiophores of *A. niger* were characterized by possessing small vesicles with few phialides and occasionally conidia were observed to develop from modified hyphal tips. It was suggested that these modifications in conidiophore complexity might represent a partial switch-on of the conidiophore mechanism, also perhaps suggesting that the conidiophore could be bypassed and conidia produced from more simplified structures.

During the tower fermenter studies by Broderick (1981), modified reproductive structures were frequently observed with *A. niger* and *A. ochraceus*. Under some experimental conditions conidiophore heads were very similar to subaerial structures, with only a slight reduction in phialide numbers. In other cases single phialides were developed directly upon hyphal tips. Both types of structures often occurred side by side on the same hyphal filament. With *A. ochraceus* a common variant produced vesicles and phialides almost directly on the hyphal surface, without any conidiophore.

Conidiating structures under tower fermenter conditions were primarily formed at the edge of the mycelial floc, the center remaining vegetative. The simplified structures were the most rapidly formed in cultures exhibiting complex and simplified forms. Broderick (1981) has suggested that the simplified conidial production process responding to restricted growth is a "panic" step undertaken by the organism to ensure survival should the nutrient limitation prove too severe for vegetative growth. Under less nutrient-demanding conditions the fungus can expend the energy to synthesize the complex conidiophores which can then produce large numbers of conidia.

III. THALLIC CONIDIOGENESIS

A. Morphology

Cole and Samson (1979) recognized two basic kinds of thallic conidiogenesis; a holothallic form and a thallic-arthric form. In holothallic conidium ontogeny a terminal or intercalary segment of a fertile hypha is converted into a thick-walled and usually pigmented propagule. In this mode of conidium development all wall layers of the conidiogenous cell are involved in the formation of the conidium wall. Examples of this kind of conidiogenesis can be found in Cole and Samson (1979) and include several *Microsporum* and *Trichophyton* species.

The thallic-arthric conidium ontogeny is characterized by conversion and disarticulation of a preexisting, determinate hyphal element (i.e., a hyphal element whose extension growth has ceased at or before the onset of conidiogenesis). Cole and Samson (1979) described two kinds of thallic-arthric ontogeny: holoarthric and enteroarthric, and these two are distinguished primarily on the basis of wall differentiation. In the holoarthric mode all wall layers of the conidiogenous cell are involved in the formation of the conidial wall, whereas in the enteroarthric mode the outer wall of the conidiogenous cell usually separates from the newly formed wall of the conidium.

Secession of thallic-arthric conidia may take place by the fission of double septa or by fracture or lysis of the walls of the adjacent cells. Fission is defined as "the kind of conidium secession which involves separation of the components of a double septum" (Kendrick, 1971c). Fracture is defined as the "kind of conidium secession which involves the rupture of the wall of an adjacent vegetative or degenerated cell at some point removed from the septum". This kind of secession depends on the application of some external mechanical stress (Kendrick, 1971c). Lysis is defined as "the kind of conidium secession which involves the dissolution of the wall of an adjacent degenerated cell". The dissolution may sometimes be autolytic and sometimes carried out by the enzymes of other microorganisms in the natural surroundings. In the Deuteromycotina fission seems to be the most common method by which the conidia are released (Kendrick, 1971c).

Illustrative examples of the two modes of thallic-arthric conidiogenesis have been given by Cole (1975) and Cole and Samson (1979). *Geotrichum candidum* serves as a good example of holoarthric development, and *Sporendonema purpurascens* as an example of enteroarthric development.

The conversion of fertile hyphae, in which apical growth has ceased, into conidia in *G. candidum* is initiated by septation that initially occurs in an acropetal order, but which later becomes random (Cole and Kendrick, 1969). The inner wall normally becomes thickened. When the arthroconidia separate at the double septa, all the wall layers of the original hyphae are incorporated in the wall of the conidium.

Conidiogenesis in *S. purpurascens* is likewise initiated by cessation of hyphal tip growth and formation of septa, at first acropetally and later randomly along the length of the hypha (Cole, 1975). However, thereafter, *S. purpurascens* shows the characteristics of enteroarthric development by converting certain hyphal compartments into conidia while other cells undergo autolytic processes. Analysis of wall differentiation during enteroarthric development shows that those cells which become arthroconidia develop an inner, thick, tertiary wall layer which also forms the end walls of the conidium. Those cells which do not differentiate into conidia degenerate and function in the secession of the conidia. The result of this developmental process is the formation of short cylindrical conidia, which are enveloped by the partially degenerated wall of the fertile hyphae.

If thallic-arthric conidiogenesis is followed in shake flask cultures or fermenters, the breaking apart of the hyphae will start very soon after formation of the first conidial septa. This is probably due to mechanical stress and will result in hyphal fragments which will septate and fragment further until the final conidial morphology has been reached (Kier et al., 1976; Quinn et al., 1981).

Thorough descriptions of the thallic-arthric conidium ontogeny in the Deuteromycotina can be found in Kendrick (1971b) and Cole and Samson (1979).

B. Formation of Arthroconidia

Geotrichum candidum will routinely sporulate asexually by the formation of arthroconidia (Carmichael, 1957). Only in one occasion has a perfect state been reported and *G. candidum* was described as the imperfect form of *Endomyces geotrichum* (Butler and Petersen, 1972). This fungus is well suited for studies of thallic conidiation, as it grows rapidly both on solid media and in submerged cultures (doubling time, 1.1 h at 30°C). In submerged cultures *G. candidum* grows in a homogeneous form (nonpellet), allowing culture turbidity to be used as a measure of growth (Trinci, 1972; Kier et al., 1976). Arthroconidia are released by cleavage of the septa (Trinci and Collinge, 1973). The first visible sign of sporulation is septation, which starts when tip growth of vegetative hyphae is arrested (Cole and Kendrick, 1969), and observations, at least on cultures on solid media, show that septation occurs initially in an acropetal order, but later septa are laid down randomly (Cole and Kendrick, 1969). In submerged shake cultures septation is followed by a release of much longer hyphal fragments or filaments which septate further into arthroconidia (Cotter and Niederpruem, 1971; Kier et al., 1976).

The first sign of fission is seen as dark triangles in the septum wall. The dark zone extends centripetally as dark lines bisecting the septal wall and the arthroconidium or "fragment" is released. Trinci and Collinge (1974) have presented a model for the release of arthroconidia (Fig. 2) in which they claimed that septa were formed that appeared to be continuous, with a new wall layer formed beneath the existing hyphal wall. Arthrospores normally contain two to three nuclei per spore in *G. candidum*, but the content can vary from one to seven nuclei per spore. By comparison, the number of nuclei per compartment in the hyphae is 19 ± 5 in apical compartments and 7 ± 2 in intercalary compartments (Fiddy and Trinci, 1976).

Carmichael (1957 observed that a culture of *G. candidum* in a peptone broth consisted of hyphae, typical arthrospores, and larger, rounded, thick-walled cells. Duran et al. (1973) called these thick-walled spores "yeastlike forms" and showed that these were formed under certain conditions, as, for example, low pH values. Trinci and Collinge (1974) showed a difference in the appearance of arthroconidia

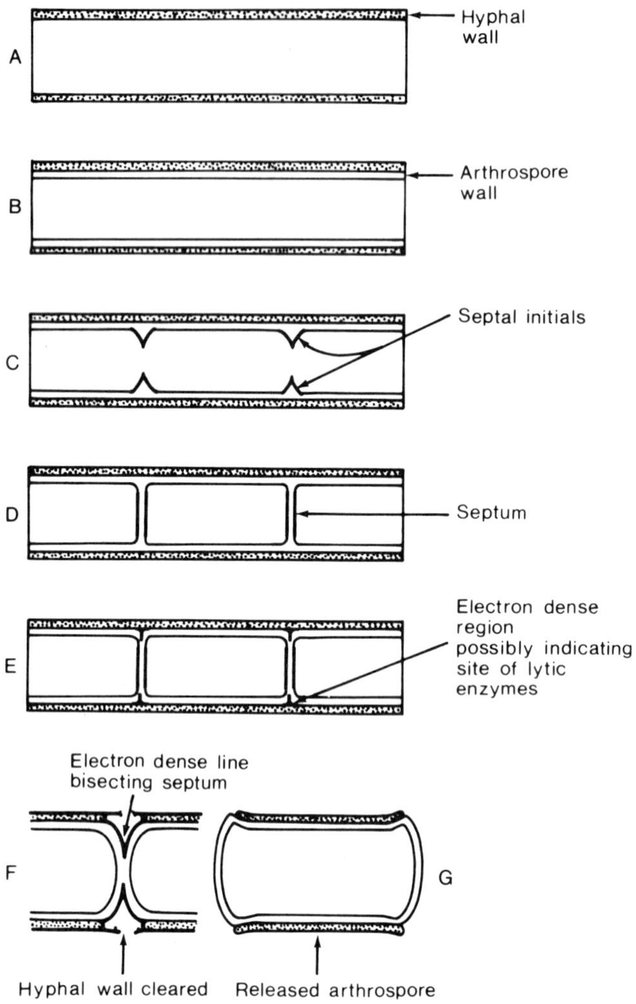

FIG. 2 Diagrammatic representation of the stages involved in arthroconidiogenesis in *Geotrichum candidum*. (Redrawn from Trinci and Collinge, 1974.)

formed in nitrogen- and carbon-starved cultures. The conidia formed in the nitrogen-starved cultures were shorter and had a greater wall thickness. Kier et al. (1976) were able to produce separately the two different forms of arthroconidia by changing the glucose concentration in simple synthetic medium. By that procedure it was possible to iso-

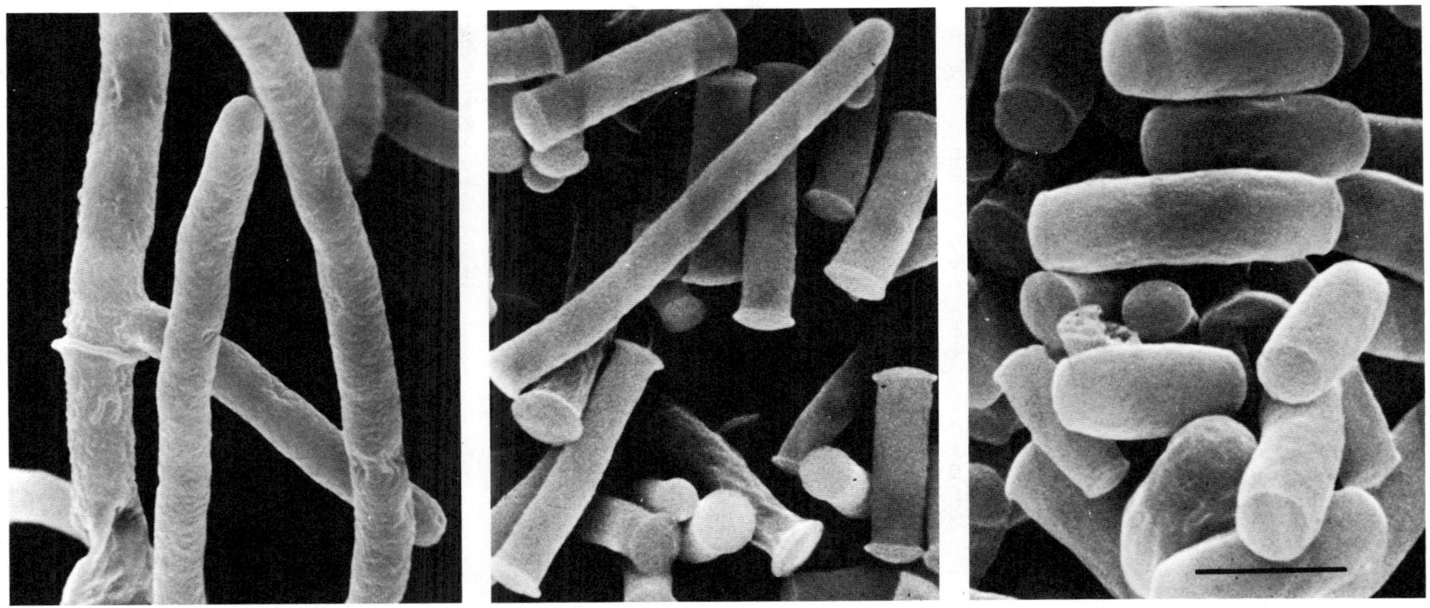

(a) (b) (c)

FIG. 3 Scanning electron micrographs of the three morphological stages of *Geotrichum candidum* (bar = 5 μm): (a) mycelium from exponential growth phase, (b) cylindrical arthroconidia, and (c) ellipsoidal arthroconidia. (From Olesen and Kier, 1981.)

late the two forms of conidia for further investigations. Allermann et al. (1978) named the two forms of arthroconidia "cylindrical" and "ellipsoidal" types.

The differences in morphology of hyphae and cylindrical and ellipsoidal conidia are shown by the scanning electron micrographs in Fig. 3, which clearly show that the cylindrical type resembles hyphal fragments, whereas the ellipsoidal type looks more developed. A difference in macromolecular composition between the two arthroconidia forms was demonstrated by Allermann et al. (1978) and it was shown that mycelium and the cylindrical conidia were almost alike in composition; in contrast, the ellipsoidal conidia had lower protein and RNA contents, but a higher carbohydrate content. Isolated cell walls had similar macromolecular compositions (Kier et al., 1980). However, the dry weight of the wall of an ellipsoidal arthroconidium was about three times that of cylindrical forms, while the content of carbohydrates in the protoplasm was about five times higher in the ellipsoidal conidia than in the cylindrical conidia. The ellipsoidal conidia differ from the cylindrical conidia in having a thicker and denser cell wall and a denser cytoplasm with carbohydrate reserve material. The main carbohydrate reserve material has recently been shown by Olesen and Kier (1981) in transmission electron micrographs by cytochemical methods to be glycogen, which was confirmed by Olsen and Allermann (unpublished results) by glycogen isolation methods that showed a glycogen content of about 20% of the dry weight in ellipsoidal arthroconidia. Duran et al. (1973) have shown that one of the differences between their "yeastlike form" and arthrospores was cell wall thickness. Steele and Fraser (1973) have also shown glycogen deposits in arthroconidia. Cotter and Niederpruem (1971) have demonstrated accumulation of mannitol and arabitol in *G. candidum* during conidiation, and Steele (1972) and da Costa and Niederpruem (1980) have shown that sugar alcohols are depleted during germination of arthroconidia.

C. Environmental-Nutritional Induction of Arthroconidiogenesis

It has long been recognized that the morphogenetic change from the mycelial phase to the sporulation phase in a fungus is caused by the environmental-nutritional state. Arthroconidiogenesis is no exception, but whether conidiogenesis is induced by carbon limitation or by some other unfavorable limitation is not clear. Trinci and Collinge (1974) showed that nitrogen starvation of *G. candidum* cultures resulted in a heavy sporulation period, with almost the entire mycelium converting to arthroconidia, whereas carbon-starved cultures consisted of short filaments. A similar picture is seen from the work of Kier et al. (1976). If the initial glucose concentration was only 0.1%, sporulation or fragmentation took place only in the stationary phase, resulting in cylindrical arthroconidia very much like the short filaments described

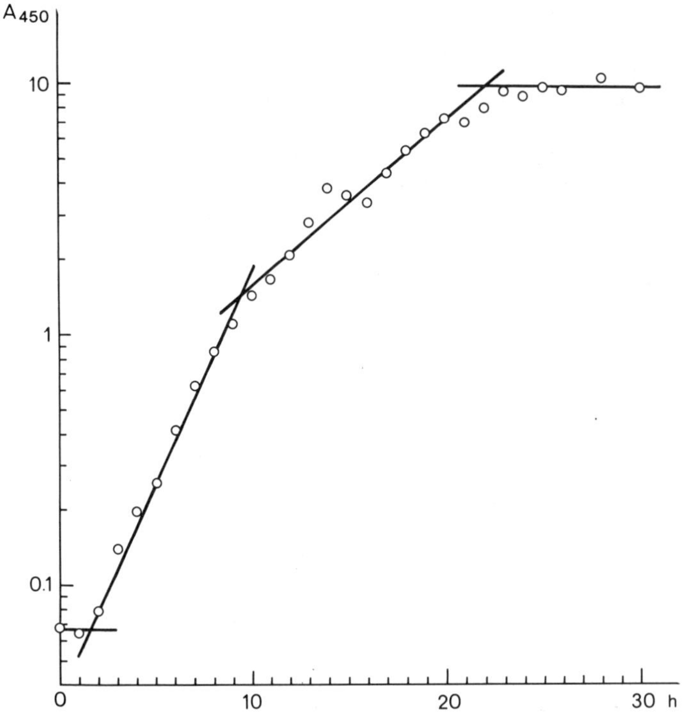

FIG. 4 Growth of *Geotrichum candidum* in 0.5% glucose culture at 25°C in shake culture. (From Kier et al., 1976.)

by Trinci and Collinge (1974). On the other hand, when the initial glucose concentration was 0.5%, without changing the other constituents of the medium, conidiogenesis took place during a second, slower growth phase (Fig. 4), resulting in ellipsoidal arthroconidia much like arthrospores found by Trinci and Collinge (1974) in nitrogen-starved cultures. Floto (1980) later showed that the growth of the second, slower growth and sporulation phase was most likely unbalanced owing to iron limitation. Duran et al. (1973) showed that the formation of their yeastlike form was caused by unfavorable conditions for growth, for example, low pH values. With *Trichophyton mentagrophytes*, Emyanitoff and Hashimoto (1979) concluded that under conditions conducive to arthrosporogenesis the growth rate was lower than that observed under nonpermissive conditions. It is likely that arthrosporogenesis in general is induced by conditions resulting in unbalanced growth. In *G. candidum* ellipsoidal arthroconidia are formed during unbalanced growth. The ellipsoidal conidia have a thick wall and a dense cytoplasm with carbohydrate reserves (glycogen and/or polyols). Carbon exhaustion results in cylindrical arthroconidia, which could well be considered as hyphal filaments or hyphal fragments, as the wall has the same thickness as the hyphal wall (Olesen and Kier, 1981) and contains very little carbohydrate reserves. A correlation between nutrition, growth rate, and sporogenesis has also been reported by Robinson and Smith (1976) for *G. candidum*. Sporulation in the continuous cultivation of *G. candidum* has been investigated by several authors (Fiddy and Trinci, 1975; Robinson and Smith, 1976; Schneider and Andersen, 1980; Quinn et al., 1981). The main conclusion of these studies has been that arthroconidia are formed mainly at low dilution rates and that the lower the dilution rate, the more spherical or ellipsoidal the conidia.

Park and Robinson (1969) suggested that induction of sporulation in *G. candidum* was due to a diffusible fungal metabolite, a sporulation factor. It is possible that the suggested sporulation factor could also be the factor responsible for self-inhibition of spore germination. Floto (1980) has shown that cultures of *G. candidum* grown in the presence of activated charcoal are unable to sporulate, although the conditions for sporulation are good (unbalanced growth), and this would further support the hypothesis that a sporulation factor is necessary to induce arthroconidiogenesis. Floto (1980) also showed that respiration was inhibited in a suspension of ellipsoidal arthroconidia and that the respiratory inhibitor could be washed off and become bound to charcoal. Katayama and Marumo (1978) have isolated a sporogenic substance, sclerosporin = R(-)-glycerol monolinolate, which was able to induce formation of arthroconidia in *Sclerotinia fructicola*.

IV. BLASTOARTHROCONIDIA

In some species both blastic and arthric processes participate in conidiogenesis. An illustrative example of this is the macroconidiation of the Monilia state of *Neurospora crassa* (for examples and references see Turian, 1976, 1977; Cole and Samson, 1979).

Experimental situations have been devised which first limit *Neurospora* to produce vegetative hyphae (submerged growth). Then by various techniques aerial hyphae can be generated which function as macroconidiophores. The tips of these conidiophores swell and bud apically. The buds enlarge into the first so-called proconidia (blastoconidia), which in turn bud other presumptive macroconidia. Finally, septa are laid down which will separate and disarticulate adjacent conidia into forms that are ontogenetically considered as blastoarthroconidia.

V. CHLAMYDOSPORES

The participants of the Kananaskis Conference (Kendrick, 1971a) ended up with a rather broad definition of the term *chlamydospore*: it was described as a "thick-walled, thallic, terminal or intercalary spore." There seem to be only minor differences between intercalary chlamydospores and some arthroconidia. If the arthroconidium definition were not restricted to those instances where a total conversion of a hypha takes place, the intercalary chlamydospores could be included with the arthroconidia. Chlamydospores do not possess a differentiated mechanism for secession, being usually liberated after dissolution of the hypha. To many mycologists the term *chlamydospore* is connected to its function rather than to its development, and the term is often used for a resistant, resting spore which allows the fungus to survive under unfavorable conditions. Chlamydospores can be found widely throughout the fungi.

A. The Chlamydospore Wall

Perhaps the most important ultrastructural aspect of the chlamydospore is the nature of the wall. The presence of a thick wall is probably one of the factors which protect chlamydospores against lysis and desiccation under unfavorable conditions. Although the information on the cell wall of *Fusarium* is better than for many other fungi, knowledge of wall structure and composition is still incomplete. A number of reports on the ultrastructure of the chlamydospore wall can be found in the literature (Stevenson and Becker, 1972, 1979; Griffiths, 1973; van Eck, 1978; Schneider and Wardrop, 1979). The wall of the mature chlamydospore of *Fusarium oxysporum* is multilayered, varying between 0.50 and 0.57 μm in total thickness (Stevenson and Becker, 1972). However, some controversy exists as to the number of layers and the nature of the final chlamydospore wall (Griffiths, 1973; Stevenson and Becker, 1979; Schneider and Wardrop, 1979). The chemistry of the wall from macroconidia and conidial chlamydospores of *Fusarium solani* formed in vitro has been investigated (van Eck, 1978). Only quantitative differences in the chemical composition of macroconidial and chlamydospore walls were measured. Chlamydospore walls were more resistant to enzymic degradation than microconidial walls, where chitinase, β-(1-3)-glucanase, and combinations of these enzymes were applied. The results indicated that chitin and β-(1-3)-glucan are essential polymeric wall components of macroconidia and chlamydospores of *F. solani*. The enhanced resistance of the chlamydospore wall to the enzymes was ascribed to the deposition of some electron-dense material on the outer layer of the mature spore. The nature of this material was not analyzed, but it was suggested to be melanin-like.

B. Lipid Reserves in Chlamydospores

A number of workers have shown that chlamydospores are rich in lipid reserves and Turian (1966) has postulated that the metabolic pathways in prospective chlamydospores may be directed toward increased synthesis of fatty materials and wall polysaccharides. Conidia and chlamydospores of *Fusarium* sp. contain numerous reserve lipid bodies (Schneider and Seaman, 1977; Stevenson and Becker, 1979). Schneider and Seaman (1977) have proposed that the reserve lipids in the developing chlamydospores of *Fusarium sulphureum* are formed intracisternally in the endoplasmic reticulum (ER) and that the mature bodies are surrounded by a unit membrane derived from the ER.

C. Chlamydospore Formation in Defined Media

Autoinhibitors (Griffin, 1970) and inducing substances (Bandara and Wood, 1978) present in soil and soil extracts have been suggested to be involved in chlamydospore formation, and the formation of chlamydospores has often been induced by exposing macroconidia or germinated macroconidia to the presence of soil or aqueous soil extracts. Such methods are usually not desirable if biochemical techniques are to be used to obtain information on biochemical processes that can explain the induction process of sporulation and the developmental events that occur. The more recent investigations that have been directed to the use of defined systems for the production of chlamydospores in high quantities is therefore important.

Abundant hyphal chlamydospore production by *F. oxysporum* has been obtained in a basal two-salt solution of KH_2PO_4 and $MgSO_4 \cdot 7H_2O$ amended with either glucose or $MgCO_3$ (Qureshi and Page, 1970). In this experiment chlamydospore production was stimulated by carbon source levels from 0.125 to 2 mg/liter, whereas levels above 2 mg/liter repressed spore induction in the basal salt solution. A modification of this method was used by Stevenson and Becker (1972) for rapid reproducible induction of large quantities of chlamydospores in a number of strains of *F. oxysporum*. Chlamydospores were induced by transferring washed, pregerminated macro- or microconidia to shallow layers of the two-salt sporulation solution with 3 mg of $MgCO_3$ per liter. Chlamydospore morphogenesis was evident by 12 h, and mature spores could be harvested at the end of 4 days. Production of only chlamydospores was achieved by pregrowing the conidia to a point before which the mycelium becomes committed to conidial formation. At this point the mycelium shows no cytological evidence of spore formation, and therefore this point can only be determined by trial and error. Such a developmental stage has also been described for *Candida albicans* (Dujardin et al., 1980) and was named "readiness." At this stage pseudomycelial thalli from cultures of *C. albicans* on rice cream decoction with glucose were able to produce chlamydospores in water without further stimulus by the sporulation medium. This stage of readiness is reached earlier at high inoculum sizes than at low inoculum sizes. Under their conditions of pregrowth and washing, Stevenson and Becker (1972) found abundant production of chlamydospores at glucose levels of 0-80 mg/liter, and high yields could even be obtained by transferring the washed pregerminated conidia to distilled water. They suggested that the production of chlamydospores in the two-salt solution without added carbon source depends on the status of the endogenous reserves in the inoculum; thus a starved inoculum may have endogenous reserves that are too low to allow for chlamydospore formation in the absence of a carbon source in the medium.

Production of large quantities of conidial chlamydospores from macroconidia of *F. sulphureum* has been obtained in distilled water (Schneider and Seaman, 1974). Differentiation of conidial cells into chlamydospores in this fungus depended upon the age and concentration of the conidia. Conidia from 30- to 38-day-old colonies on potato sucrose agar produced chlamydospores within 6 days in distilled water, whereas conidia from five-day-old colonies did not form chlamydospores, even after 40 days. At low concentrations (5×10^2-10^3 conidia per milliliter) conidia from *F. sulphureum* germinated in distilled water but did not form chlamydospores. At higher concentrations germination was inhibited and conidia differentiated into conidial chlamydospores. In contrast, conidia of *F. oxysporum* at similar concentrations in distilled water germinated and produced hyphal chlamydospores. The long period required for the aging and conversion of conidia can be avoided (Barran et al., 1977). Chlamydospores of *F. sulphureum* were formed within 72 h by incubating macroconidia at 37°C in stationary culture in a medium containing mannitol and inorganic salts. The conversion process was dependent on the inoculum level, the temperature, aeration, and the presence of an external source of nitrogen and carbon. This method for chlamydospore production has been scaled up to produce large numbers of chlamydospores for chemical analyses of cell walls (Barran et al., 1977).

D. Temperature

The use of temperature to impose stress on a fungus seems to be a useful method of inducing chlamydospore formation. In *F. sulphureum* formation of chlamydospores was induced by a combination of low oxygen tension and a temperature of 37°C, which was above the maximum growth temperature of the organism (Barran et al., 1977). In *Phytophthora lateralis*, Englander and Roth (1980) found that the optimum temperature for chlamydospore production on V8 sterol agar was 24-25°C, which was approximately the maximum temperature for linear growth and dry weight increase in this medium.

VI. MICROCYCLE CONIDIATION

In the normal developmental pattern of filamentous fungi conidiospores will germinate to form a vegetative mycelium which, in turn, under suitable environmental conditions, will generally produce asexual reproductive stages forming new conidiospores. However, under certain combinations of limiting growth parameters it is possible to induce conidiation in germination tubes than can prematurely function as conidiophores, for example, *A. niger* (Anderson and Smith, 1971b, 1972), *N. crassa* (Rossier et al., 1975), Cortat and Turian, 1974), *Penicillium urticae* (Sekiguchi et al., 1975), *Paecilomyces variotii* (Anderson et al., 1978), *G. candidum* (Park and Robinson, 1969; K. Allermann and J. Olsen, unpublished results), *Trichoderma harzianum* (Zuber and Turian, 1981), and *Acremonium diospyri* (Saviour, 1981) (Fig. 5).

This phenomenon has been termed *microcycle conidiation* and is defined as the recapitulation of conidiation following conidial germination without an intervening phase of mycelial growth. In practice, microcycle conidiation results in the production of a morphologically distinct unit comprising the parent conidium (generally, but not always, enlarged) connected to the spore-bearing apparatus (which can also arise from an abbreviated germ tube) and the second-generation conidia. Microcycle conidiation has recently been extensively reviewed (Smith et al., 1981a,b).

Microcycle conidiating systems offer many advantages over conventional mycelial-derived conidiation, since they tend to eliminate the normal phase of vegetative growth and produce a cultural biomass containing a very high proportion of conidiating material. In some of the microcycle systems it is also possible to impart much greater synchrony of conidial development than has been possible with most other submerged conidiating cultures.

A. Induction of Microcycle Conidiation

Temperature manipulation in suitable culture media has been the most practiced method of inducing microcycle conidiation. However, there are many examples where microcycle conidiation can be provoked by nutritional means alone (see Smith et al., 1981a). Recently, excellent synchronous microcycle conidiation has been achieved with *T. harzianum* (Zuber and Turian, 1981). Premature phialoconidiogenesis was achieved by applying a nutritional shock to germinated conidia. Germinated conidia were transferred into a synthetic medium containing an additional 8% glucose and 0.05% KNO_3, and after 11 h over 90% of the initial conidia were undergoing microcycle conidiation.

The temperature/nutrient manipulation of the microcycle system of *A. niger* is undoubtedly the most studied of all microcycle conidiation

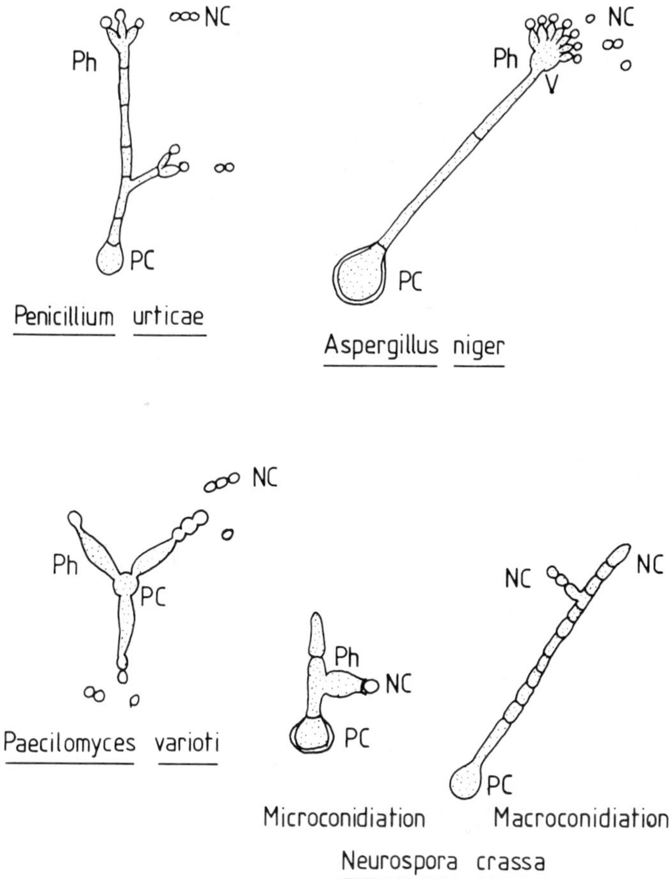

FIG. 5 Microcycle conidiation in several filamentous fungi induced by temperature and nutrient manipulations. The organisms are not drawn to scale [NC, new conidia; PC, parent conidium (enlarged); Ph, phialide; V, vesicle].

systems; in part, this is due to ease of control and the ability to produce large volumes of synchronous cultures under batch fermentation conditions.

Microcycle conidiation in *A. niger* is a two-phase system. Conidia are incubated at 44°C for 24 h, by which time the conidia increase considerably in size and complexity, achieving an optimum diameter of 20-24 μm. In the absence of the temperature limitation the medium composition would permit normal vegetative growth. The extensive increase in the size of the giant cells is an active growth process accompanied by a large increase in macromolecular synthesis and a net increase in dry weight. The ultrastructural organization of giant cells is typical in form to most fungal cells (Deans et al., 1980). Nuclear numbers increase progressively to c 16 after 24 h at 44°C. There are numerous mitochondria and abundant ribosomes. The cell walls attain a thickness of up to 2 μm in the final giant cell and have been the subject of a special study of chemical composition and associated enzymology (Deans, 1978). In particular, α-glucan was rapidly built up within the wall and later reutilized, most probably to support conidiophore outgrowth (Smith et al., 1981a). The giant cell of *A. niger* exhibits exponential growth and under optimal conditions has a specific growth rate, based on cell diameter, of 0.075 h^{-1}.

When the temperature is reduced to 30°C in *A. niger*, a thick conidiophore outgrowth is produced which is characteristic of the species. There is, however, a considerable reduction in the size of the conidiophore when compared with subaerial conidiation (Anderson and Smith, 1971b). In microcycle conidiation the conidiophore vesicle head does not form metulae or sterigmata, but directly forms phialides. Conidia produced by *A. niger* and other microcycle systems are approximately similar in size and fully capable of normal germination and vegetative outgrowth under suitable environmental conditions. The conidiophore outgrowth is derived from the inner wall of the giant cell (Deans et al., 1980) and growth is linear in nature (Smith et al., 1981b). A giant cell of *A. niger* can produce more than one conidiophore, and sometimes up to five. The considerable biomass formed by this microcycle system implies that the conidiating apparatus is continuously taking up nutrients from the medium and not merely relying on the reutilization of stored reserves in the giant cell.

Earlier studies with CO_2 additions to the air intake of the batch fermenter suggested that CO_2 was functioning as an important morphogenic agent, reducing vegetative outgrowths for the giant cells. However, careful and repeated studies by A. Al-Rawi, J. E. Smith, and B. Kristiansen (unpublished results) have failed to confirm these observations.

B. Conidiation Competence in the Microcycle of *A. niger*

In all of the identified microcyle systems the various regimes used to induce conidiation required times at least as long as those for normal conidiation. Thus conidiation competence may well be acquired during the limited growth stage of the germinating spore.

Studies with *A. niger*, using various metabolic inhibitors (Deans and Smith, 1979), have demonstrated approximate times at which the giant cells of *A. niger* complete synthesis of the nucleic acids and pro-

TABLE I Effects of the Addition of Metabolic Inhibitors at Various Stages of Giant Cell Formation on the Development of *Aspergillus niger*[a]

Time of addition (h at 44°C)	Phase reached in microcycle[b] after 12 h at 30°C				Phase reached in microcycle[b] after 24 h at 30°C			
	A	Ch	Cy	T	A	Ch	Cy	T
Control	7	7	7	7	10	10	10	10
0	2a;2a	2a;2a	2a;2a	2a;2a	2a;2a	2a;2a	2a;2a	2a;2a
2	2a;2a	2a;2a	2a;2a	2a;2a	2a;2a	2a;2a	2a;2a	2a;2a
4	2a;2a	4;2a	2a;2a	2a;2a	2a;2a	5;2a	2a;2a	2a;2a
6	2a;2a	6;2a	2a;2a	2a;2a	2a;2a	6;2a	2a;2a	2b;2a
8	2b;2b	6;2b	2b;2b	2c;2b	2b;2b	9;2b	2b;2b	2c;2b
10	6;2b	7;2b	2c;2b	6;2b	9;2b	10;2b	2b;2b	9;2b
12	6;2b	7;3	2c;2b	7;2c	9;2b	10;4	2c;2b	10;2c
14	7;2b	7;6	5;2c	7;6	10;2b	10;8	6;4	10;6
16	7;2c	7;7	7;2c	7;7	10;2c	10;9	7;6	10;9
18	7;6	7;7	7;7	7;7	10;9	10;10	10;7	10;9
20	7;7	7;7	7;7	7;7	10;9	10;10	10;8	10;9
22	7;7	7;7	7;7	7;7	10;10	10;10	10;8	10;9
24	7;7	7;7	7;7	7;7	10;10	10;10	10;8	10;10

[a]A = actinomycin D (50 μg/ml), Ch = chloramphenicol (60 μg/ml), Cy = cycloheximide (20 μg/ml), and T = tetracycline (30 μg/ml).

[b]Phases as in Fig. 1 (Deans and Smith, 1979). The number before the semicolon refers to the inhibitor being removed after 2 h of contact; the number after the semicolon refers to no inhibitor removal.

teins involved in conidiophore outgrowth. By way of adding selected inhibitors such as actinomycin D, chloramphenicol, and cycloheximide at various time intervals from spore inoculation, it was possible to demonstrate that addition up to 15-16 h completely inhibited conidiophore outgrowth. When additions were made after this time, conidiophore development was unaffected when the giant cells were later changed to the permissive 30°C (Table 1).

Thus, for *A. niger* undergoing giant cell formation, competence for conidiation starts at 15-16 h. However, when giant cells are transferred to the permissive temperature at 16 h, conidiation is poor and the outgrowths are mostly in the form of vegetative mycelium. With longer times, at 44°C there is an increasing level of conidiophore development from the giant cells, and by 20-22 h there is nearly 100% competence. Outgrowth of the conidiophore occurs after 2-4 h at the permissive temperature. These experiments would therefore suggest that competence is achieved between 15-20 h of vegetative giant cell growth, but expression of conidiation is restricted by thermal inhibition of some part of the conidiation process.

REFERENCES

Allermann, K., Floto, F., Olsen, J., Sortkjaer, O., and Kier, I. (1978). Sporulation and the macromolecular composition of the mycelium and arthrospores of *Geotrichum candidum*. *Physiol Plant.* *42*:355-358.

Anderson, J. G., and Smith, J. E. (1971a). Synchronous initiation and maturation of *Aspergillus niger* conidiophores. *Trans. Br. Mycol. Soc.* *56*:9-29.

Anderson, J. G., and Smith, J. E. (1971b). The production of conidiophores and conidia by newly germinated conidia of *Aspergillus niger* (microcycle conidiation). *J. Gen. Microbiol.* *69*:185-197.

Anderson, J. G., and Smith, J. E. (1972). The effects of elevated temperature on spore swelling and germination in *Aspergillus niger*. *Can. J. Microbiol.* *18*:289-297.

Anderson, J. G., Aryee, V., and Smith, J. E. (1978). Microcycle conidiation in *Paecilomyces variotii*. *FEMS Microbiol. Lett.* *3*:57-60.

Axelrod, D. E. (1972). Kinetics of differentiation of conidiophores and conidia by colonies of *Aspergillus nidulans*. *J. Gen. Microbiol.* *73*:181-184.

Axelrod, D. E., Gealt, M., and Pastushok, M. (1973). Gene control of developmental competence in *Aspergillus nidulans*. *Dev. Biol.* *34*:9-15.

Bandara, J. M. R. S., and Wood, R. K. S. (1978). Soil factors influencing chlamydospore formation by *Fusarium solani* f. sp. *phaseoli*. *Ann. Appl. Biol.* *90*:169-177.

Barran, L. R., Schneider, E. F., and Seaman, W. L. (1977). Requirements for the rapid conversion of macroconidia of *Fusarium sulphureum* to chlamydospores. *Can. J. Microbiol. 23*:148-151.
Broderick, A. J. (1981). Control of fungal morphology in continuous culture. Ph.D. thesis, University of Aston, Birmingham.
Broderick, A. J., and Greenshields, R. N. (1981). Sporulation of *Aspergillus niger* and *Aspergillus ochraceus* in continuous submerged liquid culture. *J. Gen. Microbiol. 126*:193-202.
Butler, E. E., and Petersen, L. J. (1972). *Endomyces geotrichum* a perfect state of *Geotrichum candidum. Mycologia 64*:365-374.
Carmichael, J. W. (1957). *Geotrichum candidum. Mycologia 49*: 820-830.
Champe, S. P., Kurtz, M. B., Yager, L. N., Butnick, N. J., and Axelrod, D. E. (1981). Spore formation in *Aspergillus nidulans*: Competence and other developmental processes. In *The Fungal Spore: Morphogenetic Controls*, G. Turian and H. R. Hohl (Eds.). Academic Press, London, pp. 255-276.
Clutterbuck, A. J. (1978). Genetics of vegetative growth and asexual reproduction. In *The Filamentous Fungi*, Vol. 3, J. E. Smith and D. R. Berry (Eds.). Edward Arnold, London, pp. 240-274.
Cochrane, V. W., and Cochrane, J. C. (1971). Chlamydospore induction in pure culture in *Fusarium solani. Mycologia 63*:462-477.
Cole, G. T. (1975). The thallic mode of condiogenesis in the Fungi Imperfecti. *Can. J. Bot. 53*:2983-3001.
Cole, G. T., and Kendrick, W. B. (1969). Conidium ontogeny in Hyphomycetes. The arthrospores of *Oidiodendron* and *Geotrichum*, and the endoarthrospores of *Sporendonema. Can. J. Bot. 47*:1773-1780.
Cole, G. T., and Samson, R. A. (1979). *Patterns of Development in Conidial Fungi*. Pitman, London.
Cortat, M., and Turian, G. (1974). Conidiation of *Neurospora crassa* in submerged culture without mycelial phase. *Arch. Microbiol. 95*:305-309.
Costa, M. S. da, and Niederpruem, D. J. (1980). Temporal accumulation of mannitol and arabitol in *Geotrichum candidum. Arch. Microbiol. 126*:57-64.
Cotter, D. A., and Niederpruem, D. J. (1971). Nutritional and temporal control of arabitol and mannitol accumulation in *Geotrichum. Arch. Microbiol. 76*:65-73.
Deans, S. G. (1978). Cell wall ontogeny during microcycle conidiation in *Aspergillus niger*. Ph.D. thesis, University of Strathclyde, Glasgow.
Deans, S. G., and Smith, J. E. (1979). Effect of metabolic inhibitors on microcycle conidiation of *Aspergillus niger. Trans. Br. Mycol. Soc. 72*:201-206.

Deans, S. G., Gull, G., and Smith, J. E. (1980). Ultrastructural changes during microcycle conidiation of *Aspergillus niger*. *Trans. Br. Mycol. Soc. 74*:493-499.

Dujardin, L., Walbaum, S., and Biguet, J. (1980). Chlamydosporulation de *Candida albicans*: Déroulement de la morphogénèse; influence de la densité d'ensemencement. *Ann. Microbiol. Inst. Pasteur 131A*:141-149.

Duran, A., Uruburu, F., and Villaneuva, J. R. (1973). Morphogenetic and nutritional studies of *Geotrichum lactis* cells. *Arch. Microbiol. 88*:245-256.

Eck, W. H. van (1978). Chemistry of cell walls of *Fusarium solani* and the resistance of spores to microbial lysis. *Soil Biol. Biochem. 10*:155-157.

Emyanitoff, R. G., and Hashimoto, T. (1979). The effects of temperature, incubation atmosphere, and medium composition on arthrospore formation in the fungus *Trichophyton mentagrophytes*. *Can. J. Microbiol. 25*:362-366.

Englander, L., and Roth, L. F. (1980). Interaction of light and sterol on sporangium and chlamydospore production by *Phytophthora lateralis*. *Phytopathology 70*:650-654.

Fiddy, C., and Trinci, A. P. J. (1975). Kinetics and morphology of glucose-limited cultures of moulds grown in a chemostat and on solid media. *Arch. Microbiol. 103*:191-197.

Fiddy, C., and Trinci, A. P. J. (1976). Nuclei, septation, branching and growth of *Geotrichum candidum*. *J. Gen. Microbiol. 97*:185-192.

Floto, F. (1980). The effect of inorganic nutrition and charcoal on the formation of arthrospores of *Geotrichum candidum*. In *Poster Abstracts Third International Fungal Spore Symposium, Gwatt, Switzerland*.

Galbraith, J. C., and Smith, J. E. (1969). Sporulation of *Aspergillus niger* in submerged liquid culture. *J. Gen. Microbiol. 59*:31-45.

Griffin, G. J. (1970). Carbon and nitrogen requirements for macroconidial germination of *Fusarium solani*: Dependence on conidial density. *Can. J. Microbiol. 16*:733-740.

Griffin, G. J. (1976). Roles of low pH, carbon and inorganic nitrogen source used in chlamydospore formation by *Fusarium solani*. *Can. J. Microbiol. 22*: 1381-1389.

Griffiths, D. A. (1973). Fine structure of the chlamydospore wall in *Fusarium oxysporum*. *Trans. Br. Mycol. Soc. 61*:1-6.

Hall, N. E. L., and Axelrod, D. E. (1978). Sporulation competence in *Aspergillus nidulans*: A role for iron in development. *Cell Differ. 7*:73-82.

Hanlin, T. (1976). Phialide and conidium development in *Aspergillus clavatus*. *Amer. J. Bot. 63*:144-155.

Hsu, S. C., and Lockwood, J. L. (1973). Chlamydospore formation in *Fusarium* in sterile salt solution. *Phytopathology 63*: 597-602.

Katayama, M., and Marumo, S. (1978). R(-)-Glycerol monolinolate, a minor sporogenic substance of *Sclerotinia fructicola*. *Agric. Biol. Chem. 42*:1431-1433.

Kendrick, W. B. (1971a). *Taxonomy of Fungi Imperfecti*. University of Toronto Press, Toronto.

Kendrick, W. B. (1971b). Arthroconidia and meristem arthroconidia. In *Taxonomy of Fungi Imperfecti*. B. Kendrick (Ed.). University of Toronto Press, Toronto, pp. 160-175.

Kendrick, W. B. (1971c). Conclusions and recommendations. In *Taxonomy of Fungi Imperfecti*. B. Kendrick (Ed.). University of Toronto Press, Toronto, pp. 253-262.

Kier, I., Allermann, K., Floto, F., Olsen, J., and Sortkjaer, O. (1976). Changes of exponential growth rates in relation to differentiation of *Geotrichum candidum* in submerged culture. *Physiol. Plant. 38*:6-12.

Kier, I., Floto, F., Olsen, J., and Allermann, K. (1980). Macromolecular composition of the wall and protoplasm of mycelia and arthrospores of *Geotrichum candidum*. *Trans. Br. Mycol. Soc. 75*:355-361.

Kurtz, M. B., and Champe, S. P. (1979). Genetic control of transport loss during development of *Aspergillus nidulans*. *Dev. Biol. 70*:82-88.

Leach, C. M., Fullerton, R. A., and Young, K. (1977). Northern Leaf Blight of maize in New Zealand: Relationship of *Drechslera turcia* airspora to factors influencing sporulation, conidium development, and chlamydospore formation. *Phytopathology 67*: 629-636.

Lovett, J. S. (1975). Growth and differentiation in the water mold *Blastocladiella emersonii*: Cytodifferentiation and the role of ribonucleic acid and protein synthesis. *Bacteriol. Rev. 39*: 345-406.

Martinelli, S. D. (1976). Conidiation of *Aspergillus nidulans* in submerged culture. *Trans. Br. Mycol. Soc. 67*:121-128.

Ng, A. M. L., Smith, J. E., and McIntosh, A. F. (1973). Conidiation of *Aspergillus niger* in continuous culture. *Arch. Microbiol. 88*:119-126.

Olesen, P., and Kier, I. (1981). Optimal preparation of mycelium and two types of arthrospores of *Geotrichum candidum*: A SEM, TEM and cytochemical study. *Scandem 81*, Stockholm, June 1981.

Park, D., and Robinson, P. M. (1969). Sporulation in *Geotrichum candidum*. *Trans. Br. Mycol. Soc. 52*:213-222.

Pastushok, M., and Axelrod, D. E. (1976). Effect of glucose, ammonium and media maintenance on the time of conidiophore initiation by surface colonies of *Aspergillus nidulans*. *J. Gen. Microbiol. 94*:221-224.

Pitt, D., and Poole, P. C. (1981). Calcium-induced conidiation in *Penicillium notatum* in submerged culture. *Trans. Br. Mycol. Soc. 76*:219-230.

Quinn, J. P., Patton, A. M., and Marchant, R. (1981). Sporulation of *Geotrichum candidum* in submerged culture. *Trans. Br. Mycol. Soc.* 77:627-635.

Qureshi, A. A., and Page, O. T. (1970). Observations on chlamydospore production by *Fusarium* in a two-salt solution. *Can. J. Microbiol. 16*:29-32.

Righelato, R. C., Trinci, A. P. J., Pirt, S., and Peat, A. (1968). The influence of maintenance energy and growth rate on the metabolic activity, morphology and conidiation of *Penicillium chrysogenum*. *J. Gen. Microbiol. 50*:399-412.

Robinson, P. M., and Smith, J. M. (1976). Morphogenesis and growth kinetics of *Geotrichum candidum* in continuous culture. *Trans. Br. Mycol. Soc. 66*:413-420.

Rossier, C., Tan That, T. C., and Turian, G. (1977). Microcyclic microconidiation in *Neurospora crassa*. *Exp. Mycol. 1*:52-62.

Saviour, R.J. (1981). Microcycle condiation by *Acremonium diospyri* in submerged culture. FEMS Microbiol. Lett. 12:287-293.

Schneider, E. F., and Seaman W. L. (1974). Development of conidial chlamydospores of *Fusarium sulphureum* in distilled water. *Can. J. Microbiol. 20*:247-254.

Schneider, E. F., and Seaman, W. L. (1977). Ontogeny of lipid bodies in the endoplasmic reticulum of *Fusarium sulphureum*. *Can. J. Microbiol. 23*:190-196.

Schneider, E. F., and Wardrop, A. B. (1979). Ultrastructural studies on the cell walls in *Fusarium sulphureum*. *Can. J. Microbiol. 25*:75-85.

Schneider, I., and Andersen, J. R. (1980). Glucose-limited continuous cultivation of *Geotrichum candidum* in a chemostat. In *Continuous Cultivation of Microorganisms*. B. Sikyta, Z. Fencl, and V. Polacek (Eds.). Prague, pp. 697-687.

Sekiguchi, J., Gaucher, G. M., and Costerton, J. W. (1975). Microcycle conidiation in *Penicillium urticae*. *Can. J. Microbiol. 21*:2048-2058.

Smith, J. E. (1978). Asexual sporulation in filamentous fungi. In *The Filamentous Fungi*, Vol. 3, J. E. Smith and D. R. Berry (Eds.). Edward Arnold, London, pp. 214-239.

Smith, J. E., and Anderson, J. G. (1973). Differentiation in the Aspergilli. *Symp. Soc. Gen. Microbiol. 23*:295-337.

Smith, J. E., Anderson, J. G., Deans, S. G., and Berry, D. R. (1981a). Biochemistry of microcycle conidiation. In *Biology of*

Conidial Fungi, Vol. 2, G. T. Cole and B. Kendrick (Eds.). Academic Press, New York, 329-356.

Smith, J. E., and Galbraith, J. C. (1971). Biochemical and physiological aspects of differentiation in the fungi. *Adv. Microbiol. Physiol.* 5:45-134.

Smith, J. E., Anderson, J. G., Kristiansen, B., Al-Rawi, A., and Yahya, G. (1981b). Microcycle conidiation. In *The Fungal Spore: Morphogenetic Controls*, G. Turian and H. R. Hohl (Eds.), Academic Press, London, pp. 627-650.

Steele, S. D. (1972). Sugars and sugar alcohols in relation to life cycle phases of *Geotrichum candidum*. *Trans. Br. Mycol. Soc.* *59*:502-506.

Steele, S. D., and Fraser, T. W. (1973). Ultrastructural changes during germination of *Geotrichum candidum* arthrospores. *Can. J. Microbiol.* *19*:1031-1034.

Stevenson, I. L., and Becker, S. A. W. E. (1972). The fine structure and development of chlamydospores of *Fusarium oxysporum*. *Can. J. Microbiol.* *18*:997-1002.

Stevenson, I. L., and Becker, S. A. W. E. (1979). The fine structure of mature and germinating chlamydospores of *Fusarium oxysporum*. *Can. J. Microbiol.* *25*:808-817.

Trinci, A. P. J. (1972). Culture turbidity as a measure of mould growth. *Trans. Br. Mycol. Soc.* *58*:467-473.

Trinci, A. P. J., and Collinge, A. J. (1974). Spore formation in nitrogen and carbon starved cultures of *Geotrichum candidum* and *Mucor racemosus*. *Trans. Br. Mycol. Soc.* *62*:351-358.

Turian, G. (1966). Morphogenesis in Ascomycetes. In *The Fungi, The Fungal Organism*, Vol. 2. G. C. Ainsworth, and A. S. Sussman (Eds.). Academic Press, New York, pp. 339-385.

Turian, G. (1974). Sporogenesis in fungi. *Annu. Rev. Phytopathol.* *12*:129-137.

Turian, G. (1975). Differentiation in *Allomyces* and *Neurospora*. *Trans. Br. Mycol. Soc.* *64*:367-380.

Turian, G. (1976). Spores in Ascomycetes, their controlled differtiation. In *The Fungal Spore: Form and Function*. D. J. Weber and W. M. Hess (Eds.). Wiley, New York, pp. 715-788.

Turian, G. (1977). Fungal Differentiation. In *Biotechnology and Fungal Differentiation*. J. Meyrath and J. D. Bu'Lock (Eds.). Academic Press, London, pp. 1-15.

Zuber, J., and Turian, G. (1981). Induction of premature phialoconidiogenesis on germinated conidia of *Trichoderma harzianum*. *Trans. Br. Mycol. Soc.* *76*:433-440.

15

Fungal Pheromones

HERMAN van den ENDE *University of Amsterdam, Amsterdam, The Netherlands*

I. INTRODUCTION

One of the means by which cells communicate with each other is by diffusible metabolites, the structure and composition of which represent a message, telling the addressed cell where to move, how to behave, and so on. One is inclined to entitle such metabolites "hormones" or "pheromones", dependent on whether the communicating cells belong to the same or different individuals. This distinction is quite inadequate when applied to fungal systems, since a particular substance, secreted by one part of thallus A may affect another part of thallus A, as well as neighboring thallus B. This is especially evident in bisexual systems. I will, however, follow the general trend in the literature and speak of pheromones in all cases where secreted specific substances are involved in intercellular communication.

Particularly from the developmental point of view, sex pheromones deserve special attention. They induce or promote the formation of specialized sex organs; they promote contacts between gametes by imposing oriented growth or movements, or by endowing them with specific agglutinative properties; and, finally, they prepare the mating cells for the final act of fusion, which certainly is a dramatic event in the cell's existence. In this chapter I will deal with some representative sexual systems which have received considerable attention in the recent literature. There are not many of them, compared to the large number of fungal species in which sex pheromones have been demonstrated more or less directly to play a role. This may be partly due to the fact that a high degree of specificity of the message is often paralleled by equal sensitivity of the responding individual and, consequently, low steady state concentrations of the pheromone concerned. Moreover, sexual interaction may be restricted to a small part of the life cycle and there often exist mechanisms to wipe out the traces of such interactions in order to allow the resumption of vegetative growth. Thus the investigator must be able, and willing, to assay and characterize very small amounts of substances of sometimes bizarre composition.

I have made a further discrimination in favor of those pheromones whose molecular structure has been elucidated. Clearly, any study on the primary mode of action of such substances and on their role in the control of developmental processes requires a knowledge of their chemical structure.

Outstanding reviews on this subject, also covering the older literature, have been produced by Raper (1952), Gooday (1974), and Kochert (1978).

II. *ALLOMYCES*

In the gametophytic phase of the aquatic fungus *Allomyces*, gametes are produced which are unicellular and flagellated and that lack a cell wall. Both male and female gametes are motile, but the latter are rather sluggish and generally become motionless soon after release. They attract male gametes by the secretion of a pheromone called *sirenin*. Both types of gametes fuse to form motile biflagellate zygotes which ultimately settle down and develop into the diploid sporophytic phase. The ultrastructural aspects of plasmogamy and karyogamy have been described by Pommerville and Fuller (1976).

The production of a chemotactic agent was first demonstrated by Machlis and Rawitscher-Kunkel (1967), who showed that male gametes are attracted to and remain in the vicinity of a dialysis membrane retaining freshly released female gametes. Mass production of the responsible agent was accomplished owing to the fact that hybrids of *Allomyces macrogynus* and *Allomyces arbuscula* were obtained which were largely male or female and thus had a seemingly diecious phenotype (Machlis, 1958; Ronne and Olson, 1976). This is important, because a monoecious *Allomyces* strain normally contains male and female gametangia in equal numbers and the secreted pheromone is rapidly consumed by male gametes (Machlis, 1973).

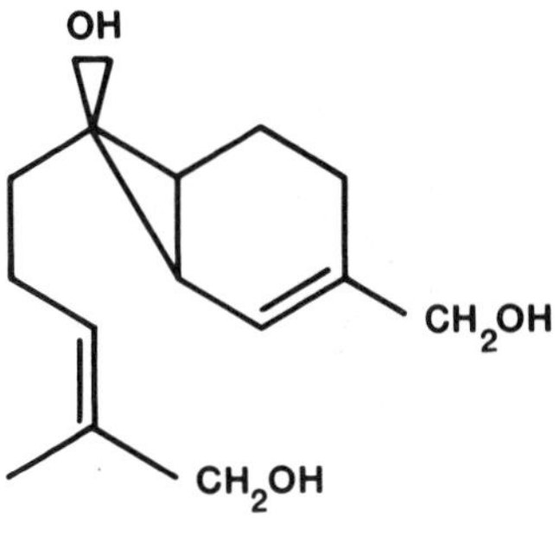

FIG. 1 Structure of sirenin.

The identification of sirenin was carried out by Machlis et al. (1968) and Nutting et al. (1968). Its structure, shown in Fig. 1, was confirmed (e.g., Plattner and Rapoport, 1971). Only the L form appeared to be active (Machlis, 1973). It showed biological activity in the range of 10^{-10} to 10^{-5} M (Carlile and Machlis, 1965).

Some evidence has been obtained that male gametes produce an antagonist of sirenin, the significance of which is uncertain (Klapper and Klapper, 1977).

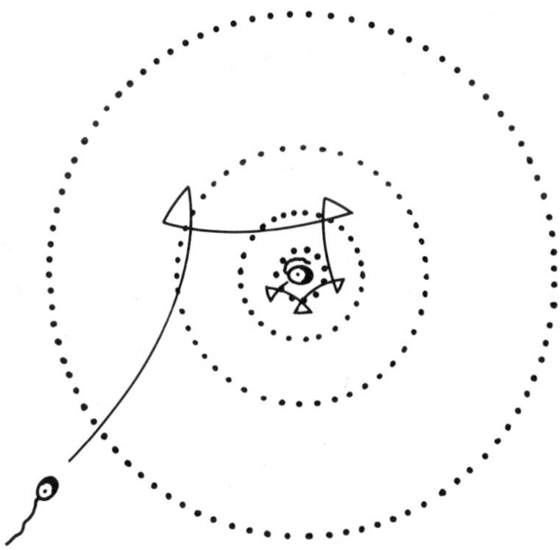

FIG. 2 Idealized motility pattern of male gametes approaching a source of sirenin. The concentration gradient of sirenin is indicated by the circles. (Redrawn from Pommerville, 1981.)

In spite of the fact that sirenin as well as dense populations of male gametes are readily available, no studies on the mode of action of this pheromone have been reported. Another important point that has been neglected since Machlis' (1969) study is that zygotic cells rapidly lose their sensitivity to sirenin and become attracted toward sources of amino acids.

Interesting work has been published by Pommerville (1978) on the behavioral response of male gametes toward a concentration gradient of sirenin. By analyzing the trajectories of moving cells, he established that they have short, slightly curved swimming paths, interrupted by jerks which alter the swimming direction in a random way. However, when a cell is swimming up a gradient of sirenin, the jerking frequency is suppressed. In contrast, a direction parallel to or down a sirenin gradient results in frequent directional changes. These continue until the cell is again moving up the gradient. The net effect of this biased random motility is an efficient approach toward the sirenin source (Fig. 2). Similar behavioral responses to a gradient of a stimulus have been described in bacteria (e.g., Adler, 1975) and algae (e.g., Feinleib, 1978). It is probable that a concentration gradient is sensed in a temporal instead of a spatial way, which means that differences in pheromone concentration are measured as a function of time. Evidence for this view is that cells also show a behavioral response when they are transferred from one uniform sirenin concentration to another (see Pommerville, 1981). The obvious analogy between this behavior and that of prokaryotes, in which the molecular aspects of this time-based concentration sensing phenomenon have received considerable attention (e.g., Springer et al., 1979), should encourage further study.

III. *ACHLYA*

The majority of *Achlya* strains exhibit self-conjugation, but the relative intensities of male and female behavior may strongly differ between different strains. Thus strongly male as well as strongly female strains, aside from intermediates, can be distinguished which behave perfectly as sexual partners (Barksdale, 1969). The first visible sexual response occurring between adjacent male and female reacting mycelia is the formation of antheridial initials in the male. These hyphae are thinner than vegetative hyphae, appear to be spiral, and are thus readily distinguishable from the stout vegetative hyphae. Next, knoblike oogonial initials are produced on the female. Antheridial hyphae grow toward these oogonial initials. Some time later crosswalls delimit antheridia and oogonia and the egg cells produced in the oogonia are fertilized by penetrating antheridial cells.

Raper (1952) has emphasized a sequential pattern in sexual morphogenesis which suggests successive mutual influences of partner

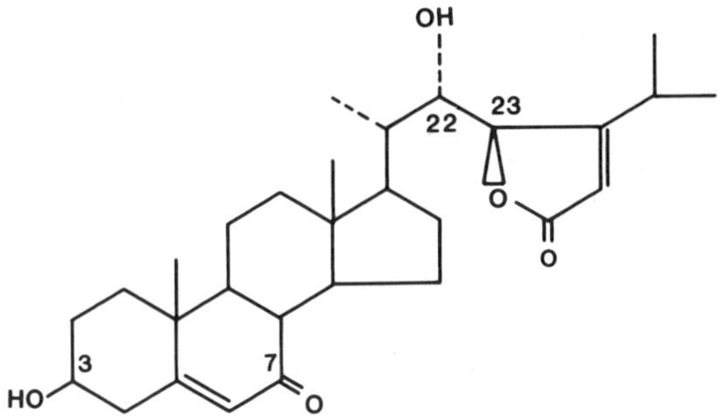

FIG. 3 Structure of antheridiol.

mycelia, involving the secretion and action of stage-characteristic pheromones: One pheromone, produced constitutively by the female mycelium, would induce antheridial initials in the male mycelium, and in response the male would produce a second pheromone, inducing the female to produce oogonial initials. By the characterization of the metabolites involved, this pattern was confirmed. Indirect evidence was provided for the assertion that the compound inducing antheridial hyphae also has an attractive influence on implicating that oogonia produce pheromone in larger amounts than vegetative mycelia, but this has never been fully investigated (Barksdale, 1963).

The antheridium-inducing compound *antheridiol* was isolated from *Achlya bisexualis* by McMorris and Barksdale in 1967, and characterized and synthesized by Arsenault et al. (1968) and Edwards et al. (1972) (Fig. 3). The configuration at C-22 and C-23 appeared to be quite critical. For example, the natural isomer (22S,23R configuration) was three orders of magnitude more active in inducing antheridial hyphae than the 22S,23S configuration (Barksdale et al. 1974). The functional groups at C-3 and C-7 seem to be relatively unimportant for biological activity.

The oogonium-inducing compound *oogoniol* was isolated and characterized by McMorris et al. (1975) from a bisexual strain of *Achlya heterosexualis* and it consisted of a mixture of three closely related compounds (Fig. 4). In addition, a varying percentage (up to 25%) of compounds with unsaturated side chain appeared to be present, the 24(28)-dehydro-oogoniols. In fact, these compounds showed a 100-fold higher oogonium-inducing activity than oogoniol, and so they might be the true hormones (Preus and McMorris, 1979).

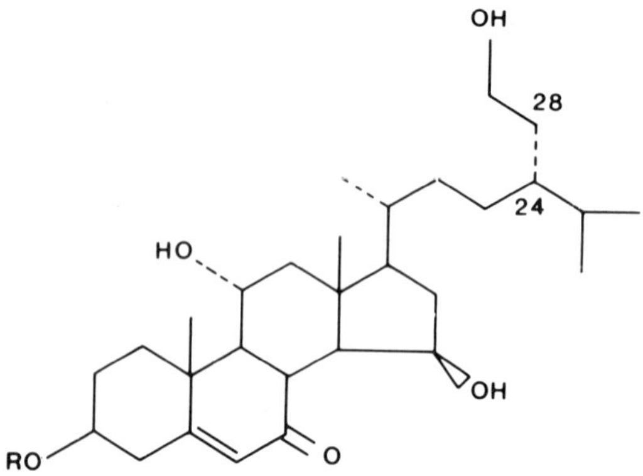

FIG. 4 Structure of the oogoniols [R = $(CH_3)_2CHCO$, CH_3CH_2CO, or CH_3CO].

It should be pointed out that the mixture of biologically active molecules was qualitatively similar, whether isolated from the bisexual *A. heterosexualis* or from the strongly male *A. ambisexualis*, strain E87, which produces the oogoniols only after treatment with antheridiol (McMorris and White, 1977). This raises the question of their biosynthesis. Presumably, fucosterol, a common sterol in these fungi, is the major precursor of both antheridiol and the oogoniols (Popplestone and Unrau, 1974; McMorris and White, 1977). Figure 5 shows the rate of incorporation of label from tritiated fucosterol into the oogoniols in male *A. ambisexualis* after antheridiol addition, while Fig. 6 shows the proposed biosynthetic routes. Thus both male and female pheromones are derived from the same precursor. It seems, then, that a similar situation exists in the differential biosynthesis of sex pheromones in *Achlya* as in the mucoraceous fungi described below. In any case, the pathway leading to the male pheromones is promoted by the female pheromone. Whether the reverse is true, that is, oogoniol stimulating antheridiol production, has not been demonstrated directly, but is borne out by the observation that oogoniol-treated mycelia have a strong attractive influence on antheridial initials, assuming still that antheridiol is the attractive agent (Barksdale, 1963). This mutually affective action of partner mycelia would neatly confirm Raper's earlier assertion that there is a sequential pattern of sexual morphogenesis.

Fucosterol has also been found to be one of the most active sterols in promoting the formation of sexual oospores in *Phytophthora* and *Pythium*. These species are unable to synthesize sterols, and this im-

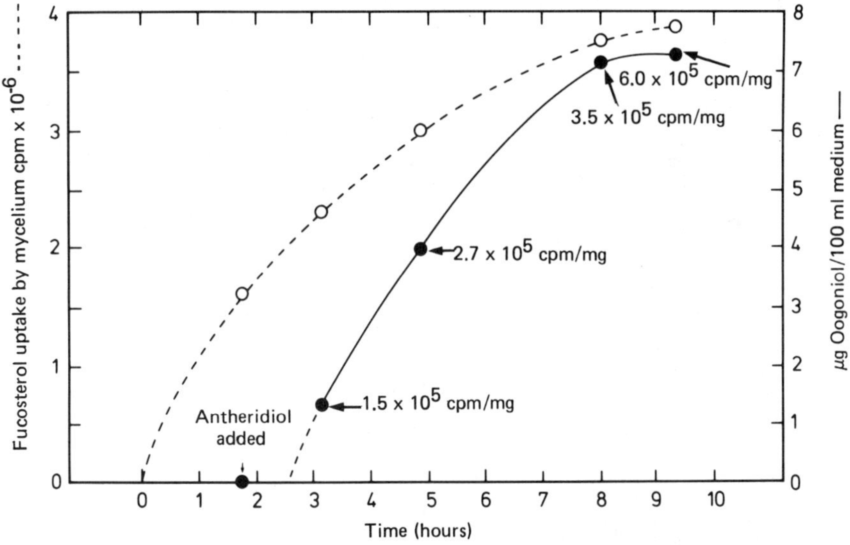

FIG. 5 Induction of oogoniol synthesis from fucosterol by antheridiol in *Achlya ambisexualis* E87. (From McMorris and White, 1977.)

plies that sterol derivatives are involved in sexual activity in these species (Gooday, 1974; Ko, 1980).

Achlya is the most primitive eukaryote in which sterols play a well-defined developmental role, with the sexual partner as a readily available target. For this reason, and in view of the analogy with mammalian steroid sex hormones, several studies have been initiated to elucidate their mode of action. For technical reasons (oogoniol is scarcely obtainable), the actions of antheridiol have mainly been taken into consideration. There are a number of responses which suggest that the de novo synthesis of special proteins is involved. In the first place, the induction of sexual organs is a clear example of sexual morphogenesis. Their shape and properties (particularly their attraction toward oogonial initials) leave little doubt as to their specialized nature. In the second place, the induction of oogoniol synthesis is subject to inhibition by cycloheximide (McMorris and White, 1977). A third action is the induction of the synthesis of cellulase, an enzyme thought to be involved in the initiation of antheridial hyphae (Nolan and Bal, 1974). Finally, antheridiol appears to induce the formation of an enzyme complex which degrades antheridiol extracellularly to an inactive metabolite (Musgrave and Nieuwenhuis, 1975). Also, this action is completely suppressed by actinomycin D and cycloheximide. At

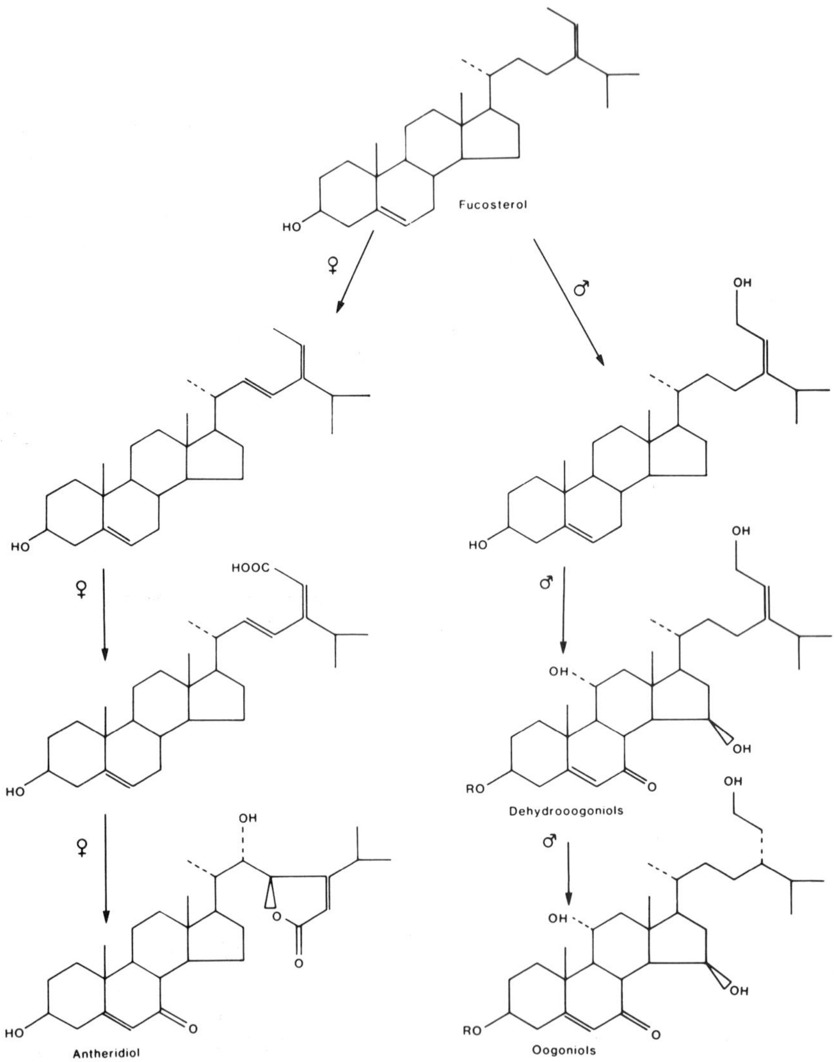

FIG. 6 Proposed biosynthetic routes of antheridiol and oogoniols. The reactions are carried out in male or female mycelia as indicated. (Modified from Popplestone and Unrau, 1974; Preus and McMorris, 1979.)

the molecular level, antheridiol treatment effects a transiently enhanced productivity of ribosomal and poly(A)-enriched RNA that precedes the formation of antheridial hyphae (Silver and Horgen, 1974; Timberlake, 1976). Sutherland and Horgen (1977), using a heterologous chromatin-*Escherichia coli* RNA-polymerase assay, demonstrated that the number of RNA polymerase binding sites and initiation sites of RNA synthesis on chromatin increased upon preincubating mycelium with antheridiol. One may expect that antheridiol-stimulated RNA synthesis would include the production of a number of specific mRNA species. However, several studies have been unsuccessful in detecting such mRNAs. Rozek and Timberlake (1980), using hybridization techniques with a broad range of sensitivities, failed to detect qualitative differences in poly(A) RNA populations derived from mycelia that were either untreated or had been exposed to antheridiol for various periods. So, if qualitative changes in mRNA production were induced by antheridiol, they must have been quite subtle. A similar conclusion was drawn by Gwynne and Brandhurst (1980), who translated poly(A)-containing RNA, isolated from mycelia at various developmental stages, in a wheat-germ cell-free system and analyzed the products by two-dimensional gel electrophoresis. They suggested that the formation of antheridial initials requires the continuous synthesis of RNA and protein, but that it also involves posttranslational modification processes of proteins. However, analyses of protein extracts by two-dimensional gel electrophoresis (Groner et al., 1976; Michalski, 1978; Gwynne and Brandhurst, 1980), although not in complete agreement, failed to reveal major differences in the protein patterns of mycelia at different developmental stages. It should also be mentioned that the experimental procedures for detecting novel mRNAs had a lower limit of sensitivity of approximately 7% of the total poly(A) RNA, and one may doubt whether antheridiol-induced qualitative changes in mRNA synthesis and/or processing should be more than that.

In passing, it should be noted that similar difficulties were encountered by Zantinge et al. (1979) in *Schizophyllum commune*. Poly(A)-containing RNA populations from mycelia with different morphological and functional characteristics (mono- and dikaryotic mycelia, regulated by two sets of incompatibility factors) did not show qualitative differences. Again, the authors suggested that transscriptional control is relatively unimportant. Thus differentiation in these fungi seems to be more difficult to tackle at the molecular level than similar processes in vertebrate systems (e.g., Woo and O'Malley, 1975), where not only quantitative but also qualitative changes in the synthesis of RNAs and proteins have been reported.

In another respect, however, the action of antheridiol seems to be reminiscent in a more encouraging way of more highly evolved systems in which a strong affinity has been demonstrated between a steroid hormone and a cytoplasmic receptor in target cells, leading to a com-

TABLE 1 The Effects of Cytosol, Antheridiol, and Cytosol-Antheridiol Mixtures on the Transcription of *Achlya* Chromatin

Chromatin preincubated with[a]	Specific radioactivity[b]
1. No additions (control)	111 ± 4
E87 cytosol	112 ± 6
Antheridiol	97 ± 2
E87 cytosol + antheridiol	415 ± 8
734 cytosol + antheridiol	117 ± 5
2. No additions (control)	82 ± 2
E87 cytosol	85 ± 1
Antheridiol	75 ± 4
E87 cytosol + antheridiol	271 ± 6

[a]1. Transcription with heterologous RNA polymerase. A total of 50 μg of chromatin DNA (strain E87) was added along with 10 μg of *E. coli* RNA polymerase to each reaction mixture. Either 2.1 nM antheridiol, 1.25-150 mg of cytosol, or a combination of both were added to each preincubation mixture. The final concentrations in the 1-ml reaction mixture were 50 mM Tris HCl (pH 8.0), 10 mM dithiothreitol, 10 mM $MgCl_2$, 20% glycerol, 0.1 M KCl, 0.2 mM ammonium sulfate, and 0.4 mM K_2HPO_4. These mixtures were preincubated for 15 min at 25°C. To begin the RNA synthetic reaction, 0.4 mM UTP, 0.4 mM ATP, 0.4 mM CTP, 0.4 mM GTP, and 10 μCi [^{3}H]UTP (specific activity, 42 Ci/mmol) were added. Each reaction mixture was then incubated at 30°C for 15 min. The reactions were terminated, the products collected, and the radioactivity determined as described by Horgen (1977).
2. Transcription with endogenous *Achlya* RNA polymerases. Preincubation mixtures were similar to those given in (1), except that 200 μg of chromatin DNA were added and *E. coli* RNA polymerase was not.
[b]In picomoles incorporated per minute per 50 μg of chromatin DNA.
Source: Modified from Horgen (1977).

plex involved in regulating template activity (Jensen and DeSombre, 1972). Horgen (1977) showed that the transcriptional activity of chromatin isolated from noninduced male cultures of *A. ambisexualis* was enhanced after preincubation with a mixture of antheridiol and a preparation of cytoplasmic components derived from a male mycelium (cytosol). When applied separately, neither the antheridiol nor the cytosol had an effect. Some results are presented in Table 1. It is also notable that the cytosol fraction from the female 734 strain had no activity, which suggests that this strain has few, if any, cytoplasmic receptors for antheridiol. However, more control experiments, also involving radioactive antheridiol of high specific activity, are required to assess and extend these results. With well-labeled antheridiol, *Achlya* could offer superior opportunities for unraveling the mechanism of differentiation at the molecular level, compared with the more sophisticated multicellular systems studied so far.

IV. ZYGOMYCETES

This large group of fungi is represented in studies of sexuality by only a few species: *Mucor mucedo*, *Blakeslea trispora*, and *Phycomyces blakesleeanus*. In fact, only Blakeslee (1904) and Burgeff (1924) have surveyed a truly representative number of species. A characteristic feature is that sexual interaction can occur, be it incompletely, between different species or even genera, as long as the partners have a different mating type. This suggests that sexual morphogenesis, as in *Achlya*, is evoked by metabolites which are common to all species concerned. This is reinforced by the hormonal studies described below.

The most frequently used organism, *M. mucedo*, forms erect sexual hyphae, called zygophores, when mycelia of both mating types (mt^+ and mt^-) approach each other. Its popularity is due to the fact that these zygophores are easily recognized in the mass of vegetative aerial hyphae. Zygophores of both mating types grow toward one another, and upon contact swell considerably and delimit gametangia which subsequently fuse and produce the zygospores. In principle, this behavior is also observed in other heterothallic and homothallic species (cf. also Schipper, 1978).

With experiments involving partner mycelia separated by a semipermeable membrane, Burgeff (1924) demonstrated that zygophore production in *M. mucedo* and several other species is controlled by low-molecular-weight substances which diffuse through the substrate or the air; subsequent work (reviewed by van den Ende, 1978) demonstrated that the *trisporic acids* as well as several trisporic acid precursors, are active zygophore-inducing agents. Figure 7 shows the structures of trisporic acids B and C which have been confirmed by total synthesis (e.g., Edwards et al., 1971; Isoe et al., 1971; Prisbylla et al., 1979).

There are a number of interesting aspects to the biosynthesis and action of the trisporic acids. It was demonstrated that they are the result of a cooperative action of both mating types. A mycelium of one mating type can only accomplish a limited part of the biosynthetic pathway and is complemented by a mycelium of the other mating type. Figure 8 shows how this complementary action is envisaged. Starting from compound I, a putative degradation product of cyclic carotenoids, ubiquitous in Mucorales (Bu'Lock et al., 1974a), mt^+ mycelia have the ability to oxidize the pro-S-methyl group at C-1 to a carboxymethyl group, which results in the formation of the methyl 4-dihydrotrispor-

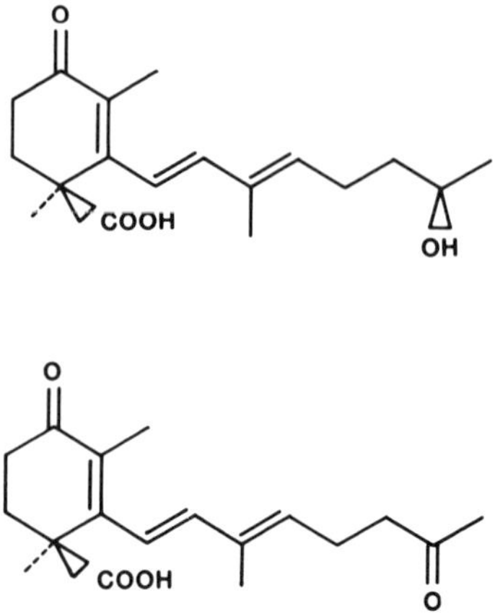

FIG. 7 Structure of the trisporic acids C (—OH) and B (= O).

ates (IV). In contrast, *mt*⁻ mycelia are able to oxidize the hydroxyl group at C-4, which leads to the trisporins (VI). Both IV and VI are end products in the single respective mycelia. However, when cultures of both mating types are mixed, both act in a complementary manner: IV is converted to V by the *mt*⁻-specific oxidation of the C-4 hydroxyl group, and subsequently hydrolyzed to VIII, while VI is converted to VIII by the corresponding oxidizing activities present in mt^+. This oxidation probably takes place via $-CH_2OH$ and $-COOH$ intermediates, similar to the oxidation of lanosterol during its conversion to cholesterol (Bechtold et al., 1972).

This concerted action of both mating types in the production of trisporic acids is also possible because the single-mating-type products, IV and VI, among other derivatives, are secreted into the culture medium (Sutter et al., 1973; Bu'Lock et al., 1974b). Their specific conversion to trisporic acids was demonstrated by Nieuwenhuis and van den Ende (1975). Two enzymes involved in the conversion of IV to trisporic acids, and/or in the formation of VI, a NADP-specific dehydrogenase and an esterase, were found to be present only in mt^- mycelia (Werkman, 1976).

FIG. 8 Proposed biosynthetic pathway leading to trisporic acids. The solid lines represent physiological reactions active in the mating types indicated; the dashed lines represent reactions postulated to be partially repressed. The structure between brackets is unstable. For side chain R, see Fig. 7; R' = $CHCHCCH_3CHCH_2OH$. (See Sutter and Whitaker, 1981a.)

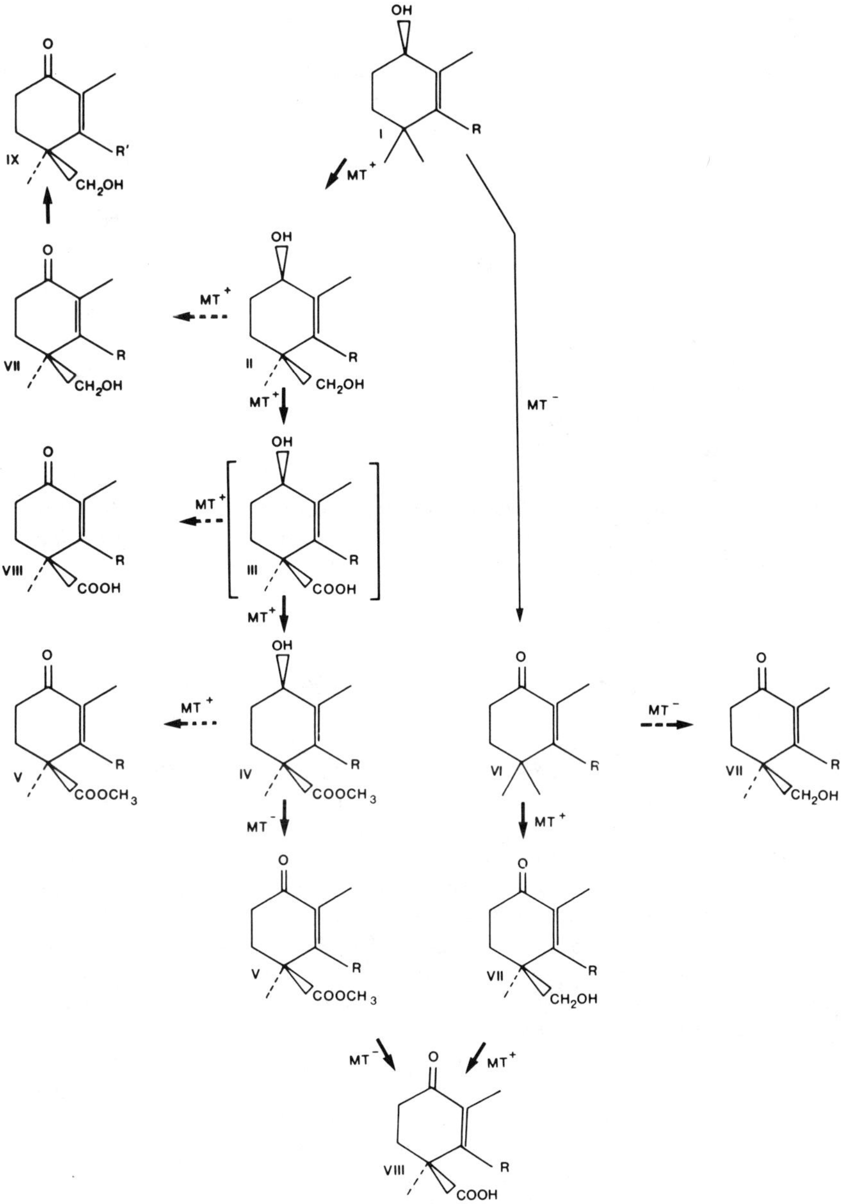
OH
I
R
MT+
MT−
O
IX
R'
CH2OH
VII
R
II
III
IV
V
VI
VIII
COOH
COOCH3

TABLE 2 Relative Zygophore-Stimulating Activities of Trisporic Acid Derivatives

Compound	Activities[a] *P. blakesleeanus* mt^+	*P. blakesleeanus* mt^-	*M. mucedo* mt^+	*M. mucedo* mt^-
Methyl 4-dihydrotrisporate B	>320	1	>8,900	4
Methyl 4-dihydrotrisporate C	>640	4	>17,900	75
Methyl 4-dihydrotrisporate C*	>6,330	11	>177,000	75
Methyl trisporate B	310	3	290	4
Methyl trisporate C	2,100	21	19,000	39
Trisporic acid B	90	90	2	1
Trisporic acid C	360	360	40	10

[a]The numbers indicate the lowest amount of a compound giving a definite positive response in bioassays for zygophores (*M. mucedo*) and enlarged zygophores (*P. blakesleeanus*), relative to the compound which was active in the lowest absolute quantity. Unity equals 27 pmol in *M. mucedo* and 745 pmol in *P. blakesleeanus*.
Source: Sutter and Whitaker (1981a).

A possible objection to the proposed exchange of metabolites might be that the extremely small quantities of the compounds involved, combined with the large dilution factor as the result of secretion into the culture medium, would make a reasonably large production of trisporic acid, as observed in *B. trispora* (40 mg/liter; Bu'Lock et al., 1972), hardly feasible. However, Werkman and van den Ende (1974) observed that the trisporic acids have a promoting action on their own biosynthesis; probably the synthesis of enzymes involved in the biosynthetic pathway is stimulated, including those which convert IV and VI, since inhibitors of protein synthesis effectively block this stimulation.

As mentioned above, a remarkable feature of trisporic acids is the ability to induce sexual organs in both mating types of *M. mucedo*. The most active isomer is a 9-cis-trisporic acid B (Bu'Lock et al., 1972). In *P. blakesleeanus*, however, the trisporic acids have very little effect. As can be seen in Table 2, trisporic acid precursors show a much higher activity in this species (Sutter and Whitaker, 1981a); also, in *M. mucedo*, these intermediates display considerable activity. The major difference in action between these components and trisporic acids is the strong mating-type specificity of the former. Particularly, the methyl 4-dihydrotrisporates (IV) are active only on *mt*⁻-mycelia while the trisporins (VI) show activity only on mt^+ mycelia. It seems, therefore, that for activity on mt^- a C-1 carboxy(methyl) group is essential, whereas for activity on mt^+ a carbonyl group at C-4 is required, the situation at C-1 being unimportant. The presence of a functional group at C-4 is required in any case, since synthetic intermediates lacking such a group show no activity (T. C. McMorris and H. van den Ende, unpublished data, 1982). Possibly these activities reflect the differential abilities of the respective mating types to convert these compounds to intracellular trisporic acids.

Another property of the trisporic acids and their derivatives, as mentioned above, is their ability to stimulate their own production. Also, in several strains they enhance carotenoid biosynthesis (see Gooday, 1978). An attractive possibility, furthermore, is that trisporic acid precursors, which are much more volatile than the trisporic acids themselves, would exert a chemotropic action on zygophores, forcing them to grow towards one another (Mesland et al., 1974). The fragility of the aerial sexual organs has so far prevented a detailed investigation of this aspect.

The differential abilities of both mating types of *M. mucedo* and related fungi to participate in trisporic acid biosynthesis has been explained by a differential repression of enzymatic functions by the mating-type alleles (Sutter and Whitaker, 1981a,b). For example, the production of an active dehydrogenase, involved in the oxidation of the C-4 hydroxyl group, would be repressed in mt^+. However, several observations indicate that this repression is incomplete. Dehydrogenase activities have been found in mt^+ mycelia of *M. mucedo*, *Mucor hiemalis*, and *P. blakesleeanus* (A. van der Does, personal communication, 1980). This corresponds with the observation by Sutter and Whitaker (1981b), that the compounds V, VIII, and maybe VII, all exerting a zygophore-inducing influence on mt^+ mycelia, also occur in culture media of mt^+, be it in small quantities. The authors suggested that oxidation at C-4 in that mating type is the result of incomplete repression of the responsible enzymes(s). This implies that the resulting metabolites must be degraded ("detoxified") to avoid the formation of zygophores by single mt^+ mycelia, for instance, by further degradation to trisporone (IX), which is inactive and which has been found in relatively large amounts in culture media of *B. trispora* mt^+.

In summary, the following simplified picture can be drawn concerning the role of pheromones in sexual development in *Mucorales*. Mycelia produce mating-type-specific factors which induce sexual organs in the other mating type, possibly by being converted to trisporic acids. The initially slight accumulation of low concentrations of trisporic acids enhances the production of pheromones, which in turn are converted to trisporic acids. In some cases (*M mucedo, B. trispora*) this amplification leads to the massive accumulation of trisporic acids. Evidently this mutual stimulation, together with the complementarity with respect to trisporic acid synthesis, results in a developmental pattern in which the activities of both participating partner mycelia are well coordinated with respect to one another.

V. YEASTS

A. Ascomycetes

When suspensions of haploid *Saccharomyces cerevisiae* cells of different mating types (*a* and α) are mixed, zygote formation by cell fusion is preceded by a number of phenomena: a temporary inhibition of cell division, the formation of cell aggregates, and the occurrence of cells with aberrant morphology ("shmoos"), together with changes in cell wall ultrastructure and composition. These phenomena can be rationalized to be functional in stimulating zygote formation, and the interesting aspect from our point of view is that they are evoked by sex pheromones. When haploid *a* cells are exposed to filtrates of an α-cell culture, cell budding is inhibited and pearlike cell deformations are induced. The reverse combination has a similar effect. Through subsequent work (reviewed by Manney et al., 1981), the responsible factors were isolated from isogenic strains and characterized. The factor derived from α-cell cultures, the α factor, consists of a group of closely related trideca- and dodecapeptides, which differ only slightly in biological activity (Stötzler et al., 1976; Fig. 9). Evidently, the α2, α3, and α4 factors are derived from the α1 compound. Chemical synthesis confirmed the structures of the peptides (Ciejek et al., 1977; Masui et al., 1977). The *a* factor is produced in much smaller quantities and is a hydrophobic peptide of unestablished structure (Manney et al., 1981).

Both the α factor as well as the *a* factor are produced consitutively by the respective mating types, although Manney and Meade (1977) suggested that *a*-factor production may be stimulated by the α factor. In this respect, the yeast system would resemble the *Achlya* and

α1 NH_2-Trp-His-Trp-Leu-Gln-Leu-Lys-Pro-Gly-Gln-Pro-Met-Tyr-COOH

α2 NH_2-His-Trp-Leu-Gln-Leu-Lys-Pro-Gly-Gln-Pro-Met-Tyr-COOH

α3 NH_2-Trp-His-Trp-Leu-Gln-Leu-Lys-Pro-Gly-Gln-Pro-Met(SO)-Tyr-COOH

α4 NH_2-His-Trp-Leu-Gln-Leu-Lys-Pro-Gly-Gln-Pro-Met(SO)-Tyr-COOH

FIG. 9 Primary structure of the α factor.

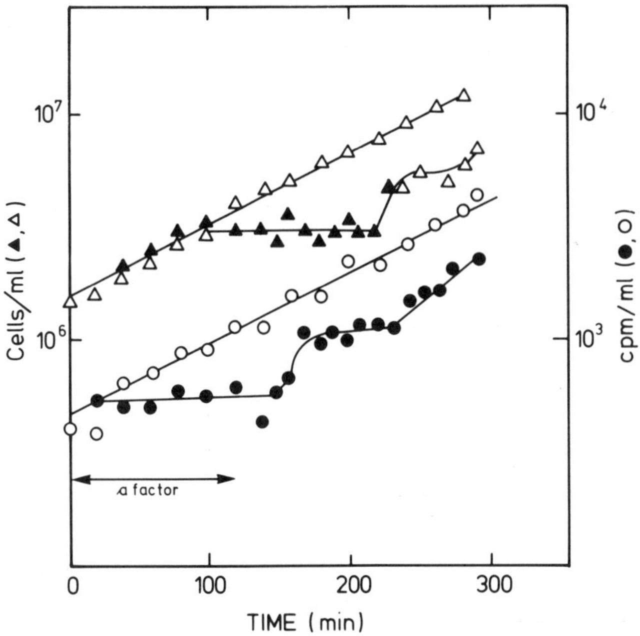

FIG. 10 Synchronization of DNA synthesis and cell division by treatment with α factor (50 unit/ml). After 120 min the factor was removed from the treated culture by filtration and the cells were washed and resuspended in the same volume of fresh medium. DNA synthesis was measured by following the incorporation of [^{14}C]-adenine into the alkali-resistant, DNase-sensitive material. The open symbols represent controls. (From Bucking-Throm et al., 1973.)

Mucor systems, where mutual stimulation of pheromone production promotes a concerted action of the sex partners.

The response of the two cell types to these factors can best be described as a temporal gamete stage during which they are made fit for conjugation and the frequency of prolonged cell contact is increased by enhanced agglutinability. The reported inhibition of budding is due to a cell cycle arrest, since the initiation of DNA synthesis is inhibited (Bucking-Throm et al., 1973). Thus the cells accumulate in the G1 phase of the cell cycle. When the blockage is released by transferring the cells to fresh medium, a synchronous pulse of DNA synthesis and cell division is observed (Fig. 10). This cell cycle arrest, as well as implying inhibition of nuclear plaque duplication (Byers and Goetsch, 1975), leads to the synchronization of both part-

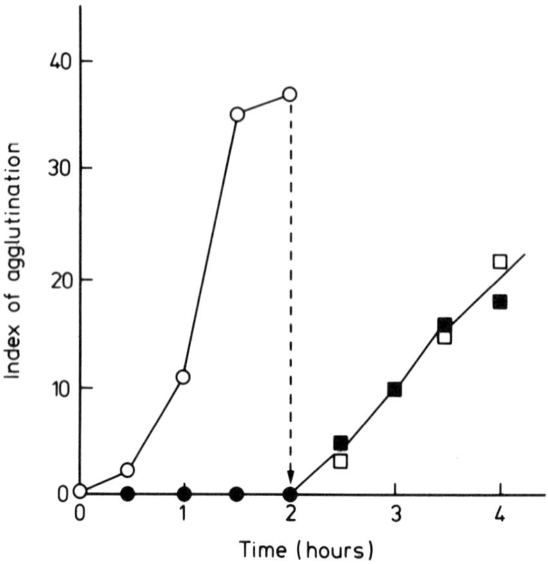

FIG. 11 Mating-type-specific agglutination between *a* and α cells of *S. cerevisiae*; ○—○, fresh mixture without cycloheximide; ●—●, fresh mixture with cycloheximide; ■—■, sonicated, 2-h agglutination mixture with cycloheximide. (From Betz et al., 1978.)

ners in a stage, which is the only one in which they are competent to mate (Sena et al., 1973).

Although DNA replication is inhibited by pheromonal action, RNA and protein synthesis are not, and this may account for a general increase in cell volume. Moreover, during this period of arrest, several activities proceed which require the synthesis of specific proteins.

One of these is the acquirement of agglutinability to a higher or lesser degree, depending on the strain. A classic example is *Hansenula wingei*, from which the surface components responsible for sex-specific agglutination have been characterized to a large extent (Yen and Ballou, 1974). Figure 11 shows the time course of agglutination, as well as evidence for the presumption that active protein synthesis is required for the acquirement of this property. Also, lomofungin, an inhibitor of RNA synthesis, appears to affect the induction of agglutinability (Fehrenbacher et al., 1978). Several authors have provided evidence that this induction is in response to the *a* and α factors (Ciejek et al., 1977; Betz et al., 1978; Fehrenbacher et al., 1978). Since the agglutination process proceeds rather slowly and is strongly affected by environmental factors, several bioassays have been devised to quantitatively measure agglutinability, for example, by measuring the change in light absorbance in yeast suspensions (Yanagishima and Yoshida, 1981) or by accelerating the process by centrifugation in conjunction with the use of complementary auxotrophic strains (Fehrenbacher et al., 1978). By these means, it appeared that the induction process itself is quite rapid, an exposure of 5 min to the pheromones being sufficient under certain circumstances to confer agglutinability upon the cells, independently of the stage of the cell cycle (Yanagishima and Yoshida, 1981).

Also, the third action of the *a* and α factors, the induction of change in cell morphology, has received considerable attention. Lipke et al. (1976) observed that it is accompanied by conspicuous changes in cell wall morphology and composition. Cells treated with α factor contained more glucan and less mannan than control cells. Also, the elongated zone appeared to contain more diffusely deposited chitin, compared to untreated cells, in which chitin incorporation in the cell wall is normally limited to the bud scars (Schekman and Brawley, 1979). These changes are reminiscent of studies in which similar cell wall modifications have been noted during the transition of the budding phase to mycelial-type phases in dimorphous yeasts (e.g., Stewart and Rogers, 1978). Electron microscopy of thin sections showed that the cell wall in the elongated zone has a diffuse outer layer and lacks the normal trilaminate appearance. Tkacz and MacKay (1979) observed that these extensions share the property of binding fluorescein isothiocyanate concanavalin A with the fusion zone in young zygotes. These results suggest that the asymmetric alterations induced by the pheromones are a prelude to, or at least could facilitate, cell wall fusion of partner cells.

Little is known about the biosynthesis and primary action of the *a* and α factors. Liao and Thorner (1980) found that the α factor inhibited an adenylate cyclase activity in yeast plasma membranes, while the addition of cyclic AMP to a cell suspension antagonized the action of α factor on *a* cells in vivo. Whether the biochemical mechanism of α-factor action implies the control of the intracellular cyclic AMP level is doubtful, however, since the observed effects were not mating-type specific. In this respect, the finding of Finkelstein and McAlister (1981) is more interesting. They discovered a ^{32}P-labeled protein which undergoes an α-factor-mediated modification. Upon pulse labeling log-phase α-factor-treated cells for 5-10 min, one labeled protein with an apparent molecular weight of 17,000 showed an alteration in electrophoretic mobility, migrating with a molecular weight of 17,500. This was observed in *a* cells only. The molecule exhibited a high turnover rate and the time course of the modification appeared to correspond nicely with the time course of cell cycle arrest and the subsequent appearance of morphologically altered cells. The relative level of modified protein decreased again when the cells resumed vegetative growth. The appearance of the modified protein was not subject to the action of inhibitors of protein synthesis, whereas cell cycle arrest and subsequent events in fact were. This implies that this phenomenon is not the consequence of sexual differentiation but, rather, could precede it.

Throm and Duntze (1970) observed that the effect of the α factor is reversible: *a* cells eventually recover from growth arrest caused by the pheromone. Studies by Chan (1977) suggested that this recovery is due to the destruction of the α factor by *a* cells. Maness and Edelman (1978) and Finkelstein and Strausberg (1979) provided evidence that removal of α-factor activity is not due to the action of an extracellular inhibitor or to binding and internalization of the molecule. Using ^{125}I- or ^{35}S-labeled α factor, they observed that only small quantities (3% or less) were bound, but that most of it was degraded to smaller fragments. This degradation was mating-type specific. Interestingly, Maness and Edelman (1978) found that this mating-type specificity was not exhibited by spheroplasts, and they suggested that the proteolytic activity responsible for α-factor breakdown is somehow masked in *a* cells and *a*/α diploids. This concurs with the fact that this proteolytic activity is cell associated; the rate of proteolysis is dependent on cell density and α-factor degradation is not observed in cell-free media. Evidently, at least in log-phase cells—reportedly, stationary cells may secrete appreciable quantities of proteolytic enzymes (Finkelstein and Strausberg, 1979)—the α-factor degrading activity takes place in some unspecified part of the cell envelope. Similar conclusions were arrived at by Ciejek and Thorner (1979), using α factor that was ^{3}H labeled at six different amino acid residues. From an analysis of the degradation products, they concluded that a primary cleavage in the degradation of the α-factor molecule exists between leucine 6 and lysine 7, yielding biologically inactive fragments. This suggests that the responsible protease is quite specific. Masui et al. (1977) reported that analogs of the α factor, in which the abovementioned residues were replaced by the corresponding D isomers, showed higher activity than the natural factor. Also, various protease inhibitors enhanced or prolonged α-factor action (e.g., Ciejek and Thorner, 1979). Thus proteolytic degradation of the α factor antagonizes its action. This might explain the transient character of the cell cycle arrest and other preconjugation cellular features.

Taken together, it seems that the *a* and α pheromones exert a number of actions on a temporary basis which, in combination, propagate conjugation. They synchronize the respective mating populations in a preconjugation-competent stage and enhance cell contact frequency by stimulating the sexual agglutinability of the cell surfaces as a first step to cell wall hydrolysis and cell fusion. It is tempting to consider the third result of pheromonal action, the occurrence of morphological modifications, as analogous to the formation of copulation tubes, observed in several basidiomycetous yeasts described below. This would at least explain why in a random aggregate of mixed *a* and α cells, cell fusion is restricted to a large extent to cell pairs only, and that multiple fusions (ménages-à-trois, or worse) occur only very rarely (Calleja et al., 1981).

A cautionary note should accompany this summary. In the above studies, often admittedly high doses of α factor were applied, sometimes resulting in negative effects on zygote production (e.g., Sena et al., 1973; Udden and Finkelstein, 1978). Furthermore, bioassays of variable design have been used to quantitatively analyze G1 arrest, as well as agglutination. Finally, owing to the α factor's enzymatic degradation in *a*-cell suspensions, its concentration in a given batch is not constant, which might affect the different actions assigned to this pheromone in various ways. This makes it very difficult to make comparisons between dose-response relationships for these actions and to judge their relative importance for zygote production and their coordination in time.

B. Basidiomycetes

In basidiomycetous yeasts, yeastlike morphology is exhibited only in the haploid stage, while the diploid stage has cells of filamentous type. Sexual reproduction of haploid cells implies cellular interaction mediated by pheromones, two examples of which have been studied in detail—*Rhodosporidium toruloides* and the related *Tremella mesenterica*.

Rhodosporidium has been investigated by Abe and coworkers (1975). They used two compatible strains *A* and *a*. Both produce pheromones which, as in ascosporous yeasts, inhibit cell budding. However, cell contacts are not promoted by sex-specific agglutination, but by the elaboration of rather long copulation tubules which (at least on agar cultures) grow tropically toward the nearest colony of the appropriate mating type (Abe et al., 1975). The situation is, as in the systems described above, of a reciprocal nature: In closely aligned streaks of compatible cells, the *a* cells are first in producing a copulation tube. Tubes of both mating types may be seen to fuse tip to tip. Thus both cell types produce a pheromone, and the orderly sequence of events suggests coordinated pheromone production. Indeed, while *A* cells produce a factor constitutively, they are stimulated to produce a tube only by culture medium of *A*-factor-treated *a* cells. This aspect has, however, not been further explored, and neither has the possibility that the induction and attraction of copulation tubes are effected by the same substance.

The *A* factor, called *rhodotorucin A*, was characterized by Kamiya et al., 1978. From 300 liters of culture filtrate, 1.6 mg of material was isolated with a biological activity of 8 μg/ml in a standard assay, involving the induction of tube formation in *a* cells. Rhodotorucin A's structure was confirmed by total synthesis (Kitada et al., 1979). It is

FIG. 12 Primary structure of rhodotorucin A.

a lipophilic peptide with a novel amino acid, S-farnesyl cysteine, at the C-terminal end of the molecule (Fig. 12). This end group is essential for biological activity, since the SH-undecapeptide showed no activity. The corresponding factor from *a* cells has not been characterized.

When *a* cells are incubated with highly purified rhodotorucin A for 5 h, inhibition of bud formation and nuclear division, as well as tube formation and stimulation of *a*-factor production, are observed to occur, suggesting that one pheromone is responsible for all these phenomena (Tsuchiya et al., 1978). The analogy between *Saccharomyces* and *Rhodosporidium* goes even further, because Tsuchiya and Fukui (1978) found that rhodotorucin A activity is rapidly inhibited in solutions containing *a* cells. However, in this case, a specific binding of the pheromone to *a* cells seems to take place (cf. Miyakawa et al., 1982).

Evidence for the involvement of pheromones in the tetrapolar *Tremella mesenterica* was obtained by Bandoni (1965) and Reid (1974). The vicinity of compatible monokaryotic cells evokes the elaboration of copulation tubules quite similar to those in *Rhodosporidium*. The structure of one of at least two postulated pheromones, tremellogen A-10, produced by cells with the *A* incompatibility factor and acting on those with the *a* genotype, was elucidated by Sakagami et al. (1979) (Fig. 13). Like rhodotorucin A, this substance inhibits nuclear and cell division and induces tube formation. The substance obtained by total synthesis appeared to have identical properties (Fujino and Kitada, 1980).

Table 3 shows the biological action of several analogs. The C-terminal methyl ester can be replaced by an amide group, but a blocked C terminus is essential for biological activity (in contrast to rhodo-

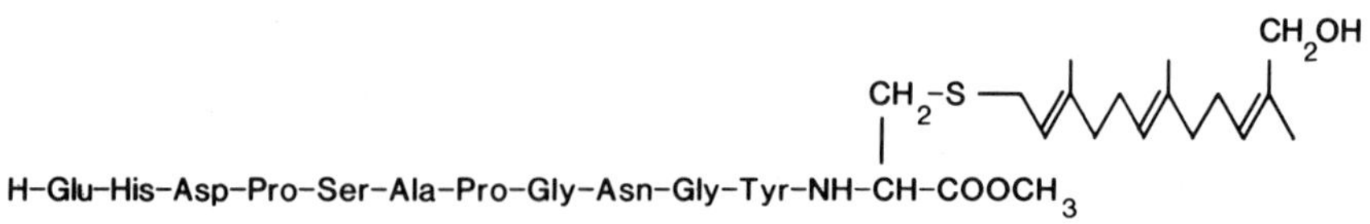

FIG. 13 Primary structure of tremellogen A-10.

TABLE 3 Biological Activity of Tremellogen A-10 Analogs

```
                                                  R
                                                  |
H-Glu-His-Asp-Pro-Ser-Ala-Pro-Gly-Asn-Gly-Tyr-Cys-X
  E   H   D   P   S   A   P   G   N   G   Y   C
```

Compound	Peptide part	-X	-R	Minimum amount inducing mating-tube formation(ng/ml)
1	E-H-D-P-S-A-P-G-N-G-Y-C (Tremerogen A-10)	$-OCH_3$	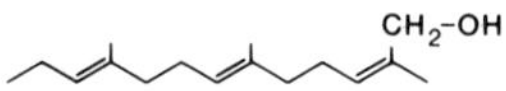	1
2	E-H-D-P-S-A-P-G-N-G-Y-NC (SH-dodecapeptide amide)	$-NH_2$		> 10.000
3	E-H-D-P-S-A-P-G-N-G-Y-C	$-NH_2$	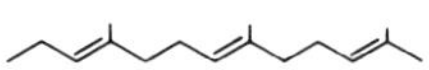	4
4	E-H-D-P-S-A-P-G-N-G-Y-C	$-OCH_3$	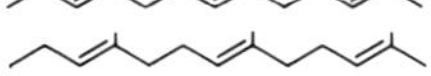	8
5	E-H-D-P-S-A-P-G-N-G-Y-C	-OH	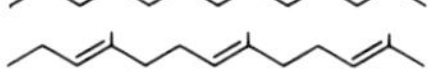	1.000
6	E-H-D-P-S-A-P-G-N-G-Y-C	$-NH_2$	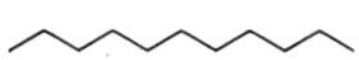	64
7	E-H-D-P-S-A-P-G-N-G-Y-C	$-NH_2$	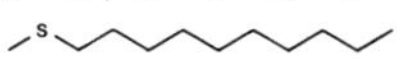	16
8	H-D-P-S-A-P-G-N-G-Y-C	$-NH_2$	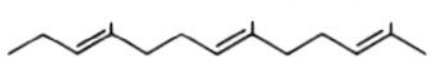	> 10.000
9	D-P-S-A-P-G-N-G-Y-C	$-NH_2$	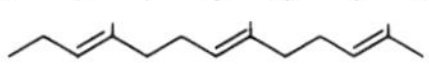	> 10.000
10	E-H-D-P-S-A-P-G-N-G-Y-C	$-NH_2$	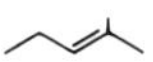	> 10.000
11	E-H-D-P-S-A-P-G-N-G-Y-C	$-NH_2$	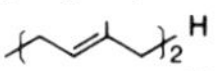	250
12	E-H-D-P-S-A-P-G-N-G-Y-C	$-NH_2$	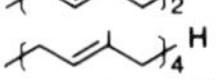	1/8
13	E-H-D-P-S-A-P-G-N-G-Y-C	$-NH_2$	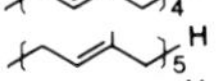	1/16
14	E-H-D-P-S-A-P-G-N-G-Y-C	$-NH_2$	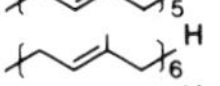	2
15	E-H-D-P-S-A-P-G-N-G-Y-C	$-NH_2$	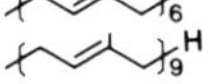	500

Source: Fujino and Kitada (1980).

torucin A). Modification of the peptide chain leads to the loss of activity, but the farnesyl group can be replaced by other (linear) isoprenoid chains, in some cases resulting in a strongly enhanced biological activity. In view of these facts, it is to be expected that more detailed studies concerning the mode of action of rhodotorucin A and tremellogen A-10 will be quite rewarding.

VI. OVERVIEW

A striking aspect of the systems described so far is the role of pheromones in the spatial and temporal coordination of the sexual process. A general pattern in heterothallic organisms in which two mating individuals are involved seems to be that one or both of them constitutively produce low levels of a signalling metabolite in small amounts. In response, the partner(s) secretes a second effective substance, which leads to an amplifying cascade of chemical interaction. There is ample evidence that a similar interactive communication also exists in homothallic mycelia, the best examples being *Achlya* (Raper, 1950) and homothallic *Mucor* sp. (Werkman, 1976). Particularly in such organisms, sexual metabolites might play their coordinative role intracellularly as well. In the coenocytic homothallic Zygomycetes, for example, one is impressed by the well-regulated spatial distribution of sexual expression within one thallus, and there is indirect evidence that signalling substances play a role in this case (Schipper and Stalpers, 1980).

The coordinative action of sex pheromones involves the timed induction of organs competent for sexual fusion. This raises the problem, especially relevant in the fungi, that they are often enclosed by a firm cell wall. A compatible contact leads to localized cell wall hydrolysis, after which the protoplasts fuse. One may ask how this hydrolysis is brought about. Is a cell wall's ability to lyse locally an intrinsic property? In other words, should the wall be looked upon as a sexual organelle, with sensing properties of its own, or is localized cell wall hydrolysis hormonally controlled? And, next, what signals cell fusion? In several organisms it is well established that the competence to fuse is an inducible property (a striking example is macrocyst formation in *Dictyostelium*, O'Day and Lewis, 1981). In *Saccharomyces*, Svoboda (1976) observed that artificially produced protoplasts of the *a* and α mating types did not produce zygotes on mixing. So, it seems that in this instance, for sexual fusion to occur, the presence of a cell wall is required. This seemingly paradoxical example is just to illustrate that we are only beginning to understand the chemical regulation of such a complex phenomenon as sexual reproduction in the fungi.

REFERENCES

Abe, K., Kusaka, J., and Fukui, S. (1975). Morphological change in the early stages of the mating process in *Rhodosporidium toruloides*. *J. Bacteriol.* *122*:710-718.

Adler, J. (1975). Chemotaxis in bacteria. In *Primitive Sensory and Communication Systems*, M. J. Carlile (Ed.). Academic, London, pp. 91-101.

Arsenault, G. P., Biemann, K., Barksdale, A. W., and McMorris, T. C. (1968). The structure of antheridiol, a sex hormone in *Achlya bisexualis*. *J. Am. Chem. Soc.* *90*:5635-5636.

Bandoni, R. J. (1965). Secondary control of conjugation in *Tremella mesenterica*. *Can. J. Bot.* *43*:627-630.

Barksdale, A. W. (1963). The role of hormone A during sexual conjugation in *Achlya ambisexualis*. *Mycologia 55*:627-632.
Barksdale, A. W. (1969). Interthallic sexual reactions in *Achlya*, a genus of the aquatic fungi. *Am. J. Bot. 47*:14-23.
Barksdale, A. W., McMorris, T.C., Seshadri, R., Arunachalam, T., Edwards, J. A., Sundeen, J., and Green, D. M. (1974). Response of *Achlya ambisexualis* E87 to the hormone antheridiol and certain other steroids. *J. Gen. Microbiol. 82*:295-299.
Bechtold, M. M. Welwicke, C. V., Cornai, K., and Gaylor, G. L. (1972). Investigation of the component reactions of oxidative sterol demethylation. *J. Biol. Chem. 247*:7650-7656.
Betz, R., Duntze, W., and Manney T. R. (1978). Mating-factor-mediated sexual agglutination in *Saccharomyces cerevisiae*. *FEMS Microbiol. Lett. 4*:107-110.
Blakeslee, A. F. (1904). Sexual reproduction in the Mucorineae. *Proc. Nat. Acad. Sci. USA 40*:205-319.
Bucking-Throm, E., Duntze, W., Hartwell, L. H., and Manney, T. R. (1973). Reversible arrest of haploid yeast cells at the initiation of DNA synthesis by a diffusible sex factor. *Exp. Cell Res. 76*:99-110.
Bu'Lock, J. D., Drake, D., and Winstanley, D. J. (1972). Specificity and transformations of the trisporic acid series of fungal sex hormones. *Phytochemistry 11*:2011-2018.
Bu'Lock, J. D., Jones, B. E., Taylor, D., Winskill, N., and Quarrie, S. A. (1974a). Sex hormones in the Mucorales. The incorporation of C20 and C18 precursors into trisporic acids. *J. Gen. Microbiol. 80*:301-306.
Bu'Lock, J. D., Winskill, N., and Jones, B. E. (1974b). Structures of the mating-type specific prohormones of *Mucorales*. *Chem. Commun.*:708-711.
Burgeff, H. (1924). Untersuchungen über Sexualität und Parasitismus bei Mucorineen. *Botan. Abh. 4*:1-35.
Byers, B., and Goetsch, L. (1975). Behavior of spindles and spindle plaques in the cell cycle and conjugation of *Saccharomyces cerevisiae*. *J. Bacteriol. 124*:511-523.
Calleja, G. B., Johnson, B. F., and Yoo, B. Y. (1981). The cell wall as sex organelle in fission yeast. In *Sexual Interactions in Eukaryotic Microbes*, D. H. O'Day and P. A. Horgen (Eds.). Academic, New York, pp. 225-261.
Carlile, M. J., and Machlis, L. (1965). The response of male gametes of *Allomyces* to the sexual hormone sirenin. *Am. J. Bot. 52*:478-483.
Chan, R. K. (1977). Recovery of *Saccharomyces cerevisiae* mating type *a* cells from G1 arrest by α factor. *J. Bacteriol. 130*:766-774.

Ciejek, E., and Thorner, J. (1979). Recovery of *S. cerevisiae a* cells from G1 arrest by α factor pheromone requires endopeptidase action. *Cell 18*:623-635.

Ciejeck, E., Thorner, J., and Geier, M. (1977). Solid phase peptide synthesis of α-factor, a yeast mating pheromone. *Biochem. Biophys. Res. Commun. 78*:952-961.

Edwards, J. A., Schwartz, V., Fajkos, J., Maddox, M. L., and Fried, J. H. (1971). Fungal sex hormones. The synthesis of (±)-7(*t*)-trisporic acid B methyl ester. The stereochemistry at C-9 of the trisporic acids. *Chem. Commun.* 292-293.

Edwards, J. A., Sundeen, J., Salmond, W., Iwadare, T., and Fried, J. H. (1972). A new synthetic route to the fungal sex hormone antheridiol and the determination of its absolute stereochemistry. *Tetrahedron Lett.*:791-794.

Ende, H. van den (1978). Sexual morphogenesis in the Phycomycetes. In *The Filamentous Fungi*, Vol. 3. J. E. Smith and D. R. Berry (Eds.). Arnold, London, pp. 257-275.

Fehrenbacher, G., Perry, K., and Thorner, J. (1978). Cell-cell recognition in *Saccharomyces cerevisiae*: Regulation of mating-specific adhesion. *J. Bacteriol. 134*:893-901.

Feinleib, M. L. (1978). Photomovement of microorganisms. *Photochem. Photobiol. 27*:849-854.

Finkelstein, D. B., and McAlister, L. (1981). α-Factor-mediated modification of a ^{32}P-labeled protein by MAT*a* cells of *Saccharomyces cerevisiae. J. Biol. Chem. 256*:2561-2566.

Finkelstein, D. B., and Strausberg, S. (1979). Metabolism of α-factor by *a* mating type cells of *Saccharomyces cerevisiae. J. Biol. Chem. 254*:796-803.

Fujino, M., and Kitada, C. (1980). Biological activity of synthetic analogs of tremellogen A-10. *Naturwissenschaften 67*:406-408.

Gooday, G. W. (1974). Fungal sex hormones. *Annu. Rev. Biochem. 43*:35-49.

Gooday, G. W. (1978). Functions of trisporic acid. *Philos. Trans. R. Soc. Lond. Ser. B. 284*:509-520.

Groner, B., Hynes, N., Sippel, A. E., and Schutz, G. (1976). Induction of specific proteins in hyphae of *Achlya ambisexualis* by the steroid hormone antheridiol. *Nature 261*:599-601.

Gwynne, D. J., and Brandhurst, B. P. (1980). Antheridiol induced differentiation of *Achlya* in the absence of detectable synthesis of new proteins. *Exp. Mycol. 4*:251-259.

Horgen, P. A. (1977). Cytosol-hormone stimulation of transcription in the aquatic fungus *Achlya ambisexualis. Biochem. Biophys. Res. Commun. 75*:1022-1028.

Isoe, S., Hayase, Y., and Sakan, T. (1971). Sexual hormones of the Mucorales. The synthesis of methyl trisporate B and C. *Tetrahedron Lett.*:3691-3694.

Jensen, E. V., and DeSombre, E. R. (1972). Mechanism of action of the female sex hormones. *Annu. Rev. Biochem. 41*: 203-230.

Kamiya, Y., Sakurai, A., Tamura, S., Takahashi, N., Abe, K, Tsuchiya, E., Fukui, S., Kitada, C., and Fujino, M. (1978). Structure of rhodotorucine A, a novel lipopeptide inducing mating tube formation on *Rhodosporidium toruloides*. *Biochem. Biophys. Res. Commun. 83*:1077-1083.

Kitada, C., Fujino, M., Kamiya, Y., Sakurai, A., Tamura, S., Takahashi, N., Tsuchiya, E., Abe, K., and Fukui, S. (1979). Synthesis of rhodotorucine A, the inducing factor of mating tube formation of *Rhodosporidium toruloides*. *Experimentia 35*:1275-1276.

Klapper, B. F., and Klapper, M. H. (1977). A natural inhibitor of sexual attraction in the water mold *Allomyces*. *Exp. Mycol. 1*:352-355.

Ko, W. H. (1980). Hormonal regulation of sexual reproduction in *Phytophthora*. *J. Gen. Microbiol. 116*:459-463.

Kochert, G. (1978). Sexual pheromones in algae and fungi. *Annu. Rev. Plant Physiol. 29*:461-486.

Liao, H., and Thorner, J. (1980). Yeast pheromone α factor inhibits adeylate cyclase. *Proc. Nat. Acad. Sci. USA 77*: 1898-1902.

Lipke, P. N., Taylor, A., and Ballou, C. E. (1976). Morphogenetic effects of α factor on *Saccharomyces cerevisiae a* cells. *J. Bacteriol. 127*:610-618.

Machlis, L. (1958). Evidence for a sexual hormone in *Allomyces*. *Physiol. Plant. 11*:181-192.

Machlis, L. (1969). Fertilization-induced chemotaxis in the zygotes of the water mould *Allomyces*. *Physiol. Plant. 22*:392-400.

Machlis, L. (1973). The chemotactic activity of various sirenins and analogues and the uptake of sirenin by the sperm of *Allomyces*. *Plant. Physiol. 52*:527-531.

Machlis, L., and Rawitscher-Kunkel, E. (1967). Mechanisms of gametic approach in plants. In *Fertilization*, Vol. 1. C. B. Metz and A. Monroy (Eds.). Academic, New York, pp. 117-161.

Machlis, L., Nutting, W. H., and Kapoport, H. (1968). The structure of sirenin. *J. Ac. Chem. Soc. 90*:1674-1676.

McMorris, T. C., and Barksdale, A. W. (1967). Isolation of a sex hormone from the water mould *Achlya bisexualis*. *Nature 215*: 320-321.

McMorris, T. C., and White, R. H. (1977). Biosynthesis of the oogoniols, steriodal sex hormones of *Achlya*: The role of fucosterol. *Phytochemistry 16*:359-362.

McMorris, T. C., Seshadri, R., Weihe, G. R., Arsenault, G. P., and Barksdale, A. W. (1975). Structures of oogoniols-1, -2, and -3, steriodal sex hormones of the water mould *Achlya*. *J. Am. Chem. Soc.* *97*:2544-2545.

Maness, P. F., and Edelman, G. M. (1978). Inactivation and chemical alteration of mating factor α by cells and spheroplasts of yeast. *Proc. Nat. Acad. Sci. USA.* *75*:1304-1308.

Manney, T. R., and Meade, J. H. (1977). Cell-cell interactions during mating in *Saccharomyces cerevisiae*. In *Microbial Interations*, J. L. Reissig (Ed.). Chapman and Hall, London, pp. 283-321.

Manney, T. R., Duntze, W., and Betz, R. (1981). The isolation characterization and physiological effects of the *Saccharomyces cerevisiae* sex pheromones. In *Sexual Interactions in Eukaryotic Microbes*, D. H. O'Day and P. A. Horgen (Eds.). Academic, New York, pp. 21-53.

Masui, Y., Chino, N., Sakaibara, S., Tanaka, T., Murakami, T., and Kita, H. (1977). Synthesis of the mating factor of *Saccharomyces cerevisiae* and its truncated peptides. The structure-activity relationship. *Biochem. Biophys. Res. Commun.* *78*:534-538.

Mesland, D. A. M., Huisman, J. G., and Ende, H. van den (1974). Volatile sexual hormones in *Mucor mucedo*. *J. Gen. Microbiol.* *80*:111-117.

Michalski, C. J. (1978). Protein synthesis during hormone stimulation in the aquatic fungus *Achlya*. *Biochem. Biophys. Res. Comm.* *84*:417-427.

Miyakawa, T., Nishikara, M., Tsuchiya, E., and Fukui, S. (1982). Role of metabolism of the mating pheromone in sexual differentiation of the Heterobasidiomycete *Rhodosporidium toruloides*. *J. Bacteriol.* *151*:1184-1194.

Musgrave, A., and Nieuwenhuis, D. (1975). Metabolism of radioactive antheridiol by *Achlya* species. *Arch. Microbiol.* *105*: 313-317.

Nieuwenhuis, M., and Ende, H. van den (1975). Sex specificity of hormone synthesis in *Mucor mucedo*. *Arch. Microbiol.* *102*: 167-169.

Nolan, R. A., and Bal, A. K. (1974). Cellulase localization in hyphae of *Achlya ambisexualis*. *J. Bacteriol.* *117*: 840-843.

Nutting, W. H., Rapoport, H., and Machlis, L. (1968). The structure of sirenin. *J. Am. Chem. Soc.* *90*:6434-6438.

O'Day, D. H., and Lewis, K. E. (1981). Pheromonal interactions during mating in *Dictyostelium*. In *Sexual Interactions in Eukaryotic Microbes*, D. H. O'Day and P. A. Horgen (Eds.). Academic, New York, pp. 199-225.

Plattner, J. J., and Rapoport, H. (1971). The synthesis of D- and L-sirenin and their absolute configurations. *J. Am. Chem. Soc.* *93*:1758-1761.

Pommerville, J. (1978). Analysis of gamete and zygote motility in *Allomyces*. *Exp. Cell Res.* *113*:161-172.
Pommerville, J. (1981). The role of sexual pheromones in *Allomyces*. In *Sexual Interactions in Eukaryotic Microbes*, D. H. O'Day and P. A. Horgen (Eds.). Academic, New York, pp. 53-72.
Pommerville, J., and Fuller, M. S. (1976). The cytology of the gametes and fertilization of *Allomyces macrogynus*. *Arch. Microbiol.* *109*:21-30.
Popplestone, C. R., and Unrau, A. M. (1974). Studies on the biosynthesis of antheridiol. *Can. J. Chem.* *52*:462-468.
Preus, M. W., and McMorris, T. C. (1979). The configuration at C-24 in oogoniol and identification of 24(28-)-dehydro-oogoniols as hormones in *Achlya*. *J. Am. Chem. Soc.* *101*:3066-3071.
Prisbylla, M. P., Takabe, K., and White, J. D. (1979). Stereospecific synthesis of (±)-trisporol B., a prohormone of *Blakeslea trispora* and a facile synthesis of (±)-trisporic acids. *J. Am. Chem. Soc.* *101*:762-763.
Raper, J. R. (1950). Sexual hormones in *Achlya*. VII. The hormonal mechanism in homothallic species. *Bot. Gaz. Chicago* *112*:1-24.
Raper, J. R. (1952). Chemical regulation of sexual processes in the thallophytes. *Bot. Rev.* *18*:447-545.
Reid, J. D. (1974). Properties of conjugation hormones (erogens) from the Basidiomycete *Tremella mesenterica*. *Can. J. Bot.* *52*:521-524.
Ronne, M., and Olson, L. W. (1976). Isolation of male strains of the aquatic phycomycete *Allomyces macrogynus*. *Hereditas* *83*:191-202.
Rozek, C. E. and Timberlake, W. E. (1980). Absence of evidence for changes in messenger RNA populations during steroid-hormone-induced cell differentiation in *Achlya*. *Exp. Mycol.* *4*:34-47.
Sakagami, Y., Isogai, A., Suzuki, A., Tamura, S., Kitada, C., and Fujino, M. (1979). Structure of tremellogen A-10, a peptidal hormone inducing conjugation tube formation in *Tremella mesenterica*. *Agric. Biol. Chem.* *43*:2643-2645.
Schekman, R., and Brawley, V. (1979). Localized deposition of chitin on the yeast cell surface in response to mating pheromone. *Proc. Nat. Acad. Sci. USA* *76*:645-649.
Schipper, M. A. A. (1978). On certain species of *Mucor* with a key to all accepted species. Studies in Mycology (Centraal bureau voor Schimmelcultures, Baarn, Holland) *17*:1-52.
Schipper, M. A. A., and Stalpers, J. A. (1980). Various aspects of the mating system in *Mucorales*. *Persoonia* *11*:53-63.

Sena, E. P., Radin, D. N., and Fogel, S. (1973). Synchronous mating in yeast. *Proc. Nat. Acad. Sci. USA 70*:1373-1377.

Silver, J. C., and Horgen, P. A. (1974). Hormonal regulation of presumptive mRNA in the fungus *Achlya ambisexualis*. *Nature 249*:252-254.

Springer, M. S., Gay, M. F., and Adler, J. (1979). Protein methylation in behavioral control mechanisms and in signal transduction. *Nature 280*:279-284.

Stewart, P. R., and Rogers, P. J. (1978). Fungal dimorphism: A particular expression of cell wall morphogenesis. In *The Filamentous Fungi*, Vol. 3. J. Smith and D. R. Berry (Eds.). E. Arnold, London, pp. 164-197.

Stötzler, D., Kiltz, H., and Duntze, W. (1976). Primary structure of α-factor peptides from *Saccharomyces cerevisiae*. *Eur. J. Biochem. 69*:397-400.

Sutherland, R. B., and Horgen, P. A. (1977). Effects of the steroid sex hormone, antheridiol, on the initiation of RNA synthesis in the simple eukaryote, *Achlya ambisexualis*. *J. Biol. Chem. 252*:8812-8820.

Sutter, R. P., and Whitaker, J. P. (1981a). Zygophore-stimulating precursors (pheromones) of trisporic acids active in (-)-*Phycomyces blakesleeanus*. *J. Bol. Chem. 256*:2334-2341.

Sutter, R. P., and Whitaker, J. P. (1981b). Sex pheromone metabolism in *Blakeslea trispora*. *Naturwissenschaften 68*:147-148.

Sutter, R. P., Capage, D. A., Harrison, T. L., and Keen, W. A. (1973). Trisporic acid biosynthesis in separate *plus* and *minus* cultures of *Blakeslea trispora*: Identification by *Mucor* assay of two mating-type-specific components. *J. Bacteriol. 114*:1074-1082.

Svoboda, A. (1976). Mating reaction in yeast protoplasts. *Arch. Microbiol. 110*:313-318.

Throm, E., and Duntze, W. (1970). Mating type-dependent inhibition of deoxyribonucleic acid synthesis in *saccharomyces cerevisiae*. *J. Bacteriol. 104*:1388-1390.

Timberlake, W. (1976). Alterations in RNA and protein synthesis associated with steroid hormone-induced sexual morphogenesis in the water mould *Achlya*. *Dev. Biol. 51*:202-214.

Tkacz, J. S., and MacKay, V. L. (1979). Sexual conjugation in yeast. *J. Cell Biol. 80*:326-333.

Tsuchiya, E., and Fukui, S. (1978). Binding of rhodotorucine A, a lipopeptidyl mating hormone, *to a cells* of *Rhodosporidium toruloides* for induction of sexual differentiation. *Biochem. Biophys. Res. Commun. 85*:473-479.

Tsuchiya, E., Abe, K., and Fukui, S. (1978). Sequence of events in early stages of sexual differentiation caused by rhodotorucin A, a fungal sex hormone: Biological determination using mating-less mutants derived from *Rhodosporidium toruloides*. *J. Gen. Appl. Microbiol. 24*:291-294.

Udden, M. M., and Finkelstein, D. B. (1978). Reaction order of *Saccharomyces cerevisiae* α-factor-mediated cell arrest and mating inhibition. *J. Bacteriol. 133*:1501-1507.

Werkman, B. A. (1976). Localization and partial characterization of a sex-specific enzyme in homothallic and heterothallic *Mucarales*. *Arch. Microbiol. 109*:209-213.

Werkman, B. A., and Ende, H. van den (1974). Trisporic acid synthesis in homothallic and heterothallic *Mucorales*. *J. Gen. Microbiol. 82*:273-278.

Woo, S. L. C., and O'Malley, B. W. O. (1975). Hormone inducible messenger RNA. *Life Sci. 17*:1039-1047.

Yanagishima, N., and Yoshida, K. (1981). Sexual interactions in *Saccharomyces cerevisiae* with special reference to the regulation of sexual agglutinability. In *Sexual Interactions in Eukaryotic Microbes* D. H. O'Day and P. A. Horgen (Eds.). Academic, New York, pp. 261-295.

Yen, P. H., and Ballou, C. E. (1974). Partial characterization of the sexual agglutination factor from *Hansenula wingei* Y-2340 type 5 cells. *Biochemistry 13*:2428-2437.

Zantinge, B., Dons, H., and Wessels, J. G. H. (1979). Comparison of poly(A)-containing RNAs in different cell types of the lower eukaryote *Schizophyllum commune*. *Eur. J. Biochem 101*: 251-260.

16

Differentiation in the Basidiomycetes

GÈRARD MANACHÈRE, JEAN-CLAUDE ROBERT, ROGER DURAND, JEAN-PAUL BRET, and MICHEL FÈVRE *Université Lyon 1, Villeurbanne, France*

I. INTRODUCTION

Theoretically, two phases can be distinguished in the formation of basidiomycete carpophores: an initial phase of fruiting release from a vegetative mycelium (initiation phase) followed by a morphogenetic phase. It is often difficult to establish the relative duration of the two phases, since they are joined to each other and progression to the second phase depends on the normal completion of the first. Nevertheless, there can be discrimination in certain cases. It is clear that, from both physiological or biotechnological aspects, it is necessary to control the development of both phases, the first being dependent on a previously suitable vegetative growth. Even when extensive mycelia are produced which are theoretically genetically capable of fruiting, many mycelia are still unable to produce carpophores. The fundamental problem to be solved is basically, How does one induce a thallus to the fruiting stage by avoiding irreversible engagement in an exclusively vegetative mode of growth? Various exogeneous and endogeneous factors are involved, the roles of which have been recently reviewed in physiological and biotechnological papers (Manachère 1977a, 1978, 1980; Moore et al., 1979; Chang and Hayes, 1978). Essentially, we will give some data here on the more important recent studies.

Many studies have been carried out on nonedible species, for example, *Schizophyllum commune* and various *Coprinus* sp., particularly *Coprinus congregatus*, which has been extensively studied in the authors' laboratory. We consider *C. congregatus* as a "model" for most studies relative to the main physiological problems of differentiation in the Basidiomycetes (Fig. 1). The choice of nonedible species as experimental systems can be explained in part on a practical basis. On the one hand, it is frequently difficult, and even impossible, to control fruiting in edible species, at least in vitro, while on the other hand, the carpophores of nonedible controllable species that have been well studied are generally of a small size and easy to cultivate in the laboratory.

II. GENERAL ASPECTS OF FRUIT BODY DEVELOPMENT

A. Morphological and Cellular Aspects of Stipe Elongation

There is a general agreement that in various Basidiomycetes the upper portion of the stipe expands to a much greater extent than the base (Bonner et al., 1956; Eilers, 1974; Cox and Niederpruem, 1975). However, in *Coprinus lagopus*, Cox and Niederpruem (1975) indicated that the most active zone of expansion was localized in the upper midregion of the stipe rather than at the top of the stipe.

Some reports on stipe elongation in *Agaricus* (Bonner et al., 1956; Hagimoto, 1964) claimed that after the fruit body is 18 mm tall, no cell division occurs in the stipe and that cell elongation alone is responsible for the increase in stipe length. By comparing increases in cell length with stipe elongation in both the wild-type and elongationless mutant *Coprinus macrorhizus*, Kamada and Takemaru (1977) concluded that stipe elongation is virtually entirely due to cell elongation.

Rapid elongation of the stipe of *Coprinus radiatus* (Eilers, 1974) and *Flammulina velutipes* (Wong and Gruen, 1977) is due to both cell

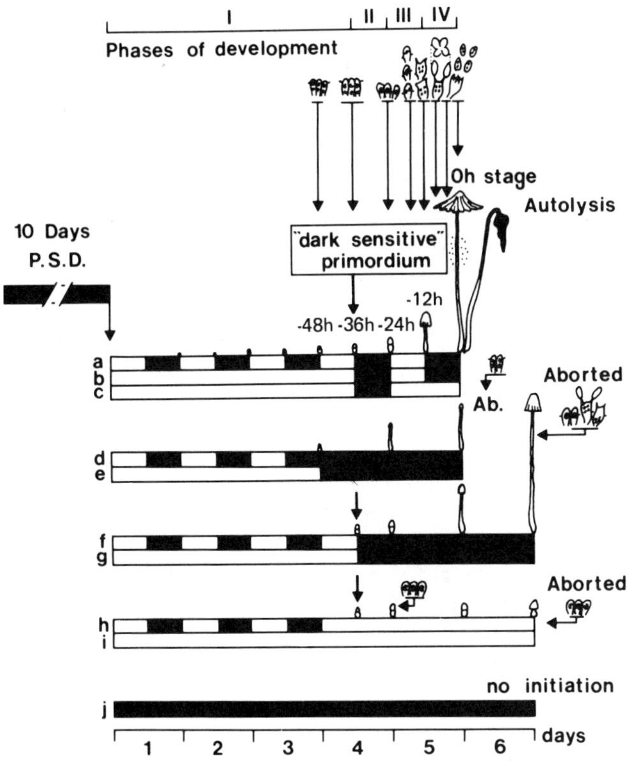

FIG. 1 Main phases of fruit body development of *Coprinus congregatus* and meiotic events as related to the photoperiodic light regime at 25°C. (Redrawn from Manachère, 1970.)

elongation and cell division. Craig et al. (1977) have reinvestigated the cellular basis for stipe elongation in *Agaricus bisporus* and have disputed the previous results obtained for this species. By measuring cell lengths at different stages of development in all regions of the stipe in *A. bisporus*, they have shown that elongations of the existing cells account for the extension of the lower and mid regions, but cell elongation and division must also occur in the upper stipe region. Using autoradiography, Craig et al. have shown that there is a nonpolar addition of new wall material in the elongating cells of the stipe and, particularly in the upper stipe, cell division can be seen to occur by the formation of new septa.

B. Role of the Mycelium and Cap on Stipe Elongation

Gruen and Wu (1972a) found that the fresh and dry weights of stipes of *F. velutipes* fruit bodies increased during the whole elongation phase and that fruit bodies grew very little if they were removed at different lengths and laid on the mycelium. These and other previous results indicated that stipe elongation requires a continuous supply of water and nutrients from the vegetative mycelium during most of the growth period and that this dependence decreases with age. The identity of the translocated nutrients was not determined, but the elongation of isolated whole fruit bodies was promoted by glucose and other low-molecular-weight carbohydrates (Gruen and Wu, 1972b). Similar results were obtained by Bret (1977a) in *C. congregatus*, where stipe elongation also depended on the presence of the vegetative mycelium at the base of the fruit body, but in *Coprinus cinereus*, the mycelium was not required during the elongation phase (Gooday, 1974). The cap, or pileus, plays an essential role in the determination of stipe elongation. In *A. bisporus* and *F. velutipes* stipe elongation depends on the pileus during most of the growth period, and growth-controlling activity originates in the lamellae (Gruen, 1963, 1969). In *C. congregatus*, the presence of the cap is necessary for complete stipe elongation until a late stage in developing fruit bodies (Bret, 1977a).

However, in *C. lagopus* (Cox and Niederpruem, 1975) and *C. cinereus* (Gooday, 1974), stipe elongation, at least in its final realization, is independent of the presence of the cap. In *Coprinus radiatus* the cap appears to be required for normal stipe development until the stipe reaches one-quarter of its final length (Eilers, 1974).

Little is known about the mechanism of action of the lamellae on stipe elongation. *Agaricus bisporus* lamellae produce a diffusate in agar which causes stipe curvature indicative of growth promotion (Urayama, 1956; Hagimoto and Konishi, 1959, 1960; Hagimoto, 1963; Gruen, 1967). Eilers (1974) reported that stipe elongation in *C. radiatus* was strongly promoted when the pileus was placed directly onto young decapitated stipes. Excised lamellae of *F. velutipes* release a diffusate in agar which enhances the elongation of decapitated stipes (Gruen, 1976). Production of this diffusate is stimulated by glucose. Tests with diffusates from young and old lamellae on stipes of different ages showed that the declining influence of lamellae in aging fruit bodies is associated both with decreasing production of the growth-promoting factor and the decreasing response of stipes to an external supply of active diffusate (Gruen, 1979). In *C. congregatus*, removal of half the cap resulted in negative curvature of the stipe when the operation was performed more than 12 h before the end of the fruit body development (Bret, 1977b). All these results imply the production of a growth-promoting factor in the cap. However, as yet there has been no successful biochemical characterization of the factor(s).

Graftings made between young basidiocarps of *Coprinus macrorhizus*, grown under different light programs, showed that a diffusible factor(s) affecting stipe elongation, basidiospore formation, and cap

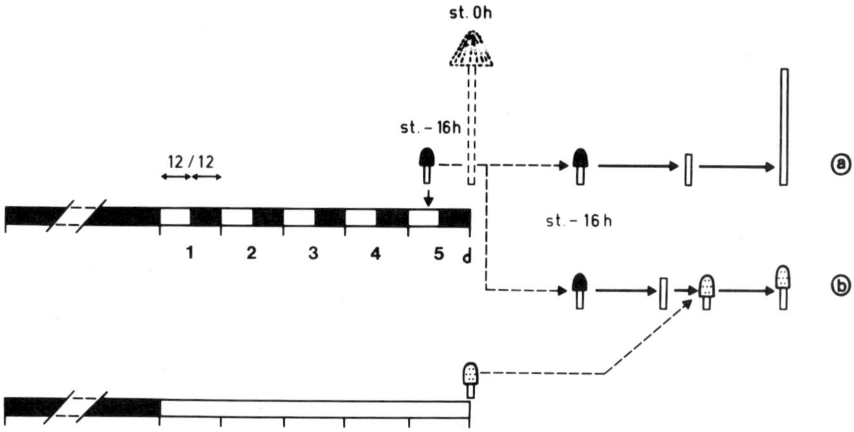

FIG. 2 Inhibition of stipe elongation of *Coprinus congregatus* by deposition of a cap from an elongationless fruit body obtained in continuous light. (After Bret in Manachère, 1978.)

autolysis was induced in the cap by the darkness phase (phase 3) during normal development of fruit bodies (Kamada and Tsuji, 1979).

Experiments consisting of joining caps and stipes of different ages allowed an evaluation of the timing and level of growth substance production in *C. congregatus*. A stimulating effect of the cap was detectable up to the time of autolysis, but maximum stimulation was only obtained when the deposited cap came from a fruit body that was 16 h before the end of its development (Bret, 1977b). In the same species, it was possible to inhibit stipe elongation using caps derived from "elongationless" fruit bodies obtained in continuous light (Fig. 2; Bret, 1978).

C. Biochemical Aspects of Stipe Elongation

The importance of rapid cell wall synthesis is obvious during carpophore development. Cell wall chitin content and chitin synthetase activity increase during stipe elongation of *C. cinereus* (Gooday, 1979). Polyoxin D, which specifically inhibits chitin synthetase activity, prevents stipe elongation in *C. cinereus* (Gooday et al;, 1976) and *A. bisporus* (Wood and Hammond, 1977).

Autoradiographic studies have shown that cell growth of expanding stipes of *C. cinereus* (Gooday, 1975) or *A. bisporus* (Craig et al., 1979) occurred by a uniform incorporation of cell wall material along the length of the cells. This intercalary elongation contrasts with the polarized tip growth of mycelial hyphae. Electron microscopic autoradiography has also located the site of chitin synthesis on the cell wall/plasma membrane region of the hyphae (Gooday, 1979; Craig et al.,

1979). No label was observed in the cytoplasm, thus excluding intracellular sites of chitin polymerization.

The chitin-synthesizing system of *C. cinereus* has been well studied by Gooday and colleagues (see Gooday, 1979). Chitin synthetase is bound to the cytoplasmic membrane. In contrast to enzymes obtained from other microorganisms, this enzyme does not exhibit zymogen activity (Cabib et al., 1979). Chitin synthetase is activated by its substrate UDP-N-acetylglucosamine and requires Mg^{2+} for activity (De Rousset-Hall and Gooday, 1975; Gooday and De Rousset-Hall, 1975). Chitin synthesis is powerfully inhibited by UDP, the second product of the enzymatic reaction. Chitin synthetase preparations exhibit membrane-bound UDPase activity capable of breaking down UDP to the much less inhibitory UMP. Chitin synthesis can be regulated by an associated uridine nucleotide cycle involving the conversion of USP to UMP in the membranes and phosphorylation of UMP to UDP Glc NAc in the cytoplasm via UDP and UTP (Gooday, 1979).

D. Biochemical Aspects of Cap Maturation

Cell wall hydrolytic enzymes—chitinase, β-1-3 glucanases, and lichenase—are produced by *Coprinus* basidiocarps (Bush, 1974; Schaefer et al., 1977; Miyake et al., 1980a). These enzymes are localized mainly in the caps, while stipes and mycelia exhibit low activities. Their synthesis occurs in the late stage of fruit body development, maximum activity being obtained when pilei start to lyse. This enzyme production correlates with fruit body morphogenesis and is independent of basidiospore formation, as sporeless mutants show the same pattern of enzyme production (Miyake et al., 1980a).

III. ROLE OF LIGHT AND TEMPERATURE

In recent years there has been little advance in understanding the influence of humidity, carbon dioxide, and nutritional factors on carpophore development, and this aspect will not be considered here. Recently, Horrière (1979) made a comparative study of the trophic requirements of some higher Basidiomycetes on synthetic media.

Thus it appears preferable to emphasize the roles of light and temperature, the two factors considered separately or interacting with each other. Of course, one must always remember that morphogenesis in Basidiomycetes is controlled by the interaction of many environmental factors (Manachère, 1980).

A. Effects of White Light

Light is required at different stages of basidiocarp production in numerous species (Durand, 1976; Tan, 1978). To date there have been no reports of normal fruiting on axenic cultures in darkness.

Generally, basidiocarp initials are successfully produced in darkness, but light is required for normal development of the primordia, for example, in *Agaricus arvensis* and *Agaricus sylvicola* (Couvy, 1974), *Panus tigrinus* (Bobbitt and Crang, 1974), and *Collybia velutipes* (Kinugawa, 1977).

Some species require light for the whole period of development. In darkness only vegetative growth occurs. Primordia of *Pleurotus ostreatus* were only formed in illuminated cultures (Eger et al., 1976). Light was required for fruit body primordium formation and initiation of pileus formation in *Favolus arcularius* (Horikoshi et al., 1974) and for the initiation and development of fertile sporophores in *Laccaria laccata* (Davis and Jong, 1976). Light was an essential factor for fruiting in *Phellinus weirii* (Li, 1979) and in the white rot species *Phanerochaete chrysosporium* (Gold and Cheng, 1979).

In many *Coprinus* species, light can act as an inducer/stimulator and as an inhibitor. In *Coprinus domesticus*, light-dark cycles induced a more rapid development of mature basidiocarps than did exposure to continuous light (Chapman and Fergus, 1973). *Coprinus lagopus* was unable to initiate meiosis at an elevated temperature (35°C) under a continuous light regime (Lu, 1972). In *C. macrorhizus*, the maturation phase was inhibited by light; continuous darkness for more than 2.5 h was required (Kamada et al., 1978). *Coprinus congregatus* has been the subject of intensive physiological studies by Manachère and co-workers (Manachère, 1970; Manachère et al., 1974; Durand, 1975; Bret, 1977a; Horrière, 1977; Robert, 1977a,b). Fruiting required successive light and dark exposures (Fig. 1). Light was needed for basidiocarp initiation, but basidiocarp maturation did not occur under continuous light at 25°C. A minimum dark requirement of 5 h suppressed photoinhibition (Robert and Durand, 1979). Darkness allowed the young basidia to enter and complete meiosis (Manachère, 1970). Following darkness, light was required to achieve normal morphogenesis.

B. Action Spectra

There have been few exact studies on the most effective wavelengths, but there is considerable evidence that many of the light-induced responses are sensitive to the blue end of the spectrum (Durand, 1976; Tan, 1978; Eger-Hummel, 1980).

Kitamoto et al. (1972) determined the action spectrum for light-induced primordium formation in *Favolus arcularius*; photoinduction was observed between 350 and 560 nm. The action spectrum for photoinduction of pileus formation in the same species showed detectable quantum effectiveness from the near-ultraviolet to the blue region of the spectrum; wavelengths above 510 nm were ineffective (Kitamoto et al., 1974). The formation of basidiocarp initials in *Psilocybe cu-*

bensis occurred only for ultraviolet and blue wavelengths, with maxima at 370, 440, and 460 nm (Badham, 1980).

In *C. congregatus*, action spectra for wavelengths between 405 and 730 nm were determined at three different isoquantal levels (2.10^{-1} to $1.3\ 10^{-3}$ nE cm^2/s). The isoquantal spectra for fruiting initiation, the photoinhibition of dark-sensitive primordia, and sporophore maturation were similar. The most effective wavelengths were in the blue part of the spectrum, at about 440-460 nm. At high irradiance levels, longer wavelengths became effective. As the same radiation could alternatively prevent the development of dark-sensitive primordia and stimulate its light-dependent phases, it has been suggested that the same photoreceptor system is effective throughout fruit body development (Durand, 1975,1976).

Recently, a photoresponse mediated by blue light has been demonstrated in *C. congregatus* (R. Durand, unpublished results). A blue light break given during a 6-h inductive night strongly inhibited fruiting; maximum inhibition occurred 3.8 h after the beginning of darkness. A well-defined action spectrum was determined for this photoperiodic effect (Fig. 3). Quantum values for a standard response (50% inhibition) were intrapolated from the quantum flux response curves. Blue light only (465 nm) appeared to be effective for light break inhibition of dark-sensitive primordia. The activity of wavelengths longer than 510 nm was due to the very small amount of scattered blue light present at high irradiance levels. This was clearly demonstrated by experiments with laser beam light, where wavelengths longer than 510 nm were ineffective.

The effect of light on *C. congregatus* is a typical "blue light response." In recent years a great variety of blue light-induced phenomena in plants, microorganisms, and animals has been studied. The similarity of the action spectra for these various responses has led to the supposition that a common photoreceptor pigment mediates the action of light in these diverse photobiological systems. The nature of the blue light photoreceptor is still an open question; nevertheless many arguments have been used to support the flavin hypothesis (Briggs, 1976; Senger, 1980).

C. Light and Temperature Interactions

In culture it can readily be demonstrated that the effect of a particular external factor on reproduction may be modified by changes in other factors. Since the pioneer work of Plunkett (1956) on *Collybia velutipes* and *Polyporus brumalis*, these interactions have received little attention. In *C. velutipes*, minimal aeration permitted primordial initiation and stipe elongation, but cap expansion was prevented in spite of the presence of light (Kinugawa, 1977). Interaction between light and the quality of the nutrient medium has been demonstrated by Tsusué (1969) and Kamada et al. (1978) in the initiation phase of *C.*

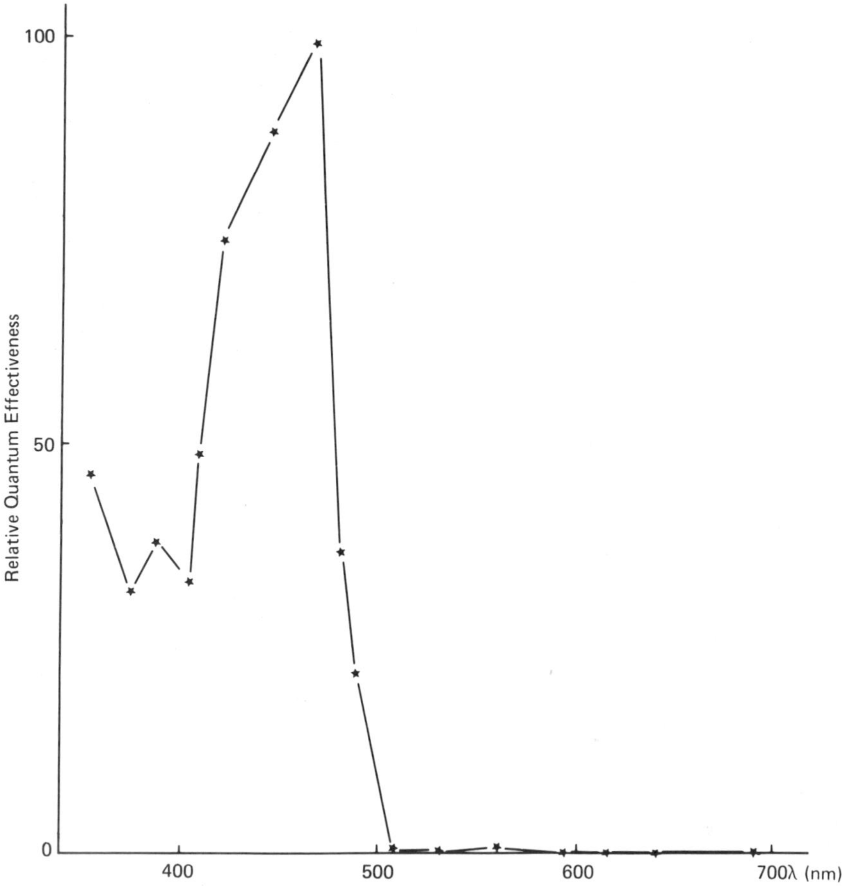

FIG. 3 Action spectrum for light break inhibition of dark-sensitive primordia of *Coprinus congregatus*. (R. Durand, unpublished results.)

macrorhizus. In *C. congregatus* the interacting effects of light and temperature have been studied. The dark requirement of the primordia could not be regarded as an absolute necessity until the temperature exceeded 17.5°C (Fig. 4). Normal development of the primordia in continuous light was obtained by lowering the basal temperature of the culture from 25 to 10°C for 6 h (Robert and Durand, 1979). Lu (1972) reported that for *C. lagopus*, meiosis was strictly dark dependent at 35°C but could proceed in continuous light at 25°C. Thus it appears that the dark requirement triggering meiosis and then basidiocarp maturation in several *Coprinus* species could be temperature de-

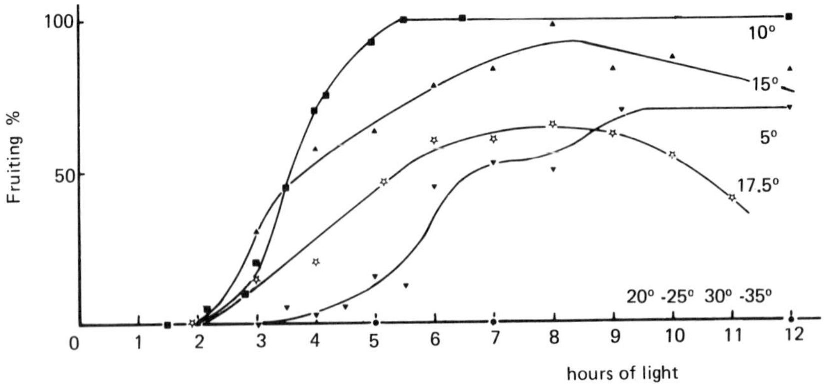

FIG. 4. Light and temperature interactions during fruiting in *Coprinus congregatus*, the effect of different temperature treatments in continuous light in place of the inductive night at the "dark-sensitive" stage. (After Robert and Durand, 1979.)

pendent. However, work is still necessary with each *Coprinus* species to confirm the similarities actually considered.

In *C. congregatus*, the interacting effects of light and temperature are dependent on the actual level of irradiance provided to the cultures. At low levels of irradiance, a slight decrease in temperature was sufficient to suppress the photoinhibition of the primordia.

IV. METABOLIC ASPECTS OF THE INITIATION PHASE

Although the results are fragmentary, there is now good evidence that passage from the vegetative phase of mycelial growth to the initiation phase of primordium formation involves important metabolic changes. These changes appear to be related to variations in the amounts of metabolites, particularly carbohydrates, in the medium and in the mycelium.

A. Changes in the Culture Medium and Fruiting Induction

Fruiting initiation of *C. congregatus* occurs when medium glucose has been almost totally removed. Fruiting is delayed if the old medium is replaced by a new one, containing glucose, before or at the moment initiation would normally occur (Robert, 1977b). Fruiting is also derepressed when glucose is partially replaced by more complex carbohydrates incorporated in the medium (Robert, 1979).

Similarly, the formation of primordia or aggregates of *A. bisporus* and *Agaricus edulis* may be regulated by a process similar to catabolite repression (Wood, 1979b). Nevertheless, developing fruit bodies of *C. velutipes* derive a part of their substrate from the glucose remaining in the medium (Kitamoto and Gruen, 1976) and fruit body initiation in *Favolus arcularius* occurs when more than half of the carbon source, which is mainly maltose, still remains in the culture filtrate (Kitamoto et al., 1978).

B. Phenoloxidases and Fruiting Induction

The increase in tyrosinase activity, parallel to the advancement of fructification, has been mentioned for diverse fungi other than Basidiomycetes.

In *A. bisporus* extracellular laccase increase in the compost (fermented straw) during the colonization phase and disappears upon development of the first flush of sporophores. An inverse relationship between the activity of cellulase (carboxymethyl cellulase) and laccase has been observed as fruiting progresses in axenic cultures, excluding the involvement of microorganisms from the compost (Wood and Goodenough, 1977). Laccase activity has been shown to be an obligatory requirement for the degradation and subsequent metabolism of lignin by basidiomycete fungi (Ander and Eriksson, 1976). The biological role of phenoloxidases has been reviewed and discussed by Molitoris (1976,1979). Increased extracellular cellulase activity at the time of fruiting may be required for the construction of a large sporophore, with the enzyme supplying the mycelium with high levels of mono- and disaccharides (Wood and Goodenough, 1977; Wood, 1979a).

In *S. commune* all tested strains that produce fruiting bodies exhibited phenoloxidase (PO) activity. Specific extracellular enzyme activity increased with the appearance of hyphal aggregates and clearly correlated with the formation of stages I–III, independently of the culture conditions, attaining a peak activity just prior to the first release of spores. Subsequent to sporulation, during the expansion stages IV-V of fruit body development, PO activity declines to barely detectable levels owing to the formation or activation of an inhibitor (Phillips and Leonard, 1976). In contrast, no evidence was found with *A. bisporus* to show that regulation of the activity could be accounted for by the production of diffusible inhibitors from the mycelium (Wood, 1980).

In vivo, substrates for PO in fungi are not known, but eight isoenzymes are activated with the initiation of fruiting in *S. commune*. Their specific contribution to the total PO activity was found to vary, with certain isoenzymes exhibiting higher peaks of activity (Phillips and Leonard, 1977). This activity is necessary but not sufficient for fruit body formation, as monokaryons, like dikaryons, may or may not exhibit PO activity if they are sterile (Leslie and Leonard, 1979).

Similarly, tyrosinase activity was produced in *C. macrorhizus* even when fruiting was inhibited in the dark or by 2-deoxyglucose (Nyunoya and Ishikawa, 1979,1980a). It was suggested that the production of tyrosinase and melanin formation may be regulated independently of fruit body formation. Induction of the enzyme appears to involve de novo protein synthesis (inhibition by cycloheximide) and is subject to catabolite repression (Nyunoya and Ishikawa, 1980a).

C. "Glucose Effect" on Fruiting and Fruiting-Inducing Substances

In *C. macrorhizus* glucose represses fruiting of both mono- and dikaryotic mycelia and acts by controlling the synthesis of inducing enzymes of cAMP metabolism. Positive correlation has been established between the level of cAMP and fruiting. Adenyl cyclase, which catalyzes the formation of cAMP from ATP, is more influenced by glucose (in vivo but not in vitro) than is phosphodiesterase, which degrades cAMP to AMP, and a cAMP binding activity that is required is also affected by glucose (Uno and Ishikawa, 1974; for a review see Ishikawa and Uno, 1977).

The highest level of adenyl cyclase has been detected in young mycelia just prior to cell aggregation, and enzyme activity decreased at the stage of fruit body maturation. The observed location of the enzyme in the particulate fractions containing plasma membranes, microsomes, and mitochondria did not change during fruit body formation under various light conditions and may be related to a possible location of a photoreceptor, since light stimulated production (Uno and Ishikawa,1975).

It has been reported (Uno and Ishikawa, 1976) that the glycogen phosphorylase of *C. macrorhizus* is activated in the presence of cAMP, the regulation of this enzyme thus resembling that of mammalian cells. In contrast, glycogen synthetase in this fungus is inhibited by the presence of cAMP and this inhibition is reversed by glucose-6-phosphate (G6P). The manner in which cAMP stimulates conversion of the enzyme remains to be elucidated (Uno and Ishikawa, 1978). The results led the authors to suggest that cAMP may regulate the amount of glycogen in mycelia by inhibition of glycogen synthetase and activation of glycogen phosphorylase.

In the light of these studies, the fruiting-inducing substances (FIS) in *C. macrorhizus* were considered to be cAMP and a cAMP-binding protein. However, Schwalb (1974b) with *S. commune*, Wood (1976,1979b) with *A. bisporus*, and von Netzer (1977) with *P. ostratus* could not find any evidence that cAMP would act as a fruiting inducer. However, Schwalb (1978a,b) discovered that cAMP could profoundly affect the morphology of developing fruit bodies; cAMP levels are elevated in *bse* (bug's ear) mutant strains and the effects of exogenous cAMP on wild strains mimicked the *bse* mutation at differing gene dosages.

A FIS analog was isolated from *A. bisporus* sporophores which induced haploid fruit body formation in *S. commune* (Rusmin and Leonard, 1975,1978): It does not appear to be cAMP, but it may be a peptide.

Some of these observations are evidently comparable to various others recently reviewed (Manachère, 1980) showing the "stimulating fruiting power" of carpophore fragments placed on a vegetative mycelium (a stimulating substance would not be specific) or on nutrient media ("regeneration power" is variable according to the maturation stage of the sporophore used as inoculum).

There are also examples of microorganisms stimulating the fruiting of higher fungi. Salas and Hancock (1972) pointed out that *Penicillium oxalicum* brings about a spectacular increase of basidiocarp production in *Mycena citricolor* when the two species are cocultivated. One (or more) basidiocarp-stimulating substance (BSS) was involved.

The positive effect of microorganisms on the release of fruiting could occur in two ways, as was hypothesized for *A. bisporus* through the release of fruiting-stimulating metabolites in the medium or via the absorption of metabolites produced by the *Agaricus* mycelium which could prevent fruiting. Wood (1976) considered the second hypothesis to be the most plausible, as fruiting was not stimulated by culture filtrates or suspensions of *Pseudomonas putida* (or other bacteria generally present). Moreover, fruiting in axenic cultures is facilitated by the presence of activated charcoal (Eger, 1961; Couvy, 1974; Long and and Jacobs, 1974).

V. METABOLIC ASPECTS OF THE MORPHOGENETIC PHASE

Once the initiation phase is accomplished, large fruit bodies develop at the expense of carbohydrates remaining in the medium or cellular constituents being stored in the mycelium and small fruit bodies. Indeed, the progression of basidiocarp morphogenesis depends on complex interactions between, for example, the medium and the vegetative mycelium, the vegetative mycelium and primordia, the developing sporophores and aborting primordia, and between the stipe and the cap of the sporophores.

A. Identity and Mobilization of Reserves

1. Soluble Carbohydrates and Glycogen

In *C. congregatus*, cultured in liquid media, a large net loss in dry matter occurring during fruit body development was due mainly to a loss in carbohydrates accumulated during vegetative growth (Fig. 5). A hot-water-soluble polysaccharide was mainly used during the late fruit body development (extension-expansion phase). The net loss was probably caused by respiration, but there may also have been some excretion of metabolites into the medium as the pH dropped simultaneously with fruit body maturation (Robert, 1977b,1979). The light-initiated primordia of *C. congregatus* require a dark period followed by a light period for maturation. Respectively, these treat-

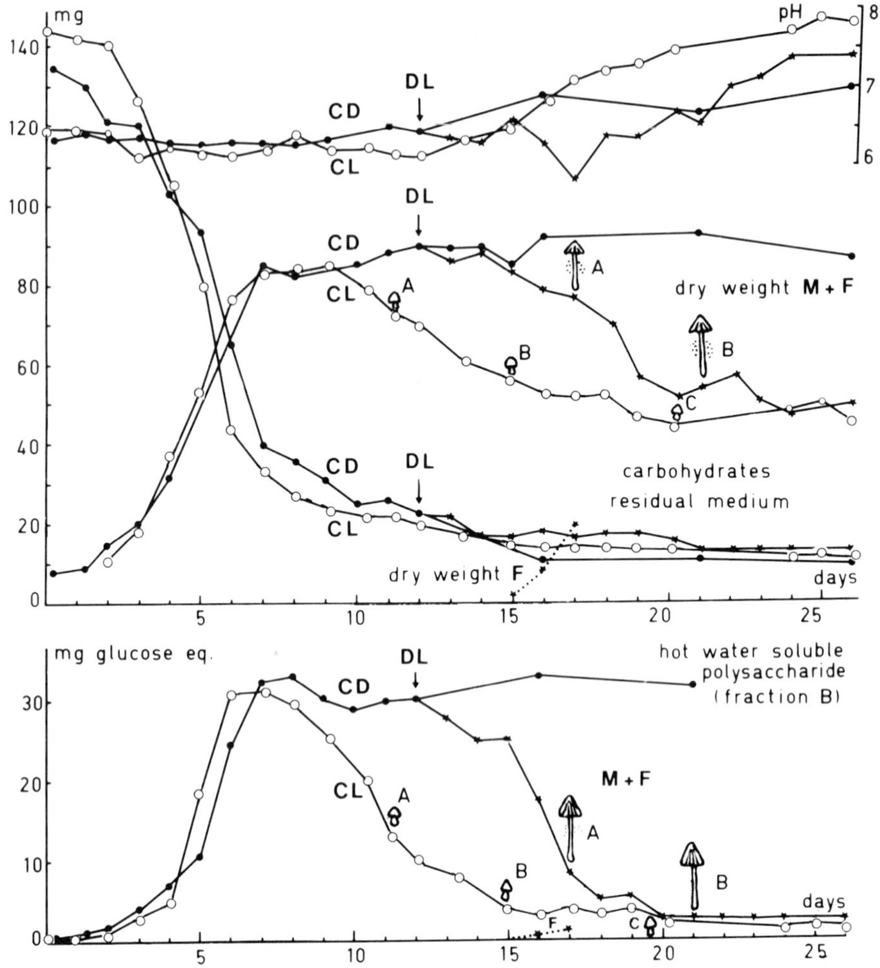

FIG. 5 Biochemical changes in cultures of *Coprinus congregatus* grown on a semidefined liquid medium [CD, continuous dark (no initiation); CL, continuous light (aborting primordia); DL, dark-light (12h, 12h Souligné) regime (sporulating fruit bodies); M + F, mycelium + fruit bodies; A, B, and C, successive fruiting flushes]. (After Robert, 1977b.)

ments allow protein and carbohydrate accumulation in the stipe, then in the cap. Alcohol-soluble carbohydrates, essentially trehalose, accumulate in the stipe if the dark period is lengthened and are probably used during light-induced cap maturation as meiosis and sporogenesis proceed. Similarly, glycogen deposition is evident in the stipe base of very young primordium initials of *C. cinereus*. Accumulation in the gill coincides with depletion of the stipe base deposit and is probably initiated at about the time karyogamy starts the meiotic division. The gill accumulations are clearly localized in the subhymenial tissue (Moore et al., 1979).

Trehalose and glycogen are, respectively, the major soluble carbohydrate and polysaccharide fractions accumulated in the mycelium and small fruit bodies (that do not develop) in *Flammulina velutipes* (Kitamoto and Gruen, 1976) and *Favolus arcularius* (Kitamoto et al., 1978). These compounds also constitute an important part of the total cell carbohydrates and appear to be reserve materials which are utilized during fruit body growth of *A. bisporus* (Hammond and Nichols, 1976; Hammond, 1979a,b). The similar levels and fluctuations of trehalose in the mycelium and the sporophore suggest that this compound is mobile (Hammond, 1979b). Translocation from the vegetative mycelium and stunted fruit bodies into the large developing ones has been suggested in *S. commune* by Wessels (1965), and recently in *F. velutipes* by Kitamoto and Gruen (1976).

2. *Cell Wall Polymers*

As cell walls are a major morphogenetic component, their synthesis and degradation can be expected to be a regulated function during development. Changes in cell wall composition have already been observed in *S. commune* during fruiting (Wessels, 1965) and have been reexamined by Schwalb (1977a). The same major polymers were found in vegetative mycelia and fruit bodies. Three glucans were identified by their relative solubilities: WSG (water-soluble glucan, mucilage, extracellular polysaccharide, β-1,3-1,6 linked), S-glucan (alkali-soluble glucan, α-1,3 linked), and R-glucan (alkali-resistant glucan, β-1, 3-1,6 linked). The difference between WSG and R-glucan appears to be the degree of branching.

Water-soluble glucan represents a large fraction of the total polysaccharide synthesized. During fruiting there is a substantial decrease in the WSG content owing to cessation of synthesis between stages II and III (Schwalb, 1977a). This appears to be due to phosphoglucomutase (PGM) inactivation by proteases, which were first observed at stage II and which increased throughout the remainder of the development, but these were not found in the vegetative mycelium. The proteases cause an increase in PGM cold lability (Schwalb, 1975) and can play a role in development.

R-Glucanase activity has already been reported during fruiting in *S. commune* (Wessels, 1965), but Schwalb (1977a) was unable to find any activity which degrades any of the cell wall polymers when the supply of glucose is not exhausted. According to Schwalb, these observations are likely to be due to starvation conditions. However, a change in resistance to enzymatic hydrolysis has been considered to explain the morphology of a "cup" mutant of *S. commune* lacking pileus growth. A recent report (Sietsma et al., 1977) indicated that this resistance of R-glucan is also involved in high-temperature (35°C) inhibition of fruiting in this species. Indeed, normal fruit bodies are formed at 35°C in an atmosphere without CO_2. The temperature effect may be due to high CO_2 tensions in the culture as a result of increased catabolism of glucose. R-Glucanase, which is glucose repressed, was produced after the glucose in the medium had been exhausted instead of the high CO_2 pressure. Thus CO_2 appears to inhibit the growth of the pilei by changing the composition of the walls in the previously formed mycelium at the time R-glucan is synthesized. The authors concluded that the glucanase-sensitive part of R-glucan may be not attached to a resistant glucan core closely associated with chitin, as mucilage is abundantly excreted in this condition and is structurally similar to the (1-3)-β-glucanase-sensitive part of the R-glucan (Sietsma et al., 1977).

Among the cell wall polymers, chitin was also reported to be used in *Favolus arcularius* (Kitamoto et al., 1978).

A working hypothesis concerning the role of bacteria in the induction of fruiting in *A. bisporus* involving induced and accelerated autolysis has been proposed by Reisinger et al. (1979). The relationship between fruiting in *A. bisporus* and the relevant microflora in the casing soil has been reviewed by Visscher (1979).

B. Enzymatic Activities and Metabolic Pathways

1. *Glucose-6-Phosphate Utilization*

The relationship between the three enzymes of glucose-6-phosphate (G6P) utilization—G6P dehydrogenase (G6P DH), phosphoglucoisomerase (PGI), and phosphoglucomutase (PGM)—was observed in *S. commune* (Schwalb, 1974a, 1978b). The change in PGM activity during fruit body development has been discussed above. This author and Moore and Ewaze (1976) have reported an initial slight increase in the G6P DH activity (related to the HMP pathway) in *C. cinereus* over the basal mycelium level which was followed by a noticeable decline as the sporophore developed. The activity of PGI (related to the EMP pathway) changed relatively much less during the same time. Moore and Ewaze (1976) reported also that G6P DH and 6-phosphogluconate dehydrogenase activities were always lower than PGI and aldolase activities in the mycelium, stipe, and cap of the sporophores.

Thus the EMP pathway represents the major route of sugar metabolism in *C. cinereus* (Moore and Ewaze, 1976). This contrasts with the *A. bisporus* sporophore in which the HMP pathway appears to be the major pathway of glucose metabolism.

Although it appeared from labeling experiments that the pathways of hexose metabolism seemed similar in the mycelium and in the sporophore (Hammond and Nichols, 1977), the results from radiorespirometry and enzyme assays were consistent with the higher activity of the HMP in the sporophore than in the mycelium (Hammond, 1977). The ratio of G6P DH to GPI increased more than fivefold from the mycelium to the fruit body (Hammond, 1979b).

Some recent papers have also been concerned with the metabolism of glucose-6-phosphate. Myunoya and Ishikawa (1980a) reported that fruit body formation in *C. macrorhizus* was inhibited by a glucose analog, 2-deoxyglucose, which is known to inhibit glycolytic enzymes such as PGI and G6P DH. Moreover, a mutant *pgi* deficient in PGI activity has been isolated (Nyunoya and Ishikawa, 1980b). This mutation leads to G6P accumulation and inhibition of fruit body formation, but the precise meaning of the mutation remains to be established.

2. *Role of the Glyoxylate Shunt of the TCA Cycle*

The glyoxylate shunt of the TCA cycle has been shown to be related to morphogenesis in some fungi. A number of enzymes of the TCA cycle have been detected in the sporophores of *C. cinereus*, but isocitrate lyase activity remained at a basal repressed level. Moreover, crude extracts of sporophore caps contained an inhibitor of this enzyme. Thus Moore and Ewaze (1976) suggested that it is most unlikely that the glyoxylate shunt plays a major function in sporophore metabolism.

3. *Evidence for the Operation of the Glutamate Decarboxylation Loop*

Activity of 2-oxoglutarate dehydrogenase could not be detected in *C. cinereus* and it was concluded that the glutamate decarboxylation loop is probably a normal part of the TCA cycle in this fungus (Moore and Ewaze, 1976). Evidence that this pathway operates in *A. bisporus* has already been presented (Piquemal et al., 1967; Piquemal, 1970) and has also been demonstrated in dormant spores (Rast et al., 1976). The primary importance of glutamate in the nitrogen metabolism of this species has been reported by Piquemal (1970). Glutamate and related compounds were also shown to be the most important amino acids in the soluble nitrogen fraction of the *Collybia velutipes* mycelium and fruit body (Naudy-de-Serres et al., 1975a,b) and at the higher levels in *Pleurotus* fruit bodies from different species (Jandaik and Rangad, 1979).

The enzymes involved in the different steps of this glutamate decarboxylation loop were studied in *A. bisporus* sporophores (Baldy, 1975,1976,1977a,b), this pathway being the major route of succinate biosynthesis. The relative importance of 4-aminobutyrate compared to 2-oxoglutarate in the respiratory metabolism has been discussed by Baldy (1977b, 1979).

4. *Reciprocal Regulation of Glutamate Dehydrogenases*

Reciprocal regulation of glutamate dehydrogenases (GDHs) was observed in dikaryotic mycelia of *C. cinereus* producing sporophores, but it could be the result of medium alterations (Stewart and Moore, 1974). The amount of GDH_{NAD} (NAD-linked enzyme) increased in all structures, the mycelium, stipe, and cap, whereas GDH_{NADP} (NADP-linked enzyme) increased only in the cap, exclusively in the gill tissue.

Regulation in the mycelia seems to depend on the action of at least two metabolites, one of carbon metabolism, possibly 2-oxoglutarate, the other of nitrogen metabolism, possibly urea. GDH_{NAD} is subject to catabolite repression and urea derepression and may be involved in ammonia assimilation; GDH_{NADP} is derepressed or induced by glucose and repressed by urea and may be reserved for specific functions, that is, a biosynthetic role.

A simple metabolite regulation model to explain the situation in the sporophore cannot be used, although GDH electrophoresis studies reveal only a single form of each GDH in the whole organism. GDH_{NADP} regulation in the cap of the sporophore is unlikely to involve spore formation as the same pattern of GDH activity is shared by fruit bodies which fail to produce spores; it may be correlated with the later metabolic changes, particularly spore dispersal and autolysis (Stewart and Moore, 1974).

Although it has been shown that the differential production of GDH_{NADP} in pileus cells is not a common phenomenon in all Basidiomycetes (Moore and Al-Gharawi, 1976), GDH_{NADP} has been reported to increase at the stage of first meiotic prophase in pileus, but not in stipe cells of dikaryotic and monokaryotic fruiting bodies, in *C. macrorhizus* (Ishikawa and Serizawa, 1978; Miyake et al., 1980a; Miyake et al., 1980b). However, as for *C. cinereus*, sporeless mutants have been able to produce approximately the same levels of enzyme, and as autolysis occurred it was concluded that the production of enzyme and basidiospore formation are independent of each other (Miyake et al., 1980a).

5. *Role of GDH and Coordinate Regulation of Enzymes*

Glutamate dehydrogenase was presumed to be a component of the glutamate decarboxylation loop in relation to an active TCA cycle in *C. cinereus* (Moore and Ewaze, 1976). However, as the specific activity of GDH_{NADP} rapidly exceeds that of the enzymes in this pathway, its function seems to contribute not only to the TCA cycle but also to other pathways. Moreover, two different GDHs appear to play this role in the stipe and the pileus.

It has also been suggested that under normal conditions of growth on complex media, nitrogen is assimilated in the form of amino acids, and GDH acts in support with transaminases to allow this process to continue, as well as supporting the urea cycle to allow the disposal of excess nitrogen (Al-Gharawi and Moore, 1977). Ewaze et al. (1978) reported on three enzymes—glutamine synthetase (GS), ornithine carbamyl transferase (OCT), and ornithine acetyl transferase (OAT)—which, like GDH_{NADP}, have high and increasing activities in the cap, but low and decreasing activities in the stipe. These enzymes contribute to arginine and urea synthesis and coordinate regulation could also be demonstrated in the mycelium grown on particular media. Thus the urea cycle is specifically amplified in the cap in response to unknown internal controls.

6. *Other Metabolic Activities*

Proteases have been involved in PGM inactivation in *S. commune*. Several chromatographically separable proteases which have also the capacity to inactivate a number of enzymes were responsible and were active in the neutral to weakly alkaline range (Schwalb, 1977b). Proteinase activities increase with the appearance of the first primordia of *P. ostreatus* induced by light (von Netzer, 1979).

Changes in intracellular protease activity during sporophore development in *A. bisporus* have also been monitored by Wood (1979a). Acid protease activity increased in the cap tissue, whereas neutral and alkaline protease activities were maximum in the stipe and lowest in the gill tissue. There they were suggested to degrade intracellular protein to be translocated as peptides or amino acids to the regions of the fruit body where protein synthesis is active, that is, in the gill tissue for spore production.

Mannitol accumulates in the *A. bisporus* sporophore, even at the aggregate stage, to become as much as 50% of the dry weight (Hammond and Nichols, 1976; Hammond, 1979a), whereas this compound does not exceed 6% of the dry weight and declines in concentration in the developing *Coprinus* sporophore (Darbyshire, 1974). Mannitol synthesis has been shown to involve reoxidation of NADPH, formed in the operation of the HMP, by reduction of fructose to mannitol in the mannitol dehydrogenase (MDH) reaction. However, MDH activity was similar in the mycelium and in the sporophore (Hammond, 1977). Ruffner et al. (1978) considered that MDH functions in *A. bisporus* as a synthesizing enzyme under physiological conditions, mannitol being used only under severe stress, such as during postharvest storage (Hammond and Nichols, 1975). Mannitol turnover was slower than that of the other labeled compounds (Hammond and Nichols, 1977), and as its concentration in the cell seems to be maintained at a relatively constant level throughout growth, Hammond (1979b) suggested that the possible role of this polyol as an osmoticant is consistent with its inertness, developing a turgor pressure in the stipe and pileus for the support and the promotion of the flow of water from the mycelium.

Arginine, alanine, and glutamate accumulate in the cap of *C. cinereus* sporophores as a result of amplified arginine biosynthesis (Ewaze et al., 1978). The accumulation of urea also is correlated with the absence of urease activity in the cap, although it occurs at high activity in the stipe and appears to be a constitutive but catabolite-repressed enzyme in the mycelium. As the concentration of urea remains virtually unchanged, a causal relationship with water influx into the developing cap tissue was suggested in this species; the inflation of the hymenium cells, paraphyses, accounting for the "umbrellalike" cap expansion that characterizes *Coprinus* sp. (Ewaze et al., 1978; Moore et al., 1979). It is interesting to note that a similar role was proposed for mannitol in the *A. bisporus* sporophore (Hammond, 1979b). The contrasting regulation of urease in the cap and in the stipe may also be related to the different morphogenetic effects of CO_2 released simultaneously with ammonia (Ewaze et al., 1978). Although not yet established for *Coprinus*, high CO_2 concentrations inhibit cap enlargement and promote stipe elongation in *A. bisporus* (Turner, 1977).

The timing of some changes in enzyme activity with sporulation in *S. commune* was determined with the use of temperature mutants (Bromberg and Schwalb, 1978) and dark inhibition of sporulation (Bromberg and Schwalb, 1976, 1977). It was interesting to note that in the case of the three enzymes, PGM, glucoamylase, and protease, which show appreciable changes in activity from the mycelium, the activity of the enzymes in the spore represents a reestablishment of the vegetative level (see earlier section on changes in PGM during early growth of the fruiting bodies; Schwalb, 1975, 1977b).

VI. THE FRUITING RHYTHM

A. Physiological Regulation

At the mycelial level, complex correlations exist other than those mentioned above. They show that the thalli can be considered as physiological units. Thus a culture of *C. congregatus* maintained at a temperature of 25° under alternating 12-h periods of light and darkness produces mature carpophores according to a characteristic rhythm; 4 or 5 days more often separates the beginning of photoperiods where the carpophores begin sporulating then ends with the autolyzing stage. Successive flushes of fructification can be observed (Fig. 6; Manachère, 1971). An analogous rhythm is known to exist in cultivated *Agaricus* and for the Gasteromycete *Sphaerobolus stellatus* (see the review by Manachère, 1980). In the case of these two fungi, as in the case of *Coprinus*, picking the primordia leads to an acceleration of the carpophore development of the following flushes, the acceleration being greater the earlier the stage during which primordia were picked (Fig. 7). It appears that during their development the buds of a

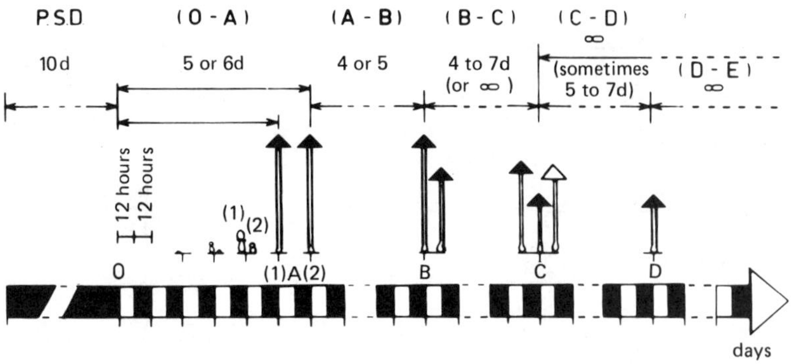

FIG. 6 Main characteristics of the fruiting rhythm of *Coprinus congregatus*. A, B, C, and D are the successive fruiting flushes. (After Manachère, 1977b.)

fruiting flush inhibit other primordia, whether or not they are destined to give carpophores of later flushes. The mechanism of inhibition is unknown. It could result from nutrient deprivation of the dominant primordia or from the production of inhibitory substances. In the particular case of *C. congregatus*, carpophore elimination, not at the primordia stage but at the maturity stage, can reduce and even prevent the production of later flushes (Manachère, 1970). It appears that once they have reached maturity, during the following 24 h, the carpophores release to the original thallus one or more substances stimulating later fruiting (Manachère, 1976,1977b). The fruiting rhythm was disturbed if mature fruit bodies were applied to the cultures before differentiation of the first flush. Fewer flushes were produced, but the yield of the first flush was enhanced and production was more synchronized, the total yield of the flushes being quite unchanged (Manachère, 1977b). The characteristics of the normal fruiting rhythm were maintained if the sporophores picked at maturity were replaced by sporophore extracts. Two different effects of these extracts were observed, on the rhythm period and on the yield of the next flush. Crude extracts from the caps stimulated earlier differentiation of the next flush, while stipe extracts, in contrast, enhanced the yield. The stimulating substance(s) was water soluble, was not destroyed by freezing or autoclaving, and had a low molecular weight (Robert, 1979, unpublished results).

B. Metabolic Aspects

Fruiting of *C. congregatus* on synthetic liquid media was correlated to the level of a hot-water-soluble polysaccharide (fraction B) accumulat-

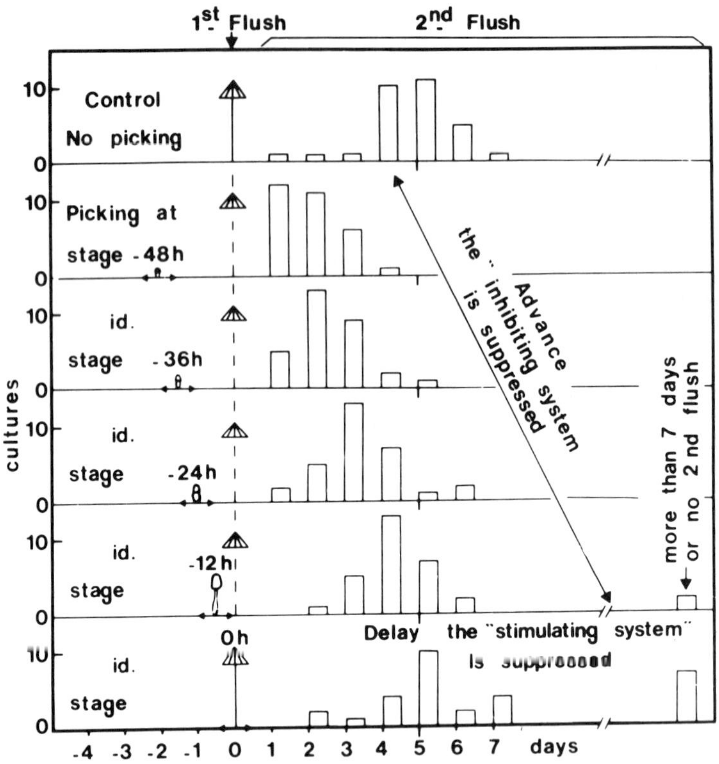

FIG. 7 "Picking effect" on the fruiting rhythm of *Coprinus congregatus*. The first flush was picked at different stages of development and differentiation of the second flush was only considered there. (Redrawn from Manachère, 1970.)

ed during vegetative growth (Robert, 1977b). The phased production of sporophores stopped when this fraction reached a minimal level and was maintained on media with a higher carbohydrate content, or when the exhausted medium was replaced by a new one (Robert, 1977b, 1979). This fundamental role of the carbon source was demonstrated, and the time at which it was incorporated interfered with the period of the fruiting rhythm. Thus a second flush could be differentiated in cultures which would otherwise fruit only once. It appears that the carbohydrate reserves of the mycelium are saved or resynthesized, allowing the differentiation of the second flush.

In *A. bisporus* a direct relationship has been observed between the yield in a flush and the mannitol level in fruit bodies from the flush (Parrish et al., 1976). Examination of sporophores throughout the cropping period showed a cyclic change in the nonstructural carbohydrates. Mean levels of mannitol were higher in the sporophores produced during a flush than in those produced in the interflush period, while trehalose and glycogen levels were significantly higher between flushes (Hammond and Nichols, 1979; Hammond, 1979b). The correlation with the yield of the flushes gives some support to the suggestion that trehalose and glycogen are used as substrates in flush growth (Hammond and Nichols, 1979). This behavior resembles that predicted by Chanter (1976) and Chanter and Thornley (1978) in their mathematical model, where a flush is triggered by the level of substrate reaching a threshold value in the mycelium.

The phased production of fruit bodies in *A. bisporus* was also shown to be associated with a phased production of ethylene from the compost, the sporocarps themselves producing no detectable ethylene (Turner et al., 1975). The activity of microorganisms present in the compost was excluded (Wood and Goodenough, 1977; Wood and Hammond, 1977; Wood, 1979a). Ethylene production was shown to originate from the mycelium and was correlated with the enlargement of fruit bodies, the fruiting structures directly influencing the metabolism of their connecting mycelia. Ethylene production was maximum when the gill color changed from pink to brown (Wood, 1979a). One must remember that a fall in the pH of the culture medium was correlated with the same phase of fruit body development in *C. congregatus* and occurred during the maturation of each flush (Robert, 1977b).

However, mycelia of *A. bisporus* grown on agar media could evolve ethylene when supplied with methionine, but other amino acids or carbohydrates had no effect. The biosynthetic pathway has not been determined. There was no evidence for a regulatory role of ethylene in the growth or development of *A. bisporus*, and no interaction with CO_2 (Ward et al., 1978). However, Visscher (1979), in a review on the effects of microflora on fruiting initiation, considered an interaction between the stimulating effect of low ethylene concentrations and the concentration of CO_2 in *A. bisporus*. The author asked to what extent this interaction is analogous to that demonstrated in higher plants and how far we can use knowledge about this antagonism to interpret mushroom induction.

VI. CONCLUSIONS

The object of this paper has not been to emphasize the possible biotechnological applications of physiological studies conducted on the differentiation of both edible and nonedible species of macromycetes. However, most of the studies discussed above have or could have biotechnological implications. Evidently, physiological researches on the control of macromycete fruiting have to be completed by ecological and genetic studies.

The intensive development of research on the metabolic aspects of fruiting together with investigations on hypothetical "fruiting substances" must not let us overlook the multiple physiological correlations that exist at all phases of growth and development, the end result being the spectacular differentiation of a sporophore. In particular, further studies on the interactions of environmental factors must be conducted in a manner similar to those studies with light and temperature. The results that have already been achieved encourage further research, but must also attest to the difficulties that will be encountered.

REFERENCES

Al-Gharawi, A., and Moore, D. (1977). Factors affecting the amount and the activity of the glutamate dehydrogenases of *Coprinus cinereus*. *Biochim. Biophys. Acta 496*:95-102

Ander, P., and Eriksson, K. E. (1976). The importance of phenol oxidase activity in lignin degradation by the white-rot fungus *Sporotrichum pulverulentum*. *Arch. Microbiol. 109*:1-8.

Badham, E. R. (1980). The effects of light upon basidiocarp initiation in *Psilocybe cubensis*. *Mycologia 72*:136-142.

Baldy, P. (1975). Métabolisme du γ-aminobutyrate chez *Agaricus bisporus*. I. La L-glutamate-1-carboxy-lyase. *Physiol. Plant. 34*:365-372.

Baldy, P. (1976). Métabolisme du γ-aminobutyrate chez *Agaricus bisporus*. II. La γ-aminobutyrate: α-Cetoglutarate aminotransférase. *Planta 130*:275-281.

Baldy, P. (1977a). Métabolisme du γ-aminobutyrate dans les carpophores d'*Agaricus bisporus* Lge. Thèse Sciences Naturelles, Université de Toulouse, Toulouse.

Baldy, P. (1977b). Métabolisme du γ-aminobutyrate chez *Agaricus bisporus*. III. La succinate semialdéhyde: (NAD) P^+ oxydoréductase. *Physiol. Plant. 40*:91-97.

Baldy, P. (1979). La réspiration des carpophores d'*Agaricus bisporus* Lange: Activation par le γ-aminobutyrate. *Mushroom Sci. 10*:361-374.

Bobbitt, T. F., and Crang, R. E. (1974). Light effects on fruiting in *Panus tigrinus* var. *tigrinus*. *Can. J. Bot. 52*:255-257.

Bonner, J. T., Kane, K. K., and Levey, R. H. (1956). Studies on the mechanics of growth in the common mushroom, *Agaricus campestris*. *Mycologia 48*:13-19.

Bret, J. P. (1977a). Respective role of cap and mycelium on stipe elongation of *Coprinus congregatus*. *Trans. Br. Mycol. Soc. 68*:363-369.

Bret, J. P. (1977b). Contrôle de l'élongation du stipe par le chapeau dans les carpophores de *Coprinus congregatus*. *C. R. Cong. Nat. Soc. Savantes Sect. Sci. 1*:273-284.

Bret, J. P. (1978). Stimulating and inhibiting effect of the cap on stipe elongation in the fruit bodies of *Coprinus congregatus*. In *Abstracts of Congress of Federation of European Societies of Plant Physiology*. Edinburgh, 1978, pp. 110-111.

Briggs, W. R. (1976). The nature of the blue light photoreceptor in higher plants and fungi. In *Light and Plant Development*, H. Smith (Ed.). Butterworths, London, pp. 2-18.

Bromberg, S. K., and Schwalb, M. N. (1976). Studies on basidiospore development in *Schizophyllum commune*. *J. Gen. Microbiol.* *96*:409-413.

Bromberg, S. K., and Schwalb, M. N. (1977). Isolation and characterization of temperature sensitive sporulationless mutants of the Basidiomycete *Schizophyllum commune*. *Can. J. Gen. Cytol.* *19*:477-481.

Bromberg, S. K., and Schwalb, M. N. (1978). Sporulation in *Schizophyllum commune*: Changes in enzyme activity. *Mycologia* *70*:481-486.

Bush, D. A. (1974). Autolysis of *Coprinus comatus* sporophores. *Experientia 30*:984-985.

Cabib, E., Duran, A., and Bowers, B. (1979). Localized activation of chitin synthetase in the initiation of yeast septum formation. In *Fungal Walls and Hyphal Growth*. Burnett, J. H., and Trinci, A. P. J., (Eds.) Cambridge University, Cambridge, pp. 189-201.

Chang, S. T., and Hayes, W. A. (Eds.) (1978). *The Biology and Cultivation of Edible Mushrooms*. Academic, New York.

Chanter, D. O. (1976). Mathematical models in mushroom research and production. Ph.D. thesis, University of Sussex, Sussex.

Chanter, D. O., and Thornley, J. H. M. (1978). Mycelial growth and the initiation and growth of sporophores in the mushroom crop: A mathematical model. *J. Gen. Microbiol. 106*:55-65.

Chapman, E. S., and Fergus, C. L. (1973). An investigation of the effects of light on basidiocarp formation of *Coprinus domesticus*. *Mycopathol. Mycol. Appl. 51*:315-326.

Couvy, J. (1974). La fructification d'*Agaricus arvensis* Fries Schaeffer en culture pure. Action de la température et de la lumière. *Bull. Soc. Mycol. Fr. 90*:81-88.

Cox, R. J., and Niederpruem, D. J. (1975). Differentiation in *Coprinus lagopus*. III. Expansion of excised fruit-bodies. *Arch. Microbiol. 105*:257-260.

Craig, G. D., Gull, K., and Wood, D. A. (1977). Stipe elongation in *Agaricus bisporus*. *J. Gen. Microbiol. 102*:337-347.

Craig, G. D., Newsam, R. J., Gull, K., and Wood, D. A. (1979). An ultrastructural and autoradiographic study of stipe elongation in *Agaricus bisporus*. *Protoplasma 98*:15-29.

Darbyshire, J. (1974). Developmental studies on *Coprinus lagopus* (sensu Buller). Ph.D. thesis, University of Manchester, Manchester.

Davis, E. E., and Jong, S. C. (1976). Basidiocarp formation by *Laccaria laccata* in agar culture. *Mycologia 68*:211-214.

De Rousset-Hall, A., and Gooday, G. W. (1975). A kinetic study of a solubilized chitin synthetase preparation from *Coprinus cinereus*. *J. Gen. Microbiol. 89*:146-154.

Durand, R. (1975). Photomorphogenèse d'un basidiomycète *Coprinus congregatus* Bull. ex Fr.: Influences de variations quantitatives et qualitatives de la lumière sur les phases successives du développement des carpophores. Analyse du phénomène de photoreception. Thèse 3e cycle, Lyon.

Durand, R. (1976). Influence des radiations lumineuses sur les processus de reproduction des champignons; hypothèses sur l'identité des photorécepteurs. Revue bibliographique. *Mycopathologia 60*:3-16.

Eger, G. (1961). Untersuchungen über die Funktion der Deckschicht bei der Fruchtkörperbildung des Kulturchampignons, *Psalliota bispora* Lge. *Arch. Mikrobiol. 39*:313-334.

Eger, G., Gottwald, H. D., and Netzer, U. von (1976). The action of light and other factors on sporophore initiation in *Pleurotus ostreatus*. *Mushroom Sci. 9*:575-583.

Eger-Hummel, G. (1980). Blue light photomorphogenesis in mushrooms (Basidiomycetes). In *The Blue Light Syndrome*, H. Senger (Ed.). Springer-Verlag, Berlin, pp. 555-562.

Eilers, F. I. (1974). Growth regulation in *Coprinus radiatus*. *Arch. Microbiol. 96*:353-364.

Ewaze, J. O., Moore, D., and Stewart, G. R. (1978). Co-ordinate regulation of enzymes involved in ornithine metabolism and its relation to sporophore morphogenesis in *Coprinus cinereus*. *J. Gen. Microbiol. 107*:343-357.

Gold, M. H., and Cheng, T. M. (1979). Conditions for fruit body formation in the white rot Basidiomycete *Phanerochaete chrysosporium*. *Arch. Microbiol. 121*:37-41.

Gooday, G. W. (1974). Control of development of excised fruit-bodies and stipes of *Coprinus cinereus*. *Trans. Br. mycol. Soc. 62*:391-399.

Gooday, G. W. (1975). The control of differentiation in fruit bodies of *Coprinus cinereus*. *Rep. Tottori Mycol. Inst. 12*:151-160.

Gooday, G. W. (1979). Chitin synthesis and differentiation in *Coprinus cinereus*. In *Fungal Walls and Hyphal Growth*. J. H. Burnett and A. P. J. Trinci (Eds.). Cambridge University, Cambridge, pp. 203-223.

Gooday, G. W., and De Rousset-Hall, A. (1975). Properties of chitin synthetase from *Coprinus cinereus*. *J. Gen. Microbiol. 89*: 137-145.

Gooday, G. W., De Rousset-Hall, A., and Hunsley, D. (1976). Effect of polyoxin D on chitin synthesis in *Coprinus cinereus*. *Trans. Br. Mycol. Soc. 67*:193-200.

Gruen, H. E. (1963). Endogenous growth regulation in carpophore of *Agaricus bisporus*. *Plant Physiol.* *38*:652-666.
Gruen, H. E. (1967). Growth regulation in fruit bodies of *Agaricus bisporus*. *Mushroom Sci.* *6*:103-120.
Gruen, H. E. (1969). Growth and rotation of *Flammulina velutipes* fruit bodies and the dependence of stipe elongation on the cap. *Mycologia* *61*:149-166.
Gruen, H. E. (1976). Promotion of stipe elongation on *Flammulina velutipes* by a diffusate from excised lamellae supplied with nutrients. *Can. J. Bot.* *54*:1306-1315.
Gruen, H. E. (1979). Control of rapid stipe elongation by the lamellae in fruit bodies of *Flammulina velutipes*. *Can. J. Bot.* *57*:1131-1135.
Gruen, H. E., and Wu, S. H. (1972a). Promotion of stipe elongation in isolated *Flammulina velutipes* fruit bodies by carbohydrates natural extracts, and amino acids. *Can. J. Bot.* *50*:803-818.
Gruen, S. E., and Wu, S. H. (1972b). Dependance of fruit body elongation on the mycelium in *Flammulina velutipes*. *Mycologia* *64*:995-1007.
Hagimoto, H. (1963). Studies on the growth of fruit body of fungi. IV. The growth of the fruit body of *Agaricus bisporus* and the economy of the mushroom growth hormone. *Bot. Mag. Tokyo* *76*:256-263.
Hagimoto, H. (1964). On the growth of the fruit body of *Agaricus bisporus* (Lange) Sing. *Trans. Br. Mycol. Soc.* *4*:158-164.
Hagimoto, H., and Konishi, M. (1959). Studies on the growth of fruit body of fungi. I. Existence of a hormone active to the growth of fruit body in *Agaricus bisporus* (Lange) Sing. *Bot. Mag. Tokyo* *72*:359-366.
Hagimoto, H., and Konishi, M. (1960). Studies on the growth of fruit body of fungi. II. Activity and stability of the growth hormone in the fruit body of *Agaricus bisporus* (Lange) Sing. *Bot. Mag. Tokyo* *73*:283-287.
Hammond, J. B. W. (1977). Carbohydrate metabolism in *Agaricus bisporus*: Oxidative pathways in mycelium and sporophore. *J. Gen. Microbiol.* *102*:245-248.
Hammond, J. B. W. (1979a). Changes in the composition of harvested mushrooms. *Phytochemistry* *18*:415-418.
Hammond, J. B. W. (1979b). The role of non-structural carbohydrates in the life-cycle of *Agaricus bisporus*. *Mushroom Sci.* *10*:391-400.
Hammond, J. B. W., and Nichols, R. (1975). Changes in respiration and soluble carbohydrates during the post-harvest storage of mushrooms (*Agaricus bisporus*). *J. Sci. Food Agric.* *26*:835-842.

Hammond, J. B. W., and Nichols, R. (1976). Carbohydrate metabolism in *Agaricus bisporus* (Lange) Sing.: Changes in soluble carbohydrates during growth of mycelium and sporophore. *J. Gen. Microbiol. 93*:309-320.

Hammond, J. B. W., and Nichols, R. (1977). Carbohydrate metabolism in *Agaricus bisporus* (Lange) Imbach.: Metabolism of ^{14}C labelled sugars by sporophores and mycelium. *New Phytol. 79*: 315-325.

Hammond, J. B. W., and Nichols, R. (1979). Carbohydrate metabolism in *Agaricus bisporus*: Changes in non-structural carbohydrates during periodic fruiting (flushing). *New Phytol. 83*:723-730.

Horikoshi, T., Kitamoto, Y., and Kasai, Z. (1974). Effect of light on the initiation of pileus formation in a Basidiomycete *Favolus arcularius*. *Plant Cell Physiol. 15*:903-911.

Horrière, F. (1977). Recherches physiologiques sur la fructification d'un basidiomycète *Coprinus congregatus* Bull. ex Fr.: Influence de quelques facteurs nutritionnels. Thèse ingénieur-docteur, Université de Lyon I, Lyon.

Horrière, F. (1979). Etude comparative des exigences trophiques de quelques basidiomycètes supérieurs fructifiant sur milieux synthétiques. Analyse bibliographique. *Mushroom Sci. 10*:665-682.

Ishikawa, T., and Uno, I. (1977). A mechanism of fruiting body formation in Basidiomycetes. In *Growth and Differentiation in Microorganisms*. T. Ishikawa, Y. Maruyama, and H. Matsumiya (Eds.), University Park, Baltimore, Md., pp. 283-295.

Ishikawa, T., and Serizawa, H. (1978). Expression of enzyme activities during the fruiting process in *Coprinus macrorhizus*. In *Abstracts, 12th International Congress of Microbiology*, Munich, p. 82, B 11.

Jandaik, C. L., and Rangad, C. O. (1979). Biochemical changes in *Pleurotus* species with respect to different growth stages. *Mushroom Sci. 10*:419-426.

Kamada, T., and Takemaru, T. (1977). Stipe elongation during basidiocarp maturation in *Coprinus macrorhizus*: Mechanical properties of stipe cell wall. *Plant Cell Physiol. 18*:831-840.

Kamada, T., Kurita, R., and Takemaru, T. (1978). Effects of light on basidiocarp maturation in *Coprinus macrorhizus*. *Plant Cell Physiol. 19*:263-275.

Kamada, T., and Tsuji, M. (1979). Darkness-induced factor affecting basidiocarp maturation in *Coprinus macrorhizus*. *Plant Cell Physiol. 20*:1445-1448.

Kinugawa, K. (1977). *Collybia velutipes* can fruit under total darkness. *Trans. Mycol. Soc. Jap. 18*:353-356.

Kitamoto, Y., and Gruen, H. E. (1976). Distribution of cellular carbohydrates during development of the mycelium and fruit bodies of *Flammulina velutipes*. *Plant Physiol. 58*:485-491.

Kitamoto, Y., Suzuki, A., and Furukawa, S. (1972). An action spectrum for light induced primordium formation in a basidiomycete, *Favolus arcularius* Fr. Ames. *Plant Physiol 49*:338-340.

Kitamoto, Y., Horikoshi, T., and Suzuki, A. (1974). An action spectrum for photo-induction of pileus formation in a basiodiomycete, *Favolus arcularius*. *Planta 119*:81-84.

Kitamoto, Y., Terashita, T., Matsuda, S., Obata, K., Hosoi, N., Kono, M., and Ichikawa, Y. (1978). Carbohydrate metabolism in *Favolus arcularius*: Changes in cellular carbohydrates during development of the mycelium and fruit bodies. *Trans. Mycol. Soc. Jap. 19*:273-281.

Leslie, J. F., and Leonard, T. J. (1979). Monokaryotic fruiting in *Schizophyllum commune*: Phenoloxidases. *Mycologia 71*:1082-1085.

Li, C. Y. (1979). Light and temperature induced sporocarp formation of *Phellinus weirii*. *Can. J. For. Res. 9*:535-538.

Long, P. E., and Jacobs, L. (1974). Aseptic fruiting of the cultivated mushroom *Agaricus bisporus*. *Trans. Br. Mycol. Soc. 63*: 99-107.

Lu, B. C. (1972). Dark dependence of meiosis at elevated temperatures in the Basidiomycete *Coprinus lagopus*. *J. Bacteriol. 111*: 833-834.

Manachère, G. (1970). Recherches physiologiques sur la fructification de *Coprinus congregatus* Bull. ex Fr. : Action de la lumière; rythme de production de carpophores. *Ann. Sci. Nat. Bot. Biol. Vég. 11*:1-95.

Manachère, G. (1971). Research on the fruiting rhythm of a basidiomycete mushroom *Coprinus congregatus* Bull. ex Fr. *J. Interdiscip. Cycle Res. 2*:199-209.

Manachère, G. (1976). Substance(s) produite par les carpophores mûrs de *Coprinus congregatus* et régularisatrice(s) du rythme de fructification du champignon; délais nécessaires à leur libération. *Rev. Mycol. 40*:329-339.

Manachère, G. (1977a). Formation of basidiocarps. In *Biotechnology and Fungal Differentiation* J. Meyrath and J. D. Bu'Lock (Eds.) Academic, London, pp. 43-65.

Manachère, G. (1977b). Modifications du rythme de fructification de *Coprinus congregatus* par le dépôt de carpophores mûrs exogènes a la surface des cultures. *Physiol. Plant. 39*:201-205.

Manachère, G. (1978). Morphogenèse des carpophores de Basidiomycètes supérieurs. Connaissances actuelles. *Rev. Mycol. 42*: 191-252.

Manachère, G. (1980). Conditions essential for controlled fruiting of macromycetes. A review. *Trans. Br. Mycol. Soc. 75*:255-270.

Manachère, G., Robert, J. C., Bret, J. P., and Hugueney, R. (1974). Recherches physiologiques sur la fructification de

Coprinus congregatus Bull. ex Fr. perspectives nouvelles. pp. 255-283 In *Travaux Mycologiques dédiés à* R. Kühner. *Bull. Soc. Linn. Lyon.*

Miyake, H., Takemaru, T., and Ishikawa, T. (1980a). Sequential production of enzymes and basidiospore formation in fruiting bodies of *Coprinus macrorhizus*. *Arch. Microbiol.* *126*:201-205.

Miyake, H., Tanaka, K., and Ishikawa, T. (1980b). Basidiospore formation in monokaryotic fruiting bodies of a mutant strain of *Coprinus macrorhizus*. *Arch. Microbiol.* *126*:207-212.

Molitoris, H. P. (1976). *Die Laccasen des Ascomyceten* Podospora anserina. *Beiträge zur Kenntnis von Struktur und Funktion eines Systems multipler Enzyme. Biblio. Mycol.* Vol. 52, Cramer, Vaduz.

Molitoris, H. P. (1979). Wood degradation, phenoloxidases and chemotaxonomy of higher fungi. *Mushroom Sci.* *10*:243-263.

Moore, D., and Al-Gharawi, A. (1976). An elevated level of NADP-linked glutamate dehydrogenase is not a general feature of the caps of agaric sporophores. *Trans. Br. Mycol. Soc.* *66*:149-150.

Moore, D., and Ewaze, J. O. (1976). Activities of some enzymes involved in metabolism of carbohydrate during sporophore development in *Coprinus cinereus*. *J. Gen. Microbiol.* *97*:313-322.

Moore, D., Elhiti, M. Y., and Butler, R. D. (1979). Morphogenesis of the carpophore of *Coprinus cinereus*. *New Phytol.* *83*:695-722.

Naudy-de-Serres, M., Latché, J. C., and Baldy, P. (1975a). Culture de *Collybia velutipes* (Curt.) Quél. sur milieu synthétique liquide. Analyse de quelques constituants du mycélium à son premier stade de développement á l'obscurité. *C. R. Acad. Sci. Sér. D.* *281*:259-262.

Naudy-de-Serres, M., Latché, J. C., and Baldy, P. (1975b). Culture de *Collybia velutipes* (Curt.) Quél. sur milieu synthétique liquide. Répercussions des conditions d'éclairement sur les teneurs du mycélium et des carpophores, en acides aminés et amides libres, en acides aminés protéiques et en acides organiques nonvolatiles. *C. R. Acad. Sci. Sèr. D.* *281*:1377-1380.

Netzer, U. von (1977). A system for the investigation of morphogenesis in Basidiomycetes. *Abstracts IMC 2*, Tampa, Fla., pp. 474.

Netzer, U. von (1979). Investigations of primordia formation in the *Pleurotus ostreatus* dikaryon "868 × 381." *Mushroom Sci.* *10*:703-711.

Nyunoya, H., and Ishikawa, T. (1979). Control of unusual hyphal morphology in a mutant of *Coprinus macrorhizus*. *Jpn. J. Genet.* *54*:11-20.

Nyunoya, H., and Ishikawa, T. (1980a). Effect of 2-deoxy-D-glucose on the induction of tyrosinase in *Coprinus macrorhizus*. *J. Gen. Appl. Microbiol.* *26*:229-238.

Nyunoya, H., and Ishikawa, T. (1980b). Relationship between deficiency of phosphoglucose isomerase in *Coprinus macrorhizus* and fruiting body formation. *J. Bacteriol. 144*:432-434.

Parrish, G. K., Beelman, R. B., and Kneebone, L. R. (1976). Relationship between crop yield and mannitol content during the crop cycle of cultivated mushrooms. *Hortic. Sci. 11*:32.

Phillips, L. E., and Leonard, T. J. (1976). Extracellular and intracellular phenoloxidase activity during growth and development in *Schizophyllum*. *Mycologia 68*:268-276.

Phillips, L. E., and Leonard, T. J. (1977). On the oxidation of dihydroxyphenylalanine and benzidine in crude cell-free fungal extracts. *Mycologia 69*:413-416.

Piquemal, M. (1970). Sur le métabolisme de l'acide glutamique, source d'azote et de carbone pour le mycélium d'*Agaricus bisporus* Lge. Thèse Doct. Spécial. Université de Toulouse, Toulouse.

Piquemal, M., Latché, J. C., and Baldy, P. (1967). Métabolisme de l'acide glutamique 3-4-^{14}C et de l'acide γ-aminobutyrique dans les carpophores d'*Agaricus hortensis*. *C. R. Acad. Sci. Sér. D 264*:318-321.

Plunkett, B. E. (1956). The influence of factors of the aeration complex and light upon fruit body form in pure cultures of an agaric and polypore. *Ann. Bot. 20*:563-586.

Rast, D. Stäuble, E. J., and Zobrist, P. (1976). The Krebs cycle and control of dormancy in the *Agaricus bisporus* spore. *New Phytol. 76*:469-477.

Reisinger, O., Desbiens, O., and Olah, G. M. (1979). Transformation du substrat organique utilisé pour la culture industrielle d'*Agaricus bisporus*. Etude ultrastructurale préliminaire et hypothèses de travail. *Mushroom Sci. 10*:287-302.

Robert, J. C. (1977a). Fruiting of *Coprinus congregatus*: Biochemical changes in fruit bodies during morphogenesis. *Trans. Br. Mycol. Soc. 68*:379-387.

Robert, J. C. (1977b). Fruiting of *Coprinus congregatus*: Relationship to biochemical changes in the whole culture. *Trans. Br. Mycol. Soc. 68*:389-395.

Robert, J. C. (1979). Quelques aspects métaboliques de la fructification du basidiomycète *Coprinus congregatus* Bull. ex Fr. *Mushroom Sci. 10*:683-702.

Robert, J. C., and Durand, R. (1979). Light and temperature requirements during fruit body development of a basidiomycete mushroom, *Coprinus congregatus*. *Physiol. Plant. 46*:174-178.

Ruffner, H. P., Rast, D., Tobler, H., and Karesch, H. (1978). Purification and properties of mannitol dehydrogenase from *Agaricus bisporus* sporocarps. *Phytochem. 17*:865-868.

Rusmin, S., and Leonard, T. J. (1975). Biochemical induction of fruiting bodies in *Schizophyllum commune*: A bioassay and its application. *J. Gen. Microbiol. 90*:217-227.

Rusmin, S., and Leonard, T. J. (1978). Biochemical induction of fruiting in *Schizophyllum*. Isolation and preliminary purification of an inducing substance from *Agaricus bisporus* mushrooms. *Plant Physiol. 61*:538-543.

Salas, J. A., and Hancock, J. G. (1972). Production of the perfect stage of *Mycena citricolor* (Berk. and Curt.) Sacc. *Hilgardia 41*:213-234.

Schaefer, H. P., Hildreth, S. W., and Drake, F. H. (1977). Glucan and glucanase involvement in autolysis of *Coprinus cinereus* gills. In *Abstracts, Second International Mycological Congress*, Fla., p. 594.

Schwalb, M. N. (1974a). Changes in activity of enzymes metabolizing glucose-6-phosphate during development of the basidiomycete *Schizophyllum commune. Dev. Biol. 40*:84-89.

Schwalb, M. N. (1974b). Effect of adenosine 3'-5'-cyclic monophosphate on the morphogenesis of fruit bodies of *Schizophyllum commune. Arch. Microbiol. 96*:17-20.

Schwalb, M. N. (1975). Developmental control of enzymes modification during fruiting of the Basidiomycete *Schizophyllum commune. Biochem. Biophys. Res. Commun. 67*:478-482.

Schwalb, M. N. (1977a). Cell wall metabolism during fruiting of the Basidiomycete *Schizophyllum commune. Arch. Microbiol. 114*:9-12.

Schwalb, M. N. (1977b). Developmentally regulated proteases from the Basidiomycete *Schizophyllum commune. J. Biol. Chem. 252*:8435-8439.

Schwalb, M. N. (1978a). A developmental mutant affecting 3'-5'-cyclic AMP metabolism in the Basidiomycete *Schizophyllum commune. FEMS Microbiol. Lett. 3*:107-110.

Schwalb, M. N. (1978b). Regulation of fruiting. In *Genetics and Morphogenesis in the Basidiomycetes*. M. N. Schwalb and P. G. Miles (Eds.) Academic, New York, pp. 135-165.

Senger, H. (Ed.) (1980). *The Blue Light Syndrome*. Springer Verlag, Berlin.

Sietsma, J. H., Rast, D., and Wessels, J. G. H. (1977). The effect of carbon dioxide on fruiting and the degradation of a cell-wall glucan in *Schizophyllum commune. J. Gen. Microbiol. 102*:385-389.

Stewart, G. R., and Moore, D. (1974). The activity of glutamate dehydrogenases during mycelial growth and sporophore development in *Coprinus lagopus* (sensu Lewis). *J. Gen. Microbiol. 83*:73-81.

Tan, K. K. (1978). Light induced fungal development. In *The Filamentous Fungi, Developmental Mycology*, Vol. 3. J. E. Smith and D. R. Berry (Eds.). Edward Arnold, London, pp. 334-357.

Tsusué, Y. M. (1969). Experimental control of fruit-body formation in *Coprinus macrorhizus*. *Dev. Growth Differ.* *11*:164-178.

Turner, E. M. (1977). Development of excised sporocarps of *Agaricus bisporus* and its control by CO_2. *Trans. Br. Mycol. Soc.* *69*:183-186.

Turner, E. M., Wright, M., Ward, T., and Osborne, D. J. (1975). Production of ethylene and other volatiles and changes in cellulase and laccase activities during the life cycle of the cultivated mushroom, *Agaricus Bisporus*. *J. Gen. Microbiol.* *91*:167-176.

Uno, I., and Ishikawa, T. (1974). Effect of glucose on the fruiting body formation and adenosine 3'-5'-cyclic monophosphate levels in *Coprinus macrorhizus*. *J. Bacteriol.* *120*:96-100.

Uno, I., and Ishikawa, T. (1975). Subcellular localization of adenylate cyclase during the fruiting body formation in *Coprinus macrorhizus*. *J. Gen. Appl. Microbiol.* *21*:335-343.

Uno, I., and Ishikawa, T. (1976). Effect of cyclic AMP on glycogen phosphorylase in *Coprinus macrorhizus*. *Biochim. Biophys. Acta* *452*:112-120.

Uno, I., and Ishikawa, T. (1978). Effect of cyclic AMP on glycogen synthetase in *Coprinus macrorhizus*. *J. Gen. Appl. Microbiol.* *24*:193-197.

Urayama, T. (1956). Das Wuchshormon des Fruchtkörpers von *Agaricus campestris* L. (Vorlaüfige Mitterlung). *Bot. Mag.* *69*: 298-299.

Visscher, H. R. (1979). Fructification of *A. bisporus* (Lge) Imb. in relation to the relevant microflora in the casing soil. *Mushroom Sci.* *10*:641-664.

Ward, T., Turner, E. M., and Osborne, D. J. (1978). Evidence for the production of ethylene by the mycelium of *Agaricus bisporus* and its relationship to sporocarp development. *J. Gen. Microbiol.* *104*:23-30.

Wessels, J. G. H. (1965). Morphogenesis and biochemical processes in *Schizophyllum commune* Fr. *Wentia* *13*:1-113.

Wong, W. M., and Gruen, H. E. (1977). Changes in cell size and nuclear number during elongation of *Flammulina velutipes* fruit bodies. *Mycologia* 5:899-913.

Wood, D. A. (1976). Primordium formation in axenic cultures of *Agaricus bisporus* (Lange) Sing. *J. Gen. Microbiol.* *95*:313-323.

Wood, D. A. (1979a). Biochemical changes during growth and development of *Agaricus bisporus*. *Mushroom Sci.* *10*:401-417.

Wood, D. A. (1979b). Studies on primordium initiation in *Agaricus bisporus* and *Agaricus bitorquis* (syn. *edulis*). *Mushroom Sci.* *10*:565-586.

Wood, D. A. (1980). Inactivation of extracellular laccase during fruiting of *Agaricus bisporus*. *J. Gen. Microbiol.* *117*:339-345.

Wood, D. A., and Goodenough, P. W. (1977). Fruiting of *Agaricus bisporus*. Changes in extracellular enzyme activities during

growth and fruiting. *Arch. Microbiol.* *114*:161-165.
Wood, D. A., and Hammond, J. B. W. (1977). Ethylene production by axenic fruiting cultures of *Agaricus bisporus*. *Appl. Environ. Microbiol.* *34*:228-229.

17

Product Formation and Differentiation in Fungi

DAVID SHEPHERD and MARIETTE CARELS *Nestlé Products Technical Assistance Company, Ltd., Orbe, Switzerland*

I. INTRODUCTION

There exists within the different genera of the fungi a wide range of mechanisms producing varying types of reproductive survival units, each depending on the type of fungus and the particular environmental conditions.

The factors influencing the initiation of sporulation have interested researchers for many years, ever since Klebs' (1928) general observations that reproduction is initiated by factors that limit growth, observations which have since been confirmed by other workers in a continued search for the exact mechanisms of sporulation induction (Hawker, 1957; Cochrane, 1958; Turian, 1969; Carter and Bull, 1969; Smith and Galbraith, 1971; Martinelli, 1976; Lawler and Weber, 1980).

Many species of fungi also produce a wide range of secondary metabolites (Turner, 1971). Their production is also influenced by the environment and many of these conditions are similar to those which cause the induction of asexual and sexual reproduction (Woodruff, 1966; Demain, 1972; Bu'Lock, 1975; Martin and McDaniel, 1976).

Bu'Lock (1965) also pointed out the relationship that both sporulation and secondary metabolism are responses which occur after a phase of balanced growth when some limiting factor becomes operative. These manifestations of survival mechanisms by fungi in response to environmental stimuli are often treated as separate events or in isolation without regard to the other, despite all the evidence that the induction of sporulation and secondary metabolism are closely linked.

In a study of the physiology of conidia formation with strains of *Claviceps purpureus* producing alkaloids, Rehacek et al. (1974) showed a relationship between conidiation and alkaloid formation, in that cultures treated with conidiation inhibitors had increased yields of alkaloids.

Sekiguchi and Gaucher (1977) correlated fungal sporulation and secondary metabolism in *Penicillium urticae*. They observed that sporulation, like secondary biosynthesis, occurred because of nitrogen limitation and required the presence of certain enzymes for a specific and limited period of time. It was observed that patulin and its precursors were not a prerequisite to conidiogenesis in *P. urticae*.

The above-mentioned studies were carried out in shaken, submerged culture. However, other work using solid culture confirmed these observations that when conidiation is limited, more secondary metabolites are formed. Hesseltine (1972) showed that if fungi were grown on solid substrates, they sporulated and aflatoxins were formed. However, if the cultures were agitated, conidiation was strongly inhibited and larger quantities of aflatoxin were formed.

Although some investigations into the mechanisms of sporulation in strains capable of secondary metabolite production have been attempted, little detailed work has been done on the balance between the two or on the role of each in the survival of the strain.

What we will try to show in this chapter is that if these phenomena were studied together, a better understanding and control of the production of certain industrially and economically important compounds could be obtained. The absence of literature in this area explains why most of the examples used to illustrate the effect of environmental conditions on spore formation and secondary metabolite production are drawn from our own work, with 40 strains of *Monascus* in submerged, shaken culture (Carels and Shepherd, 1975, 1977, 1978, 1979).

TABLE 1 Classification of *Monascus* Strains According to Their Ability to Sporulate and Produce Secondary Metabolites

Secondary metabolites	+				−			
Sexual reproduction	+		−		+		−	
Conidiation	+	−	+	−	+	−	+	−
Number of strains	4	0	3	2	9	0	17	5

Source: Shepherd (1977).

II. STUDIES WITH *MONASCUS*

A. General Observations

From the first investigations with the 40 strains from 18 species of *Monascus* it was possible to classify the strains into six groups, depending on their ability to sporulate or produce secondary metabolites. As shown in Table 1, most of the strains normally produce little or no pigment, but the majority form conidia. There were also no strains which reproduced sexually without forming conidia at the same time.

This classification was based on studies where the strains were cultivated in media containing different nitrogen sources, which also allowed a series of general conclusions to be drawn concerning the behavior of the different strains (see Table 2).

In order to have a better idea of the mechanisms involved in the expression of the above-mentioned characteristics on the different media, more detailed studies were made.

The results of these studies on the effect of media components show quite clearly that cultural conditions play an important role in determining the type and amount of secondary metabolites produced, as well as the balance between vegetative and reproductive development.

On a rich medium containing high levels of sugar, amino acids, and vitamins, only red pigments are produced [the reasons for this have been published elsewhere and will not be discussed in detail here (Carels and Shepherd, 1977)]. However, vegetative growth is high and conidiation as well as sexual reproduction can develop, apparently contradicting other observations where nutritional limitations are required for sporulation. On media containing nitrate as the sole source of nitrogen, red and yellow pigments are formed and conidiation and sexual reproduction develop. On a medium where ammonium is the sole source of nitrogen, only yellow pigments are produced, and sexual reproduction and conidiation are greatly reduced.

TABLE 2 General Observations on the Behavior of *Monascus* Species on Media Containing Different Nitrogen Sources

1.	Nitrogen source determines pH evolution.
2.	Conidiation initiated early in growth.
3.	Nitrate as nitrogen source favors conidiation and sexual reproduction.
4.	Ammonium as nitrogen source retards conidiation and sexual reproduction.
5.	Ammonium as nitrogen source favors secondary metabolite production.
6.	Uptake of nitrogen and glucose very low at beginning of growth by strains capable of sexual reproduction.

Source: Adapted from Shepherd (1977).

When the different parameters were measured during cultivation (Carels and Shepherd, 1978), it was clear that one of the major differences between cultures grown on the different media was the changing pattern of pH, as seen in Fig. 1. It is well known that when nitrate is used as the nitrogen source, the pH rises, and it descends when ammonium is used. It also rises on the rich medium, probably owing to deamination of the amino acids.

B. Effect of pH on Secondary Metabolism and Conidiation

In order to explore the exact reasons for these observations and to distinguish between the effect of the nitrogen source and the changing pattern of pH, cultures containing nitrate and ammonium as nitrogen sources were incubated normally and in media where the pH was lowered to 2.5 (Carels and Shepherd, 1978). It is clear from Figs. 2 and 3 that pH alone was responsible for these observations, irrespective of the nitrogen source. A low pH on a nitrate medium decreased the number of conidia and increased the level of secondary metabolite production. The rising pigment production in response to reduced conidia formation due to lowering the pH is less evident on ammonium medium. This can be explained by the fact that the pH of these cultures always evolves toward acidity, which will limit the number of conidia, therefore making the differences less marked than on nitrate media, where the pH normally rises.

This confirms the results from other experiments (Larmour and

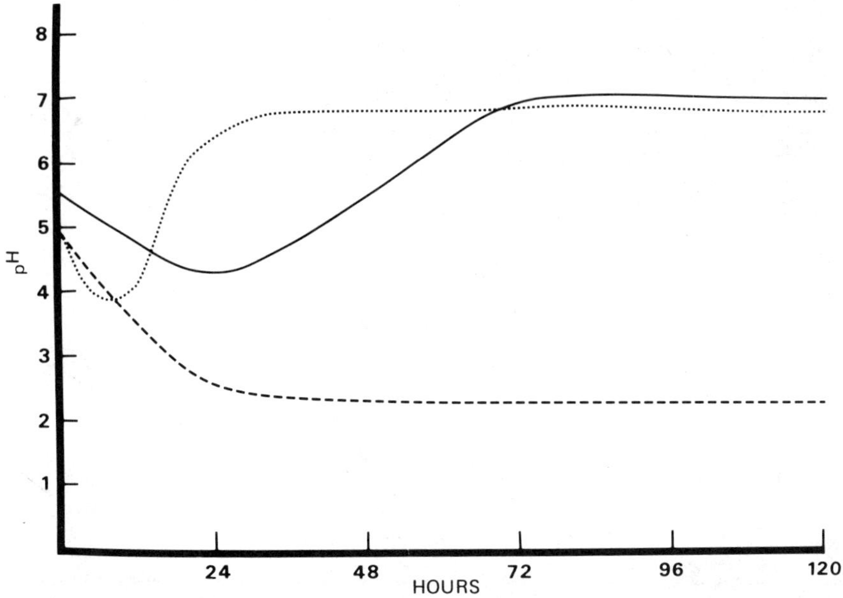

FIG. 1 Evolution of pH when *Monascus* is grown on a medium containing amino acids as the nitrogen source (————), a medium containing nitrate as the nitrogen source, (·················), and a medium containing ammonium as the nitrogen source (- - - - - - - -).

Marchant, 1977) when *Fusarium* was grown in continuous culture using different media where the pH was adjusted between 3.5 and 5.5. As the pH was increased, so did the number of conidia.

C. Effect of Amino Acids on Secondary Metabolism and Conidiation

Because of the number of spores produced on the rich medium, it was thought that there were substances present in this medium which could also be responsible for stimulating conidiation. Yeast extract was added to the nitrate and ammonium media, but the main effect was an increase in biomass formation, probably due to the presence of vitamins. However, when a mixture of the amino acids (Carels and Shepherd, 1978) present in the yeast extract was added in the same concentration, a great increase in conidiation and a decrease in pigment production were observed. When individual amino acids were added to the different media in different concentrations, no single amino acid had a significant effect on the behavior of the strains, as shown for other fungi (Dicker et al., 1969).

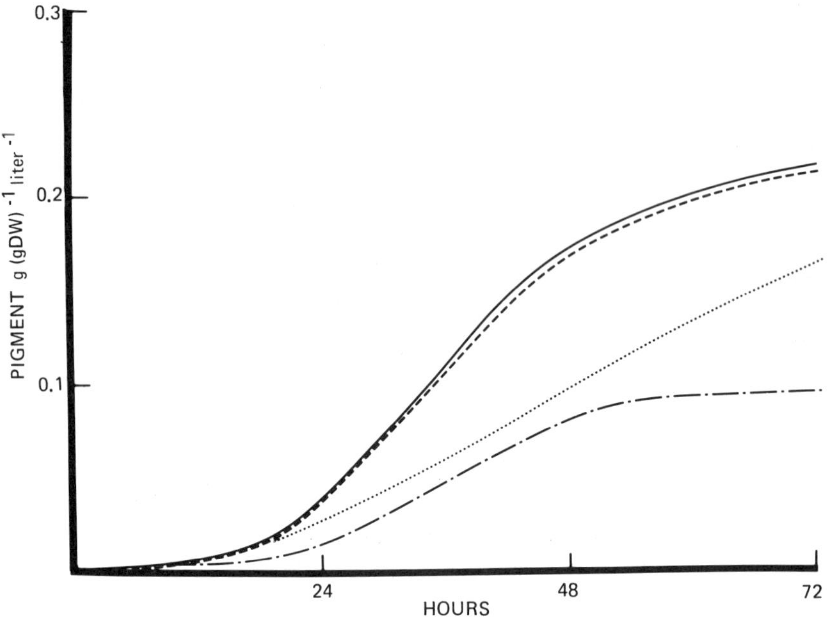

FIG. 2 Pigment production by *Monascus rubiginosus* ATCC 16367 when cultivated on a medium containing nitrate as the nitrogen source (–·–·–·–·–), the same medium where the pH was adjusted to 2.5 before inoculation (·················), a medium containing ammonia as the nitrogen source (————), and the same medium where the pH was adjusted to 2.5 before inoculation (- - - - - - - - -).

D. Effect of pH and Amino Acids on Secondary Metabolism and Conidiation

As there seem to be two possibilities for influencing conidia and secondary metabolite formation, a series of experiments were designed to investigate the effect of pH on the stimulation of conidiation by amino acid addition. As can be seen in Figs. 4 and 5, the pH effect was much stronger, because when the pH was reduced in cultures containing amino acids, conidiation was inhibited and secondary metabolite formation stimulated.

These results indicate that when growth is unrestricted, as when yeast extract is added to the medium, the fungal strains form biomass in preference to pigment and conidia. This confirms other work which has shown that some limitation of growth must occur before spores and secondary metabolites are formed (Bu'Lock, 1965; Martinelli, 1976). It would appear that the addition of amino acids to the media

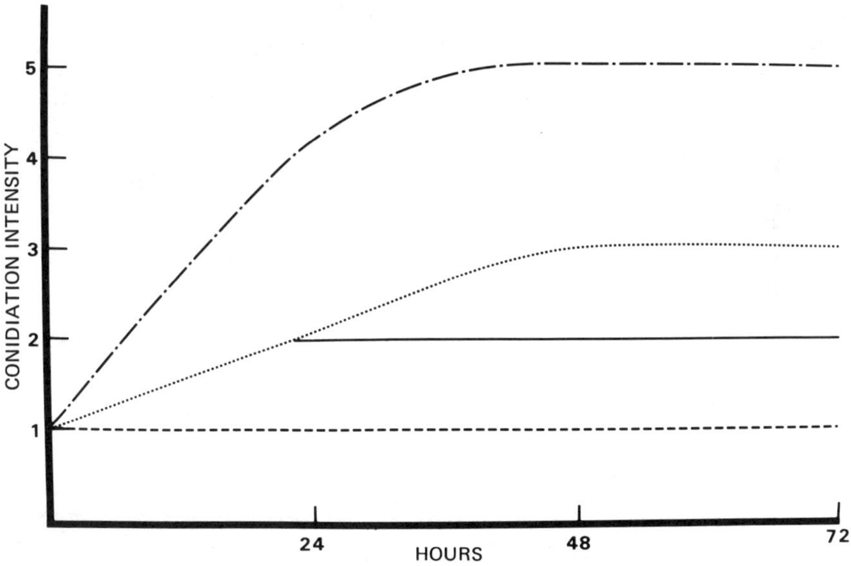

FIG. 3 Conidiation by *Monascus rubiginosus* ATCC 16367 when cultivated on a medium containing nitrate (-—·—·—), the same medium where the pH was adjusted to 2.5 before inoculation (·····················), a medium containing ammonium as the nitrogen source (————————), and the same medium where the pH was adjusted to 2.5 before inoculation (------).

is not directly responsible for the formation of an increased number of conidia, because especially on media containing nitrate the fungus has already formed large numbers of conidia before the amino acids are taken up. The inhibition of nitrate and ammonium uptake in the presence of amino acids suggests that an early acute nitrogen limitation is responsible for the induction of conidiation. As sugar uptake is not reduced, it is possible that the intermediates formed from glucose are used for the formation of conidia. The nitrogen compounds needed for conidia formation probably come from intracellular protein, since proteases are known to be stimulated during conidiation (Morton et al., 1960; Lawler and Weber, 1980). On media containing ammonium, amino acid uptake starts earlier than on nitrate medium; hence these cultures are probably less nitrogen limited, which results in fewer conidia and more biomass being formed.

The striking effects of pH on conidiation and secondary metabolite production are also reflected in nutrient uptake. At low pH, nitrate

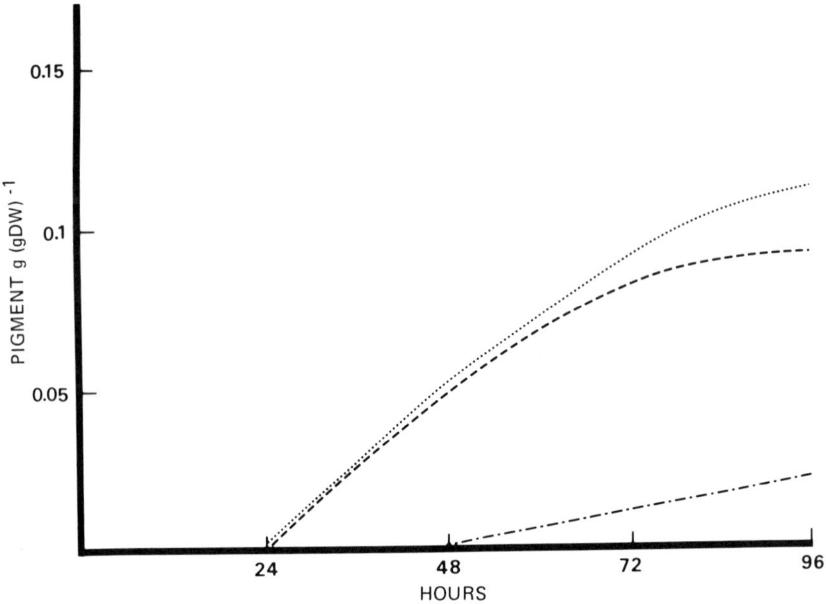

FIG. 4 Pigment production by *Monascus rubiginosus* ATCC 16367 on medium containing nitrate as the nitrogen source with added amino acids (—·—·—), the same medium where the pH was adjusted to 2.5 before inoculation (··········), a medium containing ammonium as the nitrogen source with added amino acids (no pigment was produced), and the same medium where the pH was adjusted to 2.5 before inoculation (- - - - - - -).

uptake is more efficient, whereas ammonium uptake is reduced, confirming an already known phenomenon (Morton and MacMillan, 1954). However, with amino acid uptake occurring earlier on ammonium medium, even at low pH, the intracellular nitrogen limitation is also overcome to a certain extent. Glucose uptake is always reduced when the starting pH is low, which points to a relationship between nitrogen and glucose uptake for the induction of conidiation.

E. Potassium and Phosphate Uptake

From our observations it is clear that the nitrogen source determines the evolution of pH, and this in turn directly determines conidiation and secondary metabolite production.

The stimulation of conidiation by amino acids, especially on nitrate media, is caused by the high starting pH. This was well demonstrated by the results obtained when the pH was readjusted after amino acid

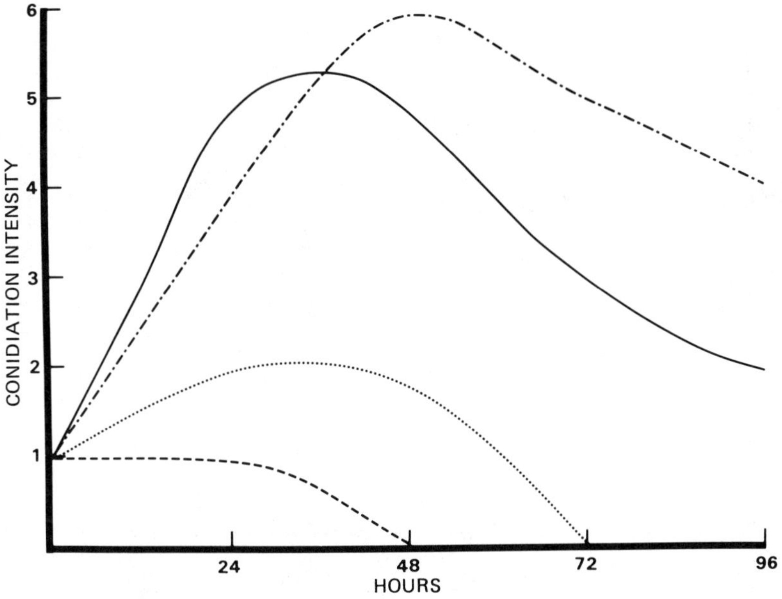

FIG. 5 Conidiation by *Monascus rubiginosus* ATCC 16367 when cultivated on medium containing nitrate as the nitrogen source with added amino acids (—·—·—), the same medium where the pH was adjusted to 2.5 before inoculation (··········), a medium containing ammonium as the nitrogen source with added amino acids (————), and the same medium where the pH was adjusted to 2.5 before inoculation (– – – – – –).

addition, because despite an uptake of amino acids, fewer conidia were formed. It is known that the intracellular pH of microorganisms does not vary greatly in response to variations of the extracellular pH (Conway and Downey, 1950; Larmour and Marchant, 1977). This being so, it is doubtful that the effects we observed are due to changes in intracellular enzymatic systems caused by the external pH. However, the pH of the medium is known to affect the transport of certain media constituents, such as inorganic phosphate and potassium, whose uptake is best when the pH is least acidic.

The results of experiments designed to investigate the effect of pH on the uptake of phosphate and potassium confirm that the phenomena observed are brought about by the differences in the uptake of certain nutrients, as summarized in Figs. 6 and 7. On ammonium medium there is a more balanced uptake of nutrients which are converted into biomass, and there is a rapid increase in growth prior to pigment production. On nitrate medium there is not a balanced growth, because

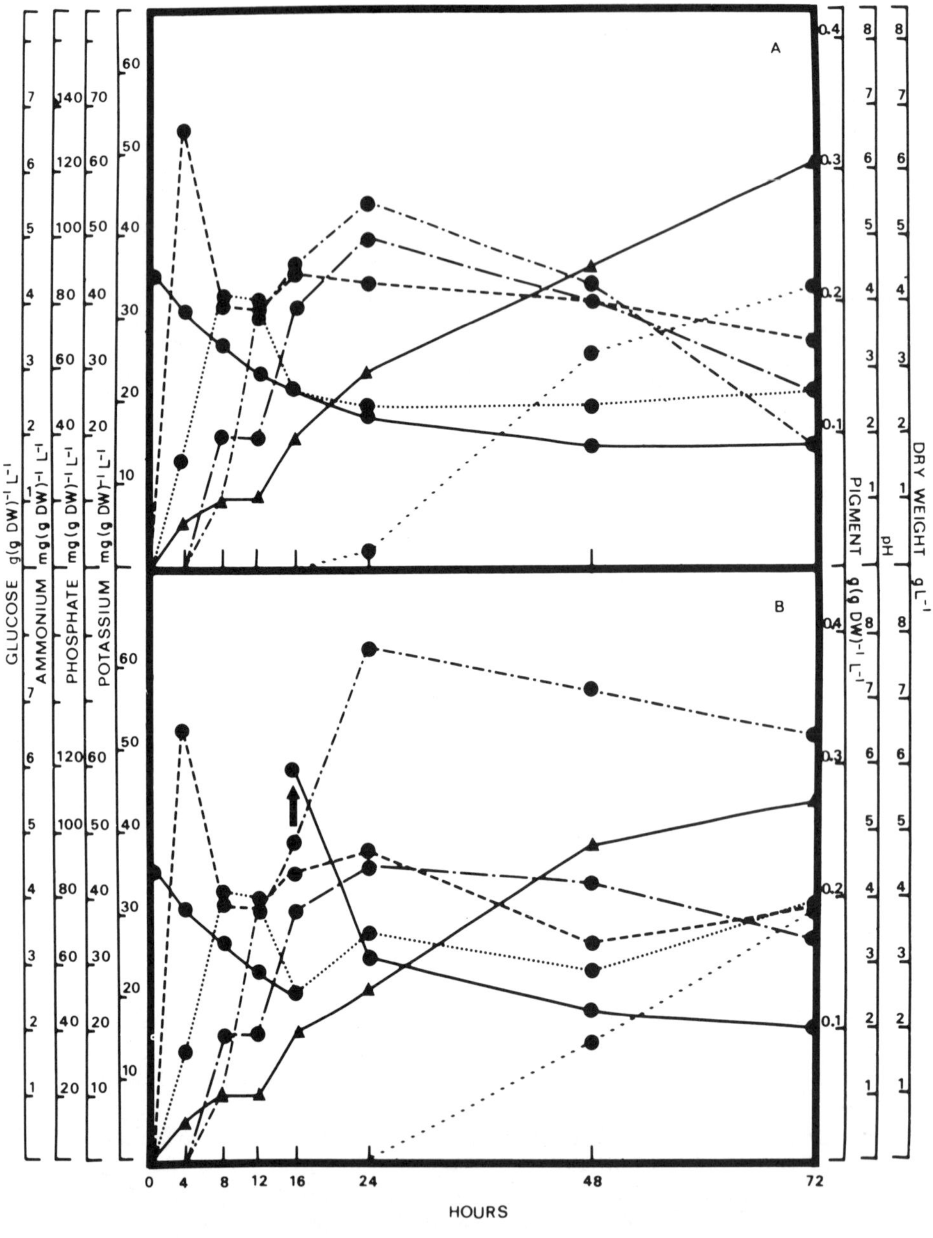

FIG. 6 Growth (dry weight) (▲——▲), pH evolution (●——●), pigment production (●·······●), and uptake of nitrate nitrogen (●– – – –●), glucose (●··········●), phosphate (●–·–·–●), and potassium (●–·——●) by *Monascus rubiginosus* ATCC 16367 when cultivated on (A) a medium containing nitrate as the nitrogen source and (B) on the

in spite of a very high initial uptake of glucose and phosphate, biomass does not increase rapidly, but conidiation is induced. The probable explanation for these observations is that there is an initial nitrogen limitation which prevents the normal functioning of the synthetic metabolic pathways, and therefore biomass formation is inhibited and glucose and phosphate are converted into storage materials needed for conidia formation.

Although the uptake of glucose and phosphate on nitrate medium is high, it is not accompanied by an equal uptake of potassium, which is necessary for the efficient functioning of the Embden-Meyerhof-Parnas pathway (Passeau and Lowry, 1963). On the ammonium medium, where more balanced growth is seen, the uptake of potassium follows that of phosphate.

These differences are confirmed when the pH of the medium is changed after 16 h. Especially when it is raised in the ammonium medium, there is an immediate increase in phosphate uptake accompanied by an increase in glucose uptake, which leads to more conidia being formed and pigment production being reduced. Potassium uptake, however, does not follow these increases.

Figures 6 and 7 also demonstrate the effect of changing the pH on secondary metabolism, because even when the pH is changed during the fermentation, the effect is immediate. Figure 8 illustrates the effect on conidiation. It can be seen that conidiation either decreases or increases within a few hours of the change in pH.

III. POSSIBLE CONTROL MECHANISMS IN *MONASCUS*

It has been demonstrated for other fungi (Smith et al., 1971) that the Embden-Meyerhof-Parnas and hexose monophosphate pathways function at all times during the development of a fungus and it is their relative contribution to the overall metabolism which changes. It is therefore clear that one does not have an all-or-nothing situation where one pathway functions and the other is shut off.

The results obtained with *Monascus* should also be interpreted in this light, because it would appear that the rate with which glucose is degraded by the different metabolic pathways determines whether biomass, conidia, or pigment are formed. At a high pH more phosphate is used which could activate the Embden-Meyerhof-Parnas pathway and eventually reduce the hexose monophosphate pathway. This in turn would limit the availability of NADPH, which is required for

---

FIG. 6 (Continued) same medium where the pH was adjusted after 16 h. The medium contained 4% glucose, 0.3% $NaNO_3$, 0.1% KH_2PO_4, 0.05% $MgSO_4 \cdot 7H_2O$, 0.05% NaCl, and 0.01% $FeSO_4 \cdot 7H_2O$. (Adapted from Carels and Shepherd, 1979.)

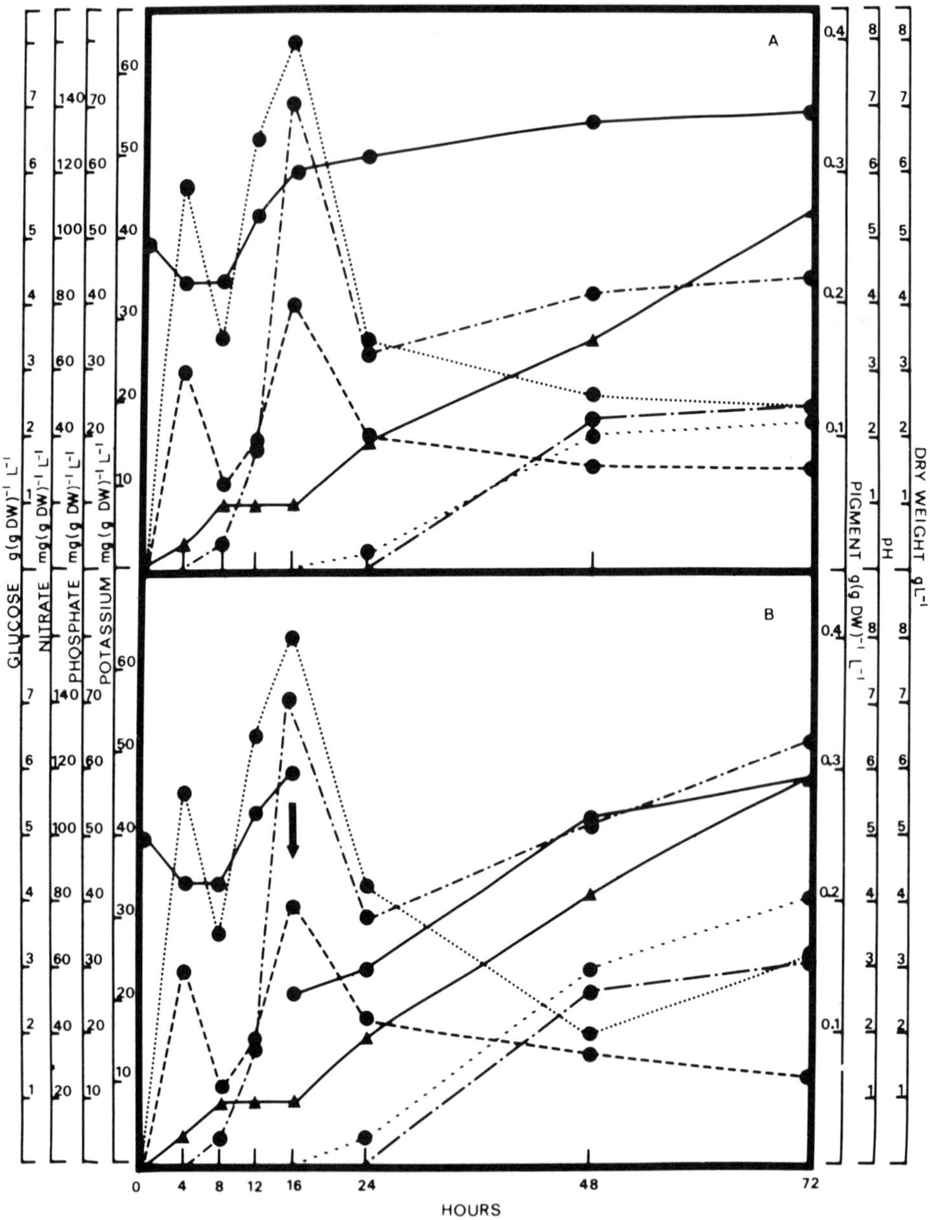

FIG. 7 Growth (dry weight) (▲———▲), pH evolution (●———●), pigment production (●·········●), and uptake of ammonium nitrogen (●– – – – –●), glucose (●···············●), phosphate (●–·–·–·●), and potassium (●—– —–●) by *Monascus rubiginosus* ATCC 16367 when cultivated on (A) a medium containing ammonium as nitrogen source and on (B) the same medium where the pH was adjusted after 16 h. The

nitrate reduction as well as for the formation of glutamic acid from α-ketoglutarate (Pateman and Cove, 1967; Kingham and Pateman, 1973). On the ammonium medium NADPH is not required for the uptake of ammonium which enters by diffusion (MacMillan, 1954) and is only necessary for the formation of glutamic acid. The pH also drops quickly, limiting phosphate uptake, which could allow a better distribution of glucose into the Embden-Meyerhof-Parnas and hexose monophosphate pathways. Such metabolic conditions would therefore always permit a more balanced growth on ammonium than on nitrate medium.

In the presence of amino acids in both media the high number of conidia could be due to an early acute intracellular nitrogen limitation. The high pH due to the addition of amino acids permits high phosphate uptake, which inhibits the hexose monophosphate pathway and stimulates the Embden-Meyerhof-Parnas pathway.

As suggested by other workers, nitrogen starvation is required for the induction of both conidia and secondary metabolite formation by fungi, which is confirmed by *Monascus*. However, the mechanisms of induction for the two systems are quite different. If culture conditions are such that the hexose monophosphate pathway is limited at the beginning of growth, acute intracellular nitrogen starvation could occur while glucose metabolism continues by the Embden-Meyerhof-Parnas pathway and conidia are therefore rapidly formed. If the hexose monophosphate pathway can function normally early during cultivation, normal growth occurs. When nitrogen is used up from the medium, excess intermediates are channeled into secondary metabolite formation. This confirms other results which suggested that in a sporulating fungus the Embden-Meyerhof-Parnas pathway is more efficient prior to the appearance of conidia (Smith et al., 1971) and that the hexose monophosphate pathway is predominant prior to secondary metabolite production (Bu'Lock et al., 1965).

IV. INDUCTION OF SEXUAL REPRODUCTION

Although the main work with *Monascus* was concerned with the industrial production of secondary metabolites, some work was carried out on strains which reproduced sexually. The results confirmed certain ideas on the control mechanisms regulating reproduction.

All the strains which reproduced sexually expressed this characteristic most efficiently on nitrate medium. On this medium the uptake of nitrate is very limited or nonexistent (Carels and Shepherd, 1977), therefore indicating that in order for sexual reproduction to be

---

FIG. 7 (Continued) medium contained 4% glucose, 0.19% NH_4Cl, 0.1% KH_2PO_4, 0.05% $MgSO_4 \cdot 7H_2O$, 0.05% NaCl, and 0.01% $FeSO_4 \cdot 7H_2O$. (Adapted from Carels and Shepherd, 1979.)

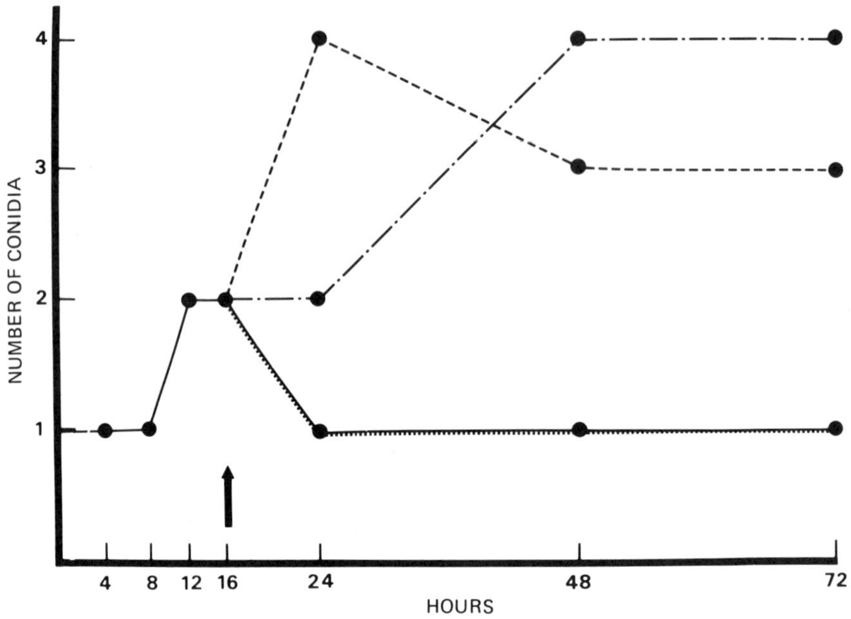

FIG. 8 Conidiation of *Monascus rubiginosus* ATCC 16367 when cultivated on a medium containing nitrate as the nitrogen source (●—·—●), the same medium where the pH was lowered after 16 h (●··········●), a medium containing ammonium as the nitrogen source (●———●), and the same medium where the pH was raised after 16 h (●- - - - -●). (See Figs. 6 and 7 for the rest of the data from these experiments.) (Data from Carels and Shepherd, 1979.)

initiated extreme physiological conditions must prevail. This idea is confirmed by the results on ammonium medium, where some nitrogen uptake occurs by diffusion and sexual reproduction, although initiated, does not develop as well.

The most striking results were those on rich media containing amino acids where sexual reproduction also occurs. However, the early uptake of amino acids, as with nitrate, is very low or nonexistent, indicating that the strains which reproduce sexually have a deficiency in their active nitrogen uptake systems. As usual, with this type of observation it is difficult to discern the exact cause; however, it was also seen with these strains that their initial uptake of glucose and phosphate is also very low. Again, the uptake of phosphate and the metabolism of glucose play important roles in the development of these strains. This was confirmed when the pH of these cultures was lowered on nitrate medium: More biomass was formed, some secondary metabolites were produced, and sexual development was limited. In parallel, nitrate, glucose, and phosphate uptake was improved.

Further confirmation of the role of phosphate was obtained when a strain which normally produced large quantities of secondary metabolites but no conidia was cultivated on nitrate medium without phosphate. Under these conditions sexual reproduction was induced.

Although more extreme physiological conditions seem to be necessary for the induction of sexual reproduction by *Monascus*, the same mechanisms appear to be operative as for the control of conidiation and secondary metabolism.

In other species of fungi the initiation of sexual reproduction seems to be more variable and the mechanisms are not as uniform as those for the induction of secondary metabolites and conidiation. This may be due to the complexity of the formation of fruiting bodies, especially in the Basidiomycetes. A detailed discussion of the mechanisms controlling sexual reproduction is beyond the scope of this chapter and numerous reviews have already been written on the subject (Raper, 1966; Turian, 1969, 1978; Smith and Galbraith, 1971; Sundberg, 1978; van den Ende, 1978).

One interesting phenomenon, however, which frequently occurs is that sexual reproduction is often linked with the formation of secondary metabolites, many of which are considered to be sex hormones (Taber, 1966; van den Ende and Stegwee, 1971; Austin et al., 1970; Gooday, 1973).

V. EFFECT OF INOCULUM

One of the most important observations to come out of the *Monascus* studies was that in well-phased cultures grown under controlled conditions the early hours of cultivation were the most critical for determining the future development of the culture. The factor which most influenced these early hours of cultivation was the inoculum.

The age of the inoculum, the medium used for its cultivation, and therefore its physiological state are of the utmost importance. If the inoculum used for the production of secondary metabolites is not in the correct physiological state, a considerable decrease in production will occur because the early hours of cultivation determine the future direction the culture will take. Other workers (Sekiguchi and Gaucher, 1977) have observed that with *P. urticae* the type of inoculum used greatly influenced the level of enzymes present at the beginning of the fermentation and hence the final level of secondary metabolites produced.

Although different levels of pigment-synthesizing enzymes were not measured in *Monascus*, it was shown that in order to obtain well-phased cultures in the production medium it was necessary to have an inoculum which was actively growing and that had no nutritional limitations (Carels and Shepherd, unpublished results). It therefore produced only vegetative biomass, very few conidia, and no secondary metabolites. The age of the inoculum was always critical, maximum efficiency only being achieved within very fine limits of 1-2 h.

These results confirm those of Calam (1976), who demonstrated the importance of the inoculum for maximum production of secondary metabolites by different microorganisms, especially the type of morphology exhibited by fungi. In later work Smith and Calam (1980) investigated inocula, used for the production of penicillin and griseofulvin, giving different yields. It was shown that biochemical factors, such as the level of enzyme activity and efficiency, were at least as important as morphology in determining yield, being carried forward from the inoculum to the production stage.

Finally, Calam and Smith (1981) investigated the time at which inoculum quality is determined and the events associated with the process. Their results are again in agreement with the observations with *Monascus*, in that the initial stage of growth is very important in fixing the future metabolic and morphological pattern of behavior of the culture.

VI. GENETIC CONSIDERATIONS

Naturally no discussion concerning the relationship between sporulation and secondary product formation can overlook the influence of the genetic makeup of the fungal organism on these parameters. As can be seen from Table 1, on any given medium the 40 different strains of *Monascus* expressed widely differing characteristics owing to their genotype. Even though these strains were easily manipulated by varying the pH of the medium, it was never possible to arrive at conditions where all the strains expressed the same phenotype or were capable of producing the same level of secondary metabolites or number of conidia.

The degree to which the genotype overcomes the effect of the environment and vice versa is very difficult to discern. However, it has already been postulated (Berry, 1975) that the efficiency with which nutrients are converted into cellular products by the organism's metabolic pathways is the sum of the genotypic and environmental controls. The balance between the two will determine the degree to which a strain can be manipulated.

At the present time it is still difficult to know exactly which genes affect the different pathways responsible for differentiation and secondary product formation; however, the considerable work carried out with *Neurospora* (Scott, 1976) would indicate that most morphological mutants are the result of enzymatic defects which cause the malfunction of major metabolic pathways.

This has recently been shown to be true for a mutant of *Streptomyces aureofaciens* which exhibited lower growth and higher tetracycline production (Colombo et al., 1981). The mutant showed a lower efficiency in the utilization of phosphate for growth and a higher cellular RNA content. The results indicate that the mutation impairs some aminoacyl RNA synthetases and could explain the lower growth in phosphate-limited conditions and the higher RNA content of the mutant.

If this is so, the strains of *Monascus* which reproduce sexually could be considered as natural morphological mutants. They seem to have defective glucose-metabolizing pathways, because in spite of ideal environmental conditions, little uptake of nutrients, especially glucose and phosphate, is observed, which creates extreme physiological conditions inside the cell.

These observations illustrate that if the genotype is such that major metabolic defects exist within a strain, then the environment is of lesser importance in controlling the type of phenotype expressed. It is therefore obvious that in order for a strain to adapt easily to a wide variety of environmental conditions, it must have a genotype which is as complete as possible and which allows the perfect functioning of the major metabolic pathways.

VII. CONCLUSIONS

Despite the fact that much of this chapter has been concerned with studies of *Monascus*, it was fortuitous to have in *Monascus* a genus that exhibits a wide range of phenotypes and species which could be manipulated using relatively simple means in order to express different phenotypes and which nevertheless allowed us to confirm many of the phenomena already described. The rapid response of the strains to pH changes allowed the elaboration of certain control mechanisms without resorting to sophisticated analytical techniques. (Certainly measurements of enzyme levels would have helped to confirm these observations). Also, for this type of study the results clearly demonstrate the great importance of studying microorganisms in controlled, well-phased cultures and, above all, the importance of studying the early developments of the culture, because it is this phase together with the physiological state of the inoculum which determines the later stages of development. Apart from being a potential producer of industrially interesting pigments, *Monascus* provides a very elegant model of how fungi can use their metabolism to survive in different environments and how quickly they can adapt.

These results provide further evidence to the idea of Bu'Lock (1975), that secondary metabolites are a means of survival for a strain, allowing metabolism to continue even though certain nutrients are lacking. *Monascus* seems to have evolved in a way whereby it can quickly change its metabolism, switching from one glycolytic pathway to another, thereby changing from sporulation to secondary metabolism and vice versa, in a very short period of time.

REFERENCES

Austin, D. J., Bu'Lock, J. D., and Drake, D. (1970). The biosynthesis of trisporic acids from β-carotene via retinol and trisporol. *Experientia 26*:348-349.

Berry, D. R. (1975). The environmental control of the physiology of filamentous fungi. In *The Filamentous Fungi, Industrial Mycology*, Vol. 1. J. E. Smith and D. R. Berry (Eds.). Edward Arnold, London, pp. 16-32.

Bu'Lock, J. D. (1965). Aspects of secondary metabolism in fungi. In *Biogenesis of Antibiotic Substances*. Z. Vanek and Z. Hostalek (Eds.). Academic, New York, pp. 61-72.

Bu'Lock, J. D. (1975). Secondary metabolism in fungi and its relationship to growth and development. In *The Filamentous Fungi, Industrial Mycology*, Vol. 1. J. E. Smith and D. R. Berry (Eds.). Edward Arnold, London, pp. 33-58.

Bu'Lock, J. D., Hamilton, D., Hulme, M. A., Powell, A. J., Smalley, H. M., Shepherd, D., and Smith, G. N. (1965). Metabolite development and secondary metabolism in *Penicillium urticae*. *Can. J. Microbiol. 2*:765-778.

Calam, C. T. (1976). Starting investigational and production cultures. *Process Biochem. 11*:7-12.

Calam, C. T., and Smith, G. M. (1981). Regulation of the biochemistry and morphology of *Penicillium chrysogenum* in relation to initial growth. *FEMS Microbiol. Lett. 10*:231-234.

Carels, M., and Shepherd, D. (1975). Sexual reproductive cycle of *Monascus* in submerged shaken culture. *J. Bacteriol. 122*:288-294.

Carels, M., and Shepherd, D. (1977). The effect of different nitrogen sources on pigment production and sporulation of *Monascus* species in submerged shaken culture. *Can. J. Microbiol. 23*:1360-1372.

Carels, M., and Shepherd, D. (1978). The effect of pH and amino acids on conidiation and pigment production of *Monascus major* ATCC 16362 and *Monascus rubiginosus* ATCC 16367 in submerged shaken culture. *Can. J. Microbiol. 24*:1346-1357.

Carels, M., and Shepherd, D. (1979). The effects of changes in pH on phosphate and potassium uptake by *Monascus rubiginosus* ATCC 16367 in submerged shaken culture. *Can. J. Microbiol. 25*:1484-1488.

Carter, B. L. A., and Bull, A. T. (1969). Studies of fungal growth and intermediary metabolism under steady and non-steady state conditions. *Biotechnol. Bioeng. 11*:785-804.

Cochrane, V. W. (1958). *Physiology of Fungi*, Wiley, New York.

Colombo, A. C., Crespi-Perellino, N., Grein, A., Minghetti, A., and Spalla, C. (1981). Metabolic and genetic aspects of the relationship between growth and tetracycline production in *Streptomyces aureofaciens*. *Biotechnol. Lett. 3*:71-76.

Conway, E. J., and Downey, M. (1950). pH values of the yeast cell. *Biochem. J. 47*:355-360.

Demain, A. L. (1972). Cellular and environmental factors affecting the synthesis and excretion of metabolites. *J. Appl. Chem. Biotechnol. 22*:345-362.

Dicker, J. W., Oulevey, N., and Turian, G. (1969). Amino acid induction of conidiation and morphological alterations in wild type and morphological mutants of *Neurospora crassa*. *Arch. Microbiol. 65*:241-257.

Ende, van den, H. (1978). Sexual morphogenesis in the Phycomycetes. In *The Filamentous Fungi, Developmental Mycology*, Vol. 3. J. E. Smith and D. R. Berry (Eds.). Edward Arnold, London, pp. 257-271.

Ende, van den, H., and Stegwee, D. (1971). Physiology of sex in Mucorales. *Bot. Rev. 37*:22-36.

Gooday, G. W. (1973). Differentiation in the Mucorales. *Symp. Soc. Gen. Microbiol. 23*:269-294.

Hawker, L. E. (1957). *Physiology of Fungi*. University of London, London.

Hesseltine, C. W. (1972). Solid state fermentation. *Biotechnol. Bioeng. 14*:517-532.

Klebs, G. (1928). *Die Bedingungen der Fortpflanzung bei Einigen Algen und Pilzen*. Fischer-Verlag, Berlin.

Kingham, J. R., and Pateman, J. A. (1973). NAD and NADP L-glutamate dehydrogenase activity and ammonium regulation in *Aspergillus nidulans*. *J. Gen. Microbiol. 78*:39-46.

Larmour, R., and Marchant, R. (1977). The induction of conidiation in *Fusarium culmorum* grown in continuous culture. *J. Gen. Microbiol. 99*:49-58.

Lawler, G. C., and Weber, D. J. (1980). Metabolism during asexual sporulation in *Rhizopus arrhizus* (Fischer). *J. Gen. Microbiol. 117*:465-474.

MacMillan, A. (1954). The entry of ammonia into fungal cells. *J. Exp. Bot. 7*:113-126.

Martin, J. F., and McDaniel, L. F. (1976). Biosynthesis of condicidin by phosphate limited resting cells of *Streptomyces griseus*. *Eur. J. Appl. Microbiol. 3*:135-144.

Martinelli, S. D. (1976). Conidiation of *Aspergillus nidulans* in submerged culture. *Trans. Br. Mycol. Soc. 67*:121-128.

Morton, A. G., and MacMillan, A. (1954). The assimilation of nitrogen from ammonium salts and nitrate by fungi. *J. Exp. Bot. 5*:232-252.

Morton, A. G., Dickerson, A. G. F., and England, D. J. F. (1960). Changes in enzyme activity of fungi during nitrogen starvation. *J. Exp. Bot. 11*:116-128.

Passeau, J. S., and Lowry, H. O. (1963). Phosphofructokinase and control of the citric acid cycle. *Biochem. Biophys. Res. Commun. 13*:372.

Pateman, J. A., and Cove, D. J. (1967). Regulation of nitrate reduction in *Aspergillus nidulans*. *Nature* (London) *215*:1234-1237.

Raper, J. R. (1966). Life cycles, basic patterns of sexuality and sexual mechanisms. In *The Fungi*, Vol. 2. G. C. Ainsworth and A. S. Sussman (Eds.). Academic, London, pp. 473-511.

Rehacek, Z., Kozova, J., Satdl, P., and Vorisek, T. (1974). The physiology of conidia formation in submerged cultures of *Claviceps purpurea* producing alkaloids. *Can. J. Microbiol. 20*:1323-1329.

Scott, W. A. (1976). Biochemical genetics of morphogenesis in *Neurospora*. *Annu. Rev. Microbiol. 30*:85-104.

Sekiguchi, J., and Gaucher, G. M. (1977). Conidiogenesis and secondary metabolism in *Penicillium urticae*. *Appl. Envir. Microbiol. 33*:147-158.

Shepherd, D. (1977). The relationships between pigment production and sporulation in *Monascus*. In *Biotechnology and Fungal Differentiation*. J. Meyrath and J. D. Bu'Lock (Eds.). Academic, London, pp. 103-118.

Smith, G. M., and Calam, C. T. (1980). Variations in inocula and their influence on the productivity of antibiotic fermentations. *Biotechnol. Lett. 2*:261-266.

Smith, J. E., and Galbraith, J. C. (1971). Biochemistry and physiological aspects of differentiation in the fungi. *Adv. Microb. Physiol. 5*:45-134.

Smith, J. E., Valenzuela-Perez, J., and Ng, W. S. (1971). Changes in activities of the Embden-Meyerhof-Parnas and pentose phosphate pathways during the growth cycle of *Aspergillus niger*. *Trans. Br. Mycol. Soc. 57*:93-101.

Sundberg, W. J. (1978). Hymenial cytodifferentiation in Basidiomycetes. In *The Filamentous Fungi, Developmental Mycology*, Vol. 3. J. E. Smith and D. R. Berry (Eds.). Edward Arnold, London, pp. 298-311.

Taber, W. A. (1966). Morphogenesis in Basidiomycetes. In *The Fungi*, Vol. 2. G. C. Ainsworth and A. S. Sussman (Eds.). Academic, London, pp. 387-412.

Turian, G. (1969). *Differenciation Fongique*. Masson, Paris.

Turian, G. (1978). Sexual morphogenesis in the Ascomycetes. In *The Filamentous Fungi, Developmental Mycology*, Vol. 3. J. E. Smith and D. R. Berry (Eds.). Edward Arnold, London, pp. 315-329.

Turner, W. B. (1971). *Fungal Metabolites*. Academic, London.

Woodruff, H. B. (1966). The physiology of antibiotic production. The role of producing organisms. *Symp. Soc. Gen. Microbiol. 16*: 22-46.

18

Genetic Aspects of Sexual Differentiation in Fungi

FRIEDHELM MEINHARDT and KARL ESSER *Lehrstuhl für Allgemeine Botanik, Ruhr-Universität, Bochum, Federal Republic of Germany*

I. INTRODUCTION

Differentiation or morphogenesis is a basic biological phenomenon common to almost all organisms. The study of the phenotypic expression of differentiation leading to detailed descriptions and taxonomic classifications of plants and animals has concerned scholars since antiquity. It is amazing that despite the great progress which has been made during the last decades in experimental biology in fields such as molecular biology, genetic engineering, nerve physiology, and so on, the whole procedure of differentiation, its initiation, and completion is still rather poorly understood. Since the beginning of this century, when Spemann (1936) started experimental research on morphogenesis, many attempts have been undertaken in both animals and plants to elucidate differentiation (Balinsky, 1970; Davidson, 1968; Fellenberg, 1980; Hadorn, 1970; Evans, 1971).

The fungi offer a wide array of developmental systems. In some cases experimental studies have led to some fundamental results, for example, the discovery that cAMP may initiate differentiation in slime molds and the genetic control of fruit body formation in higher Basidiomycetes.

However, we are far from having a substantial knowledge or even a model explaining the linkage between the genetic control of differentiation and its physiological expression at a molecular level. It is a pity that in those cases where we understand the genetic control, we lack the physiological data, and vice versa.

With this in mind, the following compilation has to be considered as an attempt to summarize the scattered genetic data not only to inform the interested reader, but also to stimulate further research in this promising area of biology. This implies that we shall not review data which deals exclusively with the environmental conditions of fungal morphogenesis. The survey of the experimental data follows the traditional classification system of fungi in order to simplify the presentation.

It should be mentioned that any study of the genetics of morphogenesis will be based on the performance and evaluation of sex and crosses. This requires a knowledge of the life cycle and of the laboratory procedures for handling specific fungi. We shall not deal here with these topics and we refer the reader to a recent review by Esser (1981). Extensive reviews dealing with fungal morphogenesis and genetics of differentiation can be found in Ainsworth and Sussman (1966), Smith and Berry (1978), Esser and Kuenen (1967), and Meyrath and Bu'Lock (1977).

II. COMPILATION OF DATA CONCERNING THE GENETICS OF SEXUAL DIFFERENTIATION

A. Myxomycetes

The cellular slime mold *Dictyostelium discoideum* (Acrasiales) is the most studied organism within this taxon. In particular, studies by Gerisch and collaborators (Gerisch, 1979; Gerisch et al., 1977, 1979; Riedel et al., 1973; Clark and Steck, 1979) have shown that cell aggregation, the prerequisite for fruit body formation, is initiated by periodic signals of cyclic adenosine monophosphate (cAMP) excreted by the cells. Its interaction with folic acid, cyclic guanidine monophosphate (cGMP), and the enzymes responsible for the synthesis of these compounds has been compared with the action of neurons in animals.

Unfortunately, the life cycle of *Dictyostelium* is not yet completely understood, especially with respect to whether a true eukaryotic sexual cycle involving karyogamy and meiosis takes place (MacInnes and Francis, 1974; Robson and Williams, 1980). Therefore the underlying genetic mechanism of this exciting example of differentiation remains obscure. There are mutants in which the transition from cell growth to aggregation is blocked (Gerisch, 1965) and where these deficiencies may be complemented by restoring the normal pathway of morphogenesis (Gerisch et al., 1975).

Representatives of the true slime molds *Physarum polycephalum* and *Didymium iridis* (Myxomycetales) have been extensively studied (see Collins, 1979). They have a well-known diploid life cycle characterized by physiological dioecism. The spores produced in typical sporangia are haploid and give rise to uninucleate halpoid amoebae of different mating types which are able to propagate asexually. After fusion of amoebae of opposite mating types, karyogamy is followed by numerous mitotic divisions, leading to the coenocytic plasmodium on which, depending on external conditions, the differentiation of sporangia takes place.

The many attempts to understand the genetic control of plasmodium formation only became successful when the so-called *colonia* strains of *P. polycephalum* were discovered which showed "haploid fruiting," that is, the amoebae aggregate to form haploid plasmodia, which also are able to differentiate sporangia and spores without karyogamy and meiosis having taken place. This phenomena is comparable to the "monokaryotic fruiting" of Basidiomycetes and shows that morphogenesis is not an integral part of the sexual cycle. Since the spores from the haploid cycle are viable, the resulting amoebae could be intercrossed with the mating types of diploid fruiters, giving diploid plasmodia with sexual fruiting. Since among the offspring of these hybrids a mendelian segregation of haploid and diploid fruiters was found, it was possible to establish a genetic system in which the expression of morphogenesis could be studied in both haploid and diploid plasmodia.

Using microdensitometry to measure the DNA content of Feulgen-stained nuclei, it was found that *colonia* strains are not self-fertile but apogamic, because plasmodia and amoebae do not differ in DNA content (Cooke and Dee, 1974). The same is true for *D. iridis* (Therrien and Collins, 1976). In a third species, *Didymium nigripes*, which is also able to form plasmodia in amoebal clones, chromosome counts led to the conclusion that amoebae and plasmodia have the same ploidy level (Dee, 1975).

Honey et al., (1979) identified a single "differentiation gene" (*dif A*) in *P. polycephalum*. Mutations of this gene inhibit plasmodia formation in strains normally producing haploid plasmodia and, subsequently, sporangia. Anderson and Dee (1977) described a strain carrying a mutation in a locus which possibly represents another differentiation gene.

B. Phycomycetes

Representatives of this taxon have long attracted the interest of research workers studying differentiation, as may be seen from numerous early reviews (Klebs, 1896; Burgeff, 1924; Couch, 1926; Kniep, 1928; Raper, 1952). These studies have revealed the existence and action of sexual hormones in both the aquatic fungi (Oomycetales;

see Mullins, 1968) and the terrestrial molds (Zygomycetales; van den Ende, 1976).

The essential feature of these sexual hormones is that they stimulate sexual morphogenesis. Thus mycelia of these fungi will not undergo any sexual differentiation unless the appropriate sexual hormone of a partner mycelium is present, irrespective of whether there is differentiation between male and female organisms (morphological dioecism) or whether they exist only as isomorphic mating types (physiological dioecism). The water mold *Achlya ambisexualis* (Saprolegniaceae) may serve as an example for the first group, and *Phycomyces blakesleeanus* (Mucoraceae) for the second. In both cases there is a series of hormonal interactions which works according to a Ping-Pong system, in that the hormone of one partner is induced concomitantly by morphological stimulation of the other partner to produce a hormonal response stimulating its differentiation.

In *Achlya* four distinct morphological phases, each initiated and controlled by one or more hormones, have been recognized (Raper, 1951).

In Mucoraceae sexual differentiation is governed by low-molecular-weight diffusable and volatile substances (Plempel, 1963a,b). Identical hormones (trisporic acids) are produced and are effective in different species (van den Ende, 1968; Austin et al., 1969).

In contrast to the detailed knowledge which has accumulated on the physiological aspects of differentiation in the Phycomycetes, the genetic data are rather scant owing in part to the following reasons. (1) It was not possible in early studies to obtain spore germination in *Achlya* under laboratory conditions, but this has now been achieved (Mullins and Raper, 1965; Barksdale, 1966). (2) The life cycle of the Oomycetales was incorrectly interpreted. It now seems clear that both the Saprolegniaceae and Phytophthoraceae are diploids. This naturally complicates the genetic analysis, because the haploid phase is restricted only to the nuclei present in the male and female sexual organs. This disadvantage has not yet been overcome, as described above for the diploid Myxomycetales. (3) Similar problems also occurred in the Mucoraceae. Germination of zygospores is very difficult to induce, and it was not known whether there was one or several meiotic divisions, because in addition to the low spore germination rate, many unexpected segregations occurred (Cerda-Olmedo, 1975).

A brief examination of the literature clearly shows how difficult it is to arrive at a sound genetic base for the physiological observations that have been recorded for many fungi.

In the offspring of crosses performed with either *Achlya, Pythium*, or *Dictyuchus*, a true mendelian segregation for males and females has never been found. In contrast, in addition to the dioecious forms, fertile monoecious forms were always obtained (Mullins and Raper 1965; Sherwood, 1969, 1971; Pratt and Green, 1973). All efforts have failed to explain this peculiar segregation pattern on the basis of multigenic control of sex determination (Lasure and Griffin, 1975). Thus at present there is no satisfying model to explain the sexual differentiation in these water molds.

This is not the case in the Mucoraceae, where at least some circumstantial evidence exists. In *Phycomyces blakesleeanus* mutants have been found that display abnormalities in carotene synthesis and aberrant sexual reactions, such as deficiencies in zygophore formation (Heisenberg and Cerda-Olmedo, 1968; Sutter, 1975). This failure to differentiate arises because the carotenes are precursors of the trisporic acids, the effective sexual hormones in the Mucoraceae (Sutter, 1975; Bu'Lock, 1976). Mutants blocked in zygospore development have also been described for *Mucor mucedo*. In this fungus trisporic acid and its precursors act as antagonists in the regulation of sexual differentiation (Wurtz and Jockusch, 1975,1978).

Recently *Phytophthora*, another Phycomycete, has become a subject of genetic studies of sexual differentiation. The complete life cycle of some species of this common plant pathogen can be observed under controlled laboratory conditions.

One of the best studied organisms is *Phytophtora drechsleri* (Khaki and Shaw, 1974; Mortimer et al., 1977). This diploid fungus is monoecious and forms male and female sex organs. However, in many isolates from Nature, owing to sexual incompatibility, no self-fertilization occurs. Cross-fertilization is possible between different mating types. Unfortunately, the genetic mechanism of this breeding system is not yet fully understood. It has been suggested that one of the two mating types of this diploid fungus is heterozygous and the other homozygous. There is also strong evidence for aneuploidy in some strains, causing self-fertility (Mortimer et al., 1977). Chang et al. (1974) studied the sexuality of *Phytophtora cinnamomii*. Some of their unexplained data fit within the interpretation of the Shaw group.

In contrast to the rather closely related representatives of Saprolegniaceae (e.g., *Achlya*), evidence for the existence of sexual hormones in *Phytophthora* has only recently been demonstrated (Ko, 1980).

C. Ascomycetes

Among the Ascomycetes sexual differentiation is present only in the filamentous forms, but not in yeasts. Mycologists and geneticists have long focused their interest on this subject, mainly because the perfect filamentous Ascomycetes produce, in contrast to higher Basidiomycetes, well-developed female and in most cases male sexual organs. Stimulated by Burgeff's (1924) observations on sexual hormones in the Mucoraceae, important emphasis has also been placed on this taxonomic group because of the demonstration of sexual hormones, as may be seen from the compilation in Hartmann's book on sexuality (1943).

A further trend also encouraged by Hartmann was to prove the existence of "relative sexuality," which was part of his dogmatic speculations of the genetic control of sexuality. However, many years later it was unequivocally shown that all known cases of "relative sexuality" in fungi could be explained by mendelian segregations of morphogenetic genes (see discussions and references in Esser, 1962).

Parallel to the progress of fungal genetics and following the *Neurospora* boom in the 1950s, genetic control of morphogenesis was studied in a number of different Ascomycetes, making partial use of the advantage to perform tetrad analysis.

In the early studies the assumption was followed that sexual chromosomes (as known for mosses, e.g., Allen, 1919; Lorber, 1927) or single genes might be responsible for the formation of male and female sex organs, as suggested for the phycomycetes (for details see Esser and Kuenen, 1967). However, this seemed improbable, because the hermaphroditic character of these haplonts was difficult to explain on a genetic basis of this kind. After some indications that a series of genes might be responsible for sexual differentiation in the plant pathogen *Glomerella cingulata* (Lucas et al., 1944; Chilton et al., 1945; Markert, 1949, 1952; McGahen and Wheeler, 1951; Wheeler and McGahen, 1952; Wheeler, 1954), a breakthrough was made in three laboratories working independently with *Sordaria.*

This coprophilous fungus is very well suited for all kinds of genetic research, because its life cycle is completed within 1 week, there is no ascospore dormancy, and asexual spores are not formed. The fungus is self-fertile, displaying apandric development; that is, there are no male sex organs. Its asci allow the analysis of ordered tetrads.

The two species studied in detail are *Sordaria macrospora* (large spores), by Esser and Straub (1958) and Heslot (1958), and *Sordaria fimicola* (rather small spores), by Carr and Olive (1959). The investigations on *Sordaria* have led to identical results, which are presented in detail using *S. macrospora* as an example.

As may be seen from Fig. 1, there are at least 15 genes needed to complete the life cycle. Two of these, *c* and *r*, are responsible for differentiation of the female sex organs. The further development of these ascogonia into protoperithecia is regulated by three more genes (*cit*, *spd*, and *p*). The remaining 10 genes are involved in the formation of fruiting bodies (perithecia): three in the stages from the dikaryotic phase formation, three in the meiotic divisions and spore production, one in the linear arrangement of ascospores, one in spore pigment development, and two in spore discharge. None of these genes are allelic.

The fact that a number of nonallelic genes affect the same developmental stage does not mean that they have the same genetic effect. On the contrary, it has been shown that such genes act at different points in the same developmental process.

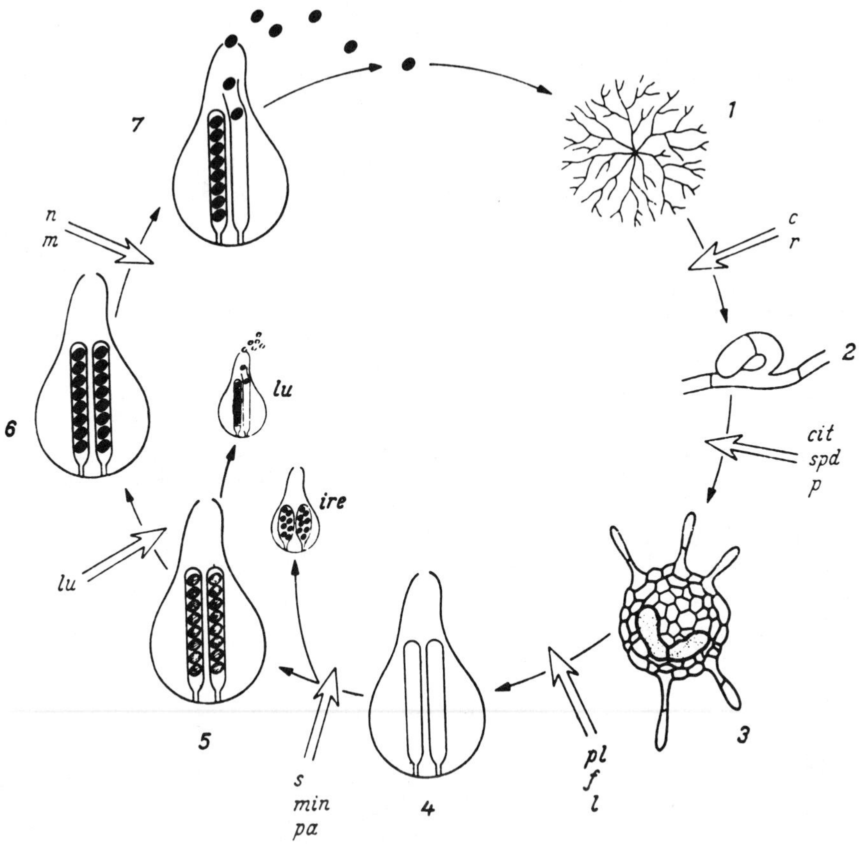

FIG. 1 Genetic control of sexual differentiation in *Sordaria macrospora*. The life cycle of this ascomycete is arbitrarily subdivided into the following developmental stages, (1) mycelium formation, (2) differentiation of the ascogonium, (3) transformation of the ascogonium into a protoperithecium, (4) the dikaryotic phase, with karyogamy and ascus formation; (5) meiosis and spore formation, (6) maturation of the ascospores, and (7) discharge of the ascospores. Male sexual organs are not produced in this apandrous fungus. The arrows with the gene symbols show at which stage of differentiation particular mutations block the developmental sequence. For further explanation see text. (Adapted from Esser and Straub, 1958.)

The mutants *min*, *pa*, and *s* all produce sterile fruiting bodies. Cytological investigation has shown that in the asci of *min* and *pa* meiosis does not occur, whereas the differentiation of asci in the *s* mutant is not blocked until after meiosis.

The expression of these morphogenetic genes naturally depends on interactions with environmental conditions. It was found that light had little influence, but this was not so for temperature. The optimal temperature for mycelial growth is 30 °C, but a slight increase to 31 °C completely blocks the sexual cycle in the wild strain at the protoperithecial stage (see stage 3 in Fig. 1). In contrast, in the mutant *colora* (not listed in Fig. 1) this arrest of differentiation occurred at 30 °C (Esser, 1980).

Following the unraveling of the genetic mechanism of sexual differentiation for *Sordaria*, it has been possible to explain the above-quoted experimental results of the Wheeler group on *Glomerella* along these lines (Wheeler et al., 1959).

This also holds true for some sporadic data derived from investigations of *Podospora anserina* (Esser, 1956; Durrens et al., 1979; Simonet and Zickler, 1978) and for *Neurospora* (Johnson, 1978, 1979; Mylek and Threlkeld, 1974; Raju and Perkins, 1978).

Thus the observations in genera that are not closely related support the conclusion that *in fungi (at least in Ascomycetes) differentiation of sexual organs and fruit bodies is under polygenic control*.

In *Podospora anserina* a single gene (*vacua*) effects transgression between the formation of female sexual organs and fruit bodies in being responsible for the formation of the ascocarp (Esser and Graw, 1980). The action of the *vacua* gene, which is also instrumental in a single mating type, is completely different from the many genes interfering with the cytology of ascus development after fertilization between + and − strains in this fungus (Simonet and Zickler, 1978); *vacua* causes a phenotype which is also completely different from those of the *barren* strains of *Neurospora crassa* (partial sterile perithecia), as described by Raju and Perkins (1978). However, threshold genes of this kind controlling the step from mycelial growth to plectenchymatic growth exist in the higher Basidiomycetes, where they effect the transgression from mycelial to stromatic growth (see below).

In contrast to this very comprehensive data, little is known about the physiological parameters and events involved in differentiation. In *S. fimicola* biotin is required for perithecial development (Barnett and Lilly, 1947). In *S. macrospora* fruit body formation depends on a specific ratio of biotin to arginine (Bahn and Hock 1974, Molowitz et al., 1976; Hock et al., 1978). Studies of the "edge effect" (accumulation of fruit bodies on the rim of the Petri dish or near mechanical obstructions) have not contributed to a better understanding of the physiological expression of the morphogenetic genes (MacDonald and Bond, 1976). In *N. crassa* there are indications that the loci involved in the morphological mutations are structural genes for specific enzymes, such as glucose-6-phosphate dehydrogenase and phosphoglucomutase (see Scott, 1976).

However, in summary, *the critical biochemical events which determine different morphogenetic steps are completely unknown.* The specific protein spectra attributed to certain stages of perithecial development (Nasrallah and Srb, 1977,1978) have a descriptive value only and cannot be linked with biochemical procedures resulting in differentiation.

D. Basidiomycetes

Sexual differentiation is found to occur only in the lower basidiomycetes (Heterobasidiomycetidae), but so far there are no genetic data which would allow an explanation of the formation of sexual organs [e.g., in the rust fungi (Uredinales)] and for further development into fruiting structures or fruit bodies on a genetic background. There is detailed knowledge available on the genetic control of breeding systems, especially for some plant pathogens within the Uredinales and Ustilaginales. Since, however, the incompatibility mechanisms are superimposed on sexual morphogenesis (Esser and Raper, 1965), an understanding of the genetic control of mating reactions does not permit an understanding of differentiation (see the reviews in Esser, 1971,1981).

In the higher basidiomycetes (Holobasidiomycetidae) sexual differentiation in the strict sense does not exist. These basidiomycetes follow, in general, a dikaryotic life cycle in which the haploid stage is restricted to the basidiospores and where the dikaryotic phase is established either already in the mature spore or immediately after spore germination via hyphal fusion (plasmogamy). In the resulting mycelia the dikaryotic phase is maintained mostly via clamp connections throughout its whole life, until karyogamy takes place in the basidia of the young fruit bodies. Since nuclear fusion is immediately followed by meiosis, the diploid phase is restricted to the basidial nucleus.

In most holobasidiomycetidae the mating reaction leading to plasmogamy is controlled by the bipolar or tetrapolar mechanism of the homogenic incompatibility system (see Esser, 1971). The genetic control of this mating system attracted the interest of geneticists rather early, ever since Kniep (1918,1920) discovered its existence. The mating mechanisms have been well documented for *Schizophyllum commune* by Raper and collaborators, Raper, 1966). They showed that both incompatibility factors (A and B) are complex genetic units, each consisting of two closely linked genes. The numerous alleles of these genes are responsible for the factor specificity in this multiallelic system. The Raper group further demonstrated that both the A and B factors controlled different steps in the sequence: hyphal fusion → nuclear exchange → nuclear migration → conjugated nuclear division

→ formation of a clamp connection. This was considered as a morphological sequence and the incompatibility factors were considered as morphogenetic traits. This is certainly justified to some extent (Day, 1960).

Parallel to Raper's work with *Schizophyllum*, comparable but less comprehensive studies were being made on a number of other basidiomycetes belonging mainly to the Poriales or Agaricales. Such data confirmed, in general, Raper's investigations in showing that in some (but not all) other species the incompatibility factors are genetically complex, but that in no way could their role in hyphal morphogenesis be shown as convincingly as in *Schizophyllum* (Esser and Kuenen, 1967; Koltin et al., 1972).

The many experimental attempts to understand the physiological expression of the incompatibility factors (e.g., Raper and Esser, 1961; Wang and Raper, 1969) have only brought about some rather speculative models (Parag, 1965; Kuhn and Parag, 1972; Ullrich, 1978).

It is somewhat surprising that the numerous studies concerning the genetics of the mating reaction were never extended to investigations of the genetics of fruit body formation. Evidently it was satisfying that once a dikaryon is established. it depends only on external conditions (such as light, temperature, and media) for fruiting to occur (Fig. 2).

The observation made in many laboratories, that dikaryons from the same species did not fruit or lost the ability to fruit in laboratory cultures, did not stimulate corresponding genetic research. The stimulus to reveal the genetic control of fruit body formation came from a rather strange phenomenon: monokaryotic fruiting. Since its first description by Wakefield (1909) this phenomenon has been observed in more than 30 species of Basidiomycetes. In this way haploid mycelia originating from a single basidiospore sometimes do form more or less normal fruit bodies, despite the fact that a plasmogamy and therefore a dikaryotic phase was never established.

A detailed genetic analysis of this phenomenon in *Polyporus ciliatus* (Stahl 1976; Stahl and Esser, 1976), *Schiz. commune* (Esser et al., 1979), and *Agrocybe aegerita* (Esser and Meinhardt, 1977; Meinhardt and Esser, 1981; Meinhardt, 1980), three not too closely related fungi, has not only unveiled the underlying mechanism of fruit body formation, but also made evident that the genes concerned control dikaryotic fruiting as well. This will be briefly exemplified using the agaric *A. aegerita*.

The fungus *A. aegerita* is of particular interest in biotechnology, because as a wood-decaying fungus, it may be used in recycling procedures to convert straw or other plant remains into protein-rich "mykofodder," while, in addition, its tasty fruit bodies can successfully compete with other edible mushrooms.

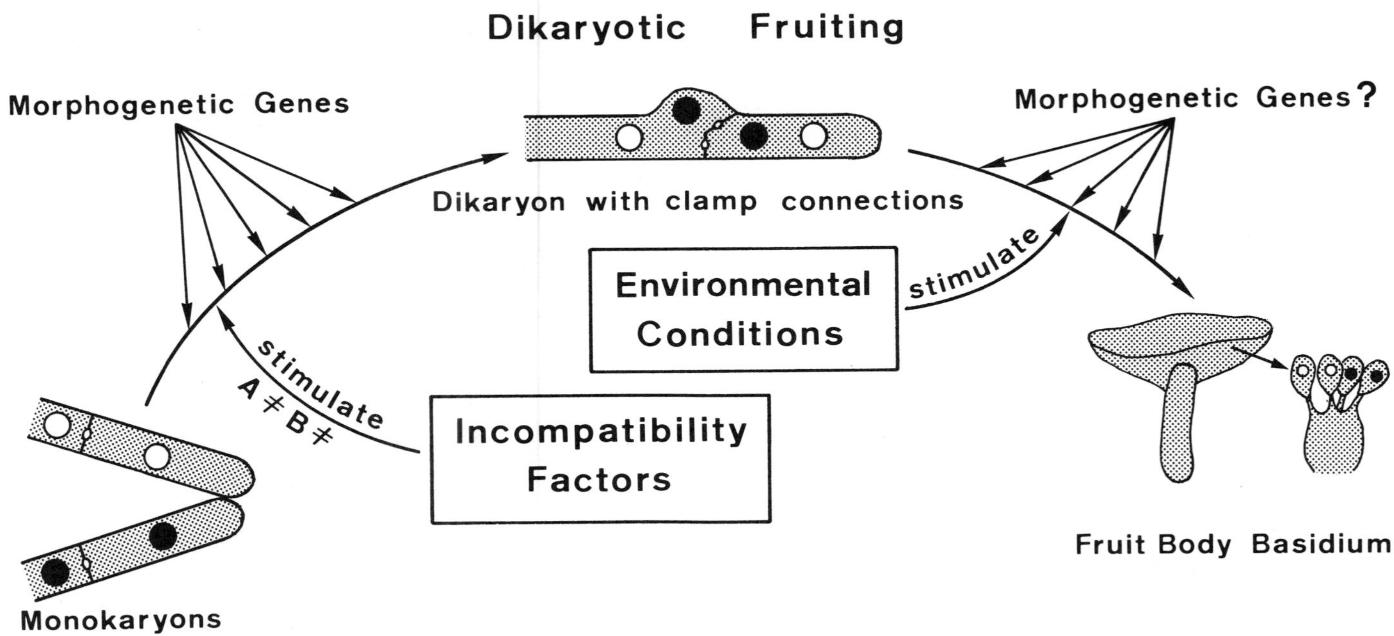

FIG. 2 Sexual cycle of a higher basidiomycete which is controlled by tetrapolar incompatibility. (Adapted from Stahl and Esser, 1976.)

If a fruit body is taken from Nature and monokaryotic mycelia are grown from single basidiospores, up to 60% of the mycelia produce fruiting initials or small but fully developed fruit bodies. The basidia of these monokaryotic fruit bodies develop two spores only. When germinated, the resulting mycelia are again monokaryons (no clamp connections!) and have the same mating type as the original monokaryon. As cytological studies have revealed in the young basidia, neither karyogamy nor meiosis takes place. Thus the formation of the monokaryotic fruit bodies is not part of a sexual cycle; in contrast, it is a form of asexual propagation.

Thus it has become evident that in higher basidiomycetes differentiation of fruit bodies does not depend on a sexual cycle.

As can be seen from Fig. 3, genetic analysis has shown that monokaryotic fruiting depends on the action of at least three genes acting sequentially: There is a threshold gene (su^+/su) which suppresses differentiation in its active form (su^+). If the allele *su* is present, differentiation may take place. The production of fruit body initials is then brought about by the gene fi^+ (fruiting initiation). The gene fb^+ (fruit body) completes the fruit body to its normal shape. Thus monokaryotic fruiting requires the genotype *su* fi^+ fb^+ to be present in a monokaryon. Any other configuration, for example, *su* fi^+ *fb* or su^+ fi^+ fb^+, leads to initials only or prevents differentiation completely.

As can be seen from Fig. 3, the three pairs of alleles are also instrumental in dikaryotic fruiting. To open the pathway of differentiation, one of the two nuclei of the dikaryon must carry the inactive allele *su* of the threshold gene (su^+/su). For further differentiation, also at least one of the active alleles (fi^+ or fb^+) needs to be present in either one of the two nuclei.

From these investigations it has become evident that the three above-mentioned genes are the hitherto unknown morphogenetic genes of the higher basidiomycetes (see Fig. 2). The failure of many laboratory cultures and even isolates taken from Nature to fruit can be explained by the absence of the active alleles of one or more of these three loci.

Various experimental data which have previously been published harmonize with the model derived from the present data.

In *Lentinus edodes* the characteristics of "sporelessness," "white pilei," and "brownish gills" are under the control of different single recessive genes. In *Coprinus macrorhizus*, Kimura and Fujio (1961) reported that the characteristic of "undeveloped fruit bodies" is controlled by a single recessive factor. In *Schiz. commune* a single recessive gene, which might be equivalent to the *fi* locus because it blocks fruiting in homozygous dikaryons, is described (Perkins and Raper, 1970), whereas many of the genes involved in the initiation of fruiting in *Schiz. commune*, as described by Leslie and Leonard (1979a,b), might affect fruiting by accident.

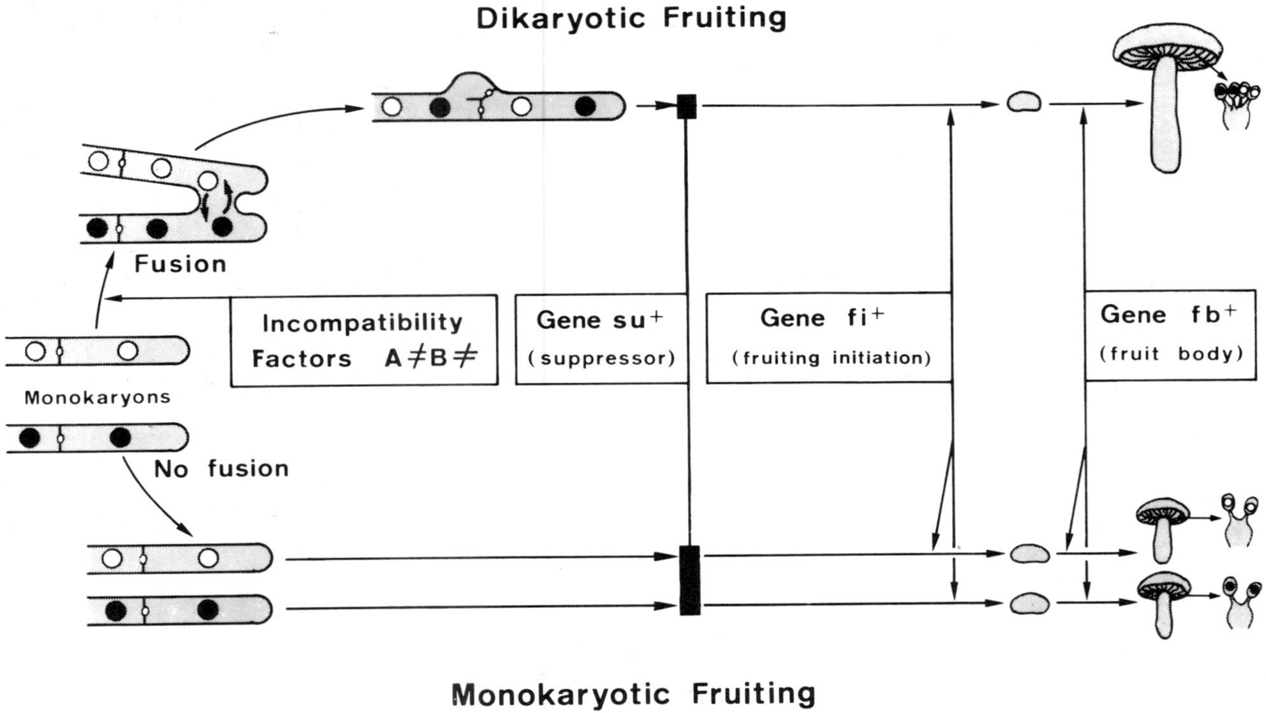

FIG. 3 Genetic control of fruit body differentiation in *Agrocybe aegerita* as an example of a higher basidiomycete. (Adapted from Meinhardt, 1980.)

In *C. macrorhizus* there exist monokaryotic mutant strains which produce four basidiospores on a basidium after a typical meiotic division (Miyake et al., 1980). At first, this appears to be in contrast to the conclusions given above, but these data can, however, be interpreted by a mutation from incompatibility to self-fertility, and it becomes evident that in higher basidiomycetes not even the meiotic event depends on the action of the incompatibility factors, as previously postulated by Raper and associates (Raper, 1966).

In contrast to the knowledge of genetic parameters of fruiting, the physiological mechanisms controlled by these genetic traits still remain obscure.

Leonard and Dick (1968) described the induction of fruiting in monokaryons of *Schiz. commune* by the imperfect fungus *Hormodendrum cladosporioides*; they isolated a fruiting-inducing substance (FIS). Leonard and Raper (1969) also proved the existence of FIS in fruit bodies of *Schiz. commune* and *Agaricus bisporus*.

Uno and Ishikawa (1973a,b) identified the FIS in *Coprinus macrorhizus* as cAMP and a protein which easily binds to cAMP. In fruiting induction cAMP seems to play the active part, because cAMP and specific enzymes involved in its metabolism are only found in strains displaying monokaryotic fruiting. A generalization of the action of cAMP is, however, not possible, as attempts to induce fruiting in *Schiz. commune* have failed (Schwalb, 1974).

The very interesting observations that the development of initials to fruit bodies depends on a reciprocal action of lytic and synthesizing enzymes such as glucanase and chitinase (Wessels, 1966,1969,1971) unfortunately lack a genetic basis.

III. CONCLUSIONS

If one surveys the entirety of the data presented in this paper, it becomes evident again that there is no general model to explain both the genetic and physiological control of sexual differentiation in fungi. This does not exist even for one class of Mycophyta. Certainly there is at least one "spotlight" in each class which gives comprehensive information about either the nature of the genetic traits responsible for morphogenesis or their physiological expression.

In the *Myxomycetes* cAMP was found to be the substance inducing fruit body formation and creating therewith the prerequisite for subsequent sexual differentiation, whereas in the *Phycomycetes* sexual differentiation was found to be governed by different hormones acting sequentially. In the *Ascomycetes*, despite a rather profound knowledge of genetics, in some organisms (e.g., *Sordaria*) and in the physiology of others (e.g., *Neurospora*) the correlation between genetic and biochemical data has not yet deciphered the molecular mechanisms of differentiation.

Finally, in the *Basidiomycetes* much emphasis has been concentrated on formal genetics and less on gene physiology, but this has unfortunately not yet revealed the gene action of a single morphogenetic gene.

However, in this context it must be stressed that the total data has revealed the existence of a hitherto unknown but evidently very general phenomenon: *The morphogenetic events connected with sexual differentiation, including fruit body formation, are not necessarily linked to the sexual cycle (karyogamy and meiosis)*. Similar or comparable modes of differentiation may also occur during asexual propagation. This indeed shows the *mutual independence of differentiation and sexuality*. This again can bring up an endless discussion along the old lines of the "chicken and the egg," namely, what was first: sex or differentiation?

REFERENCES

Ainsworth, G. C., and Sussman, A. S. (Eds.) (1966). *The Fungi*, Vol. 2. Academic, New York.

Allen, C. E. (1919). The basis of sex inheritance in *Sphaerocarpus*. *Proc. Amer. Phil. Soc. 58*:289.

Anderson, R. W., and Dee, J. (1977). Isolation and analysis of amoebal-plasmodial transition mutants in the myxomycete *Physarum polycephalum*. *Genet. Res. 29*:21-34.

Austin, D. J., Bu'Lock, J.D., and Gooday, G. W. (1969). Trisporic acids sexual hormones from *Mucor mucedo* and *Blakeslea trispora*. *Nature 223*:1178-1179.

Bahn, M., and Hock, B. (1974). Morphogenese von *Sordaria macrospora*: Induktion der Perithezeinbildung durch Arginin. *Ber. Dtsch. Bot. Ges. 87*:433-442.

Balinsky, B. I. (1970). *An Introduction to Embryology*, 3rd. ed. Saunders, Philadelphia, Pa.

Barksdale, A.W. (1966). Segregation of sex in progeny of a selfed heterozygote of *Achlya bisexualis*. *Mycologia 58*:802-804.

Barnett, H. L. and Lilly, V. G. (1947). The effects of biotin upon the formation and development of perithecia, asci, and ascospores by *Sordaria fimicola*. *Amer. J. Bot. 34*:196-204.

Bu'Lock, J. D. (1976). Cascade expression of the mating type locus in Mucorales. In *Proceedings of the Second International Symposium on the Genetics of Industrial Microorganisms*. K. D. MacDonald (Ed.). Academic, London, pp. 497-509.

Burgeff, H. (1924). Untersuchungen über Sexualität und Parasitismus bei Mucorineen. *Bot. Abh. 4*:1-135.

Carr, A. J. H., and Olive, L. S. (1959). Genetics of *Sordaria fimicola*. III. Cross-compatibility among self-fertile and self-sterile cultures. *Am. J. Bot. 46*:81-91.

Cerda-Olmedo, E. (1975). The genetics of *Phycomyces blakesleeanus*. *Genet. Res.* *25*:285-296.
Chang, S. T., Shepherd, C. J., and Pratt, B. H. (1974). Control of sexuality in *Phytophthora cinnamomii*. *Aust. J. Bot.* *22*:669-679.
Chilton, S. J. P., Lucas, G. B., and Edgerton, C. W. (1945). Genetics of Glomerella. III. Crosses with a conidial strain. *Am. J. Bot.* *32*:549-554.
Clark, R. L., and Steck, T. L. (1979). Morphogenesis in *Dictyostelium*: An orbital hypothesis. *Science* *204*:1163-1168.
Collins, O. R. (1979). Myxomycete biosystematics: Some recent developments and future research opportunities. *Bot. Rev.* *45*:146-201.
Cooke, D. J., and Dee, J. (1974). Plasmodium formation without change in nuclear DNA content in *Physarum polycephalum*. *Genet. Res.* *23*:307-317.
Couch, J.N. (1926). Heterothallism in *Dictyuchus*, a genus of the water molds. *Ann. Bot.* *40*:849-881.
Davidson, E. H., (1968). *Gene activity in early development.* Academic, New York.
Day, P. R. (1960). The structure of the A mating type locus in *Coprinus lagopus*. *Genetics* *45*:641-651.
Dee, J. (1975). Slime moulds in biological research *Sci. Prog. Oxford* *62*:523-542.
Durrens, P., Laigret, F., Labarere, J., and Bernet, J. (1979). *Podospora anserina* mutant defective in protoperithecium formation, ascospore germination and cell regeneration. *J. Bacteriol.* *140*:835-842.
Ende, H. van den (1968). Relationship between sexuality and carotene synthesis in *Blakeslea trispora*. *J. Bacteriol.* *96*:1298-1303.
Ende, H. van den (1976). *Sexual Interactions in Plants.* Academic, London.
Esser, K. (1956). Gen-Mutanten von *Podospora anserina* (CES.) REHM mit männlichem Verhalten. Naturwissenschaften *43*:284.
Esser, K. (1962). Die Genetik der sexuellen Fortpflanzung bei den Pilzen. *Biol. Zentralbl.* *81*:161-172.
Esser, K. (1971). Breeding systems in fungi and their significance for genetic recombination. *Mol. Gen. Genet.* *110*:86-100.
Esser, K. (1980). Effect of temperature and light on perithecial development in *Sordaria macrospora*. *Mycologia* *72*:619-622.
Esser, K. (1981). *Cryptogams*: Cyanobacteria, algae, fungi, lichens. Cambridge University, Cambridge.
Esser, K., and Graw, D. (1980). Homokaryotic fruiting in the bipolar-incompatible Ascomycete *Podospora anserina*. *Mycologia* *72*:534-541.
Esser, K., and Kuenen, R. (1967). *Genetics of Fungi.* Springer, Berlin.

Esser, K., and Meinhardt, F. (1977). A common genetic control of dikaryotic and monokaryotic fruiting in the Basidiomycete *Agrocybe aegerita*. *Mol. Gen. Genet.* *155*: 113-115.
Esser, K., and Raper, J. R. (Eds.) (1965). *Incompatibility in Fungi*. Springer, Berlin.
Esser, K., and Straub, J. (1958). Genetische Untersuchungen an Sordaria macrospora Auersw., Kompensation und Induktion bei genbedingten Entwicklungsdefekten. *Z. Indukt. Abstamm. Vererbungsl.* *89*:729-746.
Esser, K., Saleh, F., and Meinhardt, F. (1979). Genetics of fruit body production in higher Basidiomycetes. II. Monokaryotic and dikaryotic fruiting in *Schizophyllum commune*. *Curr. Genet.* *1*:85-88.
Evans, L. T. (1971). Flower induction and the florigen concept. *Annu. Rev. Plant. Physiol.* *22*:365-394.
Fellenberg, G. (1980). Developmental physiology. *Prog. Bot.* *42*:126-142.
Gerisch, G. (1965). Eine Mutante von *Dictyostelium minutum* mit blockierter Zentrengründung. *Z. Naturforsch.* *20*:298-301.
Gerisch, G. (1979). Control circuits in cell aggregation and differentiation of *Dictyostelium discoideum*. In *Mechanisms of Cell Change*, J. D. Ebert and T. Okada (Eds.). Wiley, New York, pp. 225-239.
Gerisch, G., Huesgen, A., and Malchow, D. (1975). Genetic control of cell differentiation and aggregation in *Dictyostelium*: The role of cyclic-AMP pulses. *Proc. FEBS Meet.* *1975*:257-267.
Gerisch, G., Maeda, Y., Malchow, D., Roos, W., Wick, U., and Wurster, B. (1977). Cyclic AMP signals and the control of cell aggregation in Dictyostelium discoideum. In *Developments and Differentiation in the Cellular Slime Moulds*, Cappuccinelli and Ashworth (Eds.). Elsevier/North-Holland, New York, pp. 105-124.
Gerisch, G., Malchow, D., Roos, W., and Wick, U. (1979). Oscillations of cyclic nucleotide concentrations in relation to the excitability of *Dictyostelium* cells. *J. Exp. Biol.* *81*:33-47.
Hadorn, E. (1970). Experimentelle Entwicklungsforschung, im besonderen an Amphibien. In *Verständliche Wissenschaft*, Vol. 77. Springer, Berlin.
Hartmann, M. (1943). *Die Sexualität*. G. Fischer, Jena.
Heisenberg, M. and Cerda-Olmedo, E. (1968). Segregation of heterokaryons in the asexual cycle of *Phycomyces*. *Mol. Gen. Genet.* *102*:187-195.
Heslot, H. (1958). Contribution a l'étude cytogénétique et génétique des Sordariacies. *Rev. Cytol. Biol. Veg.* *19*:1-209.
Hock, B., Bahn, M., Walk, R. A., and Nitschke, U. (1978). The control of fruiting body formation in the Ascomycete *Sordaria macrospora* Auersw. by regulation of hyphal development. *Planta* *141*:93-103

Honey, N. K., Poulter, R. T. M., and Teale, D. M. (1979). Genetic regulation of differentiation in *Physarum polycephalum*. *Genet. Res. 34*:131-142.

Johnson, T. E., (1978). Isolation and characteristization of perithecial development mutants in *Neurospora*. *Genetics 88*:27-47.

Johnson, T. E., (1979). A *Neurospora* mutation that arrests perithecial development as either male or female parent. *Genetics 92*:1107-1120.

Khaki, I. A., and Shaw, D. S. (1974). The inheritance of drug resistance and compatibility type in *Phytophthora drechsleri*. *Genet. Res. Cambridge, 23*:75-86.

Kimura, K., and Fujio, M. (1961). Studies on abnormal fruit-bodies of the hymenomycetous fungi. I. Undeveloped fruit-bodies of *Coprinus macrorhizus* f. *microsporus*. *Rep. Tottori Mycol. Inst. 1*:19-28.

Klebs, G. (1896). *Die Bedingungen der Fortpflanzung bei einigen Algen und Pilzen*. G. Fischer, Jena.

Kniep, H. (1918). Über morphologische und physiologische Geschlechtsdifferenzierung. *Verh. Phys. Med. Ges. N. F. 46*:1-18.

Kniep, H. (1928). *Die Sexualität der niederen Pflanzen*. Fischer, Jena.

Ko, W. H. (1980). Hormonal regulation of sexual reproduction in *Phytophthora*. *J. Gen. Microbiol. 116*:459-463.

Koltin, Y., Stamberg, J., and Lemke, P. A. (1972). Genetic structure and evolution of the incompatibility factors in higher fungi. *Bacteriol. Rev. 36*:156-171.

Kuhn, J., and Parag, Y. (1972). Protein-submit aggregation model for self-incompatibility in higher fungi: *Schizophyllum commune*. *J. Theor. Biol. 35*:77-91.

Lasure, L. L., and Griffin, D. H. (1975). Inheritance of sex in *Achlya bisexualis*. *Amer. J. Biot. 62*:216-220.

Leonard, T. J., and Dick. S. (1968). Chemical induction of haploid fruiting bodies in *Schizophyllum commune*. *Proc. Natl. Acad. Sci. 59*:745-751.

Leonard, T. J., and Raper. J. R. (1969). *Schizophyllum commune*: Gene controlling induced haploid fruiting. *Science 165*:190.

Leslie, J. F., and Leonard, T. J. (1979a). Three independent genetic systems that control initiation of a fungal fruiting body. *Mol. Gen. Genet. 171*:257-260.

Leslie, J. F., and Leonard, T. J. (1979b). Monokaryotic fruiting in *Schizophyllum commune*: Genetic control of the response to mechanical injury. *Mol. Gen. Genet. 175*:5-12.

Lucas, G. B., Chilton, J. P., and Edgerton, C. W., (1944). Genetics of *Glomerella*. I. Studies on the behavior of certain strains. *Amer. J. Bot. 31*:233-239.

MacDonald, D. J., and Bond, D. J. (1976). Genetic and environmental factors influencing the production and distribution of protoperithecia in *Sordaria brevicollis*. *J. Gen. Microbiol.* *95*:375-380.

MacInnes, M. A., and Francis, D. (1974). Meiosis in *Dictyostelium mucoroides*, *Nature* *251*:321-323.

Markert, C. L. (1949). Sexuality in the fungus *Glomerella*. *Amer. Nat.* *83*:227-231.

McGahen, J. W. and Wheeler, H. E. (1951). Genetics of glomerella. IX. Perithecial development and plasmogamy. *Amer. J. Bot.* *38*:610-617.

Meinhardt, F. (1980). Untersuchungen zur Genetik des Fortpflanzungsverhaltens und der Fruchtkörper- und Antibiotikabildung des Basidiomyceten *Agrocybe aegerita*. *Bibl. Mycol.* *75*.

Meinhardt, F., and Esser, K. (1981). Genetic studies of the Basidiomycete *Agrocybe aegerita*. II. Genetic control of fruit body formation and its practical implication. *Theor. Appl. Genet.* *60*:265-268.

Meyrath, J., and Bu'Lock, J. D. (Eds.) (1977). *Biotechnology and Fungal Differentiation*. Academic, London.

Miyake, H., Tanaka, K., and Ishikawa, T. (1980). Basidiospore formation in monokaryotic fruiting bodies of a mutant strain of *Coprinus macrorhizus*. *Arch. Microbiol.* *126*:207-212.

Molowitz, R., Bahn, M., and Hock, B. (1976). The control of fruiting body formation in the ascomycete *Sordaria macrospora* Auersw. by arginine and biotin: A two-factor analysis. *Planta* *128*:143-148.

Mortimer, A.M., Shaw, D. S., and Sansome, E. R. (1977). Genetical studies of secondary homothallism in *Phytophthora drechsleri*. *Arch. Microbiol.* *111*:255-259.

Mullins, J. T., (1968). Genetic and hormonal regulation of heterothallism in the water molds. *J. Elisha Mitchell Sci. Soc.* *84*:195-198.

Mullins, J. T., and Raper, J. R. (1965). Heterothallism in biflagellate aquatic fungi: Preliminary genetic analysis. *Science* *150*:1174-1175.

Mylek, O. M., and Threlkeld, S. F. H. (1974). A genetic study of female sterility in *Neurospora crassa*. *Genet. Res.* *24*:91-102.

Nasrallah, J. B., and Srb, A. M. (1977). Occurrence of a major protein associated with fruiting body development in *Neurospora* and related Ascomycetes. *Proc. Nat. Acad. Sci,* *74*:3831-3834.

Nasrallah, J. B., and Srb, A. M. (1978). Immunofluorescent localization of a phase-specific protein in *Neurospora tetrasperma* perithecis. *Exp. Mycol.* *2*:211-215.

Parag, Y. (1965). Common B heterokaryosis and fruiting in *Schizophyllum commune*. *Mycologia* *57*:543-561.

Perkins, J. H., and Raper, J. R. (1970). Morphogenesis in *Schizophyllum commune*. *Mol. Gen. Genet.* *106*:151-154.

Plempel, M. (1963a). Die Mucorineen-Gamone. *Naturwissenschaften* *50*:226.

Plempel, M. (1963b). Die chemischen Grundlagen der Sexualreaktion bei Zygomyceten. *Planta. 59*:492-508.

Pratt, R. G., and Green, R. J. (1973). The sexuality and population structure of *Phythium sylvaticum. Can. J. Bot. 51*:429-436.

Raju, N. B., and Perkins, D. D. (1978). Barren perithecia in *Neurospora crassa. Can. J. Genet. Cytol. 20*:41-59.

Raper, J. R. (1951). Sexual hormones in *Achlya. Am. Sci. 39*:110-121.

Raper, J. R. (1952). Chemical regulation of sexual processes in the thallophytes. *Bot. Rev. 18*:447-545.

Raper, J. R. (1966). *Genetics of Sexuality in Higher Fungi.* Ronald Press, New York.

Raper, J. R., and Esser, K. (1961). Antigenic differences due to the incompatibility factors in *Schizophyllum commune. Z. Vererbungsl. 92*:439-444.

Riedel, V., Gerisch, G., Müller, E., and Beug, H. (1973). Defective cyclic adenosin-3',5'-phosphate-phosphodiesterase regulation in morphogenetic mutants of *Dictyostelium discoideum. J. Mol. Biol. 74*:573-585.

Robson, E. G., and Williams, K. L. (1980). The mating system of the cellular slime mould *Dictyostelium discoideum. Curr. Genet. 1*:229-232.

Schwalb, M. N. (1974). Effect of adenosine 3',5'-cyclic monophosphate on the morphogenesis of fruit bodies of *Schizophyllum commune. Arch. Mikrobiol. 96*:17-20.

Scott, W. A. (1976). Biochemical genetics of morphogenesis in *Neurospora. Annu. Rev. Microbiol. 30*:85-104.

Sherwood, W. A. (1969). Sexual reactions between clonal subcultures of a strain of *Dichtyuchus monosporus. Mycologia 61*:251-263.

Sherwood, W. A. (1971). Some observations on the sexual behavior of the progeny of six isolates of *Dictyuchus monosporus, Mycologia 63*:22-30.

Simonet, J. M., and Zickler, D. (1978). Genes involved in caryogamy and meiosis in Podospora anserina. Mol. Gen. Genet. *162*:237-242.

Smith, J. E., and Berry, D. R. (Eds.) 1978. *The Filamentous Fungi,* Vol. 3. Edward Arnold, London.

Spemann, H. (1936). *Experimentelle Beiträge zu einer Theorie der Entwicklung.* Springer-Verlag, Berlin.

Stahl, U. (1976). Genetische Regulation der Fruchtkörperbildung bei höheren Basidiomyceten, monokaryotisches Fruchten bei *Polyporus ciliatus. Bibl. Mycol. 50.*

Stahl, U., and Esser, K. (1976). Genetics of fruit body production in higher Basidiomycetes. I. Monokaryotic fruiting and its correlation with dikaryotic fruiting in *Polyporus ciliatus. Mol. Gen. Genet. 148*: 183-197.

Sutter, P. R. (1975). Mutations affecting sexual development in *Phycomyces blakesleeanus*. *Proc. Nat. Acad. Sci. USA 72*:127-130.

Therrien, C. D., and Collins, O. R. (1976). Apogamic induction of haploid plasmodia in a Myxomycete *Didymium iridis*. *Dev. Biol. 49*:283-287.

Ullrich, R. C. (1978). On the regulation of gene expression: Incompatibility in *Schizophyllum*. *Genetics 88*:709-722.

Uno, I., and Ishikawa, T. (1973a). Purification and identification-of the fruiting-inducing substances in *Coprinus macrorhizus*. *J. Bacteriol. 113*:1240-1248.

Uno, I., and Ishikawa, T. (1973b). Metabolism of adenosine 3',5'-cyclic monophosphate and induction of fruiting bodies in *Coprinus macrorhizus*. *J. Bacteriol. 113*:1249-1255.

Wakefield, E. M. (1909). Über die Bedingungen der Fruchtkörperbildung sowie das Auftreten fertiler und steriler Stämme bei Hymenomyceten. *Naturwiss. Z. Forst. Landwirtsch. 7*:521-551.

Wang, C. S., and Raper, J. R. (1969). Protein specifity and sexual morphogenesis in *Schizophyllum commune*. *J. Bacteriol. 99*:291-297.

Wessels, J. G. H. (1966). Control of cell-wall glucan degradation during development in *Schizophyllum commune*. *Antonie van Leeuwenhoek J. Microbiol. Serol. 32*:341-355.

Wessels, J. G. H. (1969). A β-1, 6-glucan glucanohydrolase ininvolved in hydrolysis of cell-wall glucan in *Schizophyllum commune*. *Biochim. Biophys. Acta. 178*:191-193.

Wessels, J. G. H. (1971). Cell wall metabolism and morphogenesis in *Schizophyllum*. In *The 10th International Congress on Microbiology, Mexico*. pp. 141-146.

Wheeler, H. W. (1954). Genetics and evolution of heterothallism in *Glomerella*. *Phytopathology 44*:342-345.

Wheeler, H. E., and McGahen, J. W. (1952). Genetics of *Glomerella*. X. Genes affecting sexual reproduction. *Am. J. Bot. 39*:110-119.

Wheeler, H. E., Driver, C. H., and Campa, C. (1959). Cross- and self-fertilization in *Glomerella*. *Am. J. Bot. 46*:361-365.

Wurtz, T., and Jockusch, H. (1975). Sexual differentiation in *Mucor:* trisporic acid response mutants and mutants blocked in zygospore development. *Dev. Biol. 43*:213-222.

Wurtz, T., and Hockusch, H. (1978). Morphogenesis in *Mucor mucedo*: Mutations affecting gamone response and organ differentiation. *Mol. Gen. Genet. 159*:249-257.

19

Genetic Recombination in Fungi Following Protoplast Fusion and Transformation

JOHN F. PEBERDY *University of Nottingham, Nottingham, England*

I. INTRODUCTION

Genetic recombination is an essential feature of both sexual and parasexual reproductive processes in the fungi. The initial event in both is the fusion of normal vegetative cells or of specialized reproductive cells, with the possibility of two genetically different nuclei being brought together in a common cytoplasm. Subsequent nuclear fusion followed by meiosis in the specialized reproductive cells during the sexual process or mitotic nondisjunction, mitotic recombination, and haploidization during the parasexual process will result in recombination events.

The realization of techniques for protoplast isolation and culture during the past two decades has resulted in the further development of new technologies for genetic manipulation. In protoplast fusion a situation similar to the natural reproductive processes exists, with the bringing together of whole genomes into one cell or one mycelium. A variety of recombinant classes may arise from this association, but generally speaking, the appearance of specific types cannot be predicted. This contrasts with the other new technology, namely, transformation. In this process the aim is to produce specific recombinant clones following the introduction of specific gene(s) into a recipient organism. Theoretically, transformation offers possibilities for the cloning of genes from a variety of sources into fungi. Clearly, in the long term the two techniques will be seen as complementary, with protoplast fusion being a procedure for the broader transmission of genetic material that has been cloned into a fungus.

II. REQUIREMENTS OF THE PROTOPLAST SYSTEM FOR GENETIC STUDIES

The usefulness of protoplasts as genetic tools stems from the fact that they can be isolated in large numbers and cultured on osmotically stabilized media to induce reversion to the normal cell form. These aspects have been extensively reviewed in recent years (Peberdy, 1976,1979) and only some of the more recent observations on these aspects will be discussed here.

A. Protoplast Isolation

The release of protoplasts from fungal cells is most effectively achieved by the enzymatic digestion of the cell wall in the presence of an osmotic stabilizer. Since the first reports on protoplast isolation from yeasts (Eddy and Williamson, 1959) and filamentous fungi (Emerson and Emerson, 1958; Bachmann and Bonner, 1959) steady progress has been made in improving protoplast yields. The major factors affecting protoplast isolation, assuming that an effective enzyme is available, are the physiological age of the cells and the osmotic stabilizer used in the digestion. Recently it was shown that other factors may be important with Ca^{2+} ions causing increased protoplast production from *Aspergillus niger* (Thomas and Davis, 1980). Whether effects such as this are universal is not clear, because repeating the experiment with *Aspergillus nidulans* had no effect (R. E. Bradshaw, unpublished data).

Lytic enzymes effective against yeasts have been available from commercial sources for many years. Groups working on protoplasts from filamentous species have generally had to produce the necessary enzymes in their own laboratories. This is a time-consuming procedure and the product obtained can be very variable. One of the latest reports on this aspect described a lytic complex derived from *Oerskovia xanthenolytica* supplemented with an endogenous α-glucanase from *A. nidulans* which gave consistently higher protoplast yields from this same fungus than had been obtained in the past (van den Broek et al., 1979).

Clearly, enzymes available commercially would also be advantageous to the many workers dealing with protoplasts from filamentous fungi. A recent survey of a number of polysaccharases (Hamlyn et al., 1981) revealed the presence of side activities in some of the preparations that were highly effective against several species of fungi, including *Saccharomyces cerevisiae*. Two enzymes were of particular interest; Novozym 234 (Novo Enzyme Products, Ltd.) and Cellulase CP (John & E. Sturge, Ltd.). In some instances a single enzyme was very active (Table 1), but for other species a mixture of enzymes was best (Table 2). Novozym 234 has also been shown to be effective against yeast (Stephen and Nassim, 1981). A feature which will add to the attractiveness of these enzymes is their relatively low cost.

B. Protoplast Culture

The relative ease with which protoplasts have been adopted as tools in genetic research is in part a consequence of the extensive experimental work on protoplast culture and reversion (Peberdy, 1979; Necas, 1980). The endogenous factors which control reversion ability in a protoplast are not known. The reversion frequency is highly variable between different species and while the absence of a nucleus is a factor (Garcia Acha et al., 1966), it cannot be the only one. Issac (1978) found that only 25% of the protoplasts of *A. nidulans* lacked nuclei, yet the reversion frequencies for the strain were in the range of 20-50%, based on the total protoplast count. In *Volvariella volvacea* the reversion frequency has recently been found to be exceedingly low, at 2%, but the proportion of nucleated protoplasts is again very high (Santiago, 1982).

Clearly, the reversion frequency of a protoplast preparation is an important factor in genetic manipulation experiments. It is conceivable that in some situations, for example, interspecies crosses, differential reversion frequencies of the two strains could be used to advantage as a selective factor. Conversely, cell wall regeneration and protoplast reversion could prove to be a limiting factor in attempted crosses involving species that are widely different in their taxonomic designation. Unbalanced wall synthesis may occur in such fusion products, resulting in abnormal wall regeneration and consequently a nonviable revertant.

In transformation experiments a high reversion frequency is clearly desirable to enhance the likelihood of plasmid uptake into a viable protoplast and subsequent gene expression.

C. Protoplast Fusion Strategies

The earliest attempts to obtain fusion of fungal protoplasts embodied very simple techniques, with compaction of protoplasts by centrifugation (Ferenczy et al., 1974). The fusion frequencies obtained were very low and not too dissimilar from back-mutation frequencies of many

TABLE 1 Isolation of Protoplasts Using Commercial Polysaccharases

Enzyme	Protoplast yield $\times 10^{-6}$ ml^{-1} (±SD)			
	Acremonium chrysogenum[a]	*Aspergillus nidulans*	*Aspergillus niger*	*Penicillium chrysogenum*
Cellulase (Merck)	8.00 ± 3.27	2.07 ± 0.57	0.83 ± 0.33	NT[b]
Cellulase CP (Sturge)	133.00 ± 18.00	19.00 ± 5.00	0.94 ± 0.06	8.3
Cellulase CT (Sturge)	1.67 ± 1.03	<0.25	<0.25	<0.25
Novozym 234 (Novo)	148.00 ± 12	190 ± 53	1.57 ± 0.71	37.90 ± 5.00
Cereflo 200L (Novo)	<0.25	<0.25	<0.25	<0.25

[a]Mycelium pretreated with dithiothrietol before protoplast isolation.

[b]NT, not tested.
Source: Hamlyn et al. (1981).

TABLE 2 Synergistic Action of Commercial Polysaccharases and Protoplast Release from Aspergilli

Enzyme mixture	Protoplast yield $\times 10^{-6}$ ml^{-1} (±SD)			
	A. nidulans	*A. niger*	*A. rugulosus*	*A. ochraceous*
Novozym 234 + Cellulase CP	280.00 ± 14	17.9 ± 0.3	56.9 ± 4.2	7.63 ± 1.19
Novozym 234 + Cereflo 200L	69.7 ± 10.7	11.4 ± 0.9	*NT*	*NT*
Novozym 234 + Cereflo 200 L + Cellulase CPP	175 ± 60	25.1 ± 4.4	*NT*	13.4 ± 1.7
Cellulase CP + Cereflo 200L	17.3 ± 3.9	1.72 ± 0.75	*NT*	*NT*

Source: Hamlyn et al. (1981).

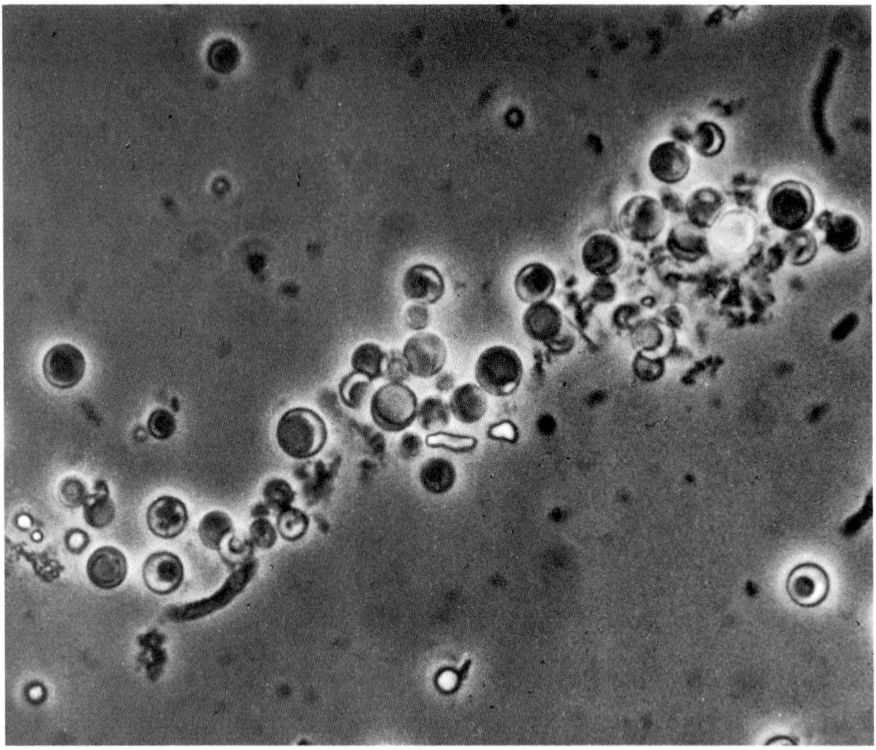

(a)

FIG. 1 Protoplasts of *Aspergillus nidulans* (a) before treatment and (b) after treatment with polyethylene glycol, showing aggregation of protoplasts.

auxotrophic markers. Improvements followed the introduction of various fusogenic agents that had been used successfully with plant protoplasts and animal cells. In the first of these (Binding and Weber, 1974), sea water or Ca^{2+} ions at high pH were shown to give enhanced fusion frequencies, but even better results followed the use of the polyethylene glycols (Anné and Peberdy, 1975,1976; Ferenczy et al., 1975a,b) (Fig. 1). The experimental design for the selection of fusion products followed a conventional type of mating experiment using complementary auxotrophic mutants. Applying this strict regime assumes that a heterokaryotic state or a diploid (or its equivalent in interspecies crosses) can be achieved. This is not always so, as was found by Hamlyn and Ball (1979) in crosses with *Acremonium chryogenum (Cephalosporium acremonium)*; a range of selective media were required to recover the total spectrum of fusion products. This broad

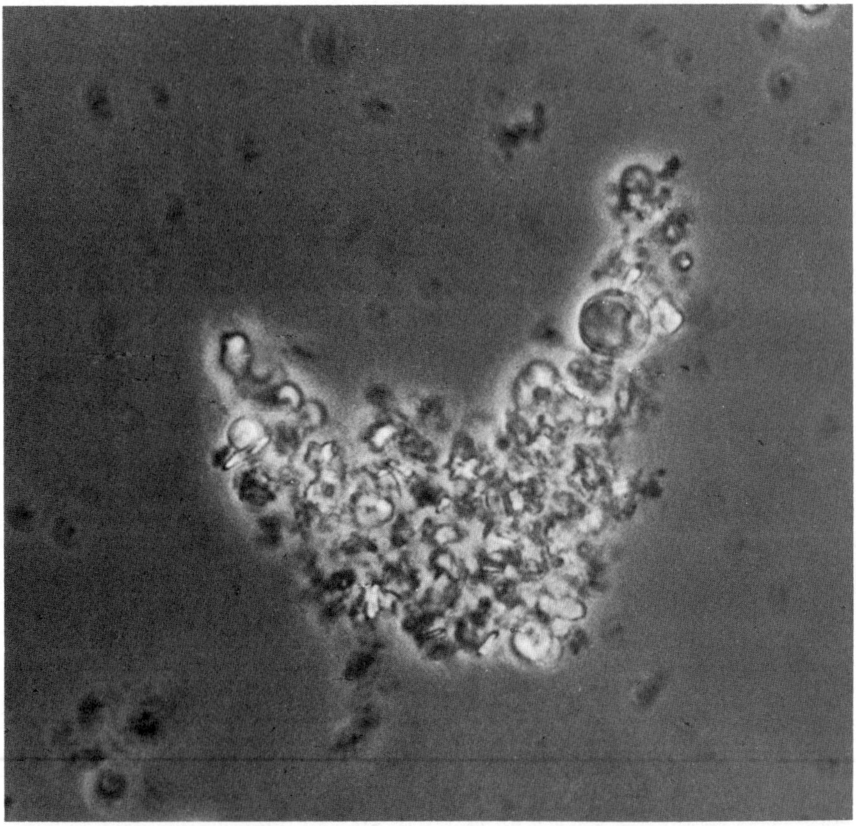

(b)

FIG. 1 (Continued)

selection approach may be applicable to interspecies crosses, but has not been reported to date.

The expansion of protoplast fusion technology into more applied areas provides a challenge to develop alternative methods for selection. The introduction of auxotrophic mutations into industrial strains is not a desirable practice in view of the likely deterioration in the productivity of the strain. Furthermore, mutant isolation is time-consuming. An interesting approach was adopted in bacterial systems in which the protoplasts from one parent were rendered nonviable prior to fusion by heat treatment (Fodor et al., 1978) or an antibiotic (Levi et al., 1977). There are no published reports on the use of these techniques with fungi; however, there would appear to be no obstacle to their application. In *Streptomyces coelicolor* Hopwood and Wright (1979) showed to form heterokaryons, one with another. Compatible strains are put that the recombination frequency was enhanced if protoplasts were irradiated with ultraviolet light to kill 99.99% before fusion. Hockney and Freeman (1980) used the same approach, with success, in an inter-species of yeasts.

III. SOMATIC HYBRIDIZATION AND RECOMBINATION FOLLOWING PROTOPLAST FUSION

Protoplast fusion has application in two aspects of somatic hybridization; in intra- and interspecies crosses. In many fungi intraspecies crosses can be easily established by conventional procedures, for example, by establishment of balanced heterokaryons via hyphal fusions and by cell fusions using compatible mating strains; however, protoplast fusion may offer advantages even with some of these strains. The method is increasingly seen as a means of establishing a genetic system in species in which a natural breeding system is unknown.

A. Intraspecies Protoplast Fusion

Protoplast fusion has been studied in a wide range of yeast species. Following the first experiments with *S. cerevisiae* (van Solingen and van der Plaat, 1977), several other studies were reported (Maráz et al., 1978; Yamamoto and Fukui, 1977; Gunge and Tamaru, 1978; Christensen, 1979; Arima and Takano, 1979; Kozhina and Cherpurnaya, 1980; Böttcher et al., 1980). Fusions were carried out between opposite mating types (α + a), as well as between identical mating types (α + α, *a* + *a*). The products of fusion crosses were normally diploid, as confirmed by DNA estimation, cell size and haploidization tests (Maráz et al., 1978), and by genetic analyses of matings with diploid tester strains (Gunge and Tamaru, 1978). Diploids obtained from crosses between identical mating types did not undergo sexual differentiation (Maráz et al., 1978). Crosses between compatible mating strains of *Schizosaccharomyces pombe* resulted in diploid formation and subsequent ascospore formation, but incompatible matings gave stable diploids (Sipiczki and Ferenczy, 1977a).

Diploid fusion products were also obtained from crosses in the basidiomycetous yeast *Rhodosporidium toruloides* (Sipiczki and Ferenczy, 1977b; and Becher and Böttcher, 1980). These isolates were generally stable, with a spontaneous mitotic segregation frequency of 10^{-3} or less. Where crosses involved mutants from different wild types, this frequency was much higher (Becher and Böttcher, 1980). Segregation of some of these diploids gave only parental genotypes, a situation also observed in *Pichia guilliermondii* (Böttcher et al., 1980). This pattern of segregation is not understood; heterokaryosis was ruled out because the cells were uninucleate. A possible cause may be that the fusion products were aneuploid. Diploid fusion products were also reported in *Saccharomycopsis lipolytica* (Stahl, 1978; Spata and Weber, 1980) and *Kluyveromyces lactis* (Morgan et al., 1977).

An exception to this pattern of diploid formation was found in *Candida tropicalis*, in which the fusion products were heterokaryons (Fournier et al., 1977; Vallin and Ferenczy, 1978). In the absence of strong selection the parental strains readily segregated, but with maintained selection a few faster-growing diploids were found.

Protoplast fusion might provide a method with which to study ploidy relationships in fungi. Using strains with common markers, selection for second- and third-generation crosses would be possible. Only one study of this interesting application has been reported to date. Klinner and colleagues (Klinner et al., 1980) set up crosses using *Pichia guilliermodii*. They recovered prototrophic progeny following a second-generation cross; however, the isolates varied in stability and DNA content and their ploidy did not appear to be clearly defined.

The events immediately following protoplast fusion still need to be resolved. It is interesting to recall the earlier reports of protoplast reversion in yeast (Necas, 1971) in this context. During the initial stage of wall regeneration about the protoplast, repeated nuclear divisions occurred and the uninucleate condition was restored at the second or third generation of bud formation. If similar events follow protoplast fusion, then a multinucleate heterokaryon would be formed. To explain the fact that fusion products are diploids, that is, colonies arising from fused protoplasts, the nuclear fusion event must occur early in the developmental sequence of wall regeneration and budding.

Intraspecies protoplast fusion has proved useful in fungi in two respects: (1) in fungi in which heterokaryosis is difficult or impossible to achieve by natural events and (2) as a tool for the study of heterokaryon incompatibility.

Attempted heterokaryosis and genetic analysis in *Acremonium chrysogenum* using hyphal anastomosis gave limited results (Nüesch et al., 1973); however, resolution of the problem was possible using protoplast fusion (Hamlyn and Ball, 1979). An important observation in the latter work was the strong indication of a transient diploid stage in the parasexual cycle of this fungus. As referred to earlier, recovery of fusion products required the use of a range of selective media. This implied that some of the fusion products were recombinants arising from the short-lived diploid phase. Similar problems in the establishment of heterokaryons were experienced with *Beauvaria tenella* (Paris, 1977). Protoplast fusion did produce heterokaryons, but these were unstable and could not be propagated by hyphal transfer.

Fusion technology has proved to be an important tool in aiding our understanding of a fundamental aspect of fungal biology, namely, heterokaryon or vegetative incompatibility. This phenomenon is best documented in *A. nidulans*, but also occurs in other species. Independent wild isolates of this fungus are grouped according to their ability in the same group, the members of which do not form vegetative heterokaryons with members of other groups. However, sexual crossing does occur between members of different groups, suggesting that other regulatory mechanisms control fusion of the specialized reproductive cells. The ability to set up sexual crosses has enabled the existence of several nuclear genes which control incompatibility, the *het* genes, to be established, but such crosses cannot provide information on the interaction of the genes, this requiring heterokaryons. Protoplast fusion between some of the incompatibility groups proved possible (Dales and Croft, 1977,1980). The fusion products were heterokaryons and somatic diploids were also obtained, making a parasexual anaysis of the *het* genes possible.

B. Interspecies Protoplast Fusion

Reported attempts to obtain hybrids between fungal species have been few in number (Burnett, 1975). Lewis (1969) made sexual crosses between *Neurospora* species and successfully recovered hybrids which were, however, sterile, and Hastie (1973) obtained hybrids in *Verticillium* using parasexual crossing methods. As protoplast fusion became a feasible technique, the possibilities for interspecies fusions and hybridization became a major aim for many groups. The forced mixing of two cytoplasms from two species as is achieved in fusion removes one possible barrier to hybridization by more natural processes. However, despite some progress, it is still not possible to define the limits to hybridization nor to consider what implications the incompatibility mechanisms referred to earlier might have for such crosses.

A problem in hybridization studies in fungi stems from the limits of taxonomy of these organisms. In many instances the morphological differences between species are few, and where such species can be hybridized using the fusion technique, questions as to true species differences are raised. For the present we have to rely on existing species designations and possibly an interest in new taxonomic approaches may develop as more hybrids are produced. Consequently, the terms *closely related* and *distally related* are used to describe the components of hybrids, but clearly these are very relative terms.

Studies on hybridization in yeast are few in comparison to the extensive research on intraspecies crosses. Whittaker and Leach (1978) isolated stable prototrophs from a cross between *K. lactis* and *Kluyveromyces fragilis* which had a larger cell volume and a higher DNA content than either of the parents. Another positive result was obtained from a cross involving *S. cerevisiae* and *Saccharomyces diastaticus* (Hockney and Freeman, 1980). Interest in this cross stemmed from a desire to develop strains of brewing yeast that could utilize the residual dextrin present in beer at the end of a normal fermentation. Problems with ploidy in *S. cerevisiae* (most brewing strains having ploidies greater than haploid) entailed some modifications to the selection procedure. Other reports have indicated less success. Sipiczki (1979) attempted a cross using *Schiz. pombe* and *Schizosaccharomyces octosporus*, but haploidization gave only the latter parent as segregants. An unsuccessful attempt at hybridization between *Hansenula* species has also been reported (Savchenko and Kapultsevich, 1980).

Attempts at intergeneric hybridization and recombination have also been disappointing. Progeny obtained following fusion of *C. tropicalis* protoplasts and *Saccharomycopsis fibuligera* protoplasts were uninucleate and had sugar assimilation patterns of one or the other parent or a combination thereof. However, these latter types proved unstable when subcultured, giving parental types (Provost et al., 1978). Spata and Weber (1980) fused protoplasts of *Saccharomycopsis lipolytica* and *Pichia guilliermondii* and Svoboda (1980) reported an attempt at hybridization between *S. cerevisiae* and *Schiz. pombe*, but in neither case were stable hybrids obtained.

Most crosses attempted with filamentous fungi have been more successful. The most detailed analysis has been made on hybrids obtained by crossing *A. nidulans* and a closely related species *Aspergillus rugulosus*. The protoplast fusion products were heterokaryons and as these were maintained under highly selective conditions, on minimal medium, vigorous growing sectors developed (Kevei and Peberdy, 1977). These sectors were characterized by a dark brown pigmentation. Similar colonies were obtained when protoplasts prepared from heterokaryons were cultured. These vigorous, brown colonies were identified hybrids (Fig. 2) and have been maintained as stable forms on minimal medium for several years (J. F. Peberdy, unpublished results). Under these conditions the hybrids produce few spores which give further hybrid colonies on germination.

Relaxation of selection by growth of the hybrids on complete medium allows spontaneous segregation to occur. In the first crosses the strains used had few genetic markers and this resulted in the majority of the segregants being parental types. However, in later work the strains had more markers and the hybrids from these give predominantly nonparental segregants (J. F. Peberdy, unpublished results). Similar patterns of segregation were observed when benomyl or other haploidizing agents were used (Kevei and Peberdy, 1979; R.E. Bradshaw, unpublished results). A much broader spectrum of segregants is obtained under these conditions (Fig. 2). The benefit of utilizing *A. nidulans* in these crosses lies in the fact that a large collection of genetically mapped strains is available. Thus it was possible to produce hybrids in which all the linkage groups of *A. nidulans* were marked genetically (Kevei and Peberdy, 1979). Analysis of the segregants obtained following growth of the hybrids on medium containing benomyl or chloral hydrate showed that the *A. nidulans* linkage groups segregated in a random manner.

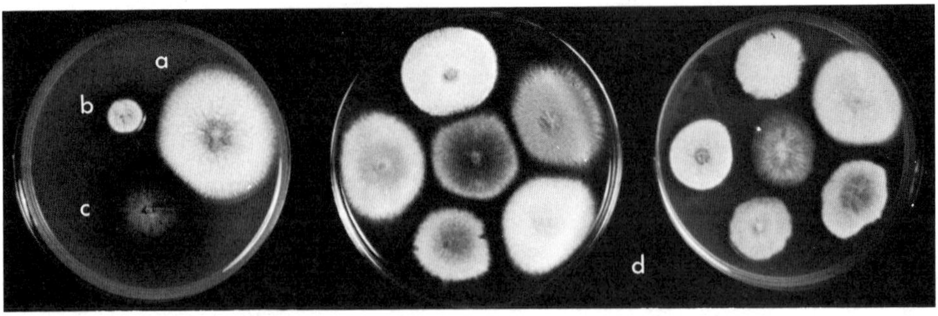

FIG. 2 Hybridization of *Aspergillus nidulans* and *Aspergillus rugulosus*: (a) *A. nidulans* parent, (b) *A. rugulosus* parent, (c) hybrid, and (d) benlate-induced segregants.

The recovery of most of the markers, possessed by the haploid parents, in the segregants clearly indicates the presence of these markers and the complementary wild-type genes in the hybrids; that is, the hybrids are heterozygous. Clearly, the markers and corresponding wild-type genes are not necessarily allelic, and it is unlikely that the loci involved have the same linkage relationships.

The segregation patterns of these hybrids show similarities to those of somatic diploids of *A. nidulans* or *A. rugulosus* and supports the view that the hybrid phenotype was the result of a fusion of nuclei of the two parents. In the earlier reports the hybrids were referred to as diploid; however, this is misleading and alloploid is a more appropriate term. Traditionally, somatic intraspecies diploids are characterized by estimates of spore DNA content. This was not feasible with the hybrids and so analysis was made on the protoplasts (Kevei and Peberdy, 1977) and this showed the hybrid nuclei to contain approximately twice as much DNA as the parental strains.

A cross using closely related *Penicillium* species gave similar results (Peberdy et al., 1977). The hybrid from *Penicillium chryosogenum* and *Penicillium cyaneofulvum* was more stable than the *Aspergillus* hybrid when selection was relaxed. Segregation occurred in the presence of haploidization agents and the whole spectrum of markers, although fewer than for the later *Aspergillus* crosses, was recovered. Somewhat later Anné and Eyssen (1978) produced similar types of hybrids with similar properties, using *Penicillium citrinum* and *P. cyaneofulvum* as parents.

This behavior contrasts with that of a hybrid produced between *P. chryosogenum* and *Penicillium roqueforti* (Anné et al., 1978). The first progeny isolated from this cross proved to be very unstable under nonselective conditions, producing only the *P. roqueforti* parent. Reevaluation of the cross resulted in the isolation of several different

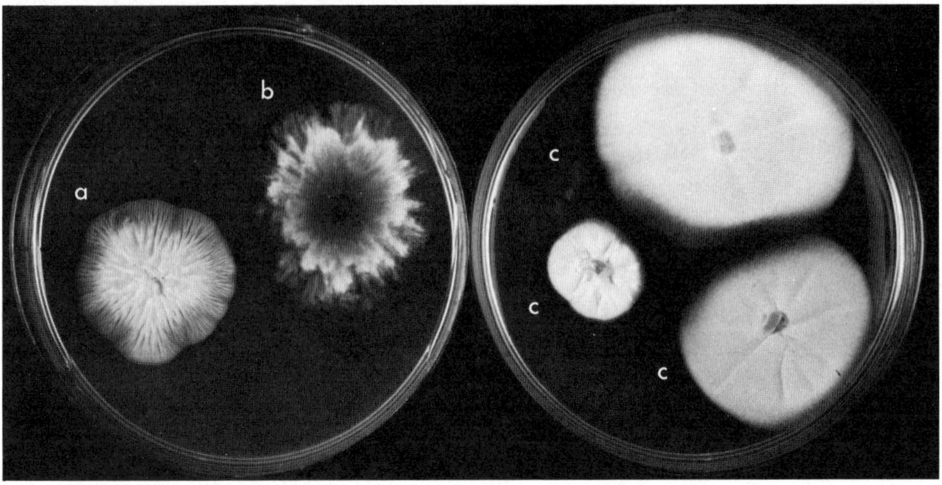

FIG. 3 Hybridization of *Penicillium chrysogenum* and *Penicillium roqueforti*: (a) *P. chrysogenum* parent, (b) *P. roqueforti* parent, and (c) hybrids with different phenotypes.

phenotypes, some as unstable as the original and others which were very stable under a wide range of growth conditions (Fig. 3; J. F. Peberdy and J. Anné, unpublished results). An isoenzyme analysis was made on some of these progenies and this showed some indications of genetic interaction (Anné and Peberdy, 1981).

Only one wide species cross in the *Aspergilli* has been described (Ferenczy, 1976), but no stable hybrid forms were obtained. The fusion progeny behaved in a manner similar to an aneuploid series, giving either parent as a segregant.

IV. TRANSFER OF CELL ORGANELLES BY PROTOPLAST FUSION

Cytoplasmic mixing and the exchange of cell organelles is an inevitable consequence of protoplast fusion and can clearly have genetic significance, as mitochondria and possibly plasmids are involved. The transmission of mitochondria and the resulting interaction has been investigated in several laboratories, with the main result being a demonstration of the restoration of respiratory deficiency in petite strains of yeast (Ferenczy and Maráz, 1977; Gunge and Tamaru, 1978). Experiments with *K. lactis* enabled a meaningful genetic analysis of the phenomenon to be carried out (Morgan et al., 1978; Allmark et al., 1978; Morgan et al., 1980). Spencer and colleagues (Spencer et al., 1980) were able to use strains with mitochondrial markers in strain improvement experiments in brewer's yeast. This approach was of particular interest, because the classical methods of breeding are not possible in this organism owing to ploidy relationships.

Evidence for recombination of mitochondrial DNA from different species of *Aspergillus* has been reported (Croft et al., 1980). The implications for other species crosses are interesting as a consequence of this result.

The uptake of isolated organelles by fungal protoplasts has received some attention. Cells of the slime mutant of *Neurospora crassa* were shown to take up spinach chloroplasts (Vasil and Giles, 1975), and yeast protoplasts behaved similarly when mixed with *Chlorella* chloroplasts (Kawakami et al., 1980). These observations provide a stimulus for further studies, particularly the uptake of nuclei, which could have interesting significance for recombination.

V. TRANSFORMATION USING PROTOPLASTS

Transformation is the process in which genetic change occurs following the uptake and expression of naked DNA by cells. It was first described in fungi in *N. crassa* several years ago (Shamoian et al., 1961), and many other studies have since been reported on the same organism (Shockley and Tatum, 1962; Mishra and Tatum, 1973; Mishra, 1976; Schablik et al., 1977) and on *A. niger* (Sen et al., 1969). There were problems in the reproducibility of some of these experiments and so far there has been no demonstration of the physical incorporation of the exogenous DNA into the genome.

In the last few years, however, significant advances have been made in the transformation of yeast using plasmid vector systems. This type of transformation has been used extensively in bacteria as a molecular cloning technique, and when used in conjunction with other genetic studies, it has provided information on the function of many genes. The transforming plasmids used in yeast are hybrid molecules comprised of bacterial plasmid DNA and yeast DNA, which includes the sequences coding the gene of interest and in most cases the whole or part of the yeast 2-μm plasmid. The hybrid plasmids are constructed by in vitro recombination methods using specific endonucleases and ligases. With the availability of these plasmids, a transformation experiment is technically relatively simple and, in essence, the same as a protoplast fusion experiment. The essential differences are that protoplasts of only one strain are used which carry a mutation of the specific gene to be cloned as a selective marker. The recipient strains used for transformation studies must have extremely low reversion frequencies ($<10^{-10}$) for the selectable gene mutation. The protoplasts are treated with PEG and the plasmid DNA is added. After a short incubation period the protoplasts are plated onto a suitable selection medium upon which the transformants develop.

In the first published experiments a *leu* 2^- strain of *S. cerevisiae* was transformed to *leu* 2^+, using an *Escherichia coli* plasmid into which yeast DNA carrying the *leu* 2^+ gene was inserted (Fig. 4a; Hinnen et al., 1978) and a similarly constructed plasmid to which the yeast 2-μm plasmid was also added (Fig. 4b; Beggs, 1978). Both plasmids could be cloned in a *leu* B^- strain of *E. coli*, because this mutation is complemented by the *leu* 2^+ gene from yeast. Although phenotypically the same, the genetic events in the transformants were very different. Hinnen et al. (1978) obtained three types of transformants, the majority of which had the whole plasmid integrated into the *leu* 2^+ region. This was a result of a recombination event between homologous sequences, resulting in a tandem duplication. These recombinants may be unstable because of further recombination in the homologous regions leading to possible exclusion of the transforming DNA. Two other types of recombinants isolated were those in which the plasmid became integrated in other regions of the yeast genome and those in which recombination and exchange of *leu* 2^+ for *leu* 2^+ occurred. The frequency of all recombinant types was 1-10 per microgram of DNA, with the latter two groups being a very small part of the total.

In the second experiment (Beggs, 1978) there was no integration of the plasmid or any part of it. This was confirmed by meiotic analysis where all four products of the ascus were either carriers or noncarriers of the plasmid. Replication of the plasmid in these transformants was autonomous. Transformants of this type are unstable, and for maintenance constant selective pressure is required. Instability is a consequence of two events; the loss of the plasmid at cell division or

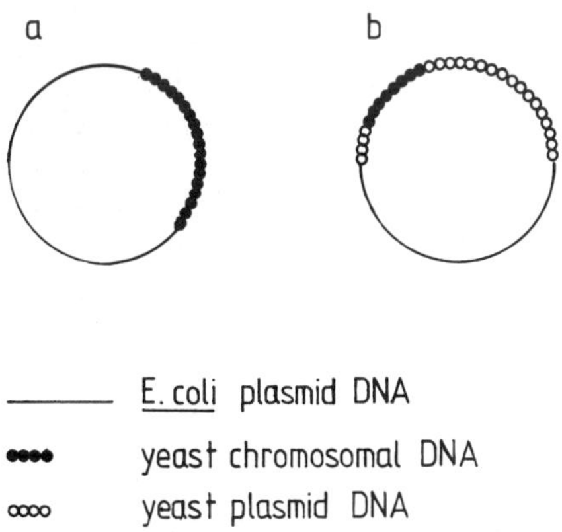

FIG. 4 Hybrid plasmids used for yeast transformation. [Data from (a) Hinnen et al., 1978, and (b) Beggs, 1978.]

loss due to recombination between the transforming plasmid and resident 2-μm plasmids in the cell (Beggs, 1978; Struhl et al., 1979). Transformation by this nonintegrating type of plasmid occurred at a much higher frequency (10^{-3}–10^{-4} transformants per microgram of DNA).

In later experiments a similarly higher frequency was observed when the hybrid plasmid included any part of the 2-μm molecule, even that lacking the replication sequence. However, it was essential for the recipient cells to be carriers of the yeast plasmid, suggesting that recombination between the transforming plasmid and a resident plasmid was a necessary event.

The integration of replication origins into transforming plasmids from other sources causes high-frequency transformation. Plasmids carrying the yeast *trp-1* region (Struhl et al., 1979; Kingsman et al., 1979) or the *arg-4* gene (Hsiao and Carbon, 1979) also included yeast chromosomal replication origins associated with each of the DNA sequences. This concept was further extended when it was demonstrated that DNA from a variety of other eukaryotic sources, including *N. crassa*, also caused the same effect as consequence of the presence of replicons (Stinchcomb, et al., 1980).

Yeast has also been shown to be a suitable organism for gene cloning and consequently furthering our understanding of gene behavior and expression (Hicks et al., 1979; Fried and Warner, 1981). However, acceptability as an organism for genetic engineering is dependent upon the ability of the cells to express genes from outside sources. The findings in this area are also encouraging. Prokaryote plasmid-borne genes (Cohen et al., 1980), transposon-borne genes (Jiminez and Davies, 1980), and chromosomal genes (Panthier et al., 1980) were found to be expressed.

Interest in transformation as a tool for gene manipulation and recombination in filamentous fungi is also very great. To date only one successful result has been published (Case et al., 1979). The *qa* 2^+ gene from *N. crassa*, which codes for catabolic dehydroquinase, was ligated into *pBR 322* to give the chimeric plasmid *pVK 88*. *Neurospora crassa qa* 2^- mutants were transformed with this plasmid, which became integrated into the fungal genome at several sites in a manner similar to that described by Hinnen et al. (1978) for *S. cerevisiae*. Very-low-frequency transformation of *N. crassa* was also obtained using linear DNA (Case et al., 1979).

VI. CONCLUSIONS

The many advances described in this chapter have taken place in the last 6 years and provide a clear statement of the impact that fungal protoplasts have provided to genetics. It is difficult at this time to predict the full extent to which protoplast fusion and transformation will develop and expand over the next few years, because we do not yet know the limits for either of the two systems.

Much has still to be unraveled regarding the sequence of events that follow protoplast fusion. Is there any possibility that these events can be controlled in such a way that the frequency of fusion or the range of species which can be hybridized can be improved? Our knowledge of protoplast physiology is also rather restricted and this could also be a limitation to future progress.

Much the same is true for transformation, but a key factor here is the question of competence, which is well known in bacteria, but does it also exist in protoplasts of eukaryotic microorganisms?

Having obtained novel organisms via these novel methods of recombination, what is their significance? Clearly, as transformation is used to manipulate a specific gene, it follows that a specific product or biochemical reaction has already been decided upon. Hybrids derived from protoplast fusion are somewhat different. The segregants derived from these hybrids represent a new mix of gene pools. This must have some biochemical significance and possibly some commercial significance as a source of novel products.

REFERENCES

Allmark, B. M., Morgan, A. J., and Whittaker, P. A. (1978). The use of protoplast fusion in demonstrating chromosomal and mitochondrial inheritance of respiratory deficiency in *Kluyveromyces lactis*, a petite negative yeast. *Mol. Gen. Genet. 159*:297-299.

Anné, J., and Eyssen, H. (1978). Isolation of inter-species hybrids of *Penicillium citrinum* and *Penicillium cyaneofulvum* following protoplast fusion. *FEMS Microbiol. Lett. 4*:87-90.

Anné, J., and Peberdy, J. F. (1975). Conditions for induced fusion of fungal protoplasts in polyethylene glycol solutions. *Arch. Microbiol. 105*:201-205.

Anné, J., and Peberdy, J. F. (1976). Induced fusion of fungal protoplasts following treatment with polyethylene glycol. *J. Gen. Microbiol. 92*:413-417.

Anné, J., and Peberdy, J. F. (1981). Characterization of interspecific hybrids between *Penicillium chrysogenum and P. cyaneofulvum* by isoenzyme analysis. *Trans. Br. Mycol. Soc. 77* (in press).

Anné, J., Eyssen, H., and De Somer, P. (1978). Somatic hybridization of *Penicillium roqueforti* and *Penicillium chrysogenum* after protoplast fusion. *Nature 262*:719-721.

Arima, K., and Takano, I. (1979). Multiple fusion of protoplasts in *Saccharomyces* yeasts. *Mol. Gen. Genet. 173*:271-277.

Bachmann, B. J., and Bonner, D. M. (1959). Protoplasts from *Neurospora crassa*. *J. Bacteriol. 78*:550-556.

Becher, D., and Böttcher, F. (1980). Hybridization of *Rhodosporidium toruloides* by protoplast fusion. In *Advances in Protoplast Research*. L. Ferenczy and G. L. Farkâs (Eds.). Akadémiai Kaidó, Budapest, pp. 105-112.

Beggs, J. D. (1978). Transformation of yeast by a replicating hybrid plasmid. *Nature 275*:104-109.

Binding, H., and Weber, H. J. (1974). The isolation, regeneration and fusion of *Phycomyces* protoplasts. *Mol. Gen. Genet. 135*:273-276.

Böttcher, F., Becher, D., Klinner, U., Samsonova, I. A., and Schilowa, B. (1980). Genetic structure of yeast hybrids constructed by protoplast fusion. In *Advances in Protoplast Research*, L. Ferenczy and G. L. Farkâs (Eds.). *Akadémiai Kaidó, Budapest*, pp. 99-104.

Brook, H. W. J. van den, Stunnenberg, H. G., and Wennekes, L. M. J. (1979). Protoplasts from *Aspergillus nidulans*. *Can. J. Microbiol. 26*:115-128.

Burnett, J. H. (1975). *Mycogenetics*, Wiley, London.

Case, M. E., Schweizer, M., Kushner, S. R., and Giles, N. H. (1979). Efficient transformation of *Neurospora crassa* by utilizing hybrid plasmid DNA. *Proc. Nat. Acad. Sci. USA 76*:5259-5263.

Christensen, B. E. (1979). Somatic hybridization in *Saccharomyces cerevisiae*: Analysis of products of protoplast fusion. *Carlsberg Res. Commun. 44*:225-233.

Cohen, J. D., Eccleshall, T. R., Needham, R. B., Federoff, H., Buchferer, B. A., and Marmur, J. (1980). Functional expression in yeast of the *Escherichia coli* plasmid gene coding for chloramphenicol acetyltransferase. *Proc. Nat. Acad. Sci. USA* 77:1078-1082.

Croft, J. H., Dales, R. B. G., Turner, G., and Earl, A. (1980). The transfer of mitochondria between species of *Aspergillus*. In *Advances in Protoplast Research*. L. Ferenczy and G. L. Farkâs (Eds.), Akadémiai Kaidó, Budapest, pp. 85-92.

Dales, R. B. G., and Croft, J. H. (1977). Protoplast fusion and the isolation of heterokaryons and diploids from vegetatively incompatible strains of *Aspergillus nidulans*. *FEMS Microbiol. Lett. 1*:201-204.

Dales, R. B. G., and Croft, J. H. (1980). Protoplast fusion and the genetical analysis of vegetative incompatibility in *Aspergillus nidulans*. In *Advances in Protoplast Research*, L. Ferenczy and G. L. Farkâs (Eds.). Adadémiai Kaidó, Budapest, pp. 73-84.

Eddy, A. A., and Williamson, D. H. (1959). Formation of aberrant cell walls and of spores by the growing of yeast protoplasts. *Nature 183*:1101-1104.

Emerson, S., and Emerson, M. R. (1958). Production, reproduction and reversion of protoplast-like structures in the osmotic strains of *Neurospora crassa*. *Proc. Nat. Acad. Sci. USA 44*:668-671.

Ferenczy, L. (1976). Some characteristics of intra- and interspecific protoplast fusion products of *Aspergillus nidulans* and *Aspergillus fumigatus*. In *Cell Genetics in Higher Plants*, D. Dudits, G. L. Farkâs, and P. Maliga (Eds.). Akadémiai Kaidó, Budapest, pp. 171-182.

Ferenczy, L., and Maráz, A. (1977). Transfer to mitochondria by protoplast fusion in *Saccharomyces cerevisiae*. *Nature 268*:524-525.

Ferenczy, L., Kevei, F., and Zsolt, J. (1974). Fusion of fungal protoplasts. *Nature 248*:793-794.

Ferenczy, L., Kevei, F., and Szegedi, M. (1975a). Increased fusion frequency of *Aspergillus nidulans* protoplasts. *Experientia 31*:50-52.

Ferenczy, L., Kevei, F., and Szegedi, M. (1975b). High-frequency fusion of fungal protoplasts. *Experientia 31*:1028-1030.

Fodor, K., Demiri, E., and Alfoldi, L. (1978). Polyethylene glycol-induced fusion of heat-inactivated and living protoplasts of *Bacillus megaterium*. *J. Bacteriol. 135*:68-70.

Fournier, P., Provost, A., Bourguignon, C., and Heslot, H. (1977). Recombination after protoplast fusion in the yeast *Candida tropicalis*. *Arch. Microbiol. 115*:143-149.

Fried, H. M., and Warner, J. R. (1981). Cloning of a yeast gene for trichodermin resistance and ribosomal protein L3. *Proc. Nat. Acad. Sci. USA 78*:238-242.

Garcia Acha, I., Lopez-Belmonte, F., and Villanueva, J. R. (1966). Regeneration of mycelial protoplasts of *Fusarium culmorum*. *J. Gen. Microbiol. 97*:169-184.

Gunge, N., and Tamaru, A. (1978). Genetic analysis of products of protoplast fusion in *Saccharomyces cerevisiae*. *Jpn. J. Genet. 53*:41-49.

Hamlyn, P. F., and Ball, C. (1979). Recombination studies with *Cephalosporium acremonium*. In *Genetics of Industrial Microorganisms*, O. K. Sebek and A. I. Laskin (Eds.). American Society for Microbiology, Washington, D. C., pp. 185-191.

Hamlyn, P. F., Bradshaw, R. E., Mellon, F. M., Santiago, C., Wilson, J. M., and Peberdy, J. F. (1981). Efficient protoplast release from fungi using commercial enzymes. *Enzyme Microb. Technol. 3* (in press).

Hastie, A. C. (1973). Hybridization of *Verticillium albo-atrum* and *Verticillium dahliae*. *Trans. Br. Mycol. Soc. 60*:511-523.

Hicks, J., Strathern, J. N., and Klar, A. J. S. (1979). Transposable mating type genes in *Saccharomyces cerevisiae*. *Nature 282*:478-483.

Hinnen, A., Hicks, J. B., and Fink, G. R. (1978). Transformation of yeast. *Proc. Nat. Acad. Sci. USA 75*:1929-1933.

Hockney, R. C., and Freeman, R. F. (1980). Construction of polysaccharide-degrading brewing yeasts by protoplast fusion.

In *Advances in Protoplast Research*, L. Ferenczy and G. L. Farkâs (Eds.). Akadémiai Kaidó, Budapest, pp. 139-144.
Hopwood, D. A., and Wright, H. M. (1979). Factors affecting recombinant frequency in protoplast fusions of *Streptomyces coelicolor*. *J. Gen. Microbiol. 111*:137-142.
Hsiao, C. -L., and Carbon, J. (1979). High frequency transformation of yeast by plasmids containing the cloned yeast *ARG4* gene. *Proc. Nat. Acad. Sci. USA. 76*:3829-3833.
Isaac, S. (1978). Ph. D. thesis, University of Nottingham; Nottingham.
Jimenez, A., and Davies, J. (1980). Expression of a transposable antibiotic resistance element in *Saccharomyces*. *Nature. 287*:869-871.
Kawakami, M., Tanaka, H., Mondo, H., Katamine, S., and Kawakami, H. (1980). Incorporation of algal thykaloid membrane and DNA in yeast protoplasts. In *Advances in Protoplast Research*. L. Ferenczy and G. L. Farkâs (Eds.). Akadémiai Kaidó, Budapest, pp. 49-54.
Kevei, F., and Peberdy, J. F. (1977). Inter-specific hybridization between *Aspergillus nidulans* and *Aspergillus rugulosus* by fusion of somatic protoplasts. *J. Gen. Microbiol. 102*:255-262.
Kevei, F., and Peberdy, J. F. (1979). Induced segregation of interspecific hybrids of *Aspergillus nidulans* and *Aspergillus rugulosus* obtained by protoplast fusion. *Mol. Gen. Genet. 170*:213-218.
Kingsman, A. J., Clark, L., Mortimer, R. K., and Carbon, J. (1979). Replication in *Saccharomyces cerevisiae* of plasmid pBR313 carrying DNA of the yeast *trpl* region. *Gene 7*:141-152.
Klinner, U., Böttcher, F., and Samsonova, I. A. (1980). Hybridization of *Pichia guilliermondii* by protoplast fusion. In *Advances in Protoplast Research,* L. Ferenczy and G. L. Farkâs, (Eds.), Akadémiai Kaidó, Budapest, pp. 113-118.
Kozhina, T. N., and Cherpurnya, O. V. (1980). Sliyanie protoplastov kav metod polucheniya gibradov u drozzhej-Sakharomicetov. *Genetika 16*:361-363.
Levi, C., Sanchez-Rivas, C., and Schaeffer, P. (1977). Further genetic studies on the fusion of bacterial protoplasts. *FEMS Microbiol. Lett. 2*:323-326.
Lewis, L. A. (1969). Genetic evidence for hybridization in an interspecific cross in the genus *Sordaria*. *J. Gen. Microbiol. 59*:359-368.
Marâz, A, Kiss, M., and Ferenczy, L. (1978). Protoplast fusion in *Saccharomyces cerevisiae* strains of identical and opposite mating types. *FEMS Microbiol. Lett. 3*:319-322.
Mishra, N. C. (1976). Episome-like behavior of DNA in transformed strains of *Neurospora crassa*. *Nature 264*:251-253.
Mishra, N. C., and Tatum, E. L. (1973). Non-mendelian inheritance

of DNA-induced inositol independence in *Neurospora crassa*. *Proc. Nat. Acad. Sci. USA 70*:3875-3879.

Morgan, A. J., Heritage, H., and Whittaker, P. A. (1977). Protoplast fusion between petite and auxotrophic mutants of petite-negative yeast *Kluyveromyces lactis*. *Microbios. Lett. 4*:103-107.

Morgan, A. J., Hall, J. L. Brunner, A., and Whittaker, P. A. (1980). Protoplast fusion in a study of mitochondrial genetics in the petite-negative yeast, *Kluyveromyces lactis*. In *Advances in Protoplast Research*, L. Ferenczy and G. L. Farkâs (Eds.). Akadémiai Kaidó, Budapest, pp.93-98.

Necas, O. (1971). Cell wall synthesis in yeast protoplasts. *Bacteriol. Rev. 35*:149-170.

Necase, O. (1980). Regeneration of protoplasts. In *Advances in Protoplast Research*, L. Ferenczy and G. L. Farkâs (Eds.). *Akadémiai Kiadó*, Budapest, pp. 151-162.

Nüesch, J., Treichler, H. J., and Liersch, M. (1973). The biosynthesis of cephalosporin C. In *Genetics of Industrial Microorganisms, Vol. 2*. Z. Vanek, Z. Hostalek, and J. Cudlin (Eds.). Elsevier, Amsterdam, pp. 309-344.

Panthier, J. -J., Fournier, P., Heslot, H., and Rambach, A. (1980). Cloned β-galactosidase gene of *Escherichia coli* is expressed in the yeast *Saccharomyces cerevisiae*. *Curr. Genet. 2*:109-113.

Paris S. (1977). Heterocaryons chez *Beauvaria tenella*. *Mycopathologia 61*:67-75.

Peberdy, J. F. (1976). Isolation and properties of protoplasts from filamentous fungi. In *Microbial and Plant Protoplasts*, J. F. Peberdy, A. H. Rose, H. J. Rogers, and E. C. Cocking, (Eds.). Academic, London, pp. 39-50.

Peberdy, J. F. (1979). Wall biogenesis by protoplasts. In *Fungal Walls and Hyphal Growth*, J. H. Burnett and A. P. J. Trinci (Eds.). Cambridge University, Cambridge, pp 49-70.

Peberdy, J. F., Eyssen, H., and Anné, J. (1977). Interspecific hybridization between *Penicillium chrysogenum* and *Penicillium cyaneofulvum* following protoplast fusion. *Mol Gen. Genet. 157*:281-284.

Provost, A., Bourguignon, C., Fournier, P., Ribet, A. M., and Heslot, H. (1978). Intergeneric hybridization in yeasts through protoplast fusion. *FEMS Microbiol. Lett. 3*:309-312.

Santiago, C. M. (1982). Ph.D. thesis, University of Nottingham, Nottingham, England.

Savchenko, G. V., and Kapultsevich, Yu. G. (1980). Hybridization of yeast from the genus *Hansenula* by protoplast fusion. In *Advances in Protoplast Research*. L. Ferenczy and G. L. Farkâs (Eds.). Akadémiai Kaidó, Budapest, pp. 125-130.

Schablik, M., Szabolcs, M., Kiss, A., Aradi, J., Zsindely, A., and Szabo, G. (1977). Conditions of transformation by DNA of *Neurospora crassa*. *Acta. Biol. Acad. Sci. Hung. 28*:273-279.

Sen, K., Nandi, P., and Mishra, A. K. (1969). Transformation of a nutritionally deficient mutant of *Aspergillus niger*. *J. Gen. Microbiol.* *55*:195-200.

Shamoian, C. A., Canazelli, A., and Melrose, J. (1961). Back mutation of a *Neurospora crassa* mutant by a nucleic acid complex from the wild strain. *Biochim. Biophys. Acta* *47*:208-211.

Shockley, T. E., and Tatum, E. L. (1962). A search for genetic transformation in *Neurospora crassa*. *Biochim. Biophys. Acta* *61*:567-572.

Sipiczki, M. (1979). Interspecific protoplast fusion in fission yeasts. *Curr. Microbiol.* *3*:37-40.

Sipiczki, M., and Ferenczy, L. (1977a). Protoplast fusion of *Schizosaccharomyces pombe* auxotrophic mutants of identical mating type. *Mol, Gen. Genet.* *151*:77-81.

Sipiczki, M., and Ferenczy, L. (1979b). Fusion of *Rhodosporidium (Rhodotorula)* protoplasts. *FEMS Microbiol. Lett* *2*:203-205.

Solingen, P. van, and Plaat, J. B. van der (1977). Fusion of yeast spheroplasts. *J. Bacteriol.* *130*:946-947.

Spata, L., and Weber, H. (1980). A study on protoplast fusion and parasexual hybridization of alkane utilizing yeasts. In *Advances in Protoplast Research*, L. Ferenczy and G. L. Farkâs (Eds.). Akadémiai Kaidó, Budapest, pp. 131-138.

Spencer, J. F. T., Land, P., and Spencer, D. M. (1980). The use of mitochondrial mutants in the isolation of hybrids involving industrial yeast strains. *Mol, Gen. Genet.* *178*:651-654.

Stahl, U. (1978). Zygote formation and recombination between like mating types in the yeast *Saccharomycopsis lipolytica* by protoplast fusion. *Mol. Gen. Genet.* *160*:111-113.

Stephen, E. R., and Nassim, A. (1981). Production of protoplasts in different yeasts by mutanase. *Can. J. Microbiol.* *27*:550-553.

Stinchcomb, D. T., Thomas, M., Kelly, J., Selker, E., and Davies, R. W. (1980). Eukaryotic DNA segments capable of autonomous replication in yeast. *Proc. Nat. Acad. Sci. USA* *77*:4559-4563.

Struhl, K., Stinchcomb, D. T., Scherer, S., and Davies, R. W. (1979). High frequency transformation of yeast: Autonomous replication of hybrid DNA molecules. *Proc. Nat. Acad. Sci. USA* *76*:1035-1039.

Svoboda, A. (1980). Intergenetic fusion of yeast protoplasts: *Saccharomyces cerevisiae and Schizosaccharomyces pombe*. In *Advance in Protoplast Research*. L. Ferenczy and G. L. Farkâs (Eds.). Akadémiai Kaidó, Budapest, pp. 119-124.

Thomas, K. R., and Davies, B. D., (1980). The effect of calcium on protoplast release from species of *Aspergillus*. *Microbios.* *28*:69-80.

Vallin, C., and Ferenczy, L. (1978). Diploid formation of *Candida tropicalis* via protoplast fusion. *Acta Microbiol. Acad. Sci. Hung.* *25*:209-212.

Vasil, I. K. and Giles, K. L. (1975). Induced transfer of higher plant chloroplasts into fungal protoplasts. *Science 190*:600.
Whittaker, P. A., and Leach, S. M. (1978). Interspecific hybrid production between the yeast *Kluyveromyces lactis* and *Kluyveromyces fragilis* by protoplast fusion. *FEMS Microbiol. Lett. 4*:31-34.
Yamamoto, M., and Fukui, S. (1977). Fusion of yeast protoplasts. *Agric. Biol. Chem. 41*:1829-1830.

Author Index

Italic numbers indicate the page on which the complete reference is listed

H

I

N

O

P

Q

R

U

V

Subject Index

A

D

I

K

L

M

N

O

P